Applications of Polymers in Surgery II

Edited by

Inamuddin[1], Arwa Alrooqi[2], Amal Alshehri[2], Hind Alluqmani[2], Mohammad Abu Jafar Mazumder[3,4]

[1]Department of Applied Chemistry, Zakir Husain College of Engineering and Technology, Faculty of Engineering and Technology, Aligarh Muslim University, Aligarh-202002, India

[2]Department of Chemistry, Faculty of Science, Al Baha University, Al Baha, Saudi Arabia

[3]Chemistry Department, King Fahd University of Petroleum & Minerals, Dhahran 31261, Saudi Arabia

[4]Interdisciplinary Research Center for Advanced Materials, King Fahd University of Petroleum & Minerals, Dhahran 31261, Saudi Arabia

Published by **Materials Research Forum LLC**
Millersville, PA 17551, USA

Published as part of the book series
Materials Research Foundations
Volume 172 (2025)
ISSN 2471-8890 (Print)
ISSN 2471-8904 (Online)

Print ISBN 978-1-64490-334-6
eBook ISBN 978-1-64490-335-3

Distributed worldwide by

Materials Research Forum LLC
105 Springdale Lane
Millersville, PA 17551
USA
https://www.mrforum.com

Manufactured in the United States of America
10 9 8 7 6 5 4 3 2 1

Table of Contents

Preface

As technological advancements continue to expand globally, the use of polymers in diverse applications has grown significantly, replacing traditional, often harmful, and expensive materials. Polymers, large molecules formed by the repeated addition of smaller units called monomers, exhibit remarkable versatility. They can be as flexible as rubber, as rigid as glass, or as elastic as fibers. These unique properties arise from their molecular structure and the manner in which monomers bond together. Polymers, whether natural or synthetic, permeate nearly every aspect of life, including therapeutic and biomedical applications, energy storage devices, and material sciences. Their widespread adoption is driven by distinctive and irreplaceable properties, such as a high strength-to-weight ratio, thermal stability, corrosion resistance, durability, versatility, flexibility, ease of processing, and cost-effectiveness.

This book, *Applications of Polymers in Surgery,* delves into the fascinating world of polymers, exploring their diverse applications, exceptional properties, and critical role in advancing medical surgery and innovation. It begins with a concise introduction to the role of polymers in surgical contexts, followed by an in-depth examination of their applications, ranging from tissue engineering, wound dressings, and implanted medical devices to blood vessel reconstruction, drug delivery, dentistry, and more. The book contains 19 chapters, summarized as follows:

Chapter 1: An overview of polymeric printed materials in various surgeries, focusing on 3D printing for surgical education, planning, and simulation. Applications of printing technology in surgery are discussed, highlighting challenges and opportunities for improving surgical practices.

Chapter 2: Explores the use of polymers as sealants and adhesives in surgical applications, discussing their mechanisms of action and material classifications.

Chapter 3: Highlights polymers in medical implants, including their versatility, biocompatibility, and applications in joint replacements and cardiovascular devices, along with challenges and advancements in polymeric implants.

Chapter 4: Examines polymers in cardiovascular surgery, focusing on vascular grafts, stents, heart valves, and drug delivery systems, including their advantages, limitations, and future prospects.

Chapter 5: Provides an overview of polymers in neurosurgery, highlighting their ability to mimic natural tissues for various applications.

Chapter 6: Details the use of polymers in orthopedic surgery, discussing materials, manufacturing techniques, and their benefits.

Chapter 7: Describes various polymers, including shape memory polymers, and their applications in surgeries.

Chapter 8: Focuses on the application of polymers in tissue engineering, emphasizing eco-friendly scaffold development and outlining future challenges.

Chapter 9: Explores polymer-based 3D-printed scaffolds for tissue regeneration, including bone, cartilage, and skin applications.

Chapter 10: Discusses natural and synthetic polymers in drug delivery systems, highlighting their roles in enhancing therapeutic outcomes.

Chapter 11: Examines polymers in drug delivery, with a focus on nanotechnology integration for advanced drug carriers.

Chapter 12: Explores the use of synthetic, natural, and hybrid polymers in wound care dressings, detailing their advantages and limitations.

Chapter 13: Discusses polymers in medical implants, highlighting biodegradable and non-biodegradable materials for various applications.

Chapter 14: Examines the use of polymers in retinal prostheses and their importance in addressing retinal degenerative diseases.

Chapter 15: Explores the role of polymers in vascular applications, emphasizing biocompatibility and advancements in smart polymers and nanotechnology.

Chapter 16: Highlights polymers in cosmetics and personal care products, discussing trends, environmental impact, and health implications.

Chapter 17: Examines polymers in dentistry, focusing on their role in advancing patient-centered dental care.

Chapter 18: Discusses advancements in vascular tissue engineering using polymeric materials to address current limitations in vascular reconstruction techniques.

This book serves as a comprehensive resource for healthcare professionals, researchers, and enthusiasts engaged in the development of advanced medical devices and techniques.

Materials Research Foundations 172 (2025) 1-42
https://doi.org/10.21741/9781644903353-1

Chapter 1

Printed Polymers in Surgery

Mohammad Hossain[1], Md. Abu Bin Hasan Susan[2,3*]

[1]Department of Chemistry, Jashore University of Science and Technology, Jashore- 7408, Bangladesh

[2]Department of Chemistry, University of Dhaka, Dhaka, Bangladesh

[3]Dhaka University Nanotechnology Center (DUNC), University of Dhaka, Dhaka, Bangladesh

susan@du.ac.bd

Abstract

The enhanced biological and mechanical properties, low production costs, ease of processing, and broad applicability of polymeric materials have led to a rise in their use in surgery. Several additive manufacturing (AM) techniques are employed in various kinds of surgery. For printing purposes, a variety of biodegradable and non-biodegradable polymeric compounds are referred to as bio-ink. At an astounding rate, this technique is customized for various surgeries. Anatomical models tailored to each patient and various types of implants have been printed using it. This chapter aims to provide a summary of the top polymeric printed materials utilized in liver transplantation, robotic urology, orthopedic, ear, nose, and throat (ENT), cardiac, vascular, andrological, spinal, and plastic surgeries, as well as the printing of maxillofacial prosthesis and medications. It also aims to develop novel polymers and techniques that will replicate the clinical setting.

Keywords

Surgical specialization, Patient-centered, Computed tomography, Anatomic models, Prosthesis

Abbreviation

CT	Computed tomography
AM	Additive manufacturing
DICOM	Digital imaging and communications in medicine
3DP	3D printing
ROI	Region of interest
CAD	Computer-aided design
ENT	Ear, nose, and throat
3D	Three dimensional
RP	Rapid prototyping

Materials Research Foundations 172 (2025) 1-42
https://doi.org/10.21741/9781644903353-1

CBCT	Cone-beam computed tomography
MRI	Magnetic resonance imaging
2D	Two dimensional
AP	Anterior-posterior
CTA	Computed tomography angiography

Contents

Materials Research Foundations 172 (2025) 1-42
https://doi.org/10.21741/9781644903353-1

1. Introduction

The integration of multiple technical developments has led to substantial advancements in surgery during the past few years. This covers all surgical specializations and subspecialties, including general surgery, where the use of minimally invasive surgical techniques is increasing. Developing a more targeted, patient-centered approach that would balance the recovery of a patient with upholding the security and level of care has always been the primary goal. Printing technology is one of the more recent technical advancements that has had a big impact on the manufacturing and product design industries [1]. Recent years have experienced a fast advancement in polymer printing, leading to the development of designed goods in various disciplines of study, particularly in the medical domain [2–5]. The various properties of polymer materials and processing methods make polymer printing advantageous for numerous medical specialties [6–9].

Digital three-dimensional (3D) models may be quickly transformed into tangible components through printing. Nowadays, computed tomography angiography (CTA) is a substantial imaging mode employed in plastic and reconstructive surgery [10–15]. With its many advantages, including affordability, dependability, and suitability for a wide range of compatible materials, this approach makes 3D printing (3DP) in surgery possible. "Zero lead time" elapsed from design to final production is the most significant benefit of printing technology for usage in the medical industry. Moreover, the majority of 3D models for 3DP medical applications may be created utilizing 3D surface restoration by image post-processing, making it simpler than with industrial procedures. In the medical arena, the latest direction toward "individualized" treatment is expedited by the potential for one-stop production, ranging from imaging to structure formation. Second, and perhaps most importantly for medical applications, 3DP, as an AM technology, demonstrates the attributes of "zero constraints and zero skills". This is because the shapes of the models made from medical photographs specific to a patient are usually too intricate to be produced using conventional fabrication techniques [16].

2. Role of Polymers in Surgery

Our daily lives involve the use of polymers in various ways. In surgical and biological procedures, synthetic polymers are employed. Basic medical equipment, test apparatus, and implants are examples of medical devices. These devices make considerable use of a variety of polymers due to their advantages in terms of weight, cost, and performance. As a result, depending on the structure-activity relationship, various types of polymers such as thermoplastics, gels, elastomers, and thermosets are widely utilized in various sectors including medicine and surgery. In the medical domain, biomaterials, including synthetic and natural polymers are used in the assessment, augmentation, and replacement of any component of the body [17]. The significant use of artificial biodegradable polymers starting in the 1960s entered into another dimension of revision surgery. The transition from biostable to biodegradable materials has been the subject of ongoing research [18–20]. Diverse polymeric systems with distinct characteristics are needed for diverse medicinal purposes. Therefore, multifunctional and combinatorial biodegradable polymers have become important. The chemical composition of the material, its solubility, molecular weight, surface

Materials Research Foundations 172 (2025) 1-42 https://doi.org/10.21741/9781644903353-1

energy, lubricity, water resistance, degradation, and the form and structure of the implant all have a significant impact on biocompatibility [21]. Polymeric biomaterials are widely used in simple membranes, multifilament meshes, large and tiny implants, and other applications [22].

Table 1. *Typical polymers and their characteristics in 3DP*

Polymer	Hardness (MPa)	Melting temperature (°C)	Benefit	Drawback
Polycarbonates	2250	110	Adjustable dynamics and porosity	Absorb atmospheric moisture, which has an impact on printing resistance and performance
Polyamides	10	250	Excellent flexibility, stress resistance, and stability	Most frequently utilized with selective laser sintering technology
Polylactic acid	230	175	Good mechanical characteristics at a low price	Durational biocompatibility
Polypropylene	1600	165	Weightless	Low tolerance to temperature changes and susceptibility to ultraviolet radiation
Acrylonitrile butadiene styrene	30	105	Strong and pliable	Non-biodegradable and contracts when exposed to air
Thermoplastic polyurethane	100	235	Variable rigidity	Unable to tolerate elevated temperatures
Poly caprolactone	216	60	Extended period of degradation	Superior viscoelastic and rheological characteristics when heated; inexpensive
Polyetheretherketone	3600	350	Strong and light with exceptional mechanical and heat resistance	Elevated melting temperature

Various advanced polymers such as polyetheretherketone, polyurethane, polylactic acid, polyamide, polyglycolide, polyethylene, polymethylmethacrylate, polyethylene terephthalate, polytetrafluoroethylene, polystyrene, and synthetic rubber are used in surgery [23]. Suture materials, tissue adhesives, sealants, and surgical meshes are prepared from these polymers. Numerous polymers are used in orthopedic implants, polymeric heart valves, plastics, reconstructive and cosmetic surgery, ophthalmology, and neurosurgery. The most prevalent

 Materials Research Forum LLC
https://doi.org/10.21741/9781644903353-1

biomaterials utilized in 3D printing technologies are polymers, which are available as gels, beads, and filaments for direct ink writing, sintering, and lithography [5,24]. Furthermore, they have properties of quick evaporation and variable mechanical strengths. Table 1 lists the printable inks that have been created in recent years using a variety of polymers.

A variety of changes in molar mass, crosslinking, co-polymers, and degree of crystallization make such varieties of applications conceivable. When selecting a polymer, mechanical properties like tensile strength, stiffness, and elasticity are typically important considerations. Biocompatibility should also be taken into account. These ideas will soon offer fruitful areas for study and advancement, which will be recognized as a significant breakthrough in the surgical sector.

3. An Approach to Creating 3D Models from Images

With the right software, which frequently comes with the computed tomography (CT) scanning device, replicas of the interior organs of humans may now be extracted from a high-resolution sequence of CT scan pictures. The software that is specifically designed for the correct storage, transfer, and display called Digital Imaging and Communications in Medicine (DICOM), processes images.

Improving, segmenting, quantifying, registering, and visualizing medical images are traditionally the primary processes in the process. The typical processing stages for creating a printed 3D model from tomography pictures are augmentation, segmentation, extraction model, smoothing and noise removal, slicing, and 3DP. Enhancing the image to clarify the boundaries in individual images is the first step in processing. Using filters, the original image is altered in this process. The second phase involves applying certain changes with the aim of grouping pixels in each DICOM image together that represent the same material. In a file that describes the 3D properties of an item, the third processing step is joining all the same group pixels from consecutive DICOM images. Once more, techniques for improving and eliminating noise in the final 3D model item must be used. This is put into practice in the fourth processing phase of the work. The 3D model file will be divided into successive two-dimensional (2D) layers once more in the fifth step of processing. Using the right software, this procedure, known as "slicing," divides the 3D model into 2D slices. The printing process will begin in the sixth and final step, using the right machine and materials.

4. Printing Technology: Shedding Light into the Surgical Education

As AM technology advances, new 3DP devices have been developed, producing realistic replicas that closely resemble real-life tissues in terms of properties. Because of this, numerous institutions are paying more and more attention to 3DP technology as a teaching tool for a variety of surgical training courses. In complex circumstances, it is also utilized for preoperative planning to help patients comprehend the geometry of their issues [25]. High-fidelity surgical teaching models have been created using this technology in various surgeries. A few important considerations are data preparation, choosing printing materials and technologies based on intended use, and so forth.

In order to derive dependable and accurate 3D models, gathering medical imaging data is essential. Various instruments like magnetic resonance imaging (MRI) ensure time savings and a respectable degree of fidelity in the results [26]. For optimal results, the structure must have a strong contrast with the surrounding environments [27]. Every piece of information is stored in DICOM format [26]. The segmentation process then comes next. It involves choosing the appropriate voxels to

 Materials Research Forum LLC
https://doi.org/10.21741/9781644903353-1

identify the anatomical landmark of interest and designating a region of interest (ROI). The software can segment data automatically, semiautomatically, or manually. Files with ROI can be transformed into stereolithography files, that are used to create 3D models [27]. Before printing the final version, more model designs and adjustments are required, including surface smoothing and other optimizations [27]. Undoubtedly, a correct portrayal of anatomical structure or pathologic entities is the primary determinant of the appropriateness and ultimate utility of manufactured models. In order to satisfy this basic need, this multistep process of quality assurance and control must be met [27]. In addition, postprocessing techniques will improve the 3D model provided that final control over anatomical accuracy is carried out, often following assessment by qualified experts [27]. The printing process, which is the last phase, requires careful consideration of numerous factors, including the best materials, different colors, total cost, and time constraints [27]. The procedure for producing medical 3DP is shown in Fig. 1 [28].

Rapid prototyping (RP) has unquestionably aided in the teaching of general surgery in a number of procedures, especially minimally invasive ones. Hepatobiliary surgery presents a favorable environment for the application of 3DP as a substitute method for providing hands-on instruction and visualization to meet the educational needs of medical students [29]. However, issues with the intricacy of the venous, biliary, and hepatic parenchyma anatomy as well as the requirement for anatomical landmarks must be addressed. 3DP has several training uses like laparoscopic procedures [30]. Although intraoperative ultrasonography and 2D imaging are helpful tools for hepatic surgical operations, certain limitations can be efficiently addressed by 3D hepatic models to aid surgeons in improving their hand-eye coordination and comprehension of the anatomy of the liver [31].

5. Anatomic Models

In surgery, constructing anatomic models is the most common use of 3DP. Education and preoperative planning are the two main uses for which anatomic models are 3D printed. Most of the research shows that 3DP is an effective tool for teaching surgical residents and medical students, as well as for pre-operative planning.

5.1 Anatomic Models: Pre-surgical Planning

There is a clear advantage to studying the anatomy of a patient with a life-size copy of the relevant parts rather than through computer display photographs. Several investigations assessed the value of creating models from images that are useful in surgical planning [32]. When evaluating the quality of their surgical procedure, the individuals who followed 3D models did better. So, the models help inexperienced physicians plan preoperatively more skillfully and be beneficial for developing skills outside of the operating theater. This is an overview of the pre-operative planning methods used by several surgical disciplines using anatomic models.

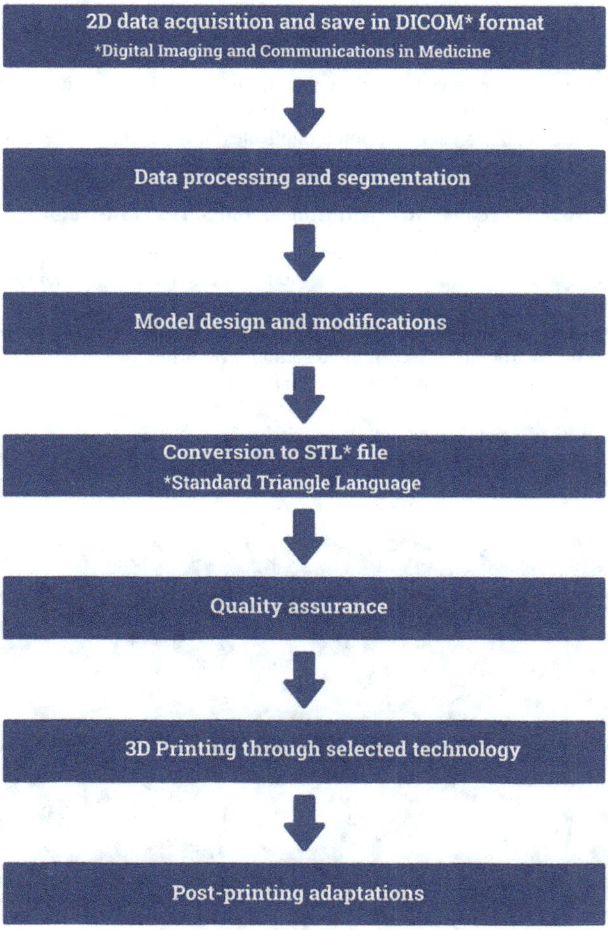

Fig. 1. Medical 3DP: A step-by-step guide. Reproduced from (Lales et al., 2020) with permission from Elsevier, Copyright [2020].

5.2 Anatomic Models: Pre-surgical Simulation

Anatomic models have been employed to conduct pre-operative mock procedures with patient-specific anatomy. When preparing surgery for a patient with an extremely complicated or uncommon surgical pathology, this can be quite helpful [33]. Pre-operative simulation in cardiothoracic surgery is used for various purposes including endovascular stenting, thoracoscopic segmentectomy, and transcatheter valve replacement [34–38]. Pediatric surgery also uses pre-operative planning in thoracoscopic baby lobectomy and laparoscopic choledochal cyst excision [39,40]. Pre-surgical simulation was deemed a useful technique here. In addition, simulations are also useful as learning tools and resources for clinical practice [33]. In order to remove a hepatoblastoma from a pediatric patient, printed duplicates of the hepatic vasculature were used [41,42].

5.3 Anatomic Models: Pre-operative Adjustment and Surgical Instrument Selection

The choice of surgical equipment and implants is physically guided and shaped by anatomic models. Fig. 2 displays a sample of the anatomic model that is made of titanium [43]. A 3DP of mandibular model tailored to the patient and a reconstruction plate model are used to pre-operatively contour the titanium plate. This is generally a beneficial procedure because of its feasibility to adjust the instrument in a controlled setting away from the operating room. This can improve implant or instrument accuracy while streamlining the process even more and cutting down on operating time. It was discovered that the intraoperative vascular and aneurysm anatomy matched the preoperatively customized patient-specific microcatheters [44]. Aneurysm models of the aortic arch have been used in vascular surgery to help select the right device for aneurysm repair [45]. Clavicular models and calcaneus models are used in orthopedic surgery [45–47].

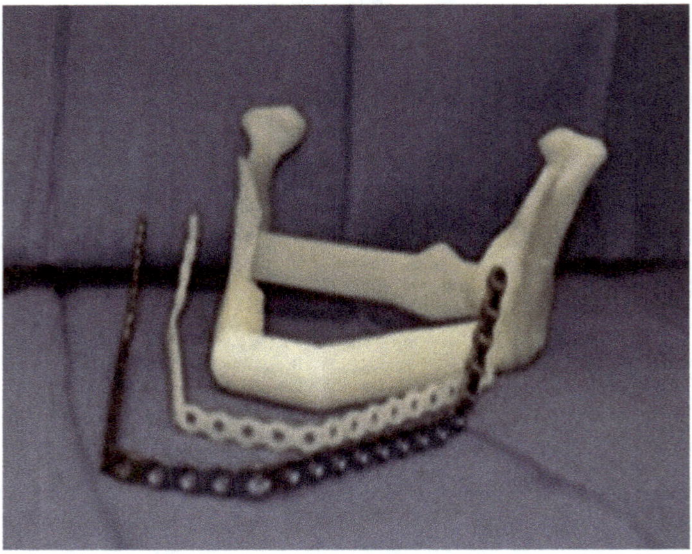

Fig. 2. *Titanium dental plate for orthodontic procedures. Reproduced from (Hoang et al., 2016) with permission from Annals of Translational Medicine, Copyright [2016].*

5.4 Anatomic Models: Education

A useful resource for medical teaching, both for resident physicians and medical students, is 3D-printed anatomic models. These physical models might be helpful when cadaveric anatomic examples are not readily available to medical students. These models are also helpful in teaching the pathology of anatomy. For instance, several types of ventricular septal abnormalities [48]. The ability to train to operate on rare surgical disorders and haptic input during real procedures has been made possible by the widespread usage of this technology.

6. Surgical Instrumentation and Guides

There are a lot of potential benefits to 3DP surgical instruments and surgical guidance. Because it makes it possible to utilize precisely tailored surgical tools and even to restore surgical apparatus in locations where obtaining it through traditional channels is not feasible. When performing front-orbital advancement surgery, neurosurgeons have discovered that using 3D-printed surgical templates is precise, user-friendly, effective, and repeatable [49]. Various common surgical equipment is printed using 3DP technology [50,51]. In addition, RP has enabled producers to design complex and unique instruments. A brand-new endoscopic equipment has been created using computer-aided design (CAD) and 3DP [52]. Not only was it possible to fabricate the gadget via 3DP, but it also allowed for rapid design modifications.

Orthopedic surgery uses printed surgical guiding templates tailored to each patient for precise osteosarcoma resection, and pronation deformity correction. This is accomplished by lumbar pedicle screw fixation, mid-cervical pedicle screw insertion, corrective osteotomy, and unicompartmental knee arthroplasties [53–57]. In orthopedic applications, using 3D-printed surgical guides has reduced intraoperative blood loss and lengthened operating times. Together with ureteral stents, a urology group has printed trocars for use in laparoscopic surgery [58]. Enhancing the prediction accuracy of MRI for prostate cancer requires an understanding of the relationship between the histology and MRI of the disease. In order to create a mold that would fit around the removed tumor, Preister *et al.* used CAD and 3DP to create an MRI dataset. A sequence of parallel slits that corresponded to individual MRI slices was included in the mold. The tumor might be sectioned after the operation in line with the mold, and the MR-histology correlation could then be ascertained by comparing the histologic sections to the pertinent MRI slice [59]. RP has been utilized in the creation of a considerable quantity of CAD-designed interocclusal wafers for orthognathic surgery, as well as maxillomandibular repositioners/splints [60–63]. These illustrations demonstrate how advancements in equipment have made orthognathic surgery increasingly more precise. 3DP has also been utilized in craniofacial surgery [64]. Furthermore, surgical guidelines tailored to each patient have been created for mandibular osteotomy and mandibular distraction [65]. Ophthalmology has also embraced 3D-printed surgical guides, which have shown to be particularly helpful in cases of technically challenging surgery (e.g., harvesting donor corneas for automated endothelial keratoplasty using Descemet stripping) [66]. A group specializing in obstetrics and gynecology printed the suction tube connector [67].

7. Issues of Ethics and Law Concerning Printing and Bioprinting

One significant advancement in contemporary biomedical science is the creation of 3DP applications. This technology facilitates the manufacture of organs as well as the manufacturing of medical equipment [68,69]. The moral implications of these applications are currently unknown.

Materials Research Foundations 172 (2025) 1-42
https://doi.org/10.21741/9781644903353-1

3DP generally has the potential to greatly address ongoing needs in modern healthcare systems, as it is anticipated to make the production process less time- and money-consuming by facilitating the supply of vital devices and supplies. As such, the new technology has an undeniable ethical worth in terms of its intended use. In particular, the development of bioprinting may spur significant advancements in transplantation and regenerative medicine. In actuality, the process of creating biological material guarantees the safety and functional appropriateness of the cloned tissues, including organs. Another issue that can be successfully addressed is the current shortage of biological resources for clinical usage. Bioprinting has the potential to herald a new era in biomedicine by producing biological materials. From an ethical standpoint, that viewpoint would inherently support our basic human right to physical integrity.

However, the basic principle applicable to any biomedical novelty does not eliminate the ethical validity of the 3DP goal. That concept states that no goal, no matter how moral, can justify sacrificing other essential principles that uphold the paramount importance of the dignity of human beings as it is represented in each individual. Stated differently, even though our goal is to advance society by creating research projects that could benefit society as a whole, there is no ethically acceptable way to accomplish this goal; the only way to do so is to adhere to all core values, which is a prerequisite for verifying the ethical soundness of the research goal. This adage is frequently stated in international instruments pertaining to biomedical law, where the rights and interests of any individual are deemed to take precedence above all [70].

In fact, there are a number of significant ethical concerns about the preservation and upholding of basic principles that must be addressed. These concerns are typically connected to certain facets of human rights.

8. Medical Printing: Quality and Safety

Various products are printed using polymers in printing technology [71]. It is anticipated that 3DP will cause a number of disruptions in the industry and healthcare sectors. The use of 3DP was limited to prototyping for more than 20 years. Medical equipment, dental tools, artificial organs for study, implants, and many more items have been produced in industries in recent years. More than 35% of surgical procedures in 2019 that required prostheses, and various types of implants relied heavily on 3DP.

The need for quality assurance and control procedures is increased by the quickly developing new in-hospital technology. Errors may happen at any time during the various stages of 3DP. These mistakes could result from human error, such as misunderstanding or the failure of other workflow-step-specific elements. Human error can be reduced or eliminated by workflow optimization and clear responsibility definition; intrinsic error can only be reduced and observed. A detailed workflow for 3DP is included in the specification of the process. The processes include choosing 3DP scenarios, acquiring images, segmenting data, engineering, 3DP, and getting everything ready for usage. For instance, according to a 2009 case study by Sampat *et al.*, a lower dosage frequency is a good substitute for a higher rate of therapeutic success [72]. The best way to enhance the result of baclofen may be to use a formulation of modified release that allows for dosage individualization.

One of the biggest regulatory challenges facing 3DP is safety. The advancement of new methods and materials has a continuous impact on 3DP. This puts both the manufactured goods and the safety of the 3D printers in danger. Safety in the context of bioprinting mostly relates to carrying out the process in non-accredited institutions. 3D bioprinting, which is based on utilizing living cells injected into the body, is yet an unproven therapeutic paradigm [73]. Since several biomedical applications are approaching the clinical trial stage, there are worries over the evaluation of human safety for disease modeling or medication testing [74,75]. The actual printing process and the printing materials themselves are two more crucial safety concerns. For example, the use of new polymers, which are occasionally combined with nanoparticles, necessitates post-marketing surveillance and registries and entails long-term dangers for implants. Furthermore, because of the higher degree of customization of the infrastructure and the possibility for greater decentralization, it might be harder for authorities and customers to confirm the safety of goods. Furthermore, regulations governing the use of living tissues in laboratories and stricter safety criteria for medical items as opposed to consumer goods are in place. The market entry of customized 3D-printed medical products is impeded by these standards. Although these guidelines may evolve to facilitate the creation of "mass customization" for various medical equipment, patient safety will remain the priority, and this will have an impact on the potential emergence of a 3DP infrastructure in the medical field [74,76]. The most important ones are those related to quality and safety since, regardless of any potential benefits of 3DP, they are values that cannot be compromised [77].

9. Economics in Printing

According to market assessments, industries and manufacturing will be significantly impacted by the advent of printing technology [78]. The production line has been made simpler by the new technology, which has removed production stages. Production is shifted closer to where demand is. The known production process is disrupted since the sale is made before the product is manufactured. By moving the production facilities closer to the client makes it possible to respond to market demands more directly [79]. When businesses move toward the more flexible paradigm of AM, the idea of building items in huge, complex facilities may become outdated [80]. Based on a model, Ye *et al.* argue that 3DP is unlikely to threaten the idea of large production capacity or decrease the flow of containers globally in the next few years [81]. Over the course of the projection period, the size of the healthcare 3DP market should be driven by rising demand for personalized 3D-printed devices and implants. According to Global Market Insights, Europe and North America account for the largest share of 3DP in the health sector globally. Three categories are used by Grand View Research to categorize the 3DP market: raw materials, applications, and regions. The 3DP sector is expected to grow at one of the fastest rates in the economy.

10. Printing and Liver Transplantation

As of right now, orthopedic, neurosurgical, cardiac, and oral and maxillofacial surgery are the surgical specialties where 3DP is being used the most [82]. Once the numerous immunological challenges associated with transplantation were overcome, a new area of medicine and liver transplantation surgery evolved quickly on a technical level. It might be stated that liver transplantation, among other solid organ transplants, is a multistep, intricate procedure requiring flawless surgery from beginning to end. Therefore, better preoperative planning, including a thorough grasp of anatomy of each patient, should lead to better results and a lower risk of surgical problems. The increase in transplantation criteria and the decrease in contraindications have nowadays made liver transplantation a routine procedure. Anatomical understanding with complicated spatial relationships and interpretation skills can be greatly improved by models. A physical model unique to each patient with liver cirrhosis is depicted in Fig. 3 and corresponds to their individual CT scans [83]. Using precise CT cross-sectional slices of the liver model can facilitate to study the structure. The anatomical structures of the liver, and the hepatic segments in particular, are extremely complex. The distribution of the several vascular and biliary conduits is staggered and interchangeable, which makes surgical anatomy challenging for novice surgeons and medical students to understand [84,85]. For such educational purposes, anatomical corrosion castings have been used extensively [86–88].

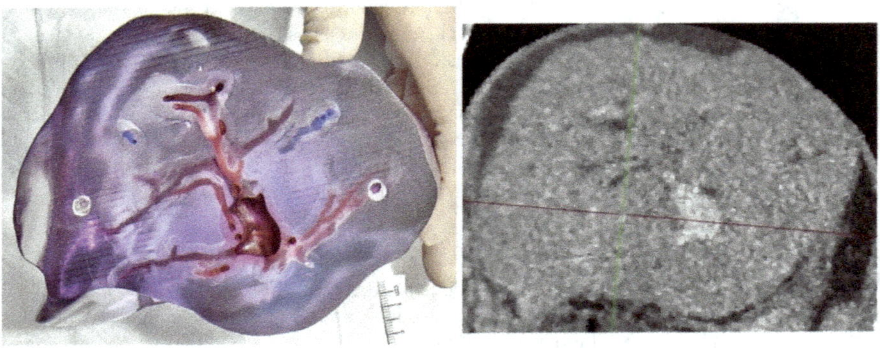

Fig. 3. *An image interpretation instruction file that correlates to the same slice of the CT scan for a cross-sectional cut in a 3D printed liver model tailored to the patient. Reproduced from (Ziogas et al., 2020) with permission from Elsevier, Copyright [2020].*

However, it is evident that the biggest obstacle facing liver transplantation today is the gap between a large number of patients waiting for a transplant and the scarcity of organ donors. The current phase of partial liver grafting has created an ideal environment for the advancement and use of 3DP, particularly in living donor segments and lobes.

11. Printing in Cardiac Surgery

The medical specialty of cardiac surgery focuses mostly on the disorders in blood arteries. The most frequent surgeries performed in this field include thoracic aortic surgery, coronary bypass, valve surgery, and other procedures for cardiac abnormalities. In addition, the entire range of the specialty includes robotic, endoscopic, minimally invasive, and endovascular operations.

Few specialties are as closely associated with technology as cardiac surgery. One of the most advanced technologies available, the extracorporeal circulation system replaces the pumping and breathing functions during surgery. To regulate perfusion intraoperatively, sophisticated monitoring tools are used. Moreover, cardiac procedures are frequently preoperatively planned using cardiac MRI and high-resolution cardiac computed tomography. The results of cutting-edge cardiac surgery technologies include various assisting devices like pacemakers. This feature, in

Materials Research Forum LLC
https://doi.org/10.21741/9781644903353-1

conjunction with the multiplicity of variations within the cardiovascular system, serves as justification for the need for preoperative 3D model study and subsequent surgical procedure design in such a broad variety of pathologies. Consequently, one of the specialties where 3DP is anticipated to advance is heart surgery.

Formation of models for anatomy instruction or display is the most common application of 3DP [89]. Therefore, 3DP can quickly illustrate a complex anatomic arrangement. Healthcare providers can learn about healthy and unhealthy structural interactions by using these models. Furthermore, they can be utilized to teach patients and the general public about the physiology and anatomy of circulatory disorders. Instructive models that illustrate congenital cardiac abnormalities, and valve operations are among the examples. A range of 3D arbitrary components, such as cardiac tissues, valves, blood arteries, and the entire heart model, have been successfully created by various tissue engineering techniques based on 3DP [90]. Therapeutic applications relating to the heart have not received much attention in recent times regarding these problems [91,92]. Fig. 4 illustrates how 3DP technology is being used to fabricate various cardiovascular system components at arbitrary gauges, from macroscale to microscale to nanoscale, to aid in the surgical procedures for a variety of cardiac problems, including myocardial infarction and other [93].

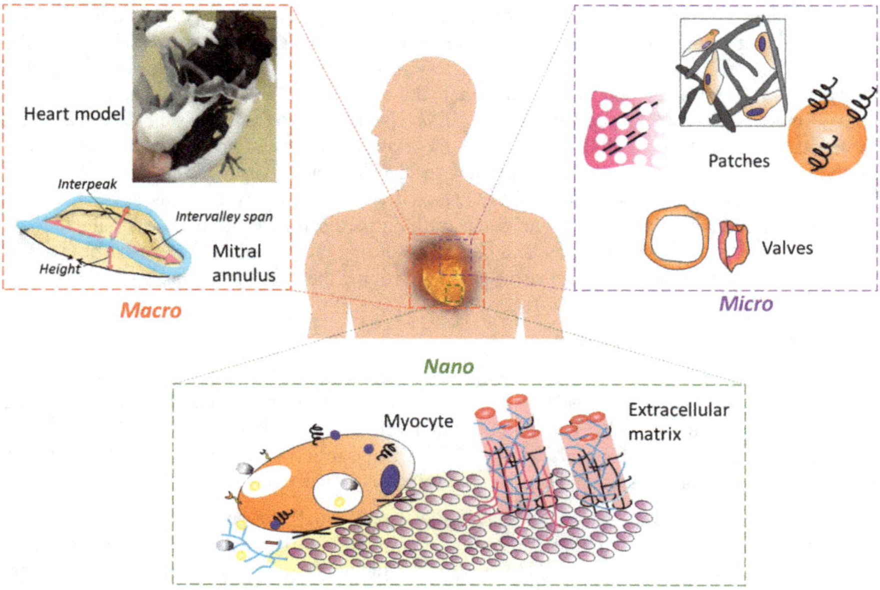

Fig. 4. *Diagrammatic representation of artificially created 3D parts or models of any size, from macro (heart model, mitral annulus, MA) to micro (cardiac patches, valves) to nano (carbon nanotubes, among other things) for cardiac surgery. Reproduced from (Kankala et al., 2017) with permission from Biofabrication, Copyright [2017].*

12. Printing in Vascular Surgery

3DP is crucial for simulating and evaluating surgical and endovascular procedures before they take place. Scientists and practicing physicians can better comprehend the anatomies of the aorta by using 3D models [94–99]. According to Ho *et al.*, vessel diameters collected from CT images before 3D printing and photos obtained from a CT scan of the model showed very slight variations in inaccuracy, guaranteeing the best accuracy [100]. It has decreased the time and frequency of complications associated with surgery. Additionally, a preoperative strategy for devices that close portosystemic shunts using 3DP has been reported [101].

Using 3DP, preoperative planning of endovascular therapy has been carried out for several aortic stenosis diseases like portal vein stenosis [102–106]. The application of models to directly observe and assess anatomical characteristics, such as the size and form of actual and pseudo lumens, may be beneficial in the aortic pathology known as aortic dissection. This ought to result in the highest caliber of clinical judgment and healthcare [100]. Medical interns and junior physicians are trained in vascular surgery using medical prototypes. Endovascular operations are known to need extreme accuracy and skills, which can be provided through training with 3D models. Practicing on cadaveric specimens is an alternate approach to endovascular surgery training, even though this approach is often impractical and costs money [107]. Geometric characteristics are also examined before turning attention to flow dynamics in vascular prototypes. Han *et al.* have demonstrated that big joining angles of capillary junctions have little effect on hemodynamic behavior [108]. Different methods of manipulating the distal artery flow resistance have been documented; these methods produce realistic models applied in innovative devices for both physiological and image-guided interventional procedure simulations [109]. Using in vitro systems, Canstein *et al.* have demonstrated that it is possible to examine the flow dynamics of the thoracic aorta [110]. Furthermore, it has been shown that combining in vivo MRI and 3DP is feasible to assess aortic abnormalities like dissection thoroughly [111]. 3DP is a technology that has several uses in vascular and endovascular surgery, including training medical students, residents, and fellows, testing possible grafts before surgery, and helping to design the procedure. This can immediately benefit patients by assisting vascular surgeons in improving the safety and quality of their work.

13. Printing in Orthopedic Surgery

A new technology called 3DP employs the right computer software to create objects based on data. The use of 3DP in medicine has grown significantly since its initial use, particularly among surgical specializations [112]. The number of studies and publications on 3DP related to orthopedic surgery is increasing gradually [113]. Advances in 3DP technology, decreased cost, and their ability to handle novel materials all contributed to this revolution.

The models are appropriate for orthopedic trauma, particularly when areas with complicated anatomy are involved. These models aid in the comprehension of specific anatomy and complex patterns of acetabular fractures [114,115]. Since its initial application in trauma patients, 3DP has spread to practically every anatomical region [116]. The techniques of reduction and fixation can be planned by surgeons. One can ascertain in advance the number of screws required, as well as their placement and sites of entrance, length, and trajectories [117–119]. When dealing with a complex fracture, 3DP is especially helpful for novice surgeons, but it is less helpful or even unneeded for seasoned surgeons when handling a straightforward fracture pattern [120]. Many surgeons value this "hands-on" approach, which allows them to practice operations on printed models at their desks, converting virtual stimulation into realistic stimulation [117]. This results in shorter operating times, less blood loss during surgery, and fewer fluoroscopies overall. Because minimally invasive procedures increase the accuracy of reduction and fixation, they can be employed even for difficult fractures. There are fewer complications, like iatrogenic nerve damage [116,120,121].

In the realm of orthopedics, spine surgery presents a possible avenue for extensive uses of 3DP. 3D models are helpful in the investigation of false articulations when employed in situations with spinal abnormalities. A safer and more precise procedure was achieved by using these models in the preoperative setting to plan the placement of pedicle screws and curve correction [122,123]. In order to treat foot abnormalities, Perthes, or Blount disease, printing technology is being used. These models can also be used to plan osteotomies in patients, whether they are pediatric or adult. Examining the abnormality will be beneficial for the surgery [124].

Improvements in 3DP technology have made it easier to create implants that are customized for each patient. The two main areas of focus are orthopedic oncology and joint arthroplasty. Despite the fact that primary, noncomplex knee arthroplasty patients can have individualized implants, these implants are more appropriate for those who have severe deformations. Examples of such cases include patients having revision arthroplasty, hip dysplasia, dwarfism, and bone tumor removal [125–127]. Standard implants are designed to fit the majority of people, but in situations where an existing implant would not be sufficient or if there are anatomical variations, a customized fit is necessary. Once a patient is chosen for a 3D-printed implant, they need to fill out a prescription paperwork explaining all the information. With mounting proof of its clinical efficacy, 3DP has been instrumental in individualized implants. Fig. 5 [121] illustrates a patient who had a major bone deformity from a car accident treated with a custom 3D-printed implant. Currently employed methods for treating major bone deficiencies brought on by trauma-induced bone loss, deformities, and nonunion include bone transfer, noncustom metal augments, vascularized bone grafts, and allograft bone repair [128]. It is possible that 3DP technology could bring about a revolution in the field of orthopedic surgery.

14. Printing in ENT Surgery

There are many articles in the literature regarding the application of 3DP in otolaryngology. These can be grouped not only by the intended application but also by the pertinent specialism (pediatric otolaryngology, rhinology, otology, and head and neck surgery/laryngology). The investigation has been carried out on otologic applications in training, mostly concerning the simulation of temporal bone surgery [129]. Using CT scan information from two selected patients, a temporal bone model was created. The ultimate assessment of the model revealed that most structures and anatomical landmarks could be replicated with satisfactory precision. However, the precision of the ossicular chain and resin inside the cells are of great concern [130]. Rose *et al.* showed that juvenile temporal bone models have satisfactory repeatability. The outcomes showed a high degree of precision and anatomic detail [131]. The correspondence between the cadaveric specimens and the affirmative responses has been provided by the trainees [132–134]. More precisely, by examining both subjective and objective techniques, Hochman *et al.* demonstrated that tactile feedback is sufficient. The experience is much more realistic because of the enhanced simulation of bone consistency made possible by material advancements [135].

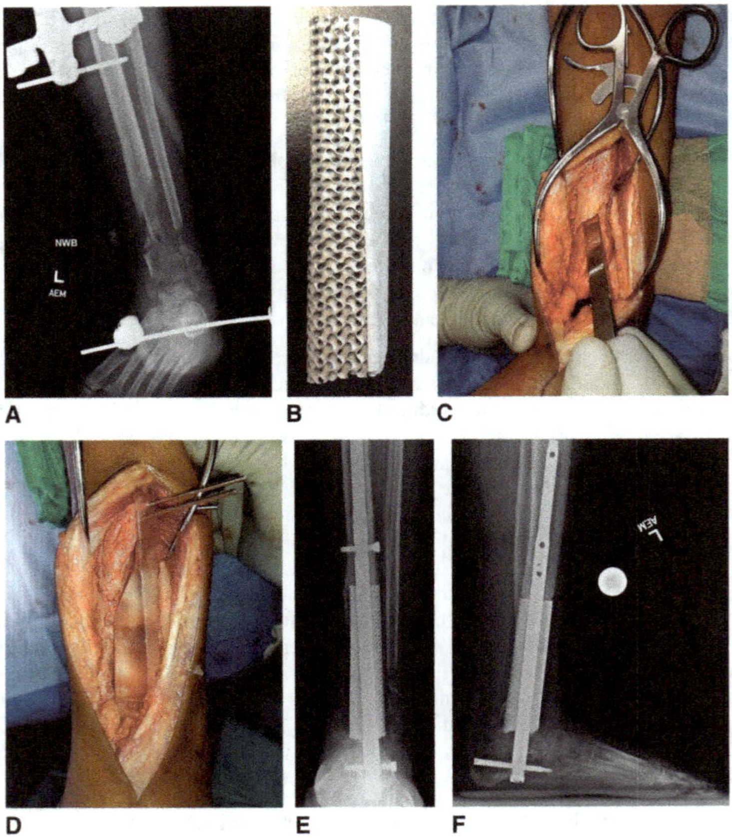

Fig. 5. *(A) Anterior-posterior (AP) radiograph of the left leg taken on a woman, age 22, who was hurt in an automobile accident. She suffered a fracture to her fibula and tibia. There is a significant bony defect where bone was lost at the scene, (B) An implant made specifically for bone deficiency using 3DP, (C) An intraoperative picture displaying bone loss and a fractured distal tibia, (D) Anatomic spacer block manufactured using 3DP for intramedullary reaming and alignment and length measurements, (E) An AP radiograph showing the 3D-printed implant successfully inserted, and (F) A lateral radiograph showing that the 3D-printed implant was successfully inserted. Reproduced from (Wixted et al., 2021) with permission from JAAOS Global Research & Reviews, Copyright [2021].*

Chiesa introduced another practical use. To manage epistaxis, Estomba *et al.* produced a 3D model that was integrated with a hydraulic system. Since the model replicated Little's area and the arteries, residents were able to practice handling anterior and posterior epistaxis [136]. Lastly, several juvenile laryngeal anomalies and a voice cord transcervical injection simulation have been presented [137–140].

15. Printing in Dental Surgery

Because of their unique biological, mechanical, and physical characteristics, synthetic polymer-based dental materials have many dental applications, including restorative, regenerative, and preventative dentistry [141]. In the domains of dentistry and maxillofacial surgery, printing technology has quickly gained traction because of its precision, efficiency, and accuracy. More precise treatment planning has been made possible by recent developments in cone-beam computed tomography (CBCT). RP polymers have been extensively used to create a variety of surgical templates [142]. Surgical templates can be produced using 3DP or milling once they are designed using polymers and pre-polymerized blocks. But compared to the latter, the former is more accurate [143,144]. Virtual simulation is crucial for orthognathic surgery, which mostly entails genioplasty. In orthognathic surgery, AM technologies based on polymers are utilized for implant trims [142]. A CT surgical guide, by definition, is a collection of instructions for surgery constructed using CBCT files [145,146]. The digital workflow is divided into three main stages: obtaining patient data, processing these data, and creating virtual plans using a particular dental CAD program [147]; and, lastly, creating the surgical guide using CAM technology (Fig. 6) [148–151].

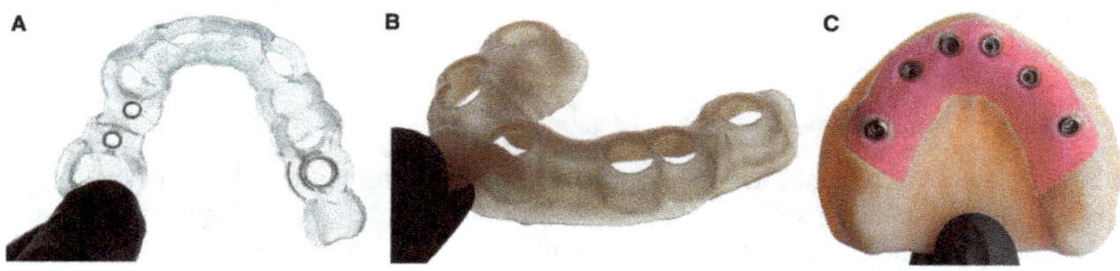

Fig. 6. *Applications of polymer 3DP in implant dentistry examples: (A) A printed surgical guide; (B) an open custom tray used for implant impression technique; and (C) a working cast of an implant-supported prosthesis. Reproduced from (Revilla-León, Sadeghpour and Özcan, 2020) with permission from Odontology, Copyright [2020].*

In the upcoming years, there will be a significant surge in dentistry for dental applications, leading to the creation of "smart" polymeric materials for use in implants and orthodontic equipment [152]. However, more advancements in processing technology and novel material compositions are required before 3D-printed polymers are widely used in all areas of dentistry.

16. Printing in Andrological Surgery

The introduction of novel biodegradable materials has led to recent advancements that have revolutionized 3DP technology in the pharmaceutical and medical industries [7]. It has completely changed the healthcare industry by making it possible to create personalized surgical instruments and biological models. Currently, various biocompatible materials are utilized in medical 3DP [153]. The development of tailored prostheses, the restoration of tissues, and the production of 3D models for pre- or postoperative simulations are among the most common therapeutic uses [4]. Despite the fact that urologists can already use 3DP, particularly for prostatic and renal surgeries [154,155].

However, comprehensive research on 3D application space in andrological surgery was not found. One potential area of use could be testicular-sparing surgery for young patients with monorchid cysts who have testicular cancer and want to preserve their fertility. For these patients, the amount of parenchyma required for reproductive function must be preserved as much as feasible. The investigation of deformities in La Peyronie disease patients may be another fascinating use. In these situations, estimating treatment options and surgical decisions helps determine the optimal surgical course of action.

When a penile prosthesis extrudes, migrates, and causes more corpora cavernosa perforations, or eventually becomes hypodimensional, 3DP can be used to design surgical procedures. In this instance, protecting the urethra and preventing iatrogenic injuries are crucial. Since surgery of this kind necessitates the highest level of precision, 3DP could improve by enhancing intervention outcomes and lowering complications.

Expenses are still a big worry whenever new technology is released. The stated price is negotiable. The price of the models is mostly fixed considering the quantity of materials used and the amount of production [156]. Large production quantities, monochromatic color, material choices, and other factors could contribute to its reduced cost [154]. Remarkable potential for treating andrological disorders as well as patient counseling has been shown by 3DP.

17. Printing in Robotic Urology

Due in part to advancements in less destructive surgery, urology has become a "hotbed" of innovation in the last few years. This has created an environment that encourages ongoing surgery. 3DP has been used by several centers worldwide to create customized physical models from radiological images of patients, which helps with informed consent procedures, improves resident teaching, and sharpens preoperative robotic surgery planning. In the discipline of urology, it was initially used as a visual aid that might help surgeons better understand kidney cancers [157].

Upon observing 3D models, surgeons can modify their surgical technique and potentially bring back a tactile/haptic feedback feature that is lacking in robotic surgeries [158–160]. To manufacture models from various materials simulation, 3DP, and polymer casting are combined [161]. Patient-specific models were created to replicate the intricate tumor architecture. This platform was built to imitate kidney texture, architecture, and perfusion. It was more detailed than von Rundstedt's fast-made, lifelike silicone models, with blood vessels that mimicked bleeding after tumor removal and the surrounding perinephric fat, gut, and musculature, enabling a fully immersive experience (Fig. 7) [162]. Particular models of kidneys are also used for pre-operative rehearsals on patient outcomes [157].

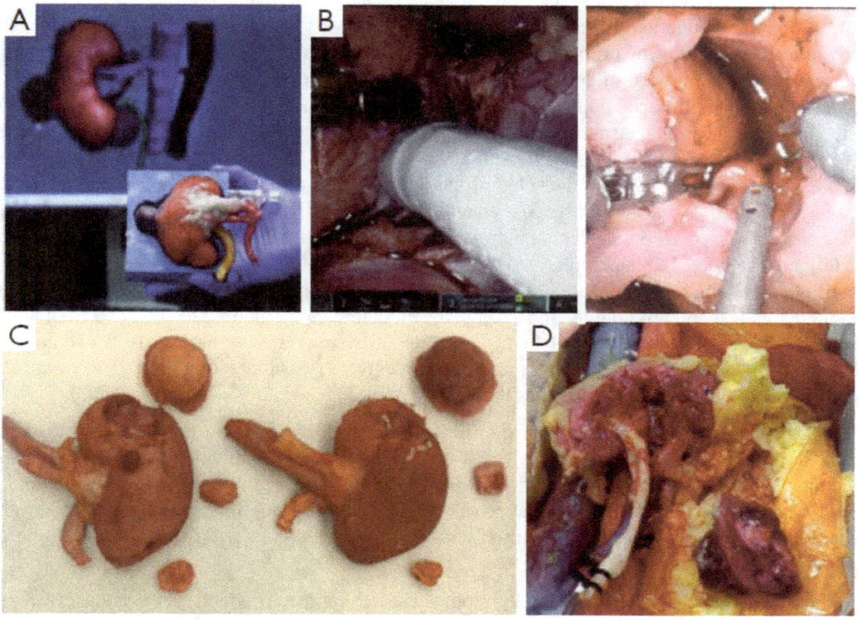

Fig. 7. *Simulation tailored to a particular patient: (A) computer design produced by segmenting the CT scan of the patient and placing a customized kidney model in a cast in the background; (B) removal of a bleeding tumor (left in a live operation, right in a practice run); (C, D) inspection of the samples after the removed lesions are depicted in simulated rehearsals. Reproduced from (Ghazi and Teplitz, 2020) with permission from Translational Andrology and Urology, Copyright [2020].*

The prediction made by Manning *et al.* that difficult surgery would eventually adopt a "practice before you play" strategy may come to pass as the discipline develops. Even for seasoned urologists, surgical rehearsal could eventually raise the bar for treatment [161]. In urology, 3DP has a promising future. In terms of surgical practice, medical education, and surgical rehearsals, it might be the next big thing.

18. Printing in Maxillofacial Surgery

Maxillofacial deformities negatively impact daily life and can result from congenital variances, trauma, or cancer. Using materials that replicate natural tissues, prosthetic healing offers both functional and cosmetic rebuilding, with polymers playing an unprecedented role in this process. The 3DP approach, where polymers are crucial, offers a quick and economical workflow process to accurately restore anatomical features unique to each patient for prosthetics.

Annually, over 69 million individuals globally experience maxillofacial deformities as a result of cancer, trauma, or inherited illnesses [163]. This entails serious negative psychological, social, and economic effects and is the primary cause of mortality and disability [164]. The complexity of the anatomy, the possibility of infection, and the disruption of important anatomical processes make reconstructing maxillofacial deformities complicated and impossible to accomplish an acceptable outcome in many cases. Thus far, polymers have been the most widely used material for prosthetics [165]. Synthetic polymers have significantly improved production, greatly resembled replacement tissues, and have been useful [166]. One of the most notable technological developments of the last few decades is digital prosthesis planning utilizing 3DP. Advanced 3D manufacturing processes, rational material design, and prosthetic methodologies have significantly impacted the traditional methods of fabricating maxillofacial prostheses. Making the most sophisticated maxillofacial prosthesis with extreme precision and customization based on individual requirements is made possible by 3DP, computer modeling, and scanning techniques. Computer-aided facilities and patient access to high-quality prostheses at a reduced cost are linked to advanced production, a realistic look, and a high degree of customization. The procedure as a whole requires less labor and expedites the healing period [167].

For maxillofacial anomalies, 3DP has significant promise for replicating tissues and anatomical features. The literature contains information on the utilization of contemporary CAD and 3DP technology in various patient-specific medical applications, ranging from design to facial prosthesis [168–170]. For maxillofacial prosthesis, there is presently no straightforward software program like DentalCAD (Exocad GmbH) with digital retention features, in contrast to intraoral rehabilitation [171].

Maxillofacial prosthetics benefit greatly from the 3DP approach, which has a good effect on emotional, psychological, and functional well-being. Treatment for facial abnormalities is challenging because of their complexity, which includes a variety of tissues and anatomical systems. The tremendous potential is determined by recent developments in 3DP technology and new perspectives on the implications of traditional and novel polymer materials, as well as their blends. With the use of this technology, patients with maxillofacial deformities can benefit from more aesthetically pleasing and practical results through the creation of personalized treatment programs and thoughtful designs.

19. Printing in Spinal Surgery

Although medical applications of this technology are still in their infancy, 3DP has the potential to completely transform healthcare due to its quick creation of bespoke shapes from a variety of materials [172,173]. Neurosurgery, specifically spinal surgery has significant potential advantages. Because of the complexity of spinal anatomy, any technology of accuracy can improve patient outcomes [172]. When performing pre-operative planning for spinal surgery, 3DP is most commonly used. A comprehensive, 3D comprehension of the disease enables more intricate planning and process simulation [172–179]. Many of the problems with standard 3D imaging are resolved when complicated disorders are assessed on a model, including the inability to accurately portray anatomy with computer skills and methods [180].

Less perioperative bleeding and faster operations are the most frequently reported benefits of improved visualization and preparation by using tailored models [181–187]. Numerous studies spanning a range of surgical procedures have demonstrated a 15–20% reduction in operation time

[188]. The main reason for the shortened operation time is realizing the problem well and the future direction of the surgical approach [188–190]. When employed in conjunction with traditional imaging for pre-operative planning, additional clinical benefits have been documented, including enhanced diagnosis, decreased fluoroscopy time, improved communication among surgical team members, and decreased screw misplacement rates [176,182]. An additional widely acknowledged advantage of 3DP models is enhanced patient education [191–193]. An intricate MRI or CT scan is far more difficult for a patient to comprehend than a physical model.

Because of the fragile nature of the surrounding anatomy, spinal surgery is inherently risky. Intraoperative guidelines that are developed using patient-specific data may be able to reduce the hazards involved with these operations [173,194]. In spine surgery, pedicle screws are a common fixation technique because, when placed correctly, they are the most effective means of stabilizing vertebrae and reducing the risk of problems [173,195–197]. To improve operation safety and monitor screw placement, computer-assisted surgery can be employed. However, because of the bone registration requirements, this technology is very costly, difficult to use, and has a lengthy learning curve that frequently necessitates hiring more staff. It also prolongs the intraoperative period [54,198,199].

These manuals could be especially helpful for more complicated anatomy. Customized guidelines are advantageous since procedural success is more difficult to obtain due to asymmetry, wide variability, and the delicate nature [200–204]. Although fixation screw placement is the primary application, the technology has been proven to be advantageous for several other operations as well. For extensive open-door laminoplasty in cervical myelopathy, Rong *et al.* used 3DP to develop guidelines [205]. 3DP technology is probably going to become commonplace and be used in more surgeries as it continues to get cheaper, faster, and more precise [206,207]. The literature and spinal surgical practice are quickly integrating this technique into their work. The method enables better patient outcomes, shorter surgical times, and improved implant properties.

20. Printing in Plastic Surgery

Contemporary imaging methods are crucial to preoperative planning in plastic surgery. Nevertheless, the limitations of traditional modalities, such as 3D reconstructions, stem from their depiction of 2D workstations. Since the development of CTA, physicians doing breast reconstructive surgery have been able to enhance clinical results by accurately and consistently choosing perforators and the best dissection technique in advance [11,208,209]. Large soft tissue defects are frequently rebuilt using perforator flap surgery following trauma or an oncologic resection. Perforator flap surgery is commonly utilized for rebuilding large soft tissue defects. With preoperative planning using CTA, reconstructive surgeons can now determine the optimal criteria. This has revolutionized the field, raised the success rate of flap transfers, and improved overall clinical results [210,211]. 3D models can provide further information beyond CTA through touch and visual assessment. A perforator flap-based restoration can be planned using a method of "reverse" model (Fig. 8) [212]. Using Osirix requires typical CTAs.

Materials Research Forum LLC
https://doi.org/10.21741/9781644903353-1

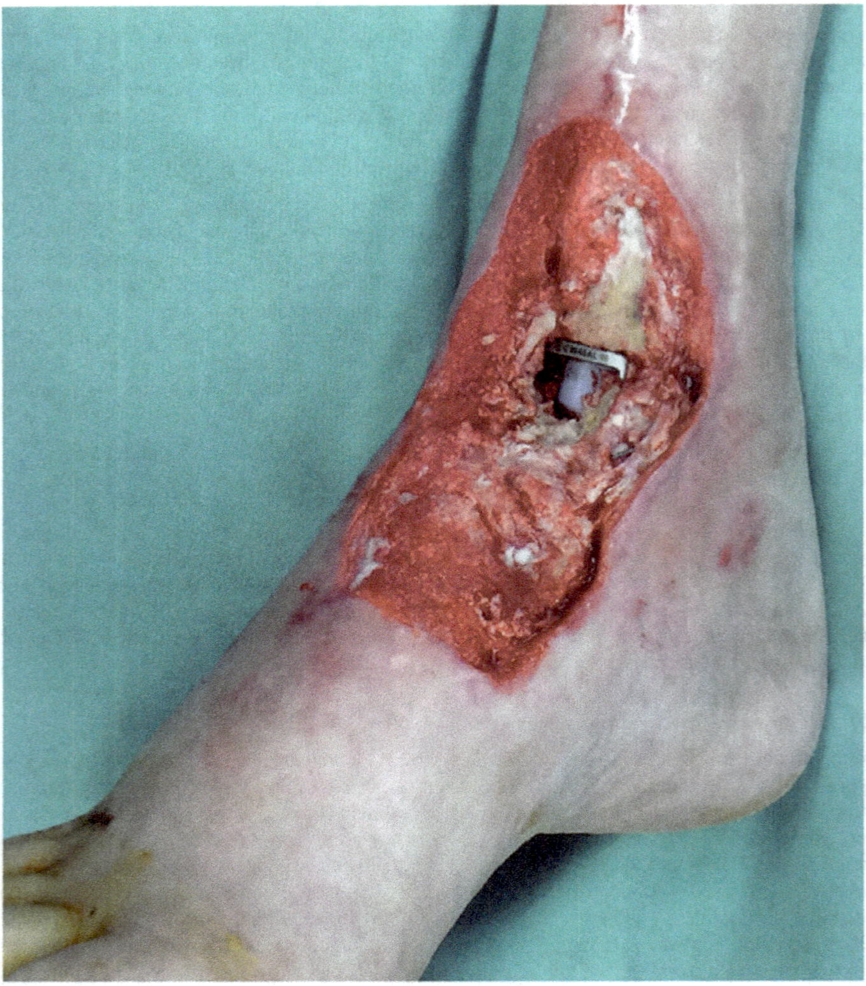

Fig. 8. *An image of the soft tissue ankle defect demonstrates the metal hardware from an earlier ankle surgery that is visible. Reproduced from (Chae et al., 2015) with permission from Microsurgery, Copyright [2015].*

Materials Research Forum LLC
https://doi.org/10.21741/9781644903353-1

Magics software (Materialise NV) helps in constructing a "reverse" model from the images (Fig. 9) [212].

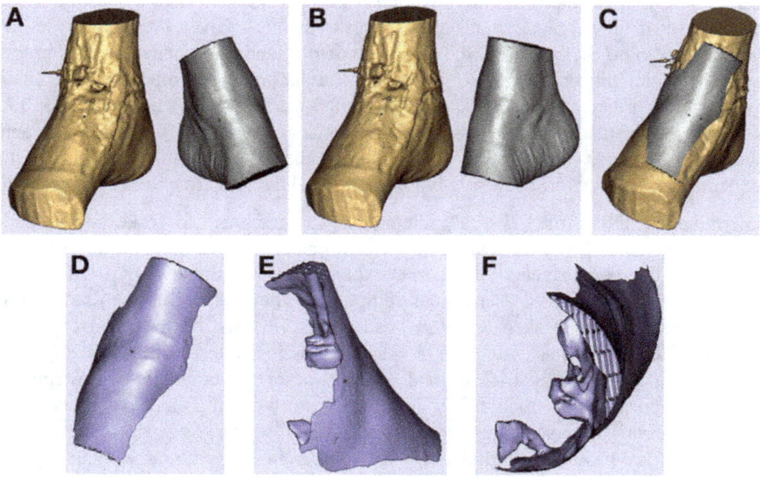

Fig. 9. *The left (normal) ankle (A) is compared to the right (pathological) ankle (B) using 3D scans. The right ankle (C) is superimposed and mirrored onto the left ankle (B). The soft tissue defect is represented as a "reverse" model by subtracting these images from one another (D-F). Reproduced from (Chae et al., 2015) with permission from Microsurgery, Copyright [2015].*

Other free open-source programs that can carry out this mirroring function are Osirix and 3D Slicer. A Cube 2 printer is used to create the "reverse" model and the diseased ankle using PLA materials. CTA is now substantial for perforator flap surgery [213]. 4D scans are also being used for creating anatomical models [214]. Using 4D CTA, researchers have assessed the dynamic flow characteristics and vascular regions of a single perforator in plastic surgery [215,216].

This helps patients to better understand the planned surgery and its possible results. This is why they can be a crucial component of informed consent. Using 3D scanning technologies like VECTRA (Canfield Imaging Systems), plastic surgeons can now precisely model possible results from a cosmetic treatment on a computer screen [217].

The sector of 3DP has experienced exponential growth and a notable decrease in cost due to the expiration of important patents. Finally, it is possible to imagine that 3DP may eventually become commonplace and serve as a vital clinical bedside instrument for plastic surgeons.

21. Challenges, Opportunities, and Limitations in Printing

The fabrication of prosthetic components, organ models, and other biomaterials is the main area of current research challenge for 3DP technology in surgery. Synthetic organ and human anatomy modeling has advanced significantly with the use of 3DP. Synthetic organ and human anatomy modeling has advanced significantly with the use of 3DP. The possibilities of eventual reproduction and tissue regeneration are becoming more and more real. New developments in 3DP help in creating models that are tailored to each patient, which offers a practical way to plan surgeries ahead of time, practice, and map spaces and times. Given the intricacy of creating a prosthetic component, SLT desktop 3DP has been determined to produce the most precise, long-lasting, and economical components for this task. Modular prostheses made to fit the anatomy and needs of a patient are created using 3D printers using a variety of polymers. Like contraceptives, prosthetics should be made specific needs and preferences of each person. Patients frequently have to travel long distances for therapy which adds to the time and financial burden for them. Consequently, among the most urgently required medical devices are braces and prosthetic limbs.

Normally, the versatility of 3DP makes it simple for designers to make adjustments. These qualities have generated a great deal of interest in medical device 3DP. More precisely, dental applications are the main drivers of the market expansion of worldwide 3DP. Dental products that can be printed are categorized as medical equipment. They can be used to make orthodontic models, replace or mend broken teeth, make dentures, crowns, bridges, and caps, as well as surgical instruments. 3DP has assured to transform medicines into prosthetic limbs, personalized on-demand drugs, patient-specific surgical models, or even human tissue. According to the literature that is currently accessible, 3DP is headed toward making this feasible, which will lead to a whole new realm of personalized medications. Therefore, it's likely that in the not-too-distant future, doctors may provide a digital file with printing instructions rather than a prescription. Although the survey notes that there are now just a few hospitals and research institutions with on-site 3DP capabilities, there should be an increase in demand for 3DP from medical personnel as additional 3DP capabilities come online. Hospitals may eventually use internal 3D printers to quickly produce body parts for trauma patients in an emergency as well as for training purposes. Today, it is already possible to artificially create breast tissue, veins, nerves, bone replacement material, and corneas using 3DP technology. It is thought to be an essential technology for the creation of whole organs or functional tissue in the future.

Despite the many benefits of 3DP, certain essential aspects of the technology still need to be significantly improved. Further manifestations in the process are required for better results. High-resolution scanners still need highly skilled personnel to operate them and are expensive. Additional disadvantages involve a lengthy scanning process that can result in mistakes because of patient movement. While multi-camera photogrammetry avoids these drawbacks, it is frequently very costly. Furthermore, the computer software that is currently on the market was not created with prosthetics in mind and is only useful for design purposes due to its extreme complexity and exclusivity for specialists. In addition, several contemporary 3D printer models are highly costly, forcing researchers to use specially made, basic 3D printers for their needs.

Conclusions

The application of 3D in surgery has expanded rapidly. This can be attributed to the remarkable versatility and vast range of possible uses of 3DP, as well as its declining cost. Given the need to use foreign software and the variety of 3DP techniques available, the path to integrating this technology into clinical practice may initially seem difficult. Including 3DP in a suitable application may be a very satisfying experience, particularly when a varied team and rapid advancements in the field are involved. We seem to be living in a dynamic, paradigm-shifting period when there is countless potential for innovation in robotic education with 3DP technology.

Conflicts of Interest

There are no conflicts to declare.

Acknowledgment

Authors acknowledge University Grants Commission of Bangladesh for financial support on a similar project.

References

[1] E. Kroll, D. Artzi, Enhancing aerospace engineering students' learning with 3D printing wind-tunnel models, Rapid Prototyp. J. 17 (2011) 393-402. https://doi.org/10.1108/13552541111156522

[2] P.F. Egan, Integrated design approaches for 3D printed tissue scaffolds: Review and outlook, Materials (Basel). 12 (2019) 2355. https://doi.org/10.3390/ma12152355

[3] T.D. Ngo, A. Kashani, G. Imbalzano, K.T.Q. Nguyen, D. Hui, Additive manufacturing (3D printing): A review of materials, methods, applications and challenges, Compos. Part B Eng. 143 (2018) 172-196. https://doi.org/10.1016/j.compositesb.2018.02.012

[4] C.L. Ventola, Medical applications for 3D printing: current and projected uses, Pharm. Ther. 39 (2014) 704.

[5] S.C. Ligon, R. Liska, J. Stampfl, M. Gurr, R. Mülhaupt, Polymers for 3D printing and customized additive manufacturing, Chem. Rev. 117 (2017) 10212-10290. https://doi.org/10.1021/acs.chemrev.7b00074

[6] P.F. Egan, K.A. Shea, S.J. Ferguson, Simulated tissue growth for 3D printed scaffolds, Biomech. Model. Mechanobiol. 17 (2018) 1481-1495. https://doi.org/10.1007/s10237-018-1040-9

[7] C. Y. Liaw, M. Guvendiren, Current and emerging applications of 3D printing in medicine, Biofabrication 9 (2017) 24102. https://doi.org/10.1088/1758-5090/aa7279

[8] J. Goole, K. Amighi, 3D printing in pharmaceutics: A new tool for designing customized drug delivery systems, Int. J. Pharm. 499 (2016) 376-394. https://doi.org/10.1016/j.ijpharm.2015.12.071

[9] I. Rubio-Pérez, A. Diaz Lantada, Surgical planning of sacral nerve stimulation procedure in presence of sacral anomalies by using personalized polymeric prototypes obtained with additive manufacturing techniques, Polymers (Basel). 12 (2020) 581. https://doi.org/10.3390/polym12030581

[10] J.M. Smit, A. Dimopoulou, A.G. Liss, C.J. Zeebregts, M. Kildal, I.S. Whitaker, A. Magnusson, R. Acosta, Preoperative CT angiography reduces surgery time in perforator flap reconstruction, J. Plast. Reconstr. Aesthetic Surg. 62 (2009) 1112-1117. https://doi.org/10.1016/j.bjps.2007.12.090

[11] W.M. Rozen, N.S. Anavekar, M.W. Ashton, D.L. Stella, D. Grinsell, R.J. Bloom, G.I. Taylor, Does the preoperative imaging of perforators with CT angiography improve operative outcomes in breast reconstruction?, Microsurg. Off. J. Int. Microsurg. Soc. Eur. Fed. Soc. Microsurg. 28 (2008) 516-523. https://doi.org/10.1002/micr.20526

[12] J. Masia, J.A. Clavero, J.R. Larranaga, X. Alomar, G. Pons, P. Serret, Multidetector-row computed tomography in the planning of abdominal perforator flaps, J. Plast. Reconstr. Aesthetic Surg. 59 (2006) 594-599. https://doi.org/10.1016/j.bjps.2005.10.024

[13] A. Alonso-Burgos, E. Garcia-Tutor, G. Bastarrika, D. Cano, A. Martinez-Cuesta, L. Pina, Preoperative planning of deep inferior epigastric artery perforator flap reconstruction with multislice-CT angiography: imaging findings and initial experience, J. Plast. Reconstr. Aesthetic Surg. 59 (2006) 585-593. https://doi.org/10.1016/j.bjps.2005.12.011

[14] J. Masia, D. Kosutic, D. Cervelli, J.A. Clavero, J.M. Monill, G. Pons, In search of the ideal method in perforator mapping: noncontrast magnetic resonance imaging, J. Reconstr. Microsurg. 26 (2010) 29-35. https://doi.org/10.1055/s-0029-1238222

[15] W.M. Rozen, D.L. Stella, J. Bowden, G.I. Taylor, M.W. Ashton, Advances in the pre-operative planning of deep inferior epigastric artery perforator flaps: Magnetic resonance angiography, Microsurg. Off. J. Int. Microsurg. Soc. Eur. Fed. Soc. Microsurg. 29 (2009) 119-123. https://doi.org/10.1002/micr.20590

[16] K. V Wong, A. Hernandez, A review of additive manufacturing, Int. Sch. Res. Not. 2012 (2012) 208760. https://doi.org/10.5402/2012/208760

[17] E. Piskin, Biodegradable polymers as biomaterials, J. Biomater. Sci. Polym. Ed. 6 (1995) 775-795. https://doi.org/10.1163/156856295X00175

[18] R. Barbucci, Integrated biomaterials science, Springer Science & Business Media, 2002. https://doi.org/10.1007/b112196

[19] J.P. Vacanti, R. Langer, Tissue engineering: the design and fabrication of living replacement devices for surgical reconstruction and transplantation, Lancet 354 (1999) S32-S34. https://doi.org/10.1016/S0140-6736(99)90247-7

[20] T. Dvir, B.P. Timko, D.S. Kohane, R. Langer, Nanotechnological strategies for engineering complex tissues, Nano-Enabled Med. Appl. (2020) 351-382. https://doi.org/10.1201/9780429399039-12

[21] D.S. Kohane, R. Langer, Polymeric biomaterials in tissue engineering, Pediatr. Res. 63 (2008) 487-491. https://doi.org/10.1203/01.pdr.0000305937.26105.e7

[22] M. Vert, Aliphatic polyesters: great degradable polymers that cannot do everything, Biomacromolecules 6 (2005) 538-546. https://doi.org/10.1021/bm0494702

[23] S. Ramakrishna, J. Mayer, E. Wintermantel, K.W. Leong, Biomedical applications of polymer-composite materials: a review, Compos. Sci. Technol. 61 (2001) 1189-1224. https://doi.org/10.1016/S0266-3538(00)00241-4

[24] E.A. Guzzi, M.W. Tibbitt, Additive manufacturing of precision biomaterials, Adv. Mater. 32 (2020) 1901994. https://doi.org/10.1002/adma.201901994

[25] M. Colaco, D.A. Igel, A. Atala, The potential of 3D printing in urological research and patient care, Nat. Rev. Urol. 15 (2018) 213-221. https://doi.org/10.1038/nrurol.2018.6

[26] F. Rengier, A. Mehndiratta, H. Von Tengg-Kobligk, C.M. Zechmann, R. Unterhinninghofen, H.-U. Kauczor, F.L. Giesel, 3D printing based on imaging data: review of medical applications, Int. J. Comput. Assist. Radiol. Surg. 5 (2010) 335-341. https://doi.org/10.1007/s11548-010-0476-x

[27] K. Townsend, T. Pietila, 3D printing and modeling of congenital heart defects: a technical review, Birth Defects Res. 110 (2018) 1091-1097. https://doi.org/10.1002/bdr2.1342

[28] G. Lales, E. Anestiadou, V. Bisbinas, J.S. Suri, G. Tsoulfas, 3D printing: shedding light into the surgical education, in: 3D Print. Appl. Med. Surg., Elsevier, 2020: pp. 21-50. https://doi.org/10.1016/B978-0-323-66164-5.00003-9

[29] E.R. Perica, Z. Sun, A systematic review of three-dimensional printing in liver disease, J. Digit. Imaging 31 (2018) 692-701. https://doi.org/10.1007/s10278-018-0067-x

[30] J.S. Witowski, J. Coles-Black, T.Z. Zuzak, M. Pędziwiatr, J. Chuen, P. Major, A. Budzyński, 3D printing in liver surgery: a systematic review, Telemed. e-Health 23 (2017) 943-947. https://doi.org/10.1089/tmj.2017.0049

[31] Y. Nishihara, Y. Isobe, Y. Kitagawa, Validation of newly developed physical laparoscopy simulator in transabdominal preperitoneal (TAPP) inguinal hernia repair, Surg. Endosc. 31 (2017) 5429-5435. https://doi.org/10.1007/s00464-017-5614-x

[32] Y. Zheng, D. Yu, J. Zhao, Y. Wu, B. Zheng, 3D printout models vs. 3D-rendered images: which is better for preoperative planning?, J. Surg. Educ. 73 (2016) 518-523. https://doi.org/10.1016/j.jsurg.2016.01.003

[33] O.C. Burdall, E. Makin, M. Davenport, N. Ade-Ajayi, 3D printing to simulate laparoscopic choledochal surgery, J. Pediatr. Surg. 51 (2016) 828-831. https://doi.org/10.1016/j.jpedsurg.2016.02.093

[34] L. Kiraly, M. Tofeig, N.K. Jha, H. Talo, Three-dimensional printed prototypes refine the anatomy of post-modified Norwood-1 complex aortic arch obstruction and allow presurgical simulation of the repair, Interact. Cardiovasc. Thorac. Surg. 22 (2016) 238-240. https://doi.org/10.1093/icvts/ivv320

[35] I. Valverde, G. Gomez, J.F. Coserria, C. Suarez-Mejias, S. Uribe, J. Sotelo, M.N. Velasco, J. Santos De Soto, A. Hosseinpour, T. Gomez-Cia, 3 D printed models for planning endovascular stenting in transverse aortic arch hypoplasia, Catheter. Cardiovasc. Interv. 85 (2015) 1006-1012. https://doi.org/10.1002/ccd.25810

[36] K.A. Barsness, D.M. Rooney, L.M. Davis, E. O'Brien, Evaluation of three sources of validity evidence for a synthetic thoracoscopic esophageal atresia/tracheoesophageal fistula repair simulator, J. Laparoendosc. Adv. Surg. Tech. 25 (2015) 599-604. https://doi.org/10.1089/lap.2014.0370

[37] T. Akiba, T. Nakada, T. Inagaki, Simulation of the fissureless technique for thoracoscopic segmentectomy using rapid prototyping, Ann. Thorac. Cardiovasc. Surg. 21 (2015) 84-86. https://doi.org/10.5761/atcs.nm.13-00322

 Materials Research Forum LLC
https://doi.org/10.21741/9781644903353-1

[38] D. Schmauss, C. Schmitz, A.K. Bigdeli, S. Weber, N. Gerber, A. Beiras-Fernandez, F. Schwarz, C. Becker, C. Kupatt, R. Sodian, Three-dimensional printing of models for preoperative planning and simulation of transcatheter valve replacement, Ann. Thorac. Surg. 93 (2012) e31-e33. https://doi.org/10.1016/j.athoracsur.2011.09.031

[39] T. Miyazaki, N. Yamasaki, T. Tsuchiya, K. Matsumoto, K. Takagi, T. Nagayasu, Airway stent insertion simulated with a three-dimensional printed airway model, Ann. Thorac. Surg. 99 (2015) e21-e23. https://doi.org/10.1016/j.athoracsur.2014.10.021

[40] K.A. Barsness, D.M. Rooney, L.M. Davis, E. O'Brien, Preliminary evaluation of a novel thoracoscopic infant lobectomy simulator, J. Laparoendosc. Adv. Surg. Tech. 25 (2015) 429-434. https://doi.org/10.1089/lap.2014.0364

[41] K. Takagi, A. Nanashima, T. Abo, J. Arai, N. Matsuo, T. Fukuda, T. Nagayasu, Three-dimensional printing model of liver for operative simulation in perihilar cholangiocarcinoma., Hepatogastroenterology. 61 (2014) 2315-2316.

[42] R. Souzaki, Y. Kinoshita, S. Ieiri, M. Hayashida, Y. Koga, K. Shirabe, T. Hara, Y. Maehara, M. Hashizume, T. Taguchi, Three-dimensional liver model based on preoperative CT images as a tool to assist in surgical planning for hepatoblastoma in a child, Pediatr. Surg. Int. 31 (2015) 593-596. https://doi.org/10.1007/s00383-015-3709-9

[43] D. Hoang, D. Perrault, M. Stevanovic, A. Ghiassi, Surgical applications of three-dimensional printing: a review of the current literature & how to get started, Ann. Transl. Med. 4 (2016). https://doi.org/10.21037/atm.2016.12.18

[44] K. Namba, A. Higaki, N. Kaneko, T. Mashiko, S. Nemoto, E. Watanabe, Microcatheter shaping for intracranial aneurysm coiling using the 3-dimensional printing rapid prototyping technology: preliminary result in the first 10 consecutive cases, World Neurosurg. 84 (2015) 178-186. https://doi.org/10.1016/j.wneu.2015.03.006

[45] M.D.B.S. Tam, S.D. Laycock, J.R.I. Brown, M. Jakeways, 3D printing of an aortic aneurysm to facilitate decision making and device selection for endovascular aneurysm repair in complex neck anatomy, J. Endovasc. Ther. 20 (2013) 863-867. https://doi.org/10.1583/13-4450MR.1

[46] K.J. Chung, D.Y. Hong, Y.T. Kim, I. Yang, Y.W. Park, H.N. Kim, Preshaping plates for minimally invasive fixation of calcaneal fractures using a real-size 3D-printed model as a preoperative and intraoperative tool, Foot Ankle Int. 35 (2014) 1231-1236. https://doi.org/10.1177/1071100714544522

[47] H.-S. Jeong, K.-J. Park, K.-M. Kil, S. Chong, H.-J. Eun, T.-S. Lee, J.-P. Lee, Minimally invasive plate osteosynthesis using 3D printing for shaft fractures of clavicles, Arch. Orthop. Trauma Surg. 134 (2014) 1551-1555. https://doi.org/10.1007/s00402-014-2075-8

[48] J.P. Costello, L.J. Olivieri, A. Krieger, O. Thabit, M.B. Marshall, S.-J. Yoo, P.C. Kim, R.A. Jonas, D.S. Nath, Utilizing three-dimensional printing technology to assess the feasibility of high-fidelity synthetic ventricular septal defect models for simulation in medical education, World J. Pediatr. Congenit. Hear. Surg. 5 (2014) 421-426. https://doi.org/10.1177/2150135114528721

[49] J. Soleman, F. Thieringer, J. Beinemann, C. Kunz, R. Guzman, Computer-assisted virtual planning and surgical template fabrication for frontoorbital advancement, Neurosurg. Focus 38 (2015) E5. https://doi.org/10.3171/2015.3.FOCUS14852

[50] W.Y. Angela, M. Khan, On-demand three-dimensional printing of surgical supplies in conflict zones, J. Trauma Acute Care Surg. 78 (2015) 201-203. https://doi.org/10.1097/TA.0000000000000481

[51] T.M. Rankin, N.A. Giovinco, D.J. Cucher, G. Watts, B. Hurwitz, D.G. Armstrong, Three-dimensional printing surgical instruments: are we there yet?, J. Surg. Res. 189 (2014) 193-197. https://doi.org/10.1016/j.jss.2014.02.020

[52] M. Maeda, N. Kanai, S. Kobayashi, T. Hosoi, R. Takagi, T. Ohki, Y. Muragaki, M. Yamato, S. Eguchi, F. Fukai, Endoscopic cell sheet transplantation device developed by using a 3-dimensional printer and its feasibility evaluation in a porcine model, Gastrointest. Endosc. 82 (2015) 147-152. https://doi.org/10.1016/j.gie.2015.01.062

[53] S. Kaneyama, T. Sugawara, M. Sumi, Safe and accurate midcervical pedicle screw insertion procedure with the patient-specific screw guide template system, Spine (Phila. Pa. 1976). 40 (2015) E341-E348. https://doi.org/10.1097/BRS.0000000000000772

[54] H. Chen, D. Wu, H. Yang, K. Guo, Clinical use of 3D printing guide plate in posterior lumbar pedicle screw fixation, Med. Sci. Monit. Int. Med. J. Exp. Clin. Res. 21 (2015) 3948. https://doi.org/10.12659/MSM.895597

[55] L. Ma, Y. Zhou, Y. Zhu, Z. Lin, Y. Wang, Y. Zhang, H. Xia, C. Mao, 3D-printed guiding templates for improved osteosarcoma resection, Sci. Rep. 6 (2016) 23335. https://doi.org/10.1038/srep23335

[56] M. Hirao, S. Ikemoto, H. Tsuboi, S. Akita, S. Ohshima, Y. Saeki, H. Yoshikawa, K. Sugamoto, T. Murase, J. Hashimoto, Computer assisted planning and custom-made surgical guide for malunited pronation deformity after first metatarsophalangeal joint arthrodesis in rheumatoid arthritis: a case report, Comput. Aided Surg. 19 (2014) 13-19. https://doi.org/10.3109/10929088.2014.885992

[57] Z. Jaffry, M. Masjedi, S. Clarke, S. Harris, M. Karia, B. Andrews, J. Cobb, Unicompartmental knee arthroplasties: robot vs. patient specific instrumentation, Knee 21 (2014) 428-434. https://doi.org/10.1016/j.knee.2013.11.017

[58] M. del Junco, Z. Okhunov, R. Yoon, R. Khanipour, S. Juncal, G. Abedi, A. Lusch, J. Landman, Development and initial porcine and cadaver experience with three-dimensional printing of endoscopic and laparoscopic equipment, J. Endourol. 29 (2015) 58-62. https://doi.org/10.1089/end.2014.0280

[59] A. Priester, S. Natarajan, J.D. Le, J. Garritano, B. Radosavcev, W. Grundfest, D.J.A. Margolis, L.S. Marks, J. Huang, A system for evaluating magnetic resonance imaging of prostate cancer using patient-specific 3D printed molds, Am. J. Clin. Exp. Urol. 2 (2014) 127.

[60] A. Shqaidef, A.F. Ayoub, B.S. Khambay, How accurate are rapid prototyped (RP) final orthognathic surgical wafers? A pilot study, Br. J. Oral Maxillofac. Surg. 52 (2014) 609-614. https://doi.org/10.1016/j.bjoms.2014.04.010

[61] S.-H. Kang, M.-K. Kim, B.C. Kim, S.-H. Lee, Orthognathic Y-splint: a CAD/CAM-engineered maxillary repositioning wafer assembly, Br. J. Oral Maxillofac. Surg. 52 (2014) 667-669. https://doi.org/10.1016/j.bjoms.2014.01.023

[62] N. Adolphs, W. Liu, E. Keeve, B. Hoffmeister, RapidSplint: virtual splint generation for orthognathic surgery-results of a pilot series, Comput. Aided Surg. 19 (2014) 20-28. https://doi.org/10.3109/10929088.2014.887778

[63] B. Li, L. Zhang, H. Sun, J. Yuan, S.G.F. Shen, X. Wang, A novel method of computer aided orthognathic surgery using individual CAD/CAM templates: a combination of osteotomy and repositioning guides, Br. J. Oral Maxillofac. Surg. 51 (2013) e239-e244. https://doi.org/10.1016/j.bjoms.2013.03.007

[64] M. Vehmeijer, M. van Eijnatten, N. Liberton, J. Wolff, A novel method of orbital floor reconstruction using virtual planning, 3-dimensional printing, and autologous bone, J. Oral Maxillofac. Surg. 74 (2016) 1608-1612. https://doi.org/10.1016/j.joms.2016.03.044

[65] A. Darwood, J. Collier, N. Joshi, W.E. Grant, V. Sauret-Jackson, R. Richards, A. Dawood, N. Kirkpatrick, Re-thinking 3D printing: a novel approach to guided facial contouring, J. Cranio-Maxillofacial Surg. 43 (2015) 1256-1260. https://doi.org/10.1016/j.jcms.2015.06.001

[66] A. Ruzza, M. Parekh, S. Ferrari, G. Salvalaio, Y. Nahum, C. Bovone, D. Ponzin, M. Busin, Preloaded donor corneal lenticules in a new validated 3D printed smart storage glide for Descemet stripping automated endothelial keratoplasty, Br. J. Ophthalmol. 99 (2015) 1388-1395. https://doi.org/10.1136/bjophthalmol-2014-306510

[67] M.L. Stitely, H. Paterson, Using three-dimensional printing to fabricate a tubing connector for dilation and evacuation, Obstet. Gynecol. 127 (2016) 317-319. https://doi.org/10.1097/AOG.0000000000001237

[68] A.B. AlAli, M.F. Griffin, P.E. Butler, Three-dimensional printing surgical applications, Eplasty 15 (2015).

[69] J. Li, M. Chen, X. Fan, H. Zhou, Recent advances in bioprinting techniques: approaches, applications and future prospects, J. Transl. Med. 14 (2016) 1-15. https://doi.org/10.1186/s12967-016-1028-0

[70] T. Goffin, P. Borry, K. Dierickx, H. Nys, Why eight EU Member States signed, but not yet ratified the Convention for Human Rights and Biomedicine, Health Policy (New. York). 86 (2008) 222-233. https://doi.org/10.1016/j.healthpol.2007.10.011

[71] L.K. Prasad, H. Smyth, 3D Printing technologies for drug delivery: a review, Drug Dev. Ind. Pharm. 42 (2016) 1019-1031. https://doi.org/10.3109/03639045.2015.1120743

[72] D. Kanters, A. de Vries, H. Boon, J. Urbach, A. Becht, H.-A. Kooistra, Quality assurance in medical 3D-printing, in: World Congr. Med. Phys. Biomed. Eng. 2018 June 3-8, 2018, Prague, Czech Repub. (Vol. 1), Springer, 2019: pp. 669-674. https://doi.org/10.1007/978-981-10-9035-6_125

[73] L. Adalbert, S.P.Y. Kanti, O. Jójárt-Laczkovich, H. Akel, I. Csóka, Expanding quality by design principles to support 3D printed medical device development following the renewed regulatory framework in Europe, Biomedicines 10 (2022) 2947. https://doi.org/10.3390/biomedicines10112947

[74] M. Kritikos, 3D Bio-Printing for Medical and Enhancement Purposes: Legal and Ethical Aspects, European Parliamentary Research Service (EPRS), 2018, (2018).

[75] F.L. Lopez, T.B. Ernest, C. Tuleu, M.O. Gul, Formulation approaches to pediatric oral drug delivery: benefits and limitations of current platforms, Expert Opin. Drug Deliv. 12 (2015) 1727-1740. https://doi.org/10.1517/17425247.2015.1060218

[76] S. Palekar, P.K. Nukala, S.M. Mishra, T. Kipping, K. Patel, Application of 3D printing technology and quality by design approach for development of age-appropriate pediatric formulation of baclofen, Int. J. Pharm. 556 (2019) 106-116. https://doi.org/10.1016/j.ijpharm.2018.11.062

[77] T. Persons, 3D Printing: Opportunities, Challenges, and Policy Implications of Additive Manufacturing, in: GAO-15-505Sp-Addictive Manuf. Forum, n. June, 2015: p. 63.

[78] C. Weller, R. Kleer, F.T. Piller, Economic implications of 3D printing: Market structure models in light of additive manufacturing revisited, Int. J. Prod. Econ. 164 (2015) 43-56. https://doi.org/10.1016/j.ijpe.2015.02.020

[79] J. Manners-Bell, K. Lyon, The implications of 3D printing for the global logistics industry, Transp. Intell. 1 (2012) 1-5.

[80] K. Cottrill, Transforming the future of supply chains through disruptive innovation, MIT Cent. Transp. Logist. Work. Pap. Spring (2011).

[81] M. Ye, The impact of 3D printing on the world container transport, (2015).

[82] T. Bartel, A. Rivard, A. Jimenez, C.A. Mestres, S. Müller, Medical three-dimensional printing opens up new opportunities in cardiology and cardiac surgery, Eur. Heart J. 39 (2018) 1246-1254. https://doi.org/10.1093/eurheartj/ehx016

[83] I.A. Ziogas, N.N. Zein, C. Quintini, C.M. Miller, G. Tsoulfas, Three-dimensional (3D) printing and liver transplantation, in: 3D Print. Appl. Med. Surg., Elsevier, 2020: pp. 97-116. https://doi.org/10.1016/B978-0-323-66164-5.00007-6

[84] D.A.B. Oliveira, R.Q. Feitosa, M.M. Correia, Segmentation of liver, its vessels and lesions from CT images for surgical planning, Biomed. Eng. Online 10 (2011) 1-23. https://doi.org/10.1186/1475-925X-10-30

[85] S.G. Farid, K.R. Prasad, G. Morris-Stiff, Operative terminology and post-operative management approaches applied to hepatic surgery: Trainee perspectives, World J. Gastrointest. Surg. 5 (2013) 146. https://doi.org/10.4240/wjgs.v5.i5.146

[86] K. Németh, R. Deshpande, Z. Máthé, A. Szuák, M. Kiss, C. Korom, Á. Nemeskéri, L. Kóbori, Extrahepatic arteries of the human liver-anatomical variants and surgical relevancies, Transpl. Int. 28 (2015) 1216-1226. https://doi.org/10.1111/tri.12630

[87] M. Uršič, M. Vrecl, G. Fazarinc, Corrosion cast study of the canine hepatic veins, Folia Morphol. (Warsz). 73 (2014) 475-481. https://doi.org/10.5603/FM.2014.0071

[88] M. Uršič, D. Ravnik, M. Hribernik, J. Pečar, J. Butinar, G. Fazarinc, Gross anatomy of the portal vein and hepatic artery ramifications in dogs: corrosion cast study, Anat. Histol. Embryol. 36 (2007) 83-87. https://doi.org/10.1111/j.1439-0264.2006.00719.x

[89] K.M. Farooqi, P.P. Sengupta, Echocardiography and three-dimensional printing: sound ideas to touch a heart, J. Am. Soc. Echocardiogr. 28 (2015) 398-403. https://doi.org/10.1016/j.echo.2015.02.005

[90] A. Shafiee, A. Atala, Printing technologies for medical applications, Trends Mol. Med. 22 (2016) 254-265. https://doi.org/10.1016/j.molmed.2016.01.003

[91] B. Mosadegh, G. Xiong, S. Dunham, J.K. Min, Current progress in 3D printing for cardiovascular tissue engineering, Biomed. Mater. 10 (2015) 34002. https://doi.org/10.1088/1748-6041/10/3/034002

[92] A.A. Giannopoulos, M.L. Steigner, E. George, M. Barile, A.R. Hunsaker, F.J. Rybicki, D. Mitsouras, Cardiothoracic applications of 3-dimensional printing, J. Thorac. Imaging 31 (2016) 253-272. https://doi.org/10.1097/RTI.0000000000000217

[93] R.K. Kankala, K. Zhu, J. Li, C.-S. Wang, S.-B. Wang, A.-Z. Chen, Fabrication of arbitrary 3D components in cardiac surgery: from macro-, micro-to nanoscale, Biofabrication 9 (2017) 32002. https://doi.org/10.1088/1758-5090/aa8113

[94] D. Shi, K. Liu, X. Zhang, H. Liao, X. Chen, Applications of three-dimensional printing technology in the cardiovascular field, Intern. Emerg. Med. 10 (2015) 769-780. https://doi.org/10.1007/s11739-015-1282-9

[95] S. Singare, Y. Liu, D. Li, B. Lu, S. He, Individually prefabricated prosthesis for maxilla reconstruction, J. Prosthodont. 17 (2008) 135-140. https://doi.org/10.1111/j.1532-849X.2007.00266.x

[96] S. Jacobs, R. Grunert, F.W. Mohr, V. Falk, 3D-Imaging of cardiac structures using 3D heart models for planning in heart surgery: a preliminary study, Interact. Cardiovasc. Thorac. Surg. 7 (2008) 6-9. https://doi.org/10.1510/icvts.2007.156588

[97] K. Subburaj, C. Nair, S. Rajesh, S.M. Meshram, B. Ravi, Rapid development of auricular prosthesis using CAD and rapid prototyping technologies, Int. J. Oral Maxillofac. Surg. 36 (2007) 938-943. https://doi.org/10.1016/j.ijom.2007.07.013

[98] R.F. Youssef, K. Spradling, R. Yoon, B. Dolan, J. Chamberlin, Z. Okhunov, R. Clayman, J. Landman, Applications of three-dimensional printing technology in urological practice, BJU Int. 116 (2015) 697-702. https://doi.org/10.1111/bju.13183

[99] P. Bangeas, G. Voulalas, K. Ktenidis, Rapid prototyping in aortic surgery, Interact. Cardiovasc. Thorac. Surg. 22 (2016) 513-514. https://doi.org/10.1093/icvts/ivv395

[100] D. Ho, A. Squelch, Z. Sun, Modelling of aortic aneurysm and aortic dissection through 3D printing, J. Med. Radiat. Sci. 64 (2017) 10-17. https://doi.org/10.1002/jmrs.212

[101] J.F.B. Chick, S.N. Reddy, C.Y. Alice, T. Kelil, R.N. Srinivasa, K.J. Cooper, W.E. Saad, Three-dimensional printing facilitates successful endovascular closure of a type II abernethy malformation using an Amplatzer Atrial Septal Occluder device, Ann. Vasc. Surg. 43 (2017) 311-e15. https://doi.org/10.1016/j.avsg.2017.02.012

[102] D. Maragiannis, M.S. Jackson, S.R. Igo, R.C. Schutt, P. Connell, J. Grande-Allen, C.M. Barker, S.M. Chang, M.J. Reardon, W.A. Zoghbi, Replicating patient-specific severe aortic

valve stenosis with functional 3D modeling, Circ. Cardiovasc. Imaging 8 (2015) e003626. https://doi.org/10.1161/CIRCIMAGING.115.003626

[103] D. Maragiannis, M.S. Jackson, S.R. Igo, S.M. Chang, W.A. Zoghbi, S.H. Little, Functional 3D printed patient-specific modeling of severe aortic stenosis, J. Am. Coll. Cardiol. 64 (2014) 1066-1068. https://doi.org/10.1016/j.jacc.2014.05.058

[104] S. Schievano, F. Migliavacca, L. Coats, S. Khambadkone, M. Carminati, N. Wilson, J.E. Deanfield, P. Bonhoeffer, A.M. Taylor, Percutaneous pulmonary valve implantation based on rapid prototyping of right ventricular outflow tract and pulmonary trunk from MR data, Radiology 242 (2007) 490-497. https://doi.org/10.1148/radiol.2422051994

[105] S. Milosavljevic, P.D. Milburn, B.W. Knox, The influence of occupation on lumbar sagittal motion and posture, Ergonomics 48 (2005) 657-667. https://doi.org/10.1080/00140130500070848

[106] H. Takao, S. Amemiya, E. Shibata, K. Ohtomo, Three-Dimensional Printing of Hollow Portal Vein Stenosis Models: A Feasibility Study., J. Vasc. Interv. Radiol. JVIR 27 (2016) 1755-1758. https://doi.org/10.1016/j.jvir.2016.05.022

[107] C. Chevallier, W. Willaert, E. Kawa, M. Centola, B. Steger, R. Dirnhofer, P. Mangin, S. Grabherr, Postmortem circulation: a new model for testing endovascular devices and training clinicians in their use, Clin. Anat. 27 (2014) 556-562. https://doi.org/10.1002/ca.22357

[108] X. Han, R. Bibb, R. Harris, Engineering design of artificial vascular junctions for 3D printing, Biofabrication 8 (2016) 25018. https://doi.org/10.1088/1758-5090/8/2/025018

[109] K. Sommer, R.L. Izzo, L. Shepard, A.R. Podgorsak, S. Rudin, A.H. Siddiqui, M.F. Wilson, E. Angel, Z. Said, M. Springer, Design optimization for accurate flow simulations in 3D printed vascular phantoms derived from computed tomography angiography, in: Med. Imaging 2017 Imaging Informatics Heal. Res. Appl., SPIE, 2017: pp. 180-191. https://doi.org/10.1117/12.2253711

[110] C. Canstein, P. Cachot, A. Faust, A.F. Stalder, J. Bock, A. Frydrychowicz, J. Küffer, J. Hennig, M. Markl, 3D MR flow analysis in realistic rapid-prototyping model systems of the thoracic aorta: comparison with in vivo data and computational fluid dynamics in identical vessel geometries, Magn. Reson. Med. An Off. J. Int. Soc. Magn. Reson. Med. 59 (2008) 535-546. https://doi.org/10.1002/mrm.21331

[111] R.R. Ahmadian, A.P. Boyd, A.M. Scotti, J.D. Collins, J.C. Carr, P.M. McCarthy, S.C. Malaisrie, A.J. Barker, M. Markl, Comprehensive evaluation of aortic disease by in-vivo 4D flow MRI and 3D printing of patient-specific models: a feasibility study, J. Cardiovasc. Magn. Reson. 18 (2016) 1-3. https://doi.org/10.1186/s12968-015-0221-2

[112] N. Martelli, C. Serrano, H. van den Brink, J. Pineau, P. Prognon, I. Borget, S. El Batti, Advantages and disadvantages of 3-dimensional printing in surgery: a systematic review, Surgery 159 (2016) 1485-1500. https://doi.org/10.1016/j.surg.2015.12.017

[113] R. Vaishya, M.K. Patralekh, A. Vaish, A.K. Agarwal, V. Vijay, Publication trends and knowledge mapping in 3D printing in orthopaedics, J. Clin. Orthop. Trauma 9 (2018) 194-201. https://doi.org/10.1016/j.jcot.2018.07.006

[114] V. Bagaria, S. Deshpande, D.D. Rasalkar, A. Kuthe, B.K. Paunipagar, Use of rapid prototyping and three-dimensional reconstruction modeling in the management of complex fractures, Eur. J. Radiol. 80 (2011) 814-820. https://doi.org/10.1016/j.ejrad.2010.10.007

[115] C. Hurson, A. Tansey, B. O'donnchadha, P. Nicholson, J. Rice, J. McElwain, Rapid prototyping in the assessment, classification and preoperative planning of acetabular fractures, Injury 38 (2007) 1158-1162. https://doi.org/10.1016/j.injury.2007.05.020

[116] H. Lal, M.K. Patralekh, 3D printing and its applications in orthopaedic trauma: a technological marvel, J. Clin. Orthop. Trauma 9 (2018) 260-268. https://doi.org/10.1016/j.jcot.2018.07.022

[117] T.M. Wong, J. Jin, T.W. Lau, C. Fang, C.H. Yan, K. Yeung, M. To, F. Leung, The use of three-dimensional printing technology in orthopaedic surgery: a review, J. Orthop. Surg. 25 (2017) 2309499016684077. https://doi.org/10.1177/2309499016684077

[118] L. Wan, X. Zhang, S. Zhang, K. Li, P. Cao, J. Li, G. Wu, Clinical feasibility and application value of computer virtual reduction combined with 3D printing technique in complex acetabular fractures, Exp. Ther. Med. 17 (2019) 3630-3636. https://doi.org/10.3892/etm.2019.7344

[119] Z. Liu, J. Jia, Y. Zhang, W. Tian, X. Jin, Y. Hu, Internal fixation of complicated acetabular fractures directed by preoperative surgery with 3D printing models, Orthop. Surg. 9 (2017) 257-260. https://doi.org/10.1111/os.12324

[120] H.J. Kang, B.S. Kim, S.M. Kim, Y.M. Kim, H.N. Kim, J.Y. Park, J.H. Cho, Y. Choi, Can preoperative 3D printing change surgeon's operative plan for distal tibia fracture?, Biomed Res. Int. 2019 (2019) 7059413. https://doi.org/10.1155/2019/7059413

[121] C.M. Wixted, J.R. Peterson, R.J. Kadakia, S.B. Adams, Three-dimensional printing in orthopaedic surgery: current applications and future developments, JAAOS Glob. Res. Rev. 5 (2021) e20. https://doi.org/10.5435/JAAOSGlobal-D-20-00230

[122] B. Garg, M. Gupta, M. Singh, D. Kalyanasundaram, Outcome and safety analysis of 3D-printed patient-specific pedicle screw jigs for complex spinal deformities: a comparative study, Spine J. 19 (2019) 56-64. https://doi.org/10.1016/j.spinee.2018.05.001

[123] W. Cho, A.V. Job, J. Chen, J.H. Baek, A review of current clinical applications of three-dimensional printing in spine surgery, Asian Spine J. 12 (2018) 171. https://doi.org/10.4184/asj.2018.12.1.171

[124] J.S. Mulford, S. Babazadeh, N. Mackay, Three-dimensional printing in orthopaedic surgery: review of current and future applications, ANZ J. Surg. 86 (2016) 648-653. https://doi.org/10.1111/ans.13533

[125] K.C. Wong, 3D-printed patient-specific applications in orthopedics, Orthop. Res. Rev. (2016) 57-66. https://doi.org/10.2147/ORR.S99614

[126] A.E.M. Eltorai, E. Nguyen, A.H. Daniels, Three-dimensional printing in orthopedic surgery, Orthopedics 38 (2015) 684-687. https://doi.org/10.3928/01477447-20151016-05

[127] A. Aimar, A. Palermo, B. Innocenti, The role of 3D printing in medical applications: a state of the art, J. Healthc. Eng. 2019 (2019) 5340616. https://doi.org/10.1155/2019/5340616

[128] C.L. Jeng, J.T. Campbell, E.Y. Tang, R.A. Cerrato, M.S. Myerson, Tibiotalocalcaneal arthrodesis with bulk femoral head allograft for salvage of large defects in the ankle, Foot Ankle Int. 34 (2013) 1256-1266. https://doi.org/10.1177/1071100713488765

[129] P. Canzi, M. Magnetto, S. Marconi, P. Morbini, S. Mauramati, F. Aprile, I. Avato, F. Auricchio, M. Benazzo, New frontiers and emerging applications of 3D printing in ENT surgery: a systematic review of the literature, Acta Otorhinolaryngol. Ital. 38 (2018) 286. https://doi.org/10.14639/0392-100X-1984

[130] P.A. Yushkevich, J. Piven, H.C. Hazlett, R.G. Smith, S. Ho, J.C. Gee, G. Gerig, User-guided 3D active contour segmentation of anatomical structures: significantly improved efficiency and reliability, Neuroimage 31 (2006) 1116-1128. https://doi.org/10.1016/j.neuroimage.2006.01.015

[131] A.S. Rose, C.E. Webster, O.L.A. Harrysson, E.J. Formeister, R.B. Rawal, C.E. Iseli, Pre-operative simulation of pediatric mastoid surgery with 3D-printed temporal bone models, Int. J. Pediatr. Otorhinolaryngol. 79 (2015) 740-744. https://doi.org/10.1016/j.ijporl.2015.03.004

[132] S.E. Mowry, H. Jammal, C. Myer IV, C.A. Solares, P. Weinberger, A novel temporal bone simulation model using 3D printing techniques, Otol. Neurotol. 36 (2015) 1562-1565. https://doi.org/10.1097/MAO.0000000000000848

[133] J.B. Hochman, C. Rhodes, D. Wong, J. Kraut, J. Pisa, B. Unger, Comparison of cadaveric and isomorphic three-dimensional printed models in temporal bone education, Laryngoscope 125 (2015) 2353-2357. https://doi.org/10.1002/lary.24919

[134] M.J. Da Cruz, H.W. Francis, Face and content validation of a novel three-dimensional printed temporal bone for surgical skills development, J. Laryngol. Otol. 129 (2015) S23-S29. https://doi.org/10.1017/S0022215115001346

[135] J.B. Hochman, J. Kraut, K. Kazmerik, B.J. Unger, Generation of a 3D printed temporal bone model with internal fidelity and validation of the mechanical construct, Otolaryngol. Neck Surg. 150 (2014) 448-454. https://doi.org/10.1177/0194599813518008

[136] C.M. Chiesa Estomba, I. González Fernández, M.Á. Iglesias Otero, How we do it: anterior and posterior nosebleed trainer, the 3D printing epistaxis project, Clin Otolaryngol 25 (2016) 1-2. https://doi.org/10.1111/coa.12711

[137] C.M. Johnson, J.T. Howell, D.J. Mettenburg, F.A. Rueggeberg, R.J. Howell, G.N. Postma, P.M. Weinberger, Mechanical modeling of the human cricoid cartilage using computer-aided design: applications in airway balloon dilation research, Ann. Otol. Rhinol. Laryngol. 125 (2016) 69-76. https://doi.org/10.1177/0003489415598999

[138] K.R. Kavanagh, V. Cote, Y. Tsui, S. Kudernatsch, D.R. Peterson, T.A. Valdez, Pediatric laryngeal simulator using 3D printed models: a novel technique, Laryngoscope 127 (2017) E132-E137. https://doi.org/10.1002/lary.26326

[139] S. Bustamante, S. Bose, P. Bishop, R. Klatte, F. Norris, Novel application of rapid prototyping for simulation of bronchoscopic anatomy, J. Cardiothorac. Vasc. Anesth. 28 (2014) 1122-1125. https://doi.org/10.1053/j.jvca.2013.08.015

Materials Research Forum LLC
https://doi.org/10.21741/9781644903353-1

[140] T.A. Ainsworth, J.B. Kobler, G.J. Loan, J.A. Burns, Simulation model for transcervical laryngeal injection providing real-time feedback, Ann. Otol. Rhinol. Laryngol. 123 (2014) 881-886. https://doi.org/10.1177/0003489414539922

[141] D. Rokaya, V. Srimaneepong, J. Sapkota, J. Qin, K. Siraleartmukul, V. Siriwongrungson, Polymeric materials and films in dentistry: An overview, J. Adv. Res. 14 (2018) 25-34. https://doi.org/10.1016/j.jare.2018.05.001

[142] H.-H. Lin, D. Lonic, L.-J. Lo, 3D printing in orthognathic surgery− A literature review, J. Formos. Med. Assoc. 117 (2018) 547-558. https://doi.org/10.1016/j.jfma.2018.01.008

[143] C.M. Cristache, S. Gurbanescu, Accuracy evaluation of a stereolithographic surgical template for dental implant insertion using 3D superimposition protocol, Int. J. Dent. 2017 (2017) 4292081. https://doi.org/10.1155/2017/4292081

[144] A. Tahmaseb, D. Wismeijer, W. Coucke, W. Derksen, Computer technology applications in surgical implant dentistry: a systematic review, Int J Oral Maxillofac Implant. 29 (2014) 25-42. https://doi.org/10.11607/jomi.2014suppl.g1.2

[145] D.A. Tyndall, J.B. Price, S. Tetradis, S.D. Ganz, C. Hildebolt, W.C. Scarfe, Position statement of the American Academy of Oral and Maxillofacial Radiology on selection criteria for the use of radiology in dental implantology with emphasis on cone beam computed tomography, Oral Surg. Oral Med. Oral Pathol. Oral Radiol. 113 (2012) 817-826. https://doi.org/10.1016/j.oooo.2012.03.005

[146] M. Klein, M. Abrams, Computer-guided surgery utilizing a computer-milled surgical template., Pract. Proced. Aesthetic Dent. PPAD 13 (2001) 165-169.

[147] M. Revilla-León, M. Sadeghpour, M. Özcan, An update on applications of 3D printing technologies used for processing polymers used in implant dentistry, Odontology 108 (2020) 331-338. https://doi.org/10.1007/s10266-019-00441-7

[148] P.B. Tardieu, L. Vrielinck, E. Escolano, Computer-assisted implant placement. A case report: treatment of the mandible., Int. J. Oral Maxillofac. Implants 18 (2003).

[149] G.A. Mandelaris, S.D. Vlk, Guided implant surgery with placement of a presurgical CAD/CAM patient-specific abutment and provisional in the esthetic zone, Compend Contin Educ Dent 35 (2014) 494-504.

[150] L. Tapie, N. Lebon, B. Mawussi, F. Duret, J.-P. Attal, Understanding dental CAD/CAM for restorations--the digital workflow from a mechanical engineering viewpoint., Int. J. Comput. Dent. 18 (2015) 21-44.

[151] S.D. Ganz, Three-dimensional imaging and guided surgery for dental implants, Dent. Clin. 59 (2015) 265-290. https://doi.org/10.1016/j.cden.2014.11.001

[152] M. Javaid, A. Haleem, 4D printing applications in medical field: a brief review, Clin. Epidemiol. Glob. Heal. 7 (2019) 317-321. https://doi.org/10.1016/j.cegh.2018.09.007

[153] C. Schubert, M.C. Van Langeveld, L.A. Donoso, Innovations in 3D printing: a 3D overview from optics to organs, Br. J. Ophthalmol. 98 (2014) 159-161. https://doi.org/10.1136/bjophthalmol-2013-304446

[154] G.E. Cacciamani, Z. Okhunov, A.D. Meneses, M.E. Rodriguez-Socarras, J.G. Rivas, F. Porpiglia, E. Liatsikos, D. Veneziano, Impact of three-dimensional printing in urology: state of the art and future perspectives. A systematic review by ESUT-YAUWP Group, Eur. Urol. 76 (2019) 209-221. https://doi.org/10.1016/j.eururo.2019.04.044

[155] E. Checcucci, D. Amparore, C. Fiori, M. Manfredi, M. Ivano, M. Di Dio, G. Niculescu, F. Piramide, G. Cattaneo, P. Piazzolla, 3D imaging applications for robotic urologic surgery: an ESUT YAUWP review, World J. Urol. 38 (2020) 869-881. https://doi.org/10.1007/s00345-019-02922-4

[156] R. Souzaki, Y. Kinoshita, S. Ieiri, N. Kawakubo, S. Obata, T. Jimbo, Y. Koga, M. Hashizume, T. Taguchi, Preoperative surgical simulation of laparoscopic adrenalectomy for neuroblastoma using a three-dimensional printed model based on preoperative CT images, J. Pediatr. Surg. 50 (2015) 2112-2115. https://doi.org/10.1016/j.jpedsurg.2015.08.037

[157] J.L. Silberstein, M.M. Maddox, P. Dorsey, A. Feibus, R. Thomas, B.R. Lee, Physical models of renal malignancies using standard cross-sectional imaging and 3-dimensional printers: a pilot study, Urology 84 (2014) 268-273. https://doi.org/10.1016/j.urology.2014.03.042

[158] M.E. Westerman, J.M. Matsumoto, J.M. Morris, B.C. Leibovich, Three-dimensional printing for renal cancer and surgical planning, Eur. Urol. Focus 2 (2016) 574-576. https://doi.org/10.1016/j.euf.2016.12.009

[159] P. Chandak, N. Byrne, H. Lynch, C. Allen, G. Rottenberg, A. Chandra, N. Raison, H. Ahmed, V. Kasivisvanathan, O. Elhage, Three-dimensional printing in robot-assisted radical prostatectomy-an Idea, Development, Exploration, Assessment, Long-term follow-up (IDEAL) Phase 2a study, BJU Int. 122 (2018) 360-361. https://doi.org/10.1111/bju.14189

[160] N. Wake, T. Rude, S.K. Kang, M.D. Stifelman, J.F. Borin, D.K. Sodickson, W.C. Huang, H. Chandarana, 3D printed renal cancer models derived from MRI data: application in pre-surgical planning, Abdom. Radiol. 42 (2017) 1501-1509. https://doi.org/10.1007/s00261-016-1022-2

[161] T. Smektala, A. Gołab, M. Królikowski, M. Słojewski, Low cost silicone renal replicas for surgical training. Technical note, Arch. Españoles Urol. 69 (2016) 434-436.

[162] A.E. Ghazi, B.A. Teplitz, Role of 3D printing in surgical education for robotic urology procedures, Transl. Androl. Urol. 9 (2020) 931. https://doi.org/10.21037/tau.2020.01.03

[163] M.C. Dewan, A. Rattani, S. Gupta, R.E. Baticulon, Y.-C. Hung, M. Punchak, A. Agrawal, A.O. Adeleye, M.G. Shrime, A.M. Rubiano, Estimating the global incidence of traumatic brain injury, J. Neurosurg. 130 (2018) 1080-1097. https://doi.org/10.3171/2017.10.JNS17352

[164] A.I.R. Maas, D.K. Menon, P.D. Adelson, N. Andelic, M.J. Bell, A. Belli, P. Bragge, A. Brazinova, A. Büki, R.M. Chesnut, Traumatic brain injury: integrated approaches to improve prevention, clinical care, and research, Lancet Neurol. 16 (2017) 987-1048.

[165] M.E. Ring, The history of maxillofacial prosthetics, Plast. Reconstr. Surg. 87 (1991) 174-184. https://doi.org/10.1097/00006534-199101000-00031

[166] R. Lanzara, M. Viswambaran, D. Kumar, Maxillofacial prosthetic materials: current status and recent advances: A comprehensive review, Int. J. Appl. Dent. Sci 7 (2021) 255-259. https://doi.org/10.22271/oral.2021.v7.i2d.1219

[167] C.H. Pritham, Fundamentals of facial prosthetics, JPO J. Prosthetics Orthot. 7 (1995) 22A. https://doi.org/10.1097/00008526-199507040-00011

[168] D. Palousek, J. Rosicky, D. Koutny, Use of digital technologies for nasal prosthesis manufacturing, Prosthet. Orthot. Int. 38 (2014) 171-175. https://doi.org/10.1177/0309364613489333

[169] E.J. Bos, T. Scholten, Y. Song, J.C. Verlinden, J. Wolff, T. Forouzanfar, M.N. Helder, P. Van Zuijlen, Developing a parametric ear model for auricular reconstruction: a new step towards patient-specific implants, J. Cranio-Maxillofacial Surg. 43 (2015) 390-395. https://doi.org/10.1016/j.jcms.2014.12.016

[170] Y.L. Yap, Y.S.E. Tan, H.K.J. Tan, Z.K. Peh, X.Y. Low, W.Y. Yeong, C.S.H. Tan, A. Laude, 3D printed bio-models for medical applications, Rapid Prototyp. J. 23 (2017) 227-235. https://doi.org/10.1108/RPJ-08-2015-0102

[171] C. Kincade, L. McHutchion, J. Wolfaardt, Digital design of patient-specific abutments for the retention of implant-retained facial prostheses, J. Prosthet. Dent. 120 (2018) 309-312. https://doi.org/10.1016/j.prosdent.2017.10.002

[172] P. Tack, J. Victor, P. Gemmel, L. Annemans, 3D-printing techniques in a medical setting: a systematic literature review, Biomed. Eng. Online 15 (2016) 1-21. https://doi.org/10.1186/s12938-015-0119-0

[173] E. Provaggi, J.J.H. Leong, D.M. Kalaskar, Applications of 3D printing in the management of severe spinal conditions, Proc. Inst. Mech. Eng. Part H J. Eng. Med. 231 (2017) 471-486. https://doi.org/10.1177/0954411916667761

[174] A.-M. Wu, Z.-X. Shao, J.-S. Wang, X.-D. Yang, W.-Q. Weng, X.-Y. Wang, H.-Z. Xu, Y.-L. Chi, Z.-K. Lin, The accuracy of a method for printing three-dimensional spinal models, PLoS One 10 (2015) e0124291. https://doi.org/10.1371/journal.pone.0124291

[175] V. Baskaran, G. Štrkalj, M. Štrkalj, A. Di Ieva, Current applications and future perspectives of the use of 3D printing in anatomical training and neurosurgery, Front. Neuroanat. 10 (2016) 69. https://doi.org/10.3389/fnana.2016.00069

[176] F. Guo, J. Dai, J. Zhang, Y. Ma, G. Zhu, J. Shen, G. Niu, Individualized 3D printing navigation template for pedicle screw fixation in upper cervical spine, PLoS One 12 (2017) e0171509. https://doi.org/10.1371/journal.pone.0171509

[177] Z. Wu, L. Huang, H. Sang, Z. Ma, S. Wan, G. Cui, W. Lei, Accuracy and safety assessment of pedicle screw placement using the rapid prototyping technique in severe congenital scoliosis, Clin. Spine Surg. 24 (2011) 444-450. https://doi.org/10.1097/BSD.0b013e318201be2a

[178] K. Mao, Y. Wang, S. Xiao, Z. Liu, Y. Zhang, X. Zhang, Z. Wang, N. Lu, Z. Shourong, Z. Xifeng, Clinical application of computer-designed polystyrene models in complex severe spinal deformities: a pilot study, Eur. Spine J. 19 (2010) 797-802. https://doi.org/10.1007/s00586-010-1359-0

[179] P.S. D'Urso, G. Askin, J.S. Earwaker, G.S. Merry, R.G. Thompson, T.M. Barker, D.J. Effeney, Spinal biomodeling, Spine (Phila. Pa. 1976). 24 (1999) 1247-1251. https://doi.org/10.1097/00007632-199906150-00013

[180] P.S. D'Urso, O.D. Williamson, R.G. Thompson, Biomodeling as an aid to spinal instrumentation, Spine (Phila. Pa. 1976). 30 (2005) 2841-2845. https://doi.org/10.1097/01.brs.0000190886.56895.3d

[181] Y.-T. Wang, X.-J. Yang, B. Yan, T.-H. Zeng, Y.-Y. Qiu, S.-J. Chen, Clinical application of three-dimensional printing in the personalized treatment of complex spinal disorders, Chinese J. Traumatol. 19 (2016) 31-34. https://doi.org/10.1016/j.cjtee.2015.09.009

[182] M. Yang, C. Li, Y. Li, Y. Zhao, X. Wei, G. Zhang, J. Fan, H. Ni, Z. Chen, Y. Bai, Application of 3D rapid prototyping technology in posterior corrective surgery for Lenke 1 adolescent idiopathic scoliosis patients, Medicine (Baltimore). 94 (2015) e582. https://doi.org/10.1097/MD.0000000000000582

[183] C. Li, M. Yang, Y. Xie, Z. Chen, C. Wang, Y. Bai, X. Zhu, M. Li, Application of the polystyrene model made by 3-D printing rapid prototyping technology for operation planning in revision lumbar discectomy, J. Orthop. Sci. 20 (2015) 475-480. https://doi.org/10.1007/s00776-015-0706-8

[184] K. Toyoda, E. Urasaki, Y. Yamakawa, Novel approach for the efficient use of a full-scale, 3-dimensional model for cervical posterior fixation: a technical case report, Spine (Phila. Pa. 1976). 38 (2013) E1357-E1360. https://doi.org/10.1097/BRS.0b013e3182a1f1bd

[185] M.T. Izatt, P.L.P.J. Thorpe, R.G. Thompson, P.S. D'Urso, C.J. Adam, J.W.S. Earwaker, R.D. Labrom, G.N. Askin, The use of physical biomodelling in complex spinal surgery, Eur. Spine J. 16 (2007) 1507-1518. https://doi.org/10.1007/s00586-006-0289-3

[186] J. Guarino, S. Tennyson, G. McCain, L. Bond, K. Shea, H. King, Rapid prototyping technology for surgeries of the pediatric spine and pelvis: benefits analysis, J. Pediatr. Orthop. 27 (2007) 955-960. https://doi.org/10.1097/bpo.0b013e3181594ced

[187] P.S. D'Urso, T.M. Barker, W.J. Earwaker, L.J. Bruce, R.L. Atkinson, M.W. Lanigan, J.F. Arvier, D.J. Effeney, Stereolithographic biomodelling in cranio-maxillofacial surgery: a prospective trial, J. Cranio-Maxillofacial Surg. 27 (1999) 30-37. https://doi.org/10.1016/S1010-5182(99)80007-9

[188] M.P. Kim, A.H. Ta, W.A. Ellsworth IV, R.A. Marco, P. Gaur, J.S. Miller, Three dimensional model for surgical planning in resection of thoracic tumors, Int. J. Surg. Case Rep. 16 (2015) 127-129. https://doi.org/10.1016/j.ijscr.2015.09.037

[189] A. Goel, B. Jankharia, A. Shah, P. Sathe, Three-dimensional models: an emerging investigational revolution for craniovertebral junction surgery, J. Neurosurg. Spine 25 (2016) 740-744. https://doi.org/10.3171/2016.4.SPINE151268

[190] J. Mizutani, T. Matsubara, M. Fukuoka, N. Tanaka, H. Iguchi, A. Furuya, H. Okamoto, I. Wada, T. Otsuka, Application of full-scale three-dimensional models in patients with rheumatoid cervical spine, Eur. Spine J. 17 (2008) 644-649. https://doi.org/10.1007/s00586-008-0611-3

[191] Y. Liew, E. Beveridge, A.K. Demetriades, M.A. Hughes, 3D printing of patient-specific anatomy: a tool to improve patient consent and enhance imaging interpretation by trainees, Br. J. Neurosurg. 29 (2015) 712-714. https://doi.org/10.3109/02688697.2015.1026799

[192] J.C. Yang, X.Y. Ma, J. Lin, Z.H. Wu, K. Zhang, Q.S. Yin, Personalised modified osteotomy using computer-aided design-rapid prototyping to correct thoracic deformities, Int. Orthop. 35 (2011) 1827-1832. https://doi.org/10.1007/s00264-010-1155-9

[193] I. Madrazo, C. Zamorano, E. Magallón, T. Valenzuela, A. Ibarra, H. Salgado-Ceballos, I. Grijalva, R.E. Franco-Bourland, G. Guízar-Sahagún, Stereolithography in spine pathology: a 2-case report, Surg. Neurol. 72 (2009) 272-275. https://doi.org/10.1016/j.surneu.2008.04.034

[194] P.A. Anderson, Clinical applications of 3D printing, Spine (Phila. Pa. 1976). 42 (2017) S30-S31. https://doi.org/10.1097/BRS.0000000000002039

[195] T. Sugawara, N. Higashiyama, S. Kaneyama, M. Sumi, Accurate and simple screw insertion procedure with patient-specific screw guide templates for posterior C1-C2 fixation, Spine (Phila. Pa. 1976). 42 (2017) E340-E346. https://doi.org/10.1097/BRS.0000000000001807

[196] T. Deng, M. Jiang, Q. Lei, L. Cai, L. Chen, The accuracy and the safety of individualized 3D printing screws insertion templates for cervical screw insertion, Comput. Assist. Surg. 21 (2016) 143-149. https://doi.org/10.1080/24699322.2016.1236146

[197] B.D. Owen, G.E. Christensen, J.M. Reinhardt, T.C. Ryken, Rapid prototype patient-specific drill template for cervical pedicle screw placement, Comput. Aided Surg. 12 (2007) 303-308. https://doi.org/10.3109/10929080701662826

[198] K. Liu, Q. Zhang, X. Li, C. Zhao, X. Quan, R. Zhao, Z. Chen, Y. Li, Preliminary application of a multi-level 3D printing drill guide template for pedicle screw placement in severe and rigid scoliosis, Eur. Spine J. 26 (2017) 1684-1689. https://doi.org/10.1007/s00586-016-4926-1

[199] S. Lu, Y.Q. Xu, Y.Z. Zhang, Y.B. Li, L.E. Xie, J.H. Shi, H. Guo, G.P. Chen, Y.B. Chen, A novel computer-assisted drill guide template for lumbar pedicle screw placement: a cadaveric and clinical study, Int. J. Med. Robot. Comput. Assist. Surg. 5 (2009) 184-191. https://doi.org/10.1002/rcs.249

[200] A.-M. Wu, S. Wang, W.-Q. Weng, Z.-X. Shao, X.-D. Yang, J.-S. Wang, H.-Z. Xu, Y.-L. Chi, The radiological feature of anterior occiput-to-axis screw fixation as it guides the screw trajectory on 3D printed models: a feasibility study on 3D images and 3D printed models, Medicine (Baltimore). 93 (2014) e242. https://doi.org/10.1097/MD.0000000000000242

[201] T. Lu, C. Liu, J. Dong, M. Lu, H. Li, X. He, Cervical screw placement using rapid prototyping drill templates for navigation: a literature review, Int. J. Comput. Assist. Radiol. Surg. 11 (2016) 2231-2240. https://doi.org/10.1007/s11548-016-1414-3

[202] K.-Y. Huang, R.-M. Lin, J.-J. Fang, A novel method of C1-C2 transarticular screw insertion for symptomatic atlantoaxial instability using a customized guiding block: A case report and a technical note, Medicine (Baltimore). 95 (2016) e5100. https://doi.org/10.1097/MD.0000000000005100

[203] S. Lu, Y.Q. Xu, G.P. Chen, Y.Z. Zhang, D. Lu, Y.B. Chen, J.H. Shi, X.M. Xu, Efficacy and accuracy of a novel rapid prototyping drill template for cervical pedicle screw placement, Comput. Aided Surg. 16 (2011) 240-248. https://doi.org/10.3109/10929088.2011.605173

[204] S. Lu, Y.Q. Xu, Y.Z. Zhang, L. Xie, H. Guo, D.P. Li, A novel computer-assisted drill guide template for placement of C2 laminar screws, Eur. Spine J. 18 (2009) 1379-1385. https://doi.org/10.1007/s00586-009-1051-4

[205] X. Rong, B.-Y. Wang, Y.-X. Deng, H. Liu, Three-dimensional printed patient-specific drilling templates for expansive open-door laminoplasty, Chin. Med. J. (Engl). 129 (2016) 2644-2646. https://doi.org/10.4103/0366-6999.192786

[206] J.O. Figueroa-Cavazos, E. Flores-Villalba, J.A. Diaz-Elizondo, O. Martínez-Romero, C.A. Rodríguez, H.R. Siller, Design Concepts of Polycarbonate-Based Intervertebral Lumbar Cages: Finite Element Analysis and Compression Testing, Appl. Bionics Biomech. 2016 (2016) 7149182. https://doi.org/10.1155/2016/7149182

[207] K. Phan, A. Sgro, M.M. Maharaj, P. D'Urso, R.J. Mobbs, Application of a 3D custom printed patient specific spinal implant for C1/2 arthrodesis, J. Spine Surg. 2 (2016) 314. https://doi.org/10.21037/jss.2016.12.06

[208] W.M. Rozen, M.W. Ashton, D.L. Stella, T.J. Phillips, D. Grinsell, G.I. Taylor, The accuracy of computed tomographic angiography for mapping the perforators of the deep inferior epigastric artery: a blinded, prospective cohort study, Plast. Reconstr. Surg. 122 (2008) 1003-1009. https://doi.org/10.1097/PRS.0b013e3181845994

[209] W.M. Rozen, M.W. Ashton, D. Grinsell, D.L. Stella, T.J. Phillips, G.I. Taylor, Establishing the case for CT angiography in the preoperative imaging of abdominal wall perforators, Microsurg. Off. J. Int. Microsurg. Soc. Eur. Fed. Soc. Microsurg. 28 (2008) 306-313. https://doi.org/10.1002/micr.20496

[210] W.M. Rozen, E. Garcia-Tutor, A. Alonso-Burgos, R. Acosta, F. Stillaert, J.L. Zubieta, M. Hamdi, I.S. Whitaker, M.W. Ashton, Planning and optimising DIEP flaps with virtual surgery: the Navarra experience, J. Plast. Reconstr. Aesthetic Surg. 63 (2010) 289-297. https://doi.org/10.1016/j.bjps.2008.10.007

[211] W.M. Rozen, D. Chubb, D. Grinsell, M.W. Ashton, Computed tomographic angiography: clinical applications, Clin. Plast. Surg. 38 (2011) 229-239. https://doi.org/10.1016/j.cps.2011.03.007

[212] M.P. Chae, F. Lin, R.T. Spychal, D.J. Hunter-Smith, W.M. Rozen, 3D-Printed haptic "Reverse" models for preoperative planning in soft tissue reconstruction: A case report, Microsurgery 35 (2015) 148-153. https://doi.org/10.1002/micr.22293

[213] W.M. Rozen, M.W. Ashton, Modifying techniques in deep inferior epigastric artery perforator flap harvest with the use of preoperative imaging, ANZ J. Surg. 79 (2009) 598-603. https://doi.org/10.1111/j.1445-2197.2009.05013.x

[214] M.P. Chae, D.J. Hunter-Smith, I. De-Silva, S. Tham, R.T. Spychal, W.M. Rozen, Four-dimensional (4D) printing: a new evolution in computed tomography-guided stereolithographic modeling. Principles and application, J. Reconstr. Microsurg. 31 (2015) 458-463. https://doi.org/10.1055/s-0035-1549006

[215] J.-Y. Nie, L.-J. Lu, X. Gong, Q. Li, J.-J. Nie, Delineating the vascular territory (perforasome) of a perforator in the lower extremity of the rabbit with four-dimensional computed tomographic angiography, Plast. Reconstr. Surg. 131 (2013) 565-571. https://doi.org/10.1097/PRS.0b013e31827c6e49

[216] S. Colohan, C. Wong, C. Lakhiani, A. Cheng, M. Maia, G. Arbique, M. Saint-Cyr, The free descending branch muscle-sparing latissimus dorsi flap: vascular anatomy and clinical applications, Plast. Reconstr. Surg. 130 (2012) 776e-787e. https://doi.org/10.1097/PRS.0b013e31826d9c5e

[217] A. Donfrancesco, P. Montemurro, P. Hedén, Three-dimensional simulated images in breast augmentation surgery: an investigation of patients' satisfaction and the correlation between prediction and actual outcome, Plast. Reconstr. Surg. 132 (2013) 810-822. https://doi.org/10.1097/PRS.0b013e3182a014cb

Applications of Polymers in Surgery II
Materials Research Foundations 172 (2025) 43-64
Materials Research Forum LLC
https://doi.org/10.21741/9781644903353-2

Chapter 2

Polymers Sealants and Adhesives in General Surgery

Y. Aylin Esquivel-Lozano[1,*], Pamela Martinez-Gomez[2], David Romero-Fierro[1], Lorena Duarte-Peña[1], and Emilio Bucio[1,*]

[1]Departamento de Química de Radiaciones y Radioquímica, Instituto de Ciencias Nucleares, Universidad Nacional Autónoma de México, Circuito Exterior, Ciudad Universitaria, Ciudad de México, 04510, México

[2]Facultad de Química. Universidad Nacional Autónoma de México, Apartado Postal 70-360. Ciudad Universitaria, Coyoacán. C.P 04510, Ciudad De México, México

* aylinesquivel95@gmail.com, ebucio@nucleares.unam.mx

Abstract

Polymers are of great importance in modern medicine due to their multiple and varied applications in this area, among which their use in the design of new adhesive materials and surgical sealants stand out. This alternative to conventional treatments offers better characteristics such as better application, prevention of infections in the treated areas and an effective hemostatic response. This chapter addresses some forms of synthesis in which natural and synthetic polymers are investigated, as well as their mechanisms of action and the different classes of sealants and adhesives that have been used.

Keywords

Polymers, Sealants, Adhesives, Hemostatic, Surgical

Contents

Materials Research Forum LLC
https://doi.org/10.21741/9781644903353-2

1. Introduction

Skin injuries caused by accidents, illnesses or surgeries are one of the most important medical conditions, whether the wound is minor or major, as in the case of patients undergoing cancer treatment or diabetics, it is a situation that without the necessary care and/or treatment can lead to a major complication. When talking about surgeries and wounds resulting from them, it is important to mention that different emergencies can occur, for example: hemorrhages, which represent a large loss of blood and can have fatal consequences for the patient.

Conventional treatments for these cases include gauze, cotton pads and suture threads, which are mainly used to keep the area covered, protect it from external agents such as dust or dirt and prevent them encountering the exposed skin of the wound. However, the susceptibility to infections is high and it results in the patient's stay in the hospital, as well as the treatments being extended and therefore the treatments also increasing the costs [1,2]. Not to mention that the healing of the surgery is complicated or is not carried out properly.

On the other hand, there are problems caused by poor wound healing and other factors such as the indiscriminate use of antibiotics, which are related to bacterial resistance. That is, commonly established antibiotics no longer have the same effect against different bacterial strains, and in severe cases the drugs are no longer effective [3]. Due to this, research has increased in the development of new antibiotics or antimicrobial agents to replace those that have lost their effect, as well as the design of new materials that help combat and reduce the effects of bacterial resistance.

One of the solutions to these problems, including bleeding, comes from the advances achieved in materials science. Great progress has been made in the design and synthesis of new and, improved

Materials Research Foundations 172 (2025) 43-64
https://doi.org/10.21741/9781644903353-2

versions of dressings that fulfill the function of protecting the wound also favoring the patient's recovery. Adhesives are materials that are sometimes referred to as "glues" because it is a way of representing their main function, which is to hold two tissues together by creating bonds strong enough to achieve this. Some of them are liquid in consistency, but when they meet the tissue or some stimulus such as a change in temperature or pH, the viscosity increases. Sealants must prevent or stop the leakage or penetration of substances, so it is common to find that sealants have homostatic properties, meaning that they are capable and efficient in stopping bleeding [4]. It is common to find that adhesives and sealants are classified as the same material, however, as already explained, each one has unique functions and characteristics that make them different, although materials that have both properties have also been obtained.

This chapter will present some adhesives and sealants made from synthetic and natural polymers that have surgical applications for deep or superficial wounds. Some classifications within these materials and the mechanism of action exhibited in terms of their adhesive or sealing effect will also be described.

2. Natural polymers as sealants and adhesives

Natural polymers are macromolecules composed of repeating units of simpler structures (monomers), linked by covalent bonds. Their origin is biological, derived from plants and animals. These polymers are primarily classified into two large groups: polysaccharides and proteins. Starch, alginate, and chitosan are notable examples of polysaccharides, while gelatin and albumin are representative examples of proteins [5,6].

Natural polymers have attracted the attention of the scientific community due to their promising potential in the field of surgery. Their biocompatibility, biodegradability and ability to stimulate tissue regeneration make them ideal candidates for various surgical applications. Polymers such as fibrin, chitosan and gelatin have proven to be particularly useful. Given their versatility, bioadhesives are being developed from a wide range of natural biopolymers that are not only biocompatible but can also confer additional properties such as hemostasis, controlled drug release and adhesion to specific tissues.

Next, we will explore specific examples of natural polymers used in surgical procedures. Figure 1 highlights the key natural polymers discussed in this chapter.

 Materials Research Forum LLC
https://doi.org/10.21741/9781644903353-2

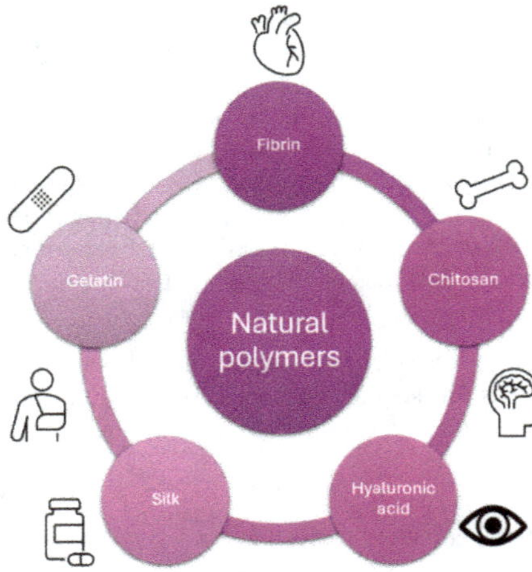

Figure 1. *Schematization of some natural polymers with medical applications.*

2.1 Fibrin

Fibrin is a biological adhesive that imitates the body's natural clotting process. It's made of human fibrinogen, bovine thrombin, and calcium chloride, which combine to create a strong, safe seal. Fibrin works by activating fibrinogen, which turns into a fibrin clot. Calcium chloride helps this clot become stronger and more stable. This stability is key for healing wounds and reducing the need for stitches.

One of the best things about fibrin glue is that your body absorbs it. This means less chance of infection and foreign reactions. It usually disappears within 10 to 14 days, matching your body's healing time. Fibrin glue can be derived from either donated blood or the patient's own blood. Fibrin glue can be applied in different ways, like using a double syringe or spraying it. This enhances its utility for many types of surgery, including heart, chest, brain, and plastic surgery [7–9].

2.2 Chitosan

Chitosan is a natural, sugar-based polysaccharide derived from the shells of crustaceans such as shrimp and crabs. Composed of repeating units of D-glucosamine (deacetylated) and N-acetyl-D-glucosamine (acetylated) as is shown in Figure 2, chitosan's structure influences its adhesive properties. The presence of D-glucosamine $(C_6H_{11}NO_5)_n$ imparts chitosan with remarkable biocompatibility and biodegradability. These properties make chitosan an excellent natural adhesive, particularly in medical applications such as wound dressings, where it not only promotes healing but also provides antimicrobial benefits. Moreover, chitosan can form strong bonds with biological tissues, further enhancing its value in biomedical applications [10].

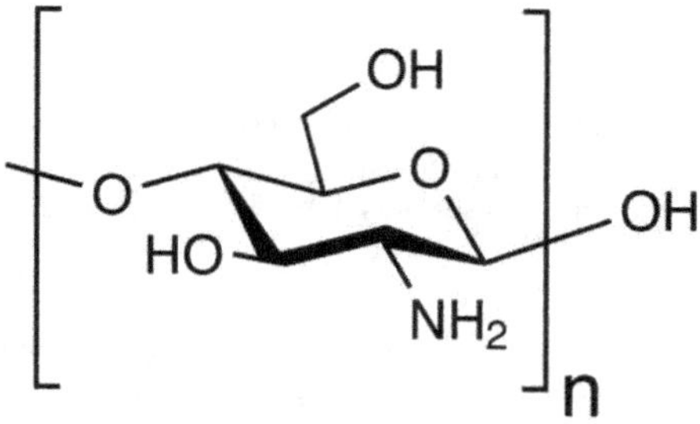

Figure 2. Chitosan structure.

Chitosan plays a fundamental role in tissue regeneration, both in wound healing and tissue engineering. Its antimicrobial and hemostatic properties create a favorable environment for healing, while its ability to act as a biocompatible scaffold promotes cell proliferation and differentiation. The slow degradation of chitosan into N-acetylglucosamine, with its anti-inflammatory properties, contributes to more efficient tissue regeneration [11,12].

2.3 Gelatin

Gelatin is a naturally occurring protein derived from collagen that is known for its ability to adhere to cells, be biocompatible, and degrade spontaneously. These characteristics make it the ideal material for a variety of biological applications, such as tissue engineering, medication delivery, and regenerative medicine [13,14].

Hydrogels based on gelatin are widely used as multifunctional adhesives for wound healing applications. These hydrogels exhibit several beneficial properties, including temperature sensitivity, mechanical flexibility, strong adhesive strength, injectability, self-healing capabilities, antibacterial activity, and the promotion of hemostasis and wound healing. Hydrogels can potentially be formulated using alternative sources of gelatin, such as mussel gelatin, waste leather, and other natural biomaterials, offering sustainable and innovative approaches to biomedical applications. [15,16].

2.4 Silk

Silk is a natural protein obtained from silkworms or spiders. Silk is known for its tensile strength and biocompatibility. Due to these properties, silk is used in tissue engineering and drug delivered systems [17,18].

Silk has been widely explored for use in mucoadhesive drug delivery systems, which enhance drug retention at the site of action. For example, silk fibroin-coated liposomes have been developed for ocular ibuprofen delivery, improving therapeutic effectiveness in eye treatments. In addition to fibroin, other silk proteins like sericin and spider silk aggregates hold great potential as natural mucoadhesive polymers, offering a biocompatible matrix for localized drug delivery to mucosal surfaces.

These silk-based systems can be combined with other polymers to create durable mucoadhesive films for transmucosal drug delivery. The adhesive properties of sericin, often discarded during

Materials Research Foundations 172 (2025) 43-64 https://doi.org/10.21741/9781644903353-2

silk processing, can also be harnessed to control the release of pharmaceuticals when blended with other materials.

Moreover, silk from the silkworm *Bombyx mori* has been utilized as a hemostatic agent due to its excellent biocompatibility and low immunogenicity. Silk is applied in various forms, including powders, sponges, films, and adhesives/sealants, for effective hemostatic applications [19,20].

2.5 Hyaluronic acid

Hyaluronic acid (HA) is a natural polysaccharide made of repeating units of disaccharides as is shown in Figure 3 (D-glucuronic and N-acetyl-D-glucosamine). It known for its wide medical applications. Its ability to form entangled structures at high molecular weights, combined with its elasticity and rheological characteristics, allows it to function effectively as a lubricating and shock-absorbing fluid. These properties are particularly beneficial in ophthalmology and orthopedics, where HA is used to protect tissues and improve joint function.

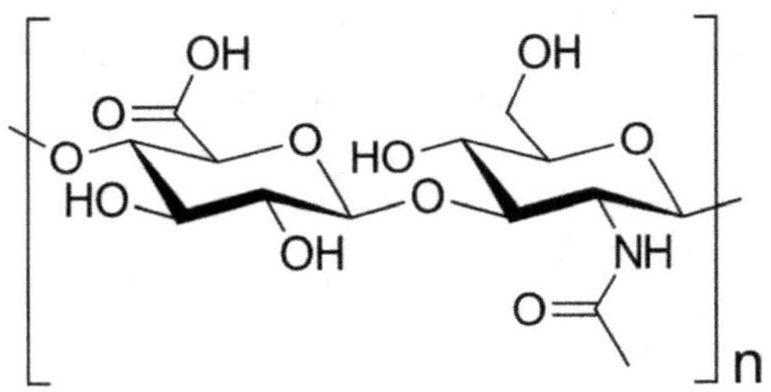

Figure 3. Hyaluronic acid structure.

HA is used in ophthalmology as a viscoelastic agent during the realization of intraocular surgeries to maintain chamber shape, to protect the cornea and reduce inflammation. It also used to treat the dye eye conditions. In orthopedics is used as a viscosupplement to improve joint function in osteoarthritis. Moreover, it may repair or protect cartilage and stimulate the production of endogenous HA [21,22].

The controlled degradation of natural polymers facilitates the gradual release of encapsulated drugs or growth factors, potentially enhancing therapeutic outcomes. Natural polymers are poised to advance surgical techniques and improve patient outcomes. Table 1 summarizes the applications of the natural polymers discussed in this section.

Materials Research Foundations 172 (2025) 43-64
https://doi.org/10.21741/9781644903353-2

Table 1. *Properties and medical applications of natural polymers.*

Biopolymer	Source	Properties	Medical Applications
Fibrin	Human blood, bovine blood	Biological adhesive, strong, safe seal, bioabsorbable	Heart, chest, brain, plastic surgery
Chitosan	Crustacean shells	Natural, sugar-based polysaccharide, biocompatible, biodegradable, antimicrobial, hemostatic	Wound dressings, tissue regeneration
Gelatin	Collagen	Natural protein, biocompatible, biodegradable, cell-adherent	Tissue engineering, drug delivery, regenerative medicine
Silk	Silkworms, spiders	Natural protein, tensile strength, biocompatible, biodegradable	Tissue engineering, drug delivery
Hyaluronic Acid	Natural polysaccharide	Biocompatible, viscoelastic, lubricating, shock-absorbing	Ophthalmology, orthopedics

3. Synthetic polymers as sealants and adhesives

In surgery applications, synthetic polymers appear as an alternative to natural polymers, when its availability is limited, offering some advantages including high thermal stability, improved mechanical properties, commercial availability, easy to be synthesized, biocompatibility and non-degradation. In this section will be discussed the most common used and promising synthetic polymers with different applications and advantages. Chemical structures of silicones, cyanoacrylates, polyethylene glycol, polyurethanes and polycarbophil are depicted in Figure 4.

Materials Research Foundations 172 (2025) 43-64
https://doi.org/10.21741/9781644903353-2

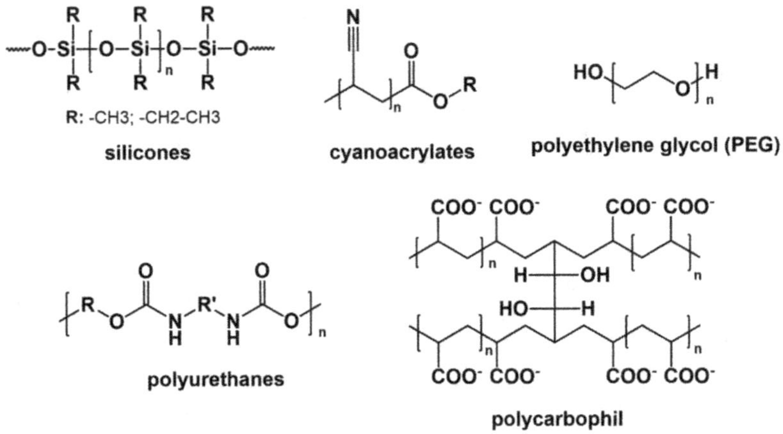

***Figure 4**. Synthetic polymers are commonly used in surgery environments as adhesives and sealants.*

3.1 Silicones

Silicones are the most common polymers used as adhesives. They consist of an organosilicon backbone, which means that they are composed of Si-O and Si-C linkages. It is considered an inorganic polymer, because it exhibits an atypical combination of an inorganic chain, being silicon atom bonded to a carbon atom through an oxygen atom, and an organic moiety, which is typically an ethyl or methyl group. This combination has a direct influence on the flexibility that silicones display allowing to adopt some configurations. This feature could be explained by the relatively long Si-O (1.64 Å) and Si-C bond (1.88 Å). For example, if the rotational energy of CH_2-CH_2 bond in polyethylene is around 13.8 kJ/mol, in the case of the Me_2Si-O bond is 3.3 kJ/mol, allowing a free rotation and flexibility of material. Other remarkable properties of silicones include their excellent resistance to chemicals and extreme weather conditions, noticeable adhesion to silicate materials and metals, heat resistance, good elastic recovery and very low shrinkage during crosslinking [23,24].

Silicone-based adhesive materials have been widely used for biomedical applications due to the relevant features of silicone chemistry, including biocompatibility, wear performance, gentleness to skin, and breathability. An especial case is the self-adhering silicones which do not require any activation immediately after application in target site, making it an interesting solution. Another benefit of silicone adhesives is the facility to remove it from skin without causing trauma or damage in a wound healing process. Silicone adhesives are used in medical devices on highly flexible catheters as well to bond silicone parts, due to having a good adhesion profile to silicone surfaces [25,26].

3.2 Cyanoacrylates

Cyanoacrylates are a family of resins derived from ethyl cyanoacrylate and related esters with application in industrial, medical and household. They are mainly composed of cyanoacrylate group which rapidly polymerizes to polycyanoacrylate (PCA) in the presence of water to form long and strong chains and reducing significantly the toxicity of monomer. Its primary use as adhesive is due to their capacity to bind to the target surface within 6 seconds with nucleophiles including water and amines and to form a strong interaction within 60 s. The interaction gives a result a degradation into cyanoacetate and formaldehyde without elimination, producing an inflammatory process. The degradation rate could be reduced by an increase in the length of the alkyl chain. If this limitation is overcome, the final PCA-based adhesive will exhibit remarkable hemostatic and antibacterial properties, whereby it can be used as microbial barrier. Other applications include hernia repair, oral approach, plastic surgery, wound healing and fistula closure [27].

To understand the mechanism of adhesion it is important to establish how CA monomers are formed. CA monomer comprises a nitrile group and an alkoxy carbonyl group, which are highly electronegative. Unsaturated carbon double bond is susceptible to Michael addition in presence of nucleophiles such as water and amines, producing an electronic rearrangement with formation of a zwitterion. This specie reacts with another accessible monomer molecule in the propagation step. During the polymerization stage, it is possible to have a covalent crosslinking between the growing polymer and amine from tissue in just a few seconds.

PCA based adhesives offers to the patient a rapid and painless application without surgery, the possibility of acting as an antimicrobial barrier, low dehiscence rates and comfortability for high patient satisfaction. In contrast, they exhibited a limited strength in the presence of fluids like water or blood, reduced mechanical properties, possible allergic pathologies, and potentially toxicity due to the presence of remaining monomers, as explained earlier [28,29].

3.3 Polyethylene glycol

Polyethylene glycol (PEG) is a synthetic polymer formed by repeating ethylene glycol repeating units. PEG is synthesized by means of polymerization of ethylene oxide and currently is commercially available in a variety of molecular weights and end group. Among its principal properties highlights their biocompatibility, easy in chemical modification, non-toxicity, water solubility and hydrophilicity. It is precisely this latest to what it is attributed to a possible mechanism of adhesion. However, in most cases it is necessary to perform some modifications to the polymer to improve the adhesive properties and establish better interactions between it and biological tissue. This modification includes the grafting of reactive moieties capable of establishing covalent bonds with proteins from tissues [23,25,27].

PEG-based adhesives have been used extensively in biomedical applications. For example, in ocular purposes, a synthetic PEG hydrogel, named ReSure®, was approved by FDA to seal corneal incisions in cataract surgery. This hydrogel is based on two main components, an *N*-hydroxysuccinimide-terminated 4-arm PEG prepolymer and a try-lisine amine crosslinker. ReSure® has shown several advantages over sutures giving to the patient a faster healing, better eye lubrication, major capability in averting fluid egress and can sustain higher intraocular pressures. Their major drawbacks lie on the requirement of mixing at least two components and a fastest application after combination. Also, this hydrogel can only be used for short periods of time, no longer than 3 days when is necessary to change it [30].

PEG-adhesives also found an application in tissue engineering. A skin patch synthesized from F127-DA hydrogel and PEG-DA as adhesive exhibited remarkable adhesive properties to different

Materials Research Foundations 172 (2025) 43-64 | https://doi.org/10.21741/9781644903353-2

animal tissues including pig skin, lung, heart, liver, kidney, and stomach, after two days of exposure. Additionally, the patch can be applied to seal punched tissues to avoid the leakage of liquid or air [31].

3.4 Polyurethanes

Polyurethanes (PUs) are a class of polymers mainly composed by two components: polyols (two or more hydroxyl groups) and isocyanates (two or more groups per molecule)- Latest groups exhibit great affinity for nucleophiles such as hydroxyl and amine groups. They are extremely versatile and can be found in the form of extremely soft elastomers to rigid and extremely hard plastics. This can be modulated and designed according to the necessities in the process of synthesis of polymer depending on the availability of functional groups with flexible or rigid groups [27].

PUs has received an increased interest in biomedical applications in some areas including catheters, stents, cardiac valves, dressings, drug delivery carriers, blood oxygenators, and tissue engineering. This interest is due to the excellent physicochemical properties of this polymer such as their mechanical flexibility, biocompatibility, wettability, high tear strength, and tailorable foam. In the field of adhesives, it is based on the high reactivity of isocyanate groups, which can be attributed to the positive charge in the carbon atom in the double system N=C=O group. Isocyanate-terminated prepolymers react with water to form a polymer network and maintain hydroxyls groups, which react covalently, via urea linkage, with amine groups from biological media [23,25]. This allows the use of PU-based adhesives in some surgery cases including renal, endocrinologic, pancreatic and colorectal.

Adhesives based on polyurethanes exhibit several advantages and disadvantages. Among the advantages it is important to highlight their biocompatibility, biodegradability, fastest reaction with amino groups present in target tissue, and excellent thermal stability at physiological temperature. However, depending on the composition of PUs, some adhesives may exhibit potential toxicity due to degradation products, containing a larger quantity of isocyanates, polyols, catalysts and crosslinkers during the polymer synthesis, and the difficulty to optimize these bioadhesives according to the target tissue [23].

3.5 Polycarbophil

Polycarbophil (PCP) is a high-molecular weight synthetic polymer which structure is based on polyacrylic acid crosslinked with polyalkenyl ethers or divinylglycol. It is a polymer with high bioadhesive properties mainly attributed to the hydrogen bond interaction between carboxylic acid and mucosal surfaces of tissues. It is water-insoluble, but in neutral pH conditions exhibits a high swelling capacity and the volume increases until of 100 times their initial volume, allowing a good adhesion with tissues [32]. It has been proved that PCP displays highest values for some properties such as swelling, humidifcation, viscosity, adhesion force in comparison with Xanthan gum, different Carbopol, chitosan, gelatin, and Acacia gum [33].

PCP exhibits applications especially in areas like mucosal, oral, and transdermal drug delivery. In mucosal drug delivery, specifically in oral application, PCP displays remarkable characteristics such as small eye irritation, long host time in the corneal surface improving bioavailability of drugs, and the possibility to be used as gels or emulsion matrix. For mucosal drug delivery. For oral drug delivery, the high hydrophilicity of PCP is a valuable aspect to increase the residence time and provide only once-daily dosing. PCP provides sustained release of drugs during long periods of time. Finally, in transdermal delivery, PCP is an alternative to transport large hydrophobic molecules such as hormones across the barrier skin by chemical bonding. All these scenarios were tested *in vitro* hence it is necessary to escalate this research level to overcome some limitations of PCP which include possible adverse reactions and possible low interactions polymer/mucus.

4. Synthesis methods of polymeric adhesives and sealants

Polymer sealants and adhesives play a crucial role in general surgery and wound healing, offering alternatives or complements to traditional sutures and staples. These materials assist in wound closure, tissue sealing, and hemostasis. The search for bioadhesives capable of withstanding application conditions and generating environments that promote cell development while ensuring wound safety by protecting it against infections or reopening is a challenge for materials science [34]. Numerous studies have been conducted on this topic, utilizing a wide variety of materials and diverse synthesis methods. The most relevant synthesis methods for adhesive hydrogels include photo-polymerization, anionic polymerization, enzymatic polymerization, polycondensation, ring-opening polymerization, and covalent crosslinking are described below and shown in Figure 5.

4.1 Photo-polymerization and photo-crosslinking

These are processes in which light, typically in the ultraviolet or visible range, is used to initiate a polymerization reaction, usually with the help of a photo-initiator. Adhesive hydrogels based on polyethylene glycol (PEG) have been synthesized using this method. PEG hydrogels are extensively researched for medical adhesives and sealants due to their ease of modification, biocompatibility, and biodegradability. PEG synthesis methods typically include free radical crosslinking and photo-polymerization; however, PEG hydrogels are often weak, and it is common to modify them with functional groups such as acrylates, thiols, or amines to enhance the strength of polymeric networks [35]. For instance, polyethylene glycol diacrylate (PEG-DA) can form hydrogels through photo-initiated crosslinking, giving rise to high adhesiveness materials. Acrylate-poloxamer F127 patches were synthesized using PEG-DA as adhesive, which was crosslinked with a LAP as a photo-initiator at a wavelength of 395 nm for 30 seconds, showing high adhesiveness to different tissues and compatibility [36]. Additionally, hydrogels of gelatin methacrylol catechol and PEG-DA crosslinked by photo-polymerization in the visible light range have been developed, presenting biocompatibility and efficiency as soft tissue sealants [37]. Other adhesives that can be produced by this method are acrylic-based adhesives, whose chain-growth polymerization may be initiated by heat, light, or chemical initiators [38]. Acrylic adhesives are mainly used on the skin since they are non-irritating and durable [39].

4.2 Anionic polymerization

This is a type of chain growth polymerization in which the initiator is an anion. Cyanoacrylates, also known as super-glues, have excellent adhesive properties and are used in different fields from medicine to industry. Cyanoacrylates are characterized by rapidly polymerizing when exposed to weak bases such as water or alcohol, even by vapors, through an anionic polymerization mechanism [40]. The electrophilic carbon of cyanoacrylate is attacked by the nucleophile, leading to the formation of a carbanion, which acts as a catalyst for the reaction by being a highly reactive

Materials Research Foundations 172 (2025) 43-64 https://doi.org/10.21741/9781644903353-2

nucleophile, which leads to the reaction propagation [41]. For medical applications, cyanoacrylates with bulky alkyl esters are preferred since they produce less irritation in the tissues [42]. Commercially used are Indermil, Dermbond, and Histoacryyl, all based on cyanoacrylates, showing improvements in healing compared with traditional sutures [43].

4.3 Enzymatic polymerization

This synthetic route involves an enzyme catalyst to carry out the polymerization or crosslinking process and is used for the manufacture of fibrin-based adhesives. Fibrin-based adhesives are biomaterials used to facilitate coagulation and wound healing, which are obtained by the combination of fibrinogen and thrombin, resulting in the instantaneous formation of a fibrin network that acts as a natural adhesive, facilitating the healing process in surgical procedures [44]. Fibrinogen and thrombin are obtained by extraction from blood plasma or by genetic engineering using recombinant production [45]. Thrombin functions as a catalyst, promoting the formation of fibrin monomers that polymerize spontaneously. Ionized calcium and aprotinin are crucial to the fibrin network formation. Calcium acts as an activator of thrombin that converts fibrinogen into fibrin and stabilizes the network. Aprotinin helps to stabilize the network by inhibiting fibrinolysis. Among the adhesives of this type on the market are Tissucal, Vivostat, and LifeSeal, which have been proven in various surgery types, such as cardiovascular, hepatic, and colorectal [46]. Furthermore, there is ongoing research on their ophthalmologic applications [47].

4.4 Polycondensation

Polycondensation is a type of polymerization reaction in which monomers bond and form mainly water as a residue; it generally involves esters and amides formation. This process is a step-growth polymerization, and it is used for the manufacture of polyurethane-based adhesives. Although these adhesives are mainly used in other areas, such as construction and crafts, they are also studied for medical applications due to their flexibility, durability, and strong adhesion [48,49]. The synthesis of polyurethane-based adhesives occurs through the formation of amides between diisocyanates and polyols. Adhesives for medical applications require biocompatible monomers, such as hexamethylene diisocyanate and polyester or polycaprolactone [50,51]. Additionally, there are studies of PEG crosslinking by polycondensation; for example, PEG-modified with succinimidyl glutarate was chemically crosslinked with PEG-modified with amines through an amidation reaction that occurs in situ and instantaneously, producing a hydrogel with high adhesiveness that could seal damaged tissues, such as pancreatic leaks [52].

4.5 Ring-opening polymerization

Is a method for synthesizing poly(lactic-co-glycolic acid) (PLGA)-based adhesives from lactic acid and glycolic acid. This polymerization process involves the conversion of a cyclic monomer into a linear polymer when it interacts with a reactive center of a molecule, which can be anionic, cationic, or radical [53]. In the case of PLGA, it is the copolymerization of two cyclic monomers, which is initiated by a catalyst, typically tin octanoate or metal derivatives; the catalyst drills the ring opening of the monomers, generating reactive centers that propagate the polymerization. The adhesive properties of PLGA are usually not as high, they can be enhanced by combining with other polymers, such as PEG, to increase this capacity [54].

4.6 Covalent crosslinking

Natural polymers, such as chitosan and collagen, have also shown potential as bioadhesives due to their biocompatibility, degradability, and high similarity to the extracellular membrane of tissues [55]. These materials exhibit adhesiveness to soft tissues since their functional groups allow interaction with the surface of the tissues. However, these materials usually have poor mechanical properties, so chemical modifications are often made to the functional groups of the chains to achieve crosslinking through covalent bonds or crosslinkers, such as glutaraldehyde, which allows the formation of Shiff bases [56,57]. Du et al. developed an adhesive hydrogel based on chitosan modified with hydrocaffeic acid, which allowed crosslinking through Michael addition and coupling reactions [58]. On the other hand, Sideek et al. synthesized chitosan hydrogels using alginate dialdehyde as a crosslinker [59].

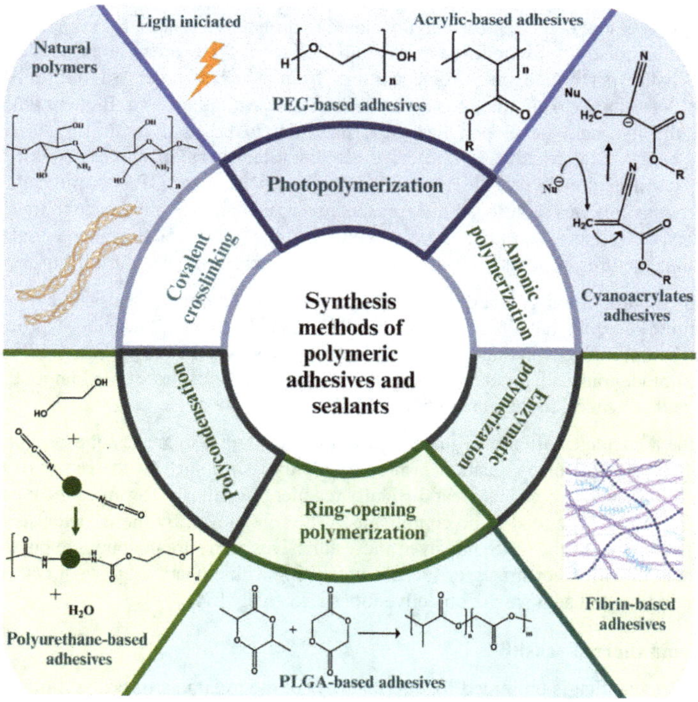

Figure 5. *Synthesis methods of polymeric adhesives and sealants.*

5. Types of adhesives and sealants

Although polymers used as adhesives and sealants can be natural or synthetic and there are different routes for obtaining the materials, they have different classifications depending on the type of material they are. Despite this, it is possible to combine them and have varied properties in the adhesives and sealants, which benefits the expansion of the possibilities of use. Below, some of the different types of adhesive and sealant polymers used in surgery will be presented.

5.1 Injectable and self-healing

Injectable hydrogels with self-healing are those that are liquid, can be stored in syringes and therefore facilitate their placement in specific sites, for this to be carried out, it is necessary that the hydrogel, upon meeting the wound, begins to present a change in its viscosity, becoming a gel. Regarding *self-healing*, this occurs when the material presents a fracture and is capable of repair itself while retaining its original properties. These represent a greater advantage over other types of adhesives or sealants because they can be molded to the shape of the wound, being a less invasive application unlike conventional materials [60].

Rosenquist, Folkesson, et al., reported an injectable hydrogel as a sealant for corneal perforations, through click chemistry with modified acetyl thiol collagen. To evaluate its adhesive and sealant effect, tests were performed on porcine corneas from which a layer was removed and then perforations were made to simulate the effects on patients with glaucoma. In these studies, it was found that the hydrogel developed has great potential to be used in the treatment of these conditions, being an alternative to corneal transplantation [61]. In another investigation, the synthesis of a self-healing hydrogel (HA-PEG) of hyaluronic acid and OPA-terminated four-armed poly(ethylene glycol) was reported, which presented a rapid and firm adherence to liver and blood vessel tissues, as well as a degradation of 6 to 22 weeks, for this reason, it is a potential alternative to conventional sutures, as it also has sealant properties to prevent leaks during surgery [62].

An injectable hydrogel must meet certain requirements, for example its consistency; before being injected it must be liquid, but when it encounters the area to be treated, the gelation process begins, which must be optimal to minimize possible adverse effects. Another important aspect to consider is the rate of biodegradability and the by-products generated, since these should not be harmful to the tissues and in general to the human body [63].

Due to all the advantages offered by injectable and self-healing polymeric adhesives and sealants, they have been sought for application in more complex issues such as the repair of the central nervous system (CNS), which has great difficulty regenerating after suffering some type of trauma and surgeries of this type tend to be complicated. The role played by these materials is to act as support in the healing of the CNS, however, these adhesives alone are not enough, but they require complementation with other strategies such as integrating with bioactive agents or cells to promote tissue regeneration and achieve the objective more effectively [64].

5.2 pH and thermo-sensitives

Another alternative that is presented for surgical uses or the treatment of post-surgical wounds by applying sealants and adhesives are those that are sensitive to stimuli, for example, to a change in temperature or pH of the tissue and even to ions. These materials have a wide application in the biomedical area because they present a change of contraction-relaxation when they are in contact with the stimulus, and it is precisely this phenomenon that allows to have a load and release of drugs for the treatment or prevention of infections caused by opportunistic bacteria [65]. However, they have another characteristic that makes them attractive for use in surgeries or wound treatments as adhesives and sealants which consists of using the contractions generated by the stimulus as an

 Materials Research Forum LLC
https://doi.org/10.21741/9781644903353-2

enhancer of the adhesiveness towards the tissues of interest [66]. For example, in research carried out by Yaodon Zhao and collaborators in 2023, a hydrogel with interpenetrated networks of *N*-dimethylbisacrylamide (NIPAM), glutaraldehyde (GTA) and hyaluronic acid (HA) was obtained where the response to temperature comes from NIPAM where it was reported that by increasing the temperature from 25 to 37 °C the adhesiveness of the hydrogel also increases. It was also reported that it is a good wound sealant through in vivo tests that also accelerates wound healing because it inhibits the inflammatory response ideal for use on wounds generated after surgery [67].

The impact of these materials is such that their use has been proposed as alternatives to conventional homeostatic treatments such as sutures or gauze, in this way polymeric materials can seal the sites where there is bleeding and protect the wound from infections and aid healing. In a recent report, a thermo- and pH-responsive hydrogel MC/CS-GA was obtained, in which chitosan-gallol (CS-GA) and methylcellulose (MC) were used, where MC is the thermo-responsive and CS-GA the pH-sensitive. The main objective of the hydrogel is to use it as an adhesive and sealant in wounds, which is also homeostatic mainly for liver tissue, which has great application potential demonstrated in vivo tests with rat liver tissue [68]. In this case, the sensitivity of the material to pH and temperature contributes to improving the adhesive capacity and mechanical properties to make a more promising hydrogel.

5.3 Sealants and adhesives with antimcrobial activity

As described so far, sealing and adhesive polymers have potential applications in modern medicine for the treatment of wounds caused by trauma or surgery; however, there is the possibility of increasing their benefits as biomaterials. This is done by adding antimicrobial agents or antibiotics, since in this way infections that can aggravate clinical cases are avoided. A bioadhesive and sealant made from gelatin, tannic acid and silver nitrate also has antimicrobial properties provided by the nanoparticles obtained through the oxidation-reduction reaction itself for the cross-linking of the hydrogel. It was reported that the material has a rapid degradation, a favorable adhesion and a good antimicrobial activity against *S. Aureus* and *E. coli* [69].

Some materials by themselves have antimicrobial activity or adhesive properties to certain surfaces, such as the case of modified chitosan (thiolted chitosan) which has been reported to increase its mucoadhesion compared to unmodified chitosan. Using this as a basis, the synthesis of thermoresponsive films of NIPAAm and thiolted chitosan was reported, in which ciprofloxacin was encapsulated as an antimicrobial and released in a controlled manner for more than 48 h [70]. Furthermore, sealants are not exclusively for use in soft tissue such as different organs in surgery, but their use has also been reported in the dental area for the treatment of caries that are regularly caused by an imbalance of bacterial biofilms. This causes countries to incur large costs of money in treatments in children and adolescents mainly. Yuyao Huang and his research group reported a resin modified with dimethylaminododecyl methacrylate (DMADDM) which is a quaternary ammonium monomer with antimicrobial properties. By modifying these resins, it was found to be a good candidate as a sealant for the treatment of caries in children since it has a long-term effect for reducing the activity of biofilms in addition to not losing its mechanical properties [71].

5.4 Mechanism of action of adhesives

So far, a broad overview has been given of the different types of polymeric sealants and adhesives that exist and some of their applications, but it is necessary to describe how the adhesion process is carried out because there are different mechanisms by which this can occur. The mechanisms for adhesion to take place are mainly through electrostatic interactions or intermolecular bonds, including hydrogen bonds, ionic crosslinking, click chemistry, to name a few, where the functional groups of the sealant or adhesive material have a favorable interaction with the functional groups of the tissue in question. In this case, we speak of a chemical adhesion where the most common reactions are Schiff bases or Michael additions[72].

For this, the hydrogel must remain perfectly attached to the area where it was applied and have good mechanical properties to maintain the integrity of the tissue; in this case, we would speak of a main mechanism of the mechanical type where the important characteristics of the material are its viscosity and that it can penetrate the tissue despite the irregularities that the application sites may have [73]. If the performance and effectiveness of the adhesive or sealant is to be strong enough to have promising results in clinical applications, it is recommended that the material present not only one mechanism of action but a combination of both, where the stability of the adhesive or sealant in the area to be treated is favored.

Conclusion

The risks in any surgical procedure of any kind are always present and even after surgery during the recovery period complications may arise. Commonly used sealants and adhesives present areas of opportunity, especially in improving the quality of the patient's recovery, as well as the surgical process itself. The polymers used to manufacture adhesives and sealants show significant progress in these areas of opportunity, since they have good biocompatibility, various forms of synthesis that allow obtaining different types of these, which implies that their application can be varied, from their use in corneal surgeries or liver surgeries. This also implies that their placement is simpler and painless, being able to access places that are very difficult with traditional materials. Depending on the polymer used, the material can have various characteristics, that is, by using natural polymers, high biodegradability and biocompatibility is ensured, minimizing the risk of adverse reactions to the new material. While synthetic polymers offer better mechanical properties that are necessary for the good performance of the adhesive or sealant. One of the great attributes of polymers used as sealants and adhesives in surgery is their versatility to be used with both properties in some cases. In addition to presenting excellent homestatic properties to stop hemorrhages.

Finally, it is necessary to continue studying the synthesis of new polymeric materials with properties as sealants or adhesives, it is also important to place greater emphasis on the degradability of the material, as well as to pay special attention to in vivo testing of the materials to have greater certainty of their good functioning and non-toxicity so that their use in humans can be scaled up and gradually replace conventional treatments.

Acknowledgments

Thanks to CONAHCyT for the doctoral scholarship provided for Yeemi Aylin Esquivel Lozano (952777), Pamela Martínez Gómez (997531) and David Romero Fierro (1175725). This work was supported by Dirección General de Asuntos del Personal Académico, Universidad Nacional Autónoma de México under Grant IN204223.

Materials Research Forum LLC
https://doi.org/10.21741/9781644903353-2

References

[1] E. Rezvani Ghomi, M. Niazi, S. Ramakrishna, The evolution of wound dressings: From traditional to smart dressings, Polym Adv Technol 34 (2023) 520–530. https://doi.org/10.1002/pat.5929

[2] A. Sanyal, S. Roy, A. Ghosh, M. Chakraborty, A. Ghosh, D. Mandal, The next frontier in hemorrhagic management: A comprehensive review on development of natural polymer-based injectable hydrogels as promising hemostatic dressings, Chemical Engineering Journal 497 (2024) 1–49. https://doi.org/10.1016/j.cej.2024.155033

[3] R. Urban-Chmiel, A. Marek, D. Stępień-Pyśniak, K. Wieczorek, M. Dec, A. Nowaczek, J. Osek, Antibiotic Resistance in Bacteria—A Review, Antibiotics 11 (2022) 2–40. https://doi.org/10.3390/antibiotics11081079

[4] M. Erdi, A. Sandler, P. Kofinas, Polymer nanomaterials for use as adjuvant surgical tools, WIREs Nanomedicine and Nanobiotechnology 15 (2023) 1–27. https://doi.org/10.1002/wnan.1889

[5] K. Vishakha S, B. Kishor D, R. Sudha S., Natural Polymers-A comprehensive Review, International Journal of Research in Pharmaceutical and Biomedical Sciences 3 (2012) 1597–1613. https://www.researchgate.net/publication/236217541

[6] S.H. Imam, C. Bilbao-Sainz, B.-S. Chiou, G.M. Glenn, W.J. Orts, Biobased adhesives, gums, emulsions, and binders: current trends and future prospects, J Adhes Sci Technol 27 (2013) 1972–1997. https://doi.org/10.1080/01694243.2012.696892

[7] M. Brennan, Fibrin glue, Blood Rev 5 (1991) 240–244. https://doi.org/10.1016/0268-960X(91)90015-5

[8] M.R. Jackson, Fibrin sealants in surgical practice: An overview, The American Journal of Surgery 182 (2001) S1–S7. https://doi.org/10.1016/S0002-9610(01)00770-X

[9] C.J. Dunn, K.L. Goa, Fibrin Sealant, Drugs 58 (1999) 863–886. https://doi.org/10.2165/00003495-199958050-00010

[10] J. Huang, Y. Liu, L. Yang, F. Zhou, Synthesis of sulfonated chitosan and its antibiofilm formation activity against E. coli and S. aureus, Int J Biol Macromol 129 (2019) 980–988. https://doi.org/10.1016/j.ijbiomac.2019.02.079

[11] F. Notario-Pérez, A. Martín-Illana, R. Cazorla-Luna, R. Ruiz-Caro, M. Veiga, Applications of Chitosan in Surgical and Post-Surgical Materials, Mar Drugs 20 (2022) 1–28. https://doi.org/10.3390/md20060396

[12] R. Franco, F. Gianfreda, M. Miranda, A. Barlattani, P. Bollero, The hemostatic properties of chitosan in oral surgery, Biomedical and Biotechnology Research Journal (BBRJ) 4 (2020) 186–188. https://doi.org/10.4103/bbrj.bbrj_43_20

[13] K. Su, C. Wang, Recent advances in the use of gelatin in biomedical research, Biotechnol Lett 37 (2015) 2139–2145. https://doi.org/10.1007/s10529-015-1907-0

[14] G.K. Saraogi, P. Gupta, U.D. Gupta, N.K. Jain, G.P. Agrawal, Gelatin nanocarriers as potential vectors for effective management of tuberculosis, Int J Pharm 385 (2010) 143–149. https://doi.org/10.1016/j.ijpharm.2009.10.004

[15] H.J. Zhang, S. Zha, X. Wang, L. Xie, S. Liang, X. Zou, X. Liu, X. Dang, X. Wang, Dual Cross-Linked Gelatin-Based Adhesive for Leather with Low Curing Temperature and

Improved Bonding Strength, ACS Appl Polym Mater 5 (2023) 10515–10525. https://doi.org/10.1021/acsapm.3c02526

[16] M. Hou, X. Wang, O. Yue, M. Zheng, H. Zhang, X. Liu, Development of a multifunctional injectable temperature-sensitive gelatin-based adhesive double-network hydrogel, Biomaterials Advances 134 (2022) 1–11. https://doi.org/10.1016/j.msec.2021.112556

[17] S.U.D. Wani, S.P. Gautam, Z.L. Qadrie, H.V. Gangadharappa, Silk fibroin as a natural polymeric based bio-material for tissue engineering and drug delivery systems-A review, Int J Biol Macromol 163 (2020) 2145–2161. https://doi.org/10.1016/j.ijbiomac.2020.09.057

[18] Y. Wang, U.-J. Kim, D.J. Blasioli, H.-J. Kim, D.L. Kaplan, In vitro cartilage tissue engineering with 3D porous aqueous-derived silk scaffolds and mesenchymal stem cells, Biomaterials 26 (2005) 7082–7094. https://doi.org/10.1016/j.biomaterials.2005.05.022

[19] Md.T. Sultan, H. Hong, O.J. Lee, O. Ajiteru, Y.J. Lee, J.S. Lee, H. Lee, S.H. Kim, C.H. Park, Silk Fibroin-Based Biomaterials for Hemostatic Applications, Biomolecules 12 (2022) 1–14. https://doi.org/10.3390/biom12050660

[20] A.E. Brooks, The Potential of Silk and Silk-Like Proteins as Natural Mucoadhesive Biopolymers for Controlled Drug Delivery, Front Chem 3 (2015) 1–8. https://doi.org/10.3389/fchem.2015.00065

[21] M. Abate, C. Schiavone, V. Salini, The Use of Hyaluronic Acid after Tendon Surgery and in Tendinopathies, Biomed Res Int 2014 (2014) 1–6. https://doi.org/10.1155/2014/783632

[22] K.L. Goa, P. Benfield, Hyaluronic Acid, Drugs 47 (1994) 536–566. https://doi.org/10.2165/00003495-199447030-00009

[23] F. Scognamiglio, A. Travan, I. Rustighi, P. Tarchi, S. Palmisano, E. Marsich, M. Borgogna, I. Donati, N. de Manzini, S. Paoletti, Adhesive and sealant interfaces for general surgery applications, J Biomed Mater Res B Appl Biomater 104 (2016) 626–639. https://doi.org/10.1002/jbm.b.33409

[24] D.M. Fitzgerald, Y.L. Colson, M.W. Grinstaff, Synthetic pressure sensitive adhesives for biomedical applications, Prog Polym Sci 142 (2023). https://doi.org/10.1016/j.progpolymsci.2023.101692

[25] S. Magalhães, L. Alves, B. Medronho, A.C. Fonseca, A. Romano, J.F.J. Coelho, M. Norgren, Brief Overview on Bio-Based Adhesives and Sealants, Polymers (Basel) 11 (2019) 1685. https://doi.org/10.3390/polym11101685

[26] T. Dunn, Adhesives, in: William Andrew Publishing, Oxford, 2015: pp. 233–238. https://doi.org/10.1016/B978-0-323-26436-5.00028-X

[27] G.-Y. Han, S.-K. Hwang, K.-H. Cho, H.-J. Kim, C.-S. Cho, Progress of tissue adhesives based on proteins and synthetic polymers, Biomater Res 27 (2023). https://doi.org/10.1186/s40824-023-00397-4

[28] P.A. Leggat, D.R. Smith, U. Kedjarune, SURGICAL APPLICATIONS OF CYANOACRYLATE ADHESIVES: A REVIEW OF TOXICITY, ANZ J Surg 77 (2007) 209–213. https://doi.org/10.1111/J.1445-2197.2007.04020.X

[29] K.L. Mittal, Progress in Adhesion and Adhesives, Progress in Adhesion and Adhesives (2015) 1–496. https://doi.org/10.1002/9781119162346

[30] E.M. Verhulst, R.R. Galarza, I.P. Herring, R.V. Ramos, Comparison of conjunctival pedicle flap to corneal adhesion achieved by Tisseel® fibrin glue, ethyl cyanoacrylate adhesive, ReSure® hydrogel sealant, and conventional suturing with 8-0 VICRYL® suture, (2023). http://hdl.handle.net/10919/113762 (accessed August 19, 2024).

[31] C. Shen, Y. Li, Q. Meng, Adhesive polyethylene glycol-based hydrogel patch for tissue repair, Colloids Surf B Biointerfaces 218 (2022). https://doi.org/10.1016/j.colsurfb.2022.112751

[32] Z. Zhu, Y. Zhai, N. Zhang, D. Leng, P. Ding, The development of polycarbophil as a bioadhesive material in pharmacy, Asian J Pharm Sci 8 (2013) 218–227. https://doi.org/10.1016/j.ajps.2013.09.003

[33] X. Tang, C. Wang, Study on the physical properties of bioadhesive polymers, Chinese Pharmaceutical Journal 39 (2005) 361–364.

[34] L. Ge, S. Chen, Recent Advances in Tissue Adhesives for Clinical Medicine, Polymers 12 (2020). https://doi.org/10.3390/polym12040939

[35] N. Artzi, T. Shazly, C. Crespo, A.B. Ramos, H.K. Chenault, E.R. Edelman, Characterization of star adhesive sealants based on PEG/dextran hydrogels., Macromolecular Bioscience 9 (2009) 754–765. https://doi.org/10.1002/mabi.200800355

[36] C. Shen, Y. Li, Q. Meng, Adhesive polyethylene glycol-based hydrogel patch for tissue repair, Colloids and Surfaces B: Biointerfaces 218 (2022) 112751. https://doi.org/https://doi.org/10.1016/j.colsurfb.2022.112751

[37] M. Ghovvati, S. Baghdasarian, A. Baidya, J. Dhal, N. Annabi, Engineering a highly elastic bioadhesive for sealing soft and dynamic tissues., Journal of Biomedical Materials Research. Part B, Applied Biomaterials 110 (2022) 1511–1522. https://doi.org/10.1002/jbm.b.35012

[38] M. Li, Y. Zhang, Y. Liu, G. Chen, N. Zhao, C. Liu, X. Li, M. Wang, J. Song, Z. Luo, A semi-interpenetrating network acrylic pressure-sensitive adhesive for efficient transdermal application with high cohesion and adhesion, Materials & Design 241 (2024) 112970. https://doi.org/https://doi.org/10.1016/j.matdes.2024.112970

[39] J.F. Kenney, T.H. Haddock, R.L. Sun, H.C. Parreira, Medical-grade acrylic adhesives for skin contact, Journal of Applied Polymer Science 45 (1992) 355–361. https://doi.org/https://doi.org/10.1002/app.1992.070450218

[40] J.K. Aronson, ed., Cyanoacrylates, in: Meyler's Side Effects of Drugs (Sixteenth Edition), Sixteenth, Elsevier, Oxford, 2016: p. 776. https://doi.org/https://doi.org/10.1016/B978-0-444-53717-1.00565-5

[41] H.S. Samuel, U. Nweke-Maraizu, E.E. Etim, Cyanoacrylate Chemistry and Polymerization Mechanisms, Progress in Chemical and Biochemical Research 7 (2024) 129–142. https://doi.org/10.48309/pcbr.2024.422869.1304

[42] P.A. Leggat, D.R. Smith, U. Kedjarune, Surgical applications of cyanoacrylate adhesives: a review of toxicity., ANZ Journal of Surgery 77 (2007) 209–213. https://doi.org/10.1111/j.1445-2197.2007.04020.x

[43] J. Ilgenfritz Neto, R.D. Aydos, I.S. Silva, L.C. Takita, A.C. de Abreu, C.A.C. Silva, K.B. Faleiros, E.M. de L. Santiago, R.T. Ramalho, Use of cyanoacrylate-based surgical adhesives associated to the macroporous tape in skin synthesis in rats., Acta Cirurgica Brasileira 34 (2019) e201900701. https://doi.org/10.1590/s0102-865020190070000001

Materials Research Forum LLC
https://doi.org/10.21741/9781644903353-2

[44] L. Yu, Z. Liu, Z. Tong, Y. Ding, Z. Qian, W. Wang, Z. Mao, Y. Ding, Sequential-Crosslinking Fibrin Glue for Rapid and Reinforced Hemostasis, Advanced Science 11 (2024) 2308171. https://doi.org/https://doi.org/10.1002/advs.202308171

[45] J.B. Cavichiolo, M. Buschle, B. Carvalho, Comparison of fibrin adhesives prepared by 3 different methods., International Archives of Otorhinolaryngology 17 (2013) 62–65. https://doi.org/10.7162/S1809-97772013000100011

[46] H.H.G. Witjes, W.A. Draaisma, M. Gagner, E.C.J. Consten, 6.637 - Staple Line Reinforcement Materials, in: P.B.T.-C.B. Ducheyne (Ed.), Elsevier, Oxford, 2011: pp. 589–592. https://doi.org/https://doi.org/10.1016/B978-0-08-055294-1.00246-4

[47] L. Fernández-Vega-Cueto, M. Persinal-Medina, N. Vázquez, M. Chacón, B. Alfonso-Bartolozzi, S. Alonso-Alonso, T. Sánchez, S. Berisa-Prado, L.M. Martínez-López, J. Merayo-Lloves, Á. Meana, A Simple, Safe, and Effective Method for Preparing Autologous Bio-Based Fibrin Glue for Ophthalmic Use., Pharmaceutics 14 (2022). https://doi.org/10.3390/pharmaceutics14112325

[48] W. Jiang, R. Hosseinpourpia, V. Biziks, S.A. Ahmed, H. Militz, S. Adamopoulos, Preparation of Polyurethane Adhesives from Crude and Purified Liquefied Wood Sawdust., Polymers 13 (2021). https://doi.org/10.3390/polym13193267

[49] G. Malucelli, A. Priola, F. Ferrero, A. Quaglia, M. Frigione, C. Carfagna, Polyurethane resin-based adhesives: curing reaction and properties of cured systems, International Journal of Adhesion and Adhesives 25 (2005) 87–91. https://doi.org/https://doi.org/10.1016/j.ijadhadh.2004.04.003

[50] S. Wendels, R. Balahura, S. Dinescu, M. Costache, L. Avérous, Synthesis and properties of biobased polyurethane tissue adhesives from bacterial polyester, Sustainable Materials and Technologies 34 (2022) e00515. https://doi.org/https://doi.org/10.1016/j.susmat.2022.e00515

[51] P. Król, Ł. Uram, B. Król, K. Pielichowska, M. Sochacka-Piętal, M. Walczak, Synthesis and property of polyurethane elastomer for biomedical applications based on nonaromatic isocyanates, polyesters, and ethylene glycol, Colloid and Polymer Science 298 (2020) 1077–1093. https://doi.org/10.1007/s00396-020-04667-8

[52] Z. Yang, X. Yu, P. Wei, Y. Huang, S. Zhou, W. Jing, Y. Zhang, L. Sun, G. Bao, X. He, B. Zhao, Tetra-armed PEG-based rapid high-adhesion, antibacterial and biodegradable pre-clinical bioadhesives for preventing pancreas leakage, Materials & Design 224 (2022) 111281. https://doi.org/https://doi.org/10.1016/j.matdes.2022.111281

[53] P.K. Deb, S.F. Kokaz, S.N. Abed, A. Paradkar, R.K. Tekade, Chapter 6 - Pharmaceutical and Biomedical Applications of Polymers, in: R.K.B.T.-B.F. of D.D. Tekade (Ed.), Advances in Pharmaceutical Product Development and Research, Academic Press, 2019: pp. 203–267. https://doi.org/https://doi.org/10.1016/B978-0-12-817909-3.00006-6

[54] H. Kamali, R. Mollaee, E. Khodaverdi, F. Hadizadeh, G.H. Zohuri, Ring-opening polymerization of poly (d,l-lactide-co-glycolide)-poly(ethylene glycol) diblock copolymer using supercritical CO2, The Journal of Supercritical Fluids 145 (2019) 133–139. https://doi.org/https://doi.org/10.1016/j.supflu.2018.12.005

[55] F. Fang, R.T.H. Linstadt, G.M. Genin, K. Ahn, S. Thomopoulos, Mechanically Competent Chitosan-Based Bioadhesive for Tendon-to-Bone Repair, Advanced Healthcare Materials 11 (2022) 2102344. https://doi.org/https://doi.org/10.1002/adhm.202102344

Materials Research Forum LLC
https://doi.org/10.21741/9781644903353-2

[56] F. Notario-Pérez, A. Martín-Illana, R. Cazorla-Luna, R. Ruiz-Caro, M.D. Veiga, Applications of Chitosan in Surgical and Post-Surgical Materials., Marine Drugs 20 (2022). https://doi.org/10.3390/md20060396

[57] H. Hamedi, S. Moradi, S.M. Hudson, A.E. Tonelli, M.W. King, Chitosan based bioadhesives for biomedical applications: A review, Carbohydrate Polymers 282 (2022) 119100. https://doi.org/https://doi.org/10.1016/j.carbpol.2022.119100

[58] X. Du, Y. Liu, H. Yan, M. Rafique, S. Li, X. Shan, L. Wu, M. Qiao, D. Kong, L. Wang, Anti-Infective and Pro-Coagulant Chitosan-Based Hydrogel Tissue Adhesive for Sutureless Wound Closure, Biomacromolecules 21 (2020) 1243–1253. https://doi.org/10.1021/acs.biomac.9b01707

[59] S.A. Sideek, H.B. El-Nassan, A.R. Fares, N.A. Elkasabgy, A.N. ElMeshad, Cross-Linked Alginate Dialdehyde/Chitosan Hydrogel Encompassing Curcumin-Loaded Bilosomes for Enhanced Wound Healing Activity, Pharmaceutics 16 (2024). https://doi.org/10.3390/pharmaceutics16010090

[60] F. Sun, Y. Bu, Y. Chen, F. Yang, J. Yu, D. Wu, An Injectable and Instant Self-Healing Medical Adhesive for Wound Sealing, ACS Appl Mater Interfaces 12 (2020) 9132–9140. https://doi.org/10.1021/acsami.0c01022

[61] J. Rosenquist, M. Folkesson, L. Höglund, J. Pupkaite, J. Hilborn, A. Samanta, An Injectable, Shape-Retaining Collagen Hydrogel Cross-linked Using Thiol-Maleimide Click Chemistry for Sealing Corneal Perforations, ACS Appl Mater Interfaces 15 (2023) 34407–34418. https://doi.org/10.1021/acsami.3c03963

[62] H. Ren, Z. Zhang, X. Cheng, Z. Zou, X. Chen, C. He, Injectable, self-healing hydrogel adhesives with firm tissue adhesion and on-demand biodegradation for sutureless wound closure, Science Advances 9 (2023) 1–18. https://doi.org/10.1126/sciadv.adh4327

[63] P. Mondal, I. Chakraborty, K. Chatterjee, Injectable Adhesive Hydrogels for Soft tissue Reconstruction: A Materials Chemistry Perspective, Chemical Record 22 (2022) 1–28. https://doi.org/10.1002/tcr.202200155

[64] C. Correia, R.L. Reis, I. Pashkuleva, N.M. Alves, Adhesive and self-healing materials for central nervous system repair, Biomaterials Advances 151 (2023) 1–17. https://doi.org/10.1016/j.bioadv.2023.213439

[65] F. Ofridam, M. Tarhini, N. Lebaz, É. Gagnière, D. Mangin, A. Elaissari, pH-sensitive polymers: Classification and some fine potential applications, Polym Adv Technol 32 (2021) 1455–1484. https://doi.org/10.1002/pat.5230

[66] E. Khadem, M. Kharaziha, H. Reza Bakhsheshi-Rad, O. Das, F. Berto, Cutting-Edge Progress in Stimuli-Responsive Bioadhesives: From Synthesis to Clinical Applications, Polymers (Basel) 14 (2022) 1–37. https://doi.org/10.3390/polym

[67] Y. Zhao, B. Yi, J. Hu, D. Zhang, G. Li, Y. Lu, Q. Zhou, Double Cross-Linked Biomimetic Hyaluronic Acid-Based Hydrogels with Thermo-Stimulated Self-Contraction and Tissue Adhesiveness for Accelerating Post-Wound Closure and Wound Healing, Adv Funct Mater 33 (2023) 1–12. https://doi.org/10.1002/adfm.202300710

[68] S.M. Hwang, E. Kim, J. Wu, M.H. Kim, H. Lee, W.H. Park, Temperature- and pH-induced dual-crosslinked methylcellulose/chitosan-gallol conjugate composite hydrogels with improved mechanical, tissue adhesive, and hemostatic properties, Int J Biol Macromol 277 (2024) 1–17. https://doi.org/10.1016/j.ijbiomac.2024.134098

[69] J. Guo, W. Sun, J.P. Kim, X. Lu, Q. Li, M. Lin, O. Mrowczynski, E.B. Rizk, J. Cheng, G. Qian, J. Yang, Development of tannin-inspired antimicrobial bioadhesives, Acta Biomater 72 (2018) 35–44. https://doi.org/10.1016/j.actbio.2018.03.008

[70] C. Radhakumary, M. Antonty, K. Sreenivasan, Drug loaded thermoresponsive and cytocompatible chitosan based hydrogel as a potential wound dressing, Carbohydr Polym 83 (2011) 705–713. https://doi.org/10.1016/j.carbpol.2010.08.042

[71] Y. Huang, H. Li, C.G. Zhu, X. Zhou, H. Wang, Q. Han, B. Ren, L. Cheng, Anti-bacterial and anti-microbial aging effects of resin-based sealant modified by quaternary ammonium monomers, J Dent 112 (2021) 1–8. https://doi.org/10.1016/j.jdent.2021.103767

[72] J. Yang, H. Yu, L. Wang, J. Liu, X. Liu, Y. Hong, Y. Huang, S. Ren, Advances in adhesive hydrogels for tissue engineering, Eur Polym J 172 (2022) 1–16. https://doi.org/10.1016/j.eurpolymj.2022.111241

[73] M.J. Sánchez-Fernández, H. Hammoudeh, R.P. Félix Lanao, M. van Erk, J.C.M. van Hest, S.C.G. Leeuwenburgh, Bone-Adhesive Materials: Clinical Requirements, Mechanisms of Action, and Future Perspective, Adv Mater Interfaces 6 (2019) 1–11. https://doi.org/10.1002/admi.201802021

Applications of Polymers in Surgery II
Materials Research Forum LLC
Materials Research Foundations 172 (2025) 65-86
https://doi.org/10.21741/9781644903353-3

Chapter 3

Polymers for Surgery Implants

Seyyed Mojtaba Mousavi[1,a*], Masoomeh Yari Kalashgrani[2], Nadiar M. Mussin[3,b], Amin Tamadon[4,c], Chin Wei Lai[5]

[1]Department of Chemical Engineering, National Taiwan University of Science and Technology, Taipei 10607, Taiwan

[2]Biotechnology Research Center, Shiraz University of Medical Sciences, Shiraz, Iran

[3]Department of Surgery No. 2, West Kazakhstan Medical University, Aktobe, Kazakhstan

[4]Department of Natural Sciences, West Kazakhstan Marat Ospanov Medical University, Aktobe, Kazakhstan

[5]Nanotechnology and Catalysis Research Centre (NANOCAT), Level 3, Block A, Institute for Advanced Studies (IAS), University Malaya (UM), 50603, Kuala Lumpur, Malaysia

[a]Kempo.smm@gmail.com, [b]nadiar_musin@zkmu.kz, [c]amintamaddon@yahoo.com

Abstract

Polymer surgery implants have emerged as a revolutionary advancement in the field of medical technology, offering innovative solutions for various surgical procedures. These implants are crafted from biocompatible and biodegradable polymers, meticulously engineered to mimic the properties of human tissues. Their unique composition allows for seamless integration into the body, minimizing the risk of rejection and promoting natural healing processes. Polymer implants have found widespread applications in orthopedics, reconstructive surgery, and tissue engineering. They serve as scaffolds for tissue regeneration, providing structural support and facilitating the growth of new cells and blood vessels. Additionally, these implants can be designed to release therapeutic agents, such as antibiotics or growth factors, directly at the implantation site, enhancing the healing process and reducing the risk of infections. The versatility of polymer implants extends beyond their structural and therapeutic capabilities. They can be tailored to specific patient needs, with customizable shapes, sizes, and degradation rates. This personalized approach ensures optimal outcomes and improved patient comfort. As research continues to advance, polymer implants hold immense potential for transforming surgical practices, offering safer, more effective, and patient-centric solutions.

Keywords

Polymer, Surgery Implant, Biodegradable Polymers, Natural and Synthetic Polymeric

Materials Research Foundations 172 (2025) 65-86
https://doi.org/10.21741/9781644903353-3

Contents

1. Introduction

Polymers, composed of repeating monomer units, have a range of unique properties such as durability, corrosion resistance, and flexibility that make them ideal for various medical applications. They have become especially favored in the development of implantable medical devices due to their excellent biocompatibility, adaptability, and quality [1]. One of the key advantages of using polymers for implants is their ability to interact effectively and safely with living tissues, providing functional integration without causing adverse reactions [2]. Additionally, their versatile nature allows for the manufacturing of polymers with a wide range of physical and mechanical characteristics, making them suitable for a broad array of implantation techniques. Key attributes of implantable polymers include biodegradability, mechanical strength, long-term stability, and biocompatibility [3]. Biocompatibility is an essential requirement for any material that comes into direct contact with bodily tissues. This characteristic ensures that the material will not trigger harmful immune responses or toxicity, promoting safe and prolonged use inside the human body [4, 5].

Medical devices made from polymers can be categorized primarily into biodegradable and non-biodegradable types, depending on their degradation characteristics. Biodegradable polymers have garnered significant attention for their potential to eliminate the need for prolonged biocompatibility assessments. These materials are designed to reside in the body for a specific timeframe and subsequently degrade naturally, eliminating the need for surgical removal. This property alleviates the requirement for a secondary surgery once the tissue has healed, making biodegradable polymers an attractive option for temporary medical applications [5-8]. The attractiveness of degradable polymers stems from their capability to replicate the characteristics of natural tissues and the extracellular matrix, alongside their straightforward processing and versatility for diverse biomedical applications [9]. These biodegradable polymer implants are employed in a variety of clinical settings, including bone, ureteral stents, cardiovascular tissue engineering, skin wound dressings, cartilage regeneration through orthopedic scaffolds, and nerve repair [10-13]. These implants offer medical professionals effective tools to aid in tissue repair and the regeneration of damaged organs [9]. One of the critical considerations in the design of biodegradable polymers is balancing the rate of degradation with the rate of tissue regeneration. If the material degrades too quickly, the defect may reoccur before full tissue repair takes place. On the other hand, if it degrades too slowly, it could interfere with the body's natural healing process [8].

Optimizing the degradation rate to match the patient's healing progress is one of the biggest challenges in biodegradable material design, especially given the variability in individual healing conditions and the extent of tissue damage. To ensure the long-term success of biodegradable biomaterials, their mechanical, physicochemical, and biological properties must remain stable throughout the degradation process. However, as the material degrades, changes in its properties and the production of degradation byproducts may impact its overall biocompatibility. These byproducts can lead to cellular changes, inflammation, or immune responses that influence the biocompatibility of the material over time [5, 14]. Therefore, understanding and controlling both the polymer's initial properties and its degradation behavior is essential for achieving a reliable and safe medical implant. The purpose of this chapter is to review polymer surgery implant and its applications in orthopedic devices, otorhinolaryngology implant, Intraocular implants, and dental implants.

2. Polymeric Materials Implants

Polymer-based biomaterials have been utilized in medical applications since the development of synthetic polymers, playing a crucial role in modern clinical practice. These materials gained widespread use due to their strength, inertness, and biocompatibility, making them ideal for replacing damaged or infected tissues, particularly in bone repair, to enhance patient outcomes. Polymer biomaterials have proven versatile, finding applications across a range of tissues, including cardiovascular, neural, musculoskeletal, and dermal tissues [15-18]. When selecting materials for biomedical implants, specific properties must be considered to ensure their suitability for medical use. These properties include surface topography, surface free energy, and the presence of functional groups that influence the interaction between the material and biological systems [19-21]. One of the key advantages of polymers is their highly tunable nature. Their chemical and physical surface behaviors can be precisely controlled, allowing for customization of bulk properties such as porosity, mechanical strength, biocompatibility, and biodegradation rates [22]. This adaptability makes polymers ideal for a variety of biomedical applications, including implants. Polymeric implants are typically small, cylindrical rods with diameters ranging from 1 to 2 mm and lengths of approximately 1 to 1.5 cm. These implants are composed of the same types of polymeric materials used in microspheres, and thus, they follow similar design considerations as discussed previously. The insertion of these implants requires relatively large needles (e.g., 16-gauge needles) or surgical incisions due to their size [23]. One of the primary advantages of polymeric implants is their ability to provide a sustained release of drugs over extended periods, ranging from several months to even years. The manufacturing process for polymeric implants often involves extrusion techniques. In brief, a mixture of the polymer is heated to a temperature slightly above its glass transition point, combined with peptide or protein particles (typically in the submicron to micron size range), and then pressed through a perforated disk using extrusion machinery. The resulting long polymer rods are cut into the desired lengths to create the final implants [24].

Materials Research Foundations 172 (2025) 65-86 https://doi.org/10.21741/9781644903353-3

2.1 Natural and Synthetic Polymeric Implants

Natural and synthetic polymers play a crucial role in the development of biomedical implants, each offering unique properties and advantages that cater to various medical applications. Natural polymers are sourced from organic materials, including plants, animals, and microorganisms. Examples of these include polysaccharides like alginate and hyaluronic acid, proteins such as collagen and silk, and bacterial polyesters. Due to their biocompatibility and biodegradability, these materials closely resemble the extracellular matrix (ECM) present in human tissues. This resemblance facilitates the seamless integration of natural polymers into biological systems, reducing the likelihood of chronic inflammatory responses and toxicity, which are often concerns with synthetic alternatives. For instance, collagen-based scaffolds are widely used in tissue engineering due to their ability to promote cell adhesion and proliferation, making them ideal for applications in wound healing and regenerative medicine [25-28]. On the other hand, synthetic polymers have gained prominence due to their versatility and ability to be engineered for specific mechanical and chemical properties. Common synthetic polymers used in implants include poly(methyl methacrylate) (PMMA), poly(lactic acid) (PLA), and poly(ethylene glycol) (PEG). These materials can be precisely tailored through various synthesis methods to achieve desired characteristics such as strength, elasticity, and degradation rates. Synthetic polymers are particularly valuable in applications requiring long-term stability and support, such as orthopedic implants where load-bearing capacity is essential [29-32]. Additionally, they can be designed to elute drugs or bioactive agents over time, enhancing therapeutic outcomes in implantable devices. The choice between natural and synthetic polymers often hinges on the specific requirements of the application. Natural polymers tend to exhibit superior biocompatibility due to their inherent biological origin; however, they may lack the mechanical strength required for certain load-bearing applications. Conversely, while synthetic polymers can be engineered for enhanced mechanical properties, they may provoke immune responses or exhibit slower degradation rates that complicate their integration into biological systems [33-35]. Donnaloja et al. [36] conducted a study on the optimization of bone scaffolds using both natural and synthetic polymers. Their analysis focuses on the current state of various widely used polymeric scaffold materials and their applications in both in vitro and in vivo models, aiming to assess their effectiveness in bone tissue engineering. The authors also discuss several strategies employed in scaffold production, including diverse material combinations, the inclusion of chemical factors, cell embedding, and advancements in manufacturing technologies [36].

2.2 Biodegradable and Non-biodegradable Polymers for Implantable Medical Devices

Biodegradable implants were designed to overcome the challenges associated with non-biodegradable implants. Composed of polymers or block copolymers, these implants degrade into smaller fragments that the body can either absorb or excrete [37, 38]. The degradation process of these materials can be controlled to regulate drug release. A significant advantage is that they do not require removal after implantation, as the body naturally degrades them. These implants can be designed similarly to monolithic or reservoir-type implants [39]. However, biodegradable implants are more complex to develop due to a narrower range of materials and stricter regulatory requirements, as the material remains in the body. The main mechanism for drug release is the degradation of the polymer matrix, which can vary between patients. Recently, there has been interest in adding chitosan to PVC polymers to improve biocompatibility for implants and scaffolds [40]. Hyper-branched polymers are also gaining popularity in biomedical applications such as drug delivery, tissue engineering, and disease detection due to their biocompatibility and ease of fabrication compared to nano-polymers [41]. Li et al. [42] investigated polymeric drug delivery systems, specifically focusing on reservoir-style implants, using both empirical and predictive models for their design. The implant examined in the study features a biodegradable polycaprolactone (PCL) membrane designed to encapsulate an active pharmaceutical ingredient (API) along with an excipient (see Figure 1). To simulate the drug release profile, various formulations of the API were evaluated in vitro under conditions that replicate the human body's environment. Upon implantation, the API dissolves and diffuses through the PCL membrane via passive diffusion. Given that PCL has a slow degradation rate, typically spanning several years, the drug release mechanism is primarily governed by diffusion rather than membrane degradation. Once the implant's reservoir reaches saturation, a stable concentration gradient is established across the membrane, resulting in zero-order release kinetics [42]. Table 1 summarizes the polymer materials used for different biomedical applications of biodegradable implant-based delivery systems.

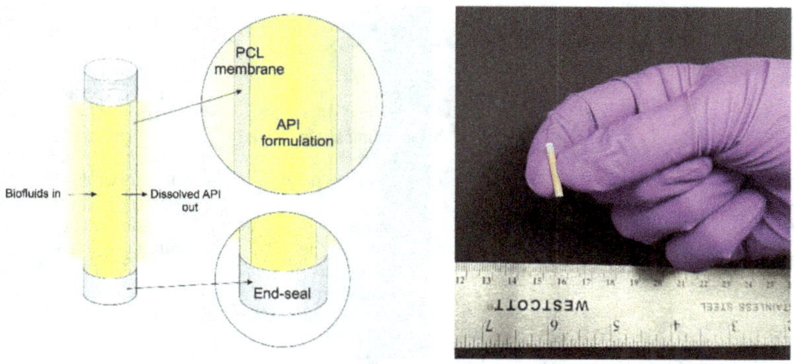

Figure 1. (Left) A schematic representation of a PCL reservoir-style implant, featuring a formulated active pharmaceutical ingredient (API) within the reservoir, with the dissolved drug gradually releasing into the surrounding medium. (Right) A digital image of a 10 mm long biodegradable implant that contains a surrogate drug formulation. Reprinted with permission from Ref. [42] Copyright © 2022 by the authors. Licensee MDPI, Basel, Switzerland. This article is an open access article distributed under the terms and conditions of the Creative Commons Attribution (CC BY) license (https://creativecommons.org/licenses/by/4.0/).

Non-biodegradable synthetic polymers are materials that resist natural decomposition, resulting in long-lasting environmental presence. This durability makes them particularly useful in various applications, especially in the medical field. Non-biodegradable polymers are often used to manufacture implants designed for long-term application in the human body. Polymers such as polyurethane and polyethylene are frequently chosen for the production of joint replacement devices, including hip and knee implants. Their widespread application is due to their high resistance to wear, long-term durability, and capacity to withstand the rigorous stress caused by daily movement and activity [43, 44]. Implants can be classified into two primary categories: monolithic and reservoir types. Monolithic implants feature a polymer matrix that uniformly disperses the drug, enabling controlled release. Conversely, reservoir-type implants contain a concentrated drug core surrounded by a permeable non-biodegradable membrane, where the membrane's thickness and permeability influence the drug release kinetics [39, 45]. Non-biodegradable implants are also extensively employed in implantable drug delivery systems, particularly for contraceptive purposes. These devices are designed for structural resilience and stability throughout their intended lifespan. However, a significant drawback of non-biodegradable implants is that they must be surgically removed once their drug load is exhausted. While the materials employed in these devices generally demonstrate favorable long-term biocompatibility, they can sometimes result in complications, including infections, tissue damage, or cosmetic concerns. Therefore, to mitigate potential adverse effects, these implants are usually extracted after their therapeutic function is fulfilled [39].

Table 1. Summary of polymer materials based on biodegradable implants.

Polymer	Method of preparation	Results	Objectives of the experiment	Ref.
Poly lactic-co-glycolic acid (PLGA)	PLGA-based in situ implant	In vitro drug released was sustained for up to 1 month	Enhance the therapeutic efficacy of Paclitaxel (PTX) while minimizing its adverse effects	[46]
Polycaprolactone–Polydimethyl Siloxane (PCL–PDMS)	Fibrous scaffolds were prepared for bone tissue regeneration using PCL–PDMS	Recovery time of 55 s; Implants were biocompatible and led to osteoblast proliferation	Performance demonstration of multifunctional shape memory polymers (SMPs) under different actuation methods	[47]
Polyglycolic acid (PGA)	————	The use of resorbable polyglycolic acid (PGA) suture in oral implantology procedures presents a good healing process and the minimum biological complications disappeared after 30 days post-surgery.	To improve oral post-surgical cycatrization with the use of resorbable polyglycolic acid (PGA) suture in patients undergoing dental implant surgery, by clinical assessment of signs of infection at 8, 15 and 30 days postsurgery	[48]
Poly(ester-anhydrides)	Polyanhydrides with alternating architectures are synthesized through the polymerization of ester-diacids, which are derived from ricinoleic acid or other hydroxy acids, in combination with anhydrides such as succinic, maleic, or phthalic anhydrides.	The findings indicate that the inclusion of hydrophobic side chains and phenyl groups, along with their proximity to anhydride bonds, plays a crucial role in enhancing stability.	Investigating the impact of ester bonds, hydrophobic side chains, and phenyl segments—along with their distances from anhydride bonds—on the stability and properties of the resulting compounds.	[49]

3. Polymeric Implants Tailored to the Patient's Specific Surgical Needs

Polymer implants have emerged as a significant advancement in the field of surgical interventions, particularly for addressing specific patient needs in reconstructive and regenerative medicine. The versatility of polymer materials allows for tailored solutions that meet the unique anatomical and physiological requirements of individual patients, enhancing surgical outcomes and patient satisfaction. These implants are designed to meet the unique anatomical and physiological needs of individual patients, enhancing surgical outcomes and improving overall patient care. The customization potential of polymers stems from their tunable mechanical properties, which allow for precise matching of stiffness and strength to that of native bone. This flexibility is particularly beneficial in applications such as craniofacial and orthopedic surgery, where the fit and functionality of an implant are crucial for successful integration and performance. The evolution of polymer implants has been driven by the need for materials that can mimic the mechanical properties and biological functions of natural tissues. Unlike traditional metal implants, polymers offer a lightweight alternative that can be engineered to possess varying degrees of stiffness and flexibility, closely resembling the properties of native bone or soft tissues. This tunability is crucial in applications such as orthopedic surgery, where the mechanical demands on implants can vary significantly based on the location and type of injury being treated [50-54]. Biodegradable polymers, such as poly(lactic acid) (PLA) and poly(caprolactone) (PCL), are frequently employed due to their capacity for gradual degradation within the body, which eliminates the necessity for a secondary surgery to remove the implant [37-39]. Additionally, one of the significant benefits of polymer implants is their potential for customization through advanced manufacturing techniques like 3D printing. This technology enables surgeons to create patient-specific implants that fit precisely into the defect site, thereby optimizing integration with surrounding tissues. The porous structure of these implants promotes cellular infiltration and vascularization, which are critical for successful healing and tissue regeneration [55, 56]. The porosity can be modified to improve the flow of nutrients and the removal of waste, thereby increasing biocompatibility and promoting better functional integration with the surrounding host tissue [57-60]. Furthermore, polymer implants can serve dual purposes beyond mere structural support; they can also function as drug delivery systems. By incorporating therapeutic agents into the polymer matrix, these implants can provide localized treatment at the site of injury or disease. This capability is particularly beneficial in managing infections or promoting healing through controlled release of growth factors or antibiotics. The design of these drug-eluting implants requires careful consideration of polymer degradation rates and drug release kinetics to ensure optimal therapeutic effects while minimizing side effects [42, 61, 62]. Despite their advantages, polymer implants are not without challenges. Issues such as mechanical strength compared to metal counterparts, potential inflammatory responses due to degradation products, and variability in patient responses to different materials must be addressed [63-65].

4. Application of Polymeric Implants in Medical

Polymeric implants have become integral in modern medicine due to their versatility and biocompatibility. These materials are utilized across various applications, including orthopedic devices, otorhinolaryngology implant, Intraocular implants, and dental implants. Polymeric materials offer several key advantages, such as ease of fabrication, flexibility, and the capacity to customize their mechanical and chemical properties for specific medical applications. For example, polymers like polyethylene and polylactic acid (PLA) are popular choices due to their biodegradability and strength, which makes them ideal for temporary implants that facilitate tissue regeneration before being absorbed by the body. Additionally, ongoing research aims to enhance the interactions between these materials and biological tissues, thereby improving osseointegration

Materials Research Foundations 172 (2025) 65-86
https://doi.org/10.21741/9781644903353-3

and minimizing inflammatory responses. Overall, the use of polymeric implants marks a significant advancement in biomedical engineering, providing innovative solutions for various medical challenges.

4.1 Orthopedic Implants

Orthopedic implants become essential for managing severe fractures that require realignment and fixation to heal properly, as well as in situations where bone fails to regenerate, leading to defects. The design of these implants must consider various factors, including material biocompatibility, mechanical and surface properties, and chemical characteristics, ensuring that the implants mimic the biomechanical properties of bone. This integration with native tissue is crucial for maintaining the implant's integrity throughout the healing period [66]. Moreover, it is vital to consider the patient's history, as certain risk factors can increase the likelihood of fractures, nonunions, or bone defects [67]. For instance, aging significantly impacts skeletal health, correlating with higher fracture rates, reduced healing capacity, and conditions such as osteoporosis and osteoarthritis [68].

Orthopedic implants play a crucial role in managing joint disorders and injuries, including conditions such as rheumatoid arthritis, post-traumatic arthritis, and osteoarthritis. These implants are classified into two main types: temporary and permanent. A comparison of the benefits and limitations of polymers used in orthopedic implants is provided in Table 2. Temporary implants, such as screws, plates, nails, pins, and wires, are typically used to support the bone during the healing process. Their design emphasizes flexibility and controlled degradation, as they are only needed for a limited time to assist in bone repair. On the other hand, permanent implants, including total hip and knee replacements, are engineered to last a lifetime. The selection of materials for these implants requires careful consideration of key factors such as resistance to corrosion, biocompatibility and wear, and mechanical strength [6, 69]. In areas of the skeleton that bear less weight, synthetic polymers, known for their extremely high molecular weight, serve as important alternatives to traditional metal implants. Various non-degradable polymers frequently employed in orthopedic applications include poly(methyl methacrylate) (PMMA), ultra-high-molecular-weight polyethylene (UHMWPE), polyurethanes (PU), and polyetheretherketone (PEEK). In addition, the U.S. Food and Drug Administration (FDA) has authorized biodegradable polymers, including polyglycolic acid (PGA), polycaprolactone (PCL), and polylactic acid (PLA) for orthopedic fixation purposes [70]. Surface modification of orthopedic implants is a critical application of polydopamine (PDA) and its derivatives, which have been shown to enhance cell-implant interactions, osteogenic activity, and osseointegration. Additionally, PDA-based materials exhibit certain in vivo biodegradability similar to eumelanin [71]. Huang et al. [72] explored the latest developments in absorbable implants within the fields of sports medicine and arthroscopic surgery. Commonly used implants in arthroscopic procedures include interference screws, endo-buttons, and suture anchors, as depicted in Figure 2i of their research [72]. Wong et al. [73] investigated the interactions between polymers and metals in their work on polymer-metal composite healthcare materials, ranging from the nano to device scale. Their study emphasizes the significant role of polymer-metal composites and nanocomposites in various biomedical applications, illustrated in Figure 2ii of their publication [73]. Durham et al. [74] utilized ion beam assisted deposition (IBAD) to apply hydroxyapatite (HA) and yttrium-stabilized zirconia (YSZ) coatings on polyether ether ketone (PEEK). In a rabbit femur implantation study, analyses using micro-computed tomography, histology, and mechanical testing revealed that the HA/YSZ

Materials Research Foundations 172 (2025) 65-86 https://doi.org/10.21741/9781644903353-3

coated implants demonstrated enhanced bone integration compared to the uncoated PEEK implants. The findings, including PEEK implants and X-ray images post-implantation, are presented in Figure 2iii of their work [74]. Additionally, Table 2 provides a summary of the advantages and disadvantages associated with various types of polymers used in orthopedic implants.

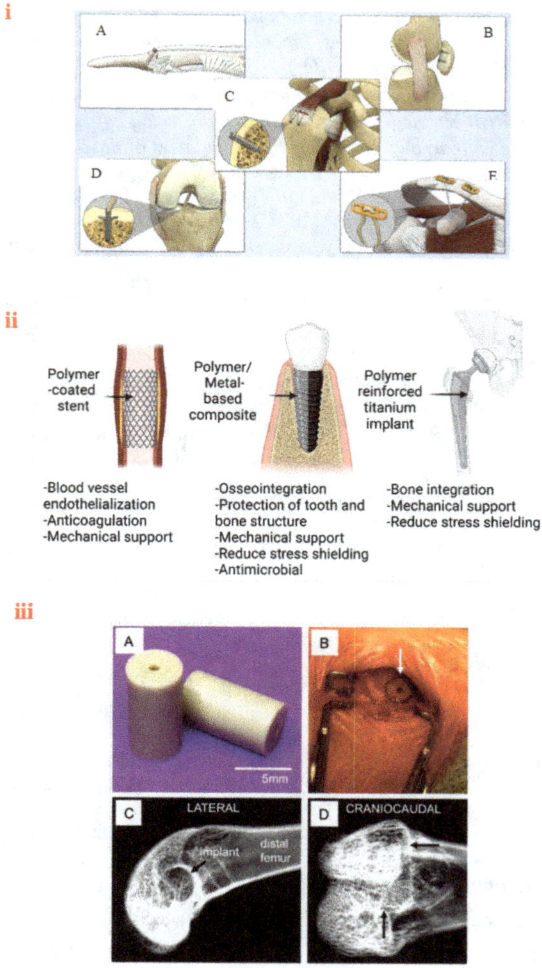

Figure 2. i) Various types of implants used in sports medicine: (A) Specialized screws for treating mallet finger injuries, (B) Screws for securing internal fixation in patellar cartilage fractures, (C) Suture anchors for repairing rotator cuff tears, (D) Interference screws used in reconstructing the anterior cruciate ligament (ACL), and (E) Endo-buttons for stabilizing acromioclavicular joint dislocations. Reprinted with permission from Ref. [72] Copyright © 2023 The Authors. Publishing services by Elsevier B.V. on behalf of KeAi Communications Co. Ltd. ii) Polymer–metal composite materials utilized in various applications such as cardiovascular stents, dental implants, and orthopedic implants. Reprinted with permission from Ref. [73] Copyright © 2022 by the authors. Licensee MDPI, Basel, Switzerland. This article is an open access article distributed under the terms and conditions of the Creative Commons Attribution (CC BY) license (https://creativecommons.org/licenses/by/4.0/). iii) (A) Dimensions of PEEK implants, (B) Surgical site showcasing an inserted implant, and (C, D) Post-implantation X-rays with arrows indicating the boundaries of the defects. Reprinted with permission from Ref. [74] Copyright © 2016 Elsevier B.V. All rights reserved. License Number: 5882550309888

Table 2. Overview of the pros and cons of various polymer types utilized in orthopedic implants.

Type of polymers	Advantages	Disadvantages	Ref.
Poly Methyl Methacrylate (PMMA)	excellent mechanical properties and rapid self-setting time	high exothermic reaction temperatures, low bioactivity, and toxicity of the monomer	[75, 76]
Polyetheretherketone (PEEK)	Elastic modulus similar to human bone, high strength, excellent wear resistance, radiolucency and biocompatibility.	bio-inertness, hydrophobicity, and susceptibility to microbial infections	[77, 78]
Ultra-High-Molecular-Weight-Polyethylene (UHMWPE)	high impact strength, high toughness, and low elastic modulus, highly withstand abrasion, and very low coefficient of friction	low tensile, transverse and compressive strengths with high creep rate	[79]
Polyurethane (PU)	Excellent chemical stability, biocompatibility and low cytotoxicity	PU degrades in the body, this may lead to insufficient contact with the surrounding tissues, thus affecting bone repair and regeneration. Also, some chemicals produced during PU degradation may cause inflammatory reactions, which may cause secondary damage to the patient.	[80]

4.2 Otorhinolaryngology Implants

The field of otolaryngology has experienced notable advancements in recent years, driven by innovations in biomaterials and implant technologies. Because otolaryngology implants interact with biological systems, they must adhere to rigorous technical standards while also addressing biological compatibility. Critical factors for the success of these implants include durability in physiological environments, appropriate biocompatibility, and mechanical functionality specific to the implant [81].

Materials Research Foundations 172 (2025) 65-86 https://doi.org/10.21741/9781644903353-3

Among the various types of implants for the ear, nose, and throat, ear implants have gained particular prominence [82]. Cochlear implants, for instance, play a crucial role in restoring hearing for individuals with severe hearing loss. In the rhinological domain, nasal packs serve as a prime example of widely utilized implants. These packs are designed to minimize discomfort following procedures like functional endoscopic sinus surgery while aiding in wound healing and preventing excessive bleeding (Figure3i) [82]. Recent studies indicate that nasal implants made from poly (L-lactide-o-D, L-lactide) are biocompatible and undergo controlled degradation within 18 months. In certain cases, complete absorption and replacement with scar tissue were observed after 24 months, while other areas were still in the absorption phase. As this process continues, the body's inflammatory response is expected to diminish over time [83].

A dexamethasone delivery system was integrated using a poly (lactic-co-glycolic acid) (PLGA) polymer matrix. This matrix consisted of a combination of polymer chains that included both free and esterified carboxyl end groups, and it was formulated without preservatives. PLGA degrades gradually into its byproducts, lactic acid and glycolic acid, through a slow breakdown process [84, 85]. Polymers like PLGA are commonly used in a range of otolaryngology devices, including corticosteroid-releasing sinus stents, voice prostheses, tympanostomy tubes, middle ear prostheses, and cochlear implants [86]. Lehner et al. [84] conducted tests on a novel biodegradable implant based on poly-(D,L-lactic-co-glycolic acid) (PLGA) for intracochlear drug delivery in animal models. The study aimed to identify the appropriate size and mechanical properties while confirming the feasibility of its administration. Researchers found that the inclusion of polyethylene glycol (PEG) as an additional excipient offered two key benefits: increased the initial rate of drug release to meet the intended profile, it enhanced the flexibility of the extrudates, and reducing the risk of cracking during bending (see Figure 3ii) [84]. In formulations without PEG, dexamethasone's release from PLGA was measured at 6.5% in the first week, rapidly increasing to nearly 50% by the second week and reaching 80% after three weeks. This release pattern is due to slow water penetration in the early phase, followed by the autocatalytic degradation of the polymer [87].

4.3 Intraocular Implants

Intraocular implants are surgically inserted into the eye to provide a controlled and sustained release of therapeutic drugs over time. Traditionally, these implants were made from non-biodegradable materials, which required surgical removal after use, leading to various associated risks. However, the advent of biodegradable polymers in the design of ocular implants has transformed their application, as they eliminate the necessity for surgical extraction [88, 89]. A significant benefit of these biodegradable polymers is their structural design, which includes deep macrovoids in the outer membrane, increasing the surface area for drug diffusion and release. While implants made from polysulfone, such as polysulfone capillary fiber implants, can be sterilized, it is crucial to recognize that they still require removal once their drug content has been depleted [90]. Additionally, while some implants can be designed as stimuli-responsive drug delivery systems, currently marketed options primarily extend the duration of drug release without altering the release rate. Numerous ongoing studies are investigating the potential of stimuli-responsive polymeric implants [91-94]. For instance, Felipe et al. [95] developed a bimatoprost implant aimed at treating glaucoma and ocular hypertension. In a 12-week study, their findings indicated that the bimatoprost implant was not inferior to timolol, suggesting its potential to enhance patient adherence and reduce the burden of glaucoma treatment [95]. Allyn et al. [96]

Materials Research Forum LLC
https://doi.org/10.21741/9781644903353-3

investigated the use of polymers in ocular drug delivery systems, highlighting how these materials enhance drug solubility, regulate pharmacokinetics, and support controlled drug release. In particular, the PLGA-based implant, Durysta®, was recently approved for the treatment of intraocular pressure (IOP) in patients diagnosed with open-angle glaucoma. Figure 4i presents a schematic diagram illustrating the implant applicator and its intended positioning, while Figure 4ii showcases several FDA-approved polymeric biomaterial products alongside their respective administration sites [96].

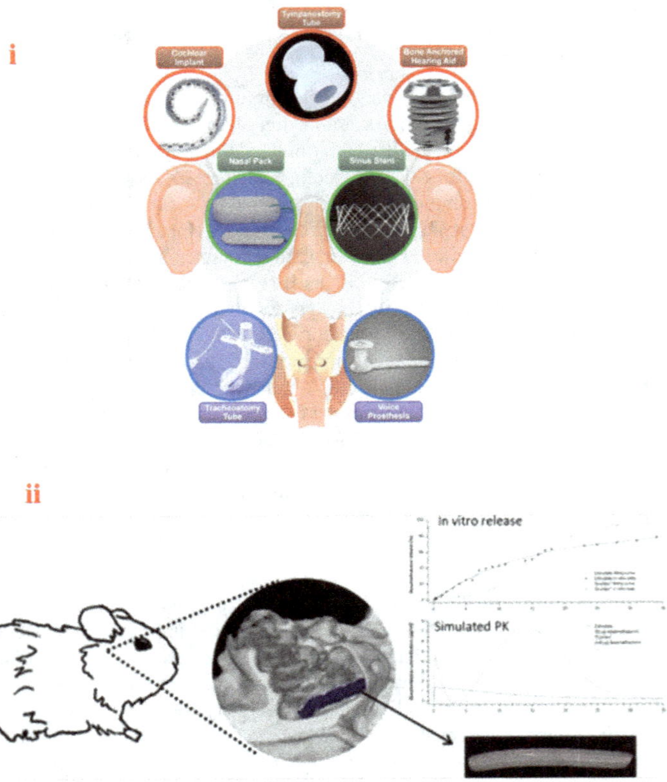

Figure 3. i) Common surgical implants utilized in otorhinolaryngology. Ear implants are marked in red, nasal implants in green, and throat implants in blue. For a color interpretation of this figure legend, please refer to the online version of this article. Reprinted with permission from Ref. [82] Copyright © 2020 Acta Materialia Inc. Published by Elsevier Ltd. All rights reserved. License Number:5882550956590 ii) Intracochlear implants made from PLGA for the release of dexamethasone. Reprinted with permission from Ref. [84] Copyright © 2019 The Authors. Published by Elsevier B.V.

Materials Research Forum LLC
https://doi.org/10.21741/9781644903353-3

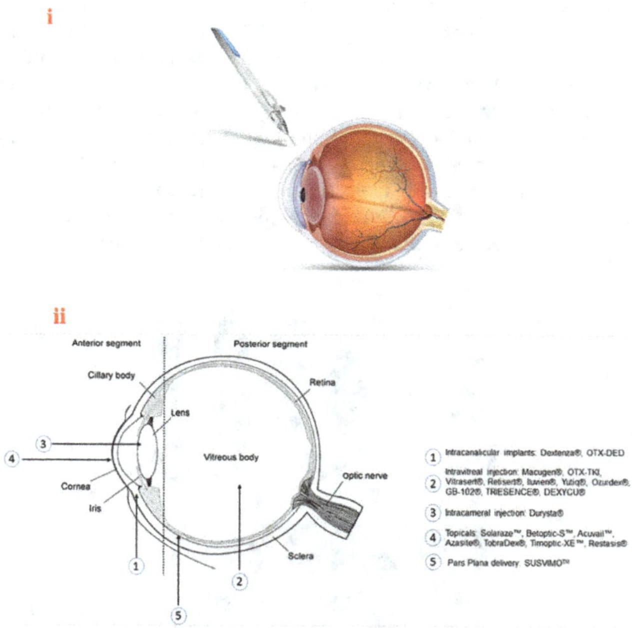

Figure 4. i) A schematic illustration of the Durysta® intracameral implant and its applicator. This device, composed of PLGA, enables the prolonged release of bimatoprost for the treatment of open-angle glaucoma, effectively overcoming some challenges associated with conventional eye drop administration. ii) The injection sites for different FDA-approved ocular drug delivery systems that utilize polymers. Reprinted with permission from Ref. [96] Copyright © 2022 Allyn, Luo, Hellwarth and Swindle-Reilly. This is an open-access article distributed under the terms of the Creative Commons Attribution License (CC BY). The use, distribution or reproduction in other forums is permitted, provided the original author(s) and the copyright owner(s) are credited and that the original publication in this journal is cited, in accordance with accepted academic practice. No use, distribution or reproduction is permitted which does not comply with these terms..

4.4 Dental Implants

A dental implant functions as an artificial tooth root designed to anchor a prosthetic tooth within the jawbone. The implant mimics the natural grooves of a tooth root and, over time, becomes integrated into the surrounding osteoid tissue, providing a stable foundation for crowns [97]. The essential components of a dental implant system include the implant itself, the abutment placed on top of the implant, and the crown. A schematic illustration depicting these various prosthetic elements is shown in Figure 5i [98]. Dental implants typically come in two shapes: screws and cylinders. The screw-shaped implants, as illustrated in Figure 5ii, have an advantage over

Materials Research Forum LLC
https://doi.org/10.21741/9781644903353-3

cylindrical implants because they more effectively transmit occlusal forces to the bone due to their design that fits snugly into the sockets. Conversely, cylindrical implants offer a smaller surface area compared to screw-type implants, which can result in reduced primary stability [99, 100]. To enhance bioactivity and enable controlled drug release, polymeric coatings have emerged as a promising strategy. Biopolymers like chitosan, cellulose, and silk fibroin-based nanomaterials are being explored for their superior bioactive and antibacterial properties as surface coatings for synthetic implants. As an example, the study by Song et al. [101] involved the use of chitosan to encapsulate Semaphorin 3A (Sema 3A), a molecule known for its bone-protective properties. This encapsulated Sema 3A was then immobilized on oxidized titanium surfaces, resulting in a prolonged release of the molecule over a span of more than two weeks [101], as illustrated in Figure 5ii(a–e).

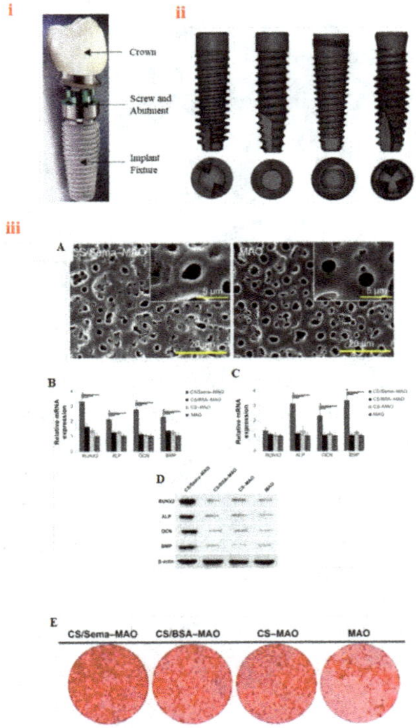

Figure 5. i) Assembly of dental implant prosthetics Reprinted. with permission from Ref. [98] Copyright © 2004 Elsevier B.V. All rights reserved. License Number:5882560475750

ii) Various types of commercially available screw implants shown from front and bottom views. Reprinted with permission from Ref. [99] Copyright © 2020 The Authors. Published by Elsevier Ltd. This is an open access article under the CC BY-NC-ND license

Materials Research Forum LLC
https://doi.org/10.21741/9781644903353-3

*(https://creativecommons.org/licenses/by-nc-nd/4.0). Selection and Peer-review under responsibility of the scientific committee of the International Conference & Exposition on Mechanical, Material and Manufacturing Technology. License Number:5882561053655. iii) Nano-chitosan-based implant surface coatings. (A) SEM pictures show the surface morphology of chitosan-semaphorin 3A-microarc oxidation (CS/Sema-MAO) against microarc oxidation (MAO). (B-E) Expression levels of osteogenic-related genes and proteins in MG63 cells cultivated on CS/Sema-MAO surfaces versus control surfaces on days 3 (B,D), 7 (C), and 21. *p < 0.05 indicates statistical significance in comparison to CS/Sema-MAO, CS-MAO, and MAO surfaces. (E) Results of the Alizarin red staining investigation. Reprinted with permission from Ref. [101].*

Conclusion

Polymer surgery implants have revolutionized the field of medical prosthetics, providing a diverse range of applications across various medical disciplines, including orthopedic devices, otorhinolaryngology implants, intraocular implants, and dental implants. These implants are engineered using advanced biocompatible polymers that are lightweight, flexible, and durable, making them ideal for both temporary and permanent use within the human body. In orthopedic devices, polymer implants such as joint replacements, bone scaffolds, and spinal cages offer improved mechanical properties that closely mimic natural bone, leading to enhanced patient outcomes and faster recovery times. Their ability to be tailored to specific needs, including the use of bioresorbable polymers for temporary scaffolds that dissolve as natural bone regenerates, has further expanded their utility in trauma and reconstructive surgeries.

In otorhinolaryngology, polymers are integral to middle ear implants, cochlear implants, and other auditory prostheses. The biocompatibility and precision in design of polymer-based implants have enhanced hearing restoration procedures, making them safer and more effective for patients with hearing impairments. Intraocular implants made from polymers, including lenses used in cataract surgery, have transformed ophthalmology by improving vision correction and providing long-term stability in the eye. The polymers used in these devices are chosen for their clarity, flexibility, and ability to remain inert within the sensitive environment of the eye, reducing the risk of complications such as infection or rejection. In dental applications, polymer implants, such as those used for dental crowns, bridges, and complete dental implants, offer aesthetic appeal, resilience, and comfort. References

Reference

[1] Seok, S., Polymer-based biocompatible packaging for implantable devices: packaging method, materials, and reliability simulation. Micromachines, 2021. 12(9): p. 1020. https://doi.org/10.3390/mi12091020

[2] Darie-Niţă, R.N., M. Râpă, and S. Frąckowiak, Special features of polyester-based materials for medical applications. Polymers, 2022. 14(5): p. 951. https://doi.org/10.3390/polym14050951

[3] Men, L., et al., Biocompatible polymers with tunable mechanical properties and conductive functionality on two-photon 3D printing. RSC advances, 2023. 13(13): p. 8586-8593. https://doi.org/10.1039/D2RA07464H

[4] Arefin, A.M., et al., Polymer 3D printing review: Materials, process, and design strategies for medical applications. Polymers, 2021. 13(9): p. 1499. https://doi.org/10.3390/polym13091499

[5] Pires, P.C., et al., Polymer-based biomaterials for pharmaceutical and biomedical applications: A focus on topical drug administration. European Polymer Journal, 2023. 187: p. 111868. https://doi.org/10.1016/j.eurpolymj.2023.111868

[6] Khan, W., et al., Implantable medical devices. Focal controlled drug delivery, 2014: p. 33-59. https://doi.org/10.1007/978-1-4614-9434-8_2

[7] Lin, S., L. Kimble, and D. Bhattacharyya, Polymer blends and composites for biomedical applications. Biomaterials for Implants and Scaffolds, 2017: p. 195-235. https://doi.org/10.1007/978-3-662-53574-5_7

[8] Suzuki, S. and Y. Ikada, Medical applications. Poly (Lactic Acid) Synthesis, Structures, Properties, Processing, Applications, and End of Life, 2022: p. 581-604. https://doi.org/10.1002/9781119767480.ch25

[9] Kirillova, A., et al., Fabrication of biomedical scaffolds using biodegradable polymers. Chemical reviews, 2021. 121(18): p. 11238-11304. https://doi.org/10.1021/acs.chemrev.0c01200

[10] Jayarama Reddy, V., et al., Nanofibrous structured biomimetic strategies for skin tissue regeneration. Wound Repair and regeneration, 2013. 21(1): p. 1-16. https://doi.org/10.1111/j.1524-475X.2012.00861.x

[11] Ravichandran, R., et al., Expression of cardiac proteins in neonatal cardiomyocytes on PGS/fibrinogen core/shell substrate for Cardiac tissue engineering. International journal of cardiology, 2013. 167(4): p. 1461-1468. https://doi.org/10.1016/j.ijcard.2012.04.045

[12] Gandhimathi, C., et al., Mimicking nanofibrous hybrid bone substitute for mesenchymal stem cells differentiation into osteogenesis. Macromolecular bioscience, 2013. 13(6): p. 696-706. https://doi.org/10.1002/mabi.201200435

[13] Naveena, N., et al., Biomimetic composites and stem cells interaction for bone and cartilage tissue regeneration. Journal of Materials Chemistry, 2012. 22(12): p. 5239-5253. https://doi.org/10.1039/c1jm14401d

[14] Nair, L.S. and C.T. Laurencin, Biodegradable polymers as biomaterials. Progress in polymer science, 2007. 32(8-9): p. 762-798. https://doi.org/10.1016/j.progpolymsci.2007.05.017

[15] Sarem, M., F. Moztarzadeh, and M. Mozafari, How can genipin assist gelatin/carbohydrate chitosan scaffolds to act as replacements of load-bearing soft tissues? Carbohydrate polymers, 2013. 93(2): p. 635-643. https://doi.org/10.1016/j.carbpol.2012.11.099

[16] Yazdanpanah, A., et al., Synthesis and Characterization of Polylactic Acid Tubular Scaffolds with Improved Mechanical Properties for Vascular Tissue Engineering. Trends in Biomaterials & Artificial Organs, 2014. 28(3).

[17] Zarrintaj, P., et al., A facile route to the synthesis of anilinic electroactive colloidal hydrogels for neural tissue engineering applications. Journal of colloid and interface science, 2018. 516: p. 57-66. https://doi.org/10.1016/j.jcis.2018.01.044

Materials Research Forum LLC
https://doi.org/10.21741/9781644903353-3

[18] Gholipourmalekabadi, M., et al., Silk fibroin/amniotic membrane 3D bi-layered artificial skin. Biomedical Materials, 2018. 13(3): p. 035003. https://doi.org/10.1088/1748-605X/aa999b

[19] Ranella, A., et al., Tuning cell adhesion by controlling the roughness and wettability of 3D micro/nano silicon structures. Acta biomaterialia, 2010. 6(7): p. 2711-2720. https://doi.org/10.1016/j.actbio.2010.01.016

[20] Hoefling, M., et al., Interaction of amino acids with the Au (111) surface: adsorption free energies from molecular dynamics simulations. Langmuir, 2010. 26(11): p. 8347-8351. https://doi.org/10.1021/la904765u

[21] Meder, F., et al., Protein adsorption on colloidal alumina particles functionalized with amino, carboxyl, sulfonate and phosphate groups. Acta biomaterialia, 2012. 8(3): p. 1221-1229. https://doi.org/10.1016/j.actbio.2011.09.014

[22] Rahmati, M., et al., Biomaterials for regenerative medicine: Historical perspectives and current trends. Cell Biology and Translational Medicine, Volume 4: Stem Cells and Cell Based Strategies in Regeneration, 2018: p. 1-19. https://doi.org/10.1007/5584_2018_278

[23] Cleland, J.L., A. Daugherty, and R. Mrsny, Emerging protein delivery methods. Current opinion in biotechnology, 2001. 12(2): p. 212-219. https://doi.org/10.1016/S0958-1669(00)00202-0

[24] Morishita, M. and N.A. Peppas, Is the oral route possible for peptide and protein drug delivery? Drug discovery today, 2006. 11(19-20): p. 905-910. https://doi.org/10.1016/j.drudis.2006.08.005

[25] Sofini, P., et al., Biomedical Applications of Natural and Synthetic Polymer based Nanocomposites. Journal of Biomaterials science. Polymer Edition, 2023: p. 1-44.

[26] Satchanska, G., S. Davidova, and P.D. Petrov, Natural and Synthetic Polymers for Biomedical and Environmental Applications. Polymers, 2024. 16(8): p. 1159. https://doi.org/10.3390/polym16081159

[27] Chahal, S., A. Kumar, and F.S.J. Hussian, Development of biomimetic electrospun polymeric biomaterials for bone tissue engineering. A review. Journal of biomaterials science, polymer edition, 2019. 30(14): p. 1308-1355. https://doi.org/10.1080/09205063.2019.1630699

[28] Rodríguez, G.R., T. Patrício, and J.D. López, Natural polymers for bone repair, in Bone Repair Biomaterials. 2019, Elsevier. p. 199-232. https://doi.org/10.1016/B978-0-08-102451-5.00008-1

[29] Hacker, M.C., J. Krieghoff, and A.G. Mikos, Synthetic polymers, in Principles of regenerative medicine. 2019, Elsevier. p. 559-590. https://doi.org/10.1016/B978-0-12-809880-6.00033-3

[30] Al-Shalawi, F.D., et al., Biodegradable synthetic polymer in orthopaedic application: A review. Materials Today: Proceedings, 2023. 74: p. 540-546. https://doi.org/10.1016/j.matpr.2022.12.254

[31] Budak, K., O. Sogut, and U. Aydemir Sezer, A review on synthesis and biomedical applications of polyglycolic acid. Journal of polymer research, 2020. 27: p. 1-19. https://doi.org/10.1007/s10965-020-02187-1

[32] Simionescu, B.C. and D. Ivanov, Natural and synthetic polymers for designing composite materials, in Handbook of bioceramics and biocomposites. 2016, Springer. p. 233-286. https://doi.org/10.1007/978-3-319-12460-5_11

[33] Prete, S., et al., Natural and synthetic polymeric biomaterials for application in wound management. Journal of Functional Biomaterials, 2023. 14(9): p. 455. https://doi.org/10.3390/jfb14090455

[34] Damodaran, V.B., D. Bhatnagar, and N.S. Murthy, Biomedical polymers: Synthesis and processing. 2016: Springer. https://doi.org/10.1007/978-3-319-32053-3

[35] Venkatraman, S., F. Boey, and L.L. Lao, Implanted cardiovascular polymers: Natural, synthetic and bio-inspired. Progress in Polymer Science, 2008. 33(9): p. 853-874. https://doi.org/10.1016/j.progpolymsci.2008.07.001

[36] Donnaloja, F., et al., Natural and synthetic polymers for bone scaffolds optimization. Polymers, 2020. 12(4): p. 905. https://doi.org/10.3390/polym12040905

[37] Claes, L. and A. Ignatius, Development of new, biodegradable implants. Der chirurg, 2002. 73: p. 990-996. https://doi.org/10.1007/s00104-002-0543-0

[38] Tian, W., et al., The influence of advancing age on implantation of drug-eluting stents. Catheterization and Cardiovascular Interventions, 2016. 88(4): p. 516-521. https://doi.org/10.1002/ccd.26333

[39] Kumar, A. and J. Pillai, Implantable drug delivery systems: An overview. Nanostructures for the engineering of cells, tissues and organs, 2018: p. 473-511. https://doi.org/10.1016/B978-0-12-813665-2.00013-2

[40] Ranjan, N., Chitosan with PVC polymer for biomedical applications: A bibliometric analysis. Materials Today: Proceedings, 2023. 81: p. 894-898. https://doi.org/10.1016/j.matpr.2021.04.274

[41] Saadati, A., M. Hasanzadeh, and F. Seidi, Biomedical application of hyperbranched polymers: Recent Advances and challenges. TrAC Trends in Analytical Chemistry, 2021. 142: p. 116308. https://doi.org/10.1016/j.trac.2021.116308

[42] Li, L., et al., Reservoir-Style Polymeric Drug Delivery Systems: Empirical and Predictive Models for Implant Design. Pharmaceuticals, 2022. 15(10): p. 1226. https://doi.org/10.3390/ph15101226

[43] Steinbüchel, A., Non-biodegradable biopolymers from renewable resources: perspectives and impacts. Current opinion in biotechnology, 2005. 16(6): p. 607-613. https://doi.org/10.1016/j.copbio.2005.10.011

[44] Lin, Q., Synthetic Non-Biodegradable Polymers, in Introduction to Biomaterials. 2006, World Scientific. p. 172-186. https://doi.org/10.1142/9789812700858_0011

[45] Martínez-Rus, F., et al., Fracture resistance of crowns cemented on titanium and zirconia implant abutments: a comparison of monolithic versus manually veneered all-ceramic systems. International Journal of Oral & Maxillofacial Implants, 2012. 27(6).

[46] Amini-Fazl, M.S., Biodegradation study of PLGA as an injectable in situ depot-forming implant for controlled release of paclitaxel. Polymer Bulletin, 2021: p. 1-14. https://doi.org/10.1007/s00289-020-03347-5

[47] Sun, L., et al., Advances in physiologically relevant actuation of shape memory polymers for biomedical applications. Polymer Reviews, 2021. 61(2): p. 280-318. https://doi.org/10.1080/15583724.2020.1825487

[48] de Romero, B.C., et al., Oral post-surgical healing with polyglycolic acid suture. Open Access Research Journal of Biology and Pharmacy, 2022. 5(2): p. 082-087. https://doi.org/10.53022/oarjbp.2022.5.2.0056

[49] Arun, Y., R. Ghosh, and A.J. Domb, Poly (ester-anhydrides) Derived from Esters of Hydroxy Acid and Cyclic Anhydrides. Biomacromolecules, 2022. 23(8): p. 3417-3428. https://doi.org/10.1021/acs.biomac.2c00542

[50] Kuznetsova, D., et al., Scaffold-and cell system-based bone grafts in tissue engineering. Medical Technologies in Medicine/Sovremennye Tehnologii v Medicine, 2014. 6(4).

[51] Guarino, V., F. Causa, and L. Ambrosio, Porosity and mechanical properties relationship in PCL porous scaffolds. Journal of Applied Biomaterials and Biomechanics, 2007. 5(3): p. 149-157.

[52] Mobaraki, M., et al., Bioinks and bioprinting: A focused review. Bioprinting, 2020. 18: p. e00080. https://doi.org/10.1016/j.bprint.2020.e00080

[53] Mora-Boza, A., et al., Glycerylphytate as an ionic crosslinker for 3D printing of multi-layered scaffolds with improved shape fidelity and biological features. Biomaterials science, 2020. 8(1): p. 506-516. https://doi.org/10.1039/C9BM01271K

[54] Ornaghi Jr, H.L., F.M. Monticeli, and L.D. Agnol, A Review on Polymers for Biomedical Applications on Hard and Soft Tissues and Prosthetic Limbs. Polymers, 2023. 15(19): p. 4034. https://doi.org/10.3390/polym15194034

[55] Al-Shalawi, F.D., et al., Biomaterials as implants in the orthopedic field for regenerative medicine: metal versus synthetic polymers. Polymers, 2023. 15(12): p. 2601. https://doi.org/10.3390/polym15122601

[56] Kovylin, R., D.Y. Aleynik, and I. Fedushkin, Modern porous polymer implants: synthesis, properties, and application. Polymer Science, Series C, 2021. 63: p. 29-46. https://doi.org/10.1134/S1811238221010033

[57] Ogay, V., et al., Progress and prospects of polymer-based drug delivery systems for bone tissue regeneration. Polymers, 2020. 12(12): p. 2881. https://doi.org/10.3390/polym12122881

[58 Kuang, T., et al., A facile approach to fabricate load-bearing porous polymer scaffolds for bone tissue engineering. Advanced Composites and Hybrid Materials, 2022. 5(2): p. 1376-1384. https://doi.org/10.1007/s42114-022-00418-1

[59] Xu, L., et al., Bioactive pore-forming bone adhesives facilitating cell ingrowth for fracture healing. Advanced Materials, 2020. 32(10): p. 1907491. https://doi.org/10.1002/adma.201907491

[60] Li, X., et al., Development of functionally graded porous titanium/silk fibroin composite scaffold for bone repair. Materials Letters, 2021. 282: p. 128670. https://doi.org/10.1016/j.matlet.2020.128670

[61] Battiston, K., et al., Polymer-free corticosteroid dimer implants for controlled and sustained drug delivery. Nature Communications, 2021. 12(1): p. 2875. https://doi.org/10.1038/s41467-021-23232-7

[62] Teo, A.J., et al., Polymeric biomaterials for medical implants and devices. ACS Biomaterials Science & Engineering, 2016. 2(4): p. 454-472. https://doi.org/10.1021/acsbiomaterials.5b00429

[63] Seal, C., K. Vince, and M. Hodgson. Biodegradable surgical implants based on magnesium alloys-A review of current research. in IOP conference series: materials science and engineering. 2009. IOP Publishing. https://doi.org/10.1088/1757-899X/4/1/012011

[64] Sumner, D., Long-term implant fixation and stress-shielding in total hip replacement. Journal of Biomechanics, 2015. 48(5): p. 797-800. https://doi.org/10.1016/j.jbiomech.2014.12.021

[65] On, S.-W., et al., Bioabsorbable osteofixation materials for maxillofacial bone surgery: A review on polymers and magnesium-based materials. Biomedicines, 2020. 8(9): p. 300. https://doi.org/10.3390/biomedicines8090300

[66] Giannoudis, P.V., T.A. Einhorn, and D. Marsh, Fracture healing: the diamond concept. injury, 2007. 38: p. S3-S6. https://doi.org/10.1016/S0020-1383(08)70003-2

[67] Andrzejowski, P. and P.V. Giannoudis, The 'diamond concept'for long bone non-union management. Journal of Orthopaedics and Traumatology, 2019. 20: p. 1-13. https://doi.org/10.1186/s10195-019-0528-0

[68] Clark, D., et al., Effects of aging on fracture healing. Current osteoporosis reports, 2017. 15: p. 601-608. https://doi.org/10.1007/s11914-017-0413-9

[69] Jin, W. and P.K. Chu, Orthopedic implants. Encyclopedia of biomedical engineering, 2019. 1(3): p. 425-439. https://doi.org/10.1016/B978-0-12-801238-3.10999-7

[70] Wang, J.L., et al., Biodegradable magnesium-based implants in orthopedics-a general review and perspectives. Advanced science, 2020. 7(8): p. 1902443. https://doi.org/10.1002/advs.201902443

[71] Jia, L., et al., Polydopamine-assisted surface modification for orthopaedic implants. Journal of orthopaedic translation, 2019. 17: p. 82-95. https://doi.org/10.1016/j.jot.2019.04.001

[72] Huang, B., et al., Absorbable implants in sport medicine and arthroscopic surgery: A narrative review of recent development. Bioactive Materials, 2024. 31: p. 272-283. https://doi.org/10.1016/j.bioactmat.2023.08.015

[73] Wong, W.-K., et al., Polymer-Metal Composite Healthcare Materials: From Nano to Device Scale. Journal of Composites Science, 2022. 6(8): p. 218. https://doi.org/10.3390/jcs6080218

[74] Durham III, J.W., et al., Hydroxyapatite coating on PEEK implants: Biomechanical and histological study in a rabbit model. Materials Science and Engineering: C, 2016. 68: p. 723-731. https://doi.org/10.1016/j.msec.2016.06.049

[75] Wei, Y., et al., Study of polymethylmethacrylate/tricalcium silicate composite cement for orthopedic application. biomedical journal, 2023. 46(3): p. 100540. https://doi.org/10.1016/j.bj.2022.05.005

[76] Ghasemi, F., et al., Different modification methods of poly methyl methacrylate (PMMA) bone cement for orthopedic surgery applications. Archives of Bone and Joint Surgery, 2023. 11(8): p. 485.

[77] Zheng, Z., et al., Strategies to improve bioactive and antibacterial properties of polyetheretherketone (PEEK) for use as orthopedic implants. Materials Today Bio, 2022. 16: p. 100402. https://doi.org/10.1016/j.mtbio.2022.100402

[78] Molinar-Díaz, J., et al., Poly-Ether-Ether-Ketone (PEEK) Biomaterials and Composites: Challenges, Progress, and Opportunities. Polymer Reviews, 2024: p. 1-39. https://doi.org/10.1080/15583724.2024.2406965

[79] Said, A., Ultra-High-Molecular-Weight-Polyethylene (UHMWPE) as Desired Polymer Material for Biomedical. Khalij-Libya Journal of Dental and Medical Research, 2022: p. 11-16. https://doi.org/10.47705/kjdmr.216103

[80] Cui, M., et al., Developments of polyurethane in biomedical applications: A review. Resources Chemicals and Materials, 2023. https://doi.org/10.1016/j.recm.2023.07.004

[81] Sternberg, K., Current requirements for polymeric biomaterials in otolaryngology. GMS current topics in otorhinolaryngology, head and neck surgery, 2009. 8.

[82] Tan, F., et al., Recent advances in the implant-based drug delivery in otorhinolaryngology. Acta Biomaterialia, 2020. 108: p. 46-55. https://doi.org/10.1016/j.actbio.2020.04.012

[83] Rippy, M.K., et al., Evaluation of absorbable PLA nasal implants in an ovine model. Laryngoscope Investigative Otolaryngology, 2018. 3(3): p. 156-161. https://doi.org/10.1002/lio2.166

[84] Lehner, E., et al., Intracochlear PLGA based implants for dexamethasone release: challenges and solutions. International Journal of Pharmaceutics: X, 2019. 1: p. 100015. https://doi.org/10.1016/j.ijpx.2019.100015

[85] Plontke, S.K., et al., Controlled release dexamethasone implants in the round window niche for salvage treatment of idiopathic sudden sensorineural hearing loss. Otology & Neurotology, 2014. 35(7): p. 1168-1171. https://doi.org/10.1097/MAO.0000000000000434

[86] Spałek, J., et al., Biocompatible materials in otorhinolaryngology and their antibacterial properties. International Journal of Molecular Sciences, 2022. 23(5): p. 2575. https://doi.org/10.3390/ijms23052575

[87] Fredenberg, S., et al., The mechanisms of drug release in poly (lactic-co-glycolic acid)-based drug delivery systems-A review. International journal of pharmaceutics, 2011. 415(1-2): p. 34-52. https://doi.org/10.1016/j.ijpharm.2011.05.049

[88] Bode, C., et al., In-situ forming PLGA implants for intraocular dexamethasone delivery. International journal of pharmaceutics, 2018. 548(1): p. 337-348. https://doi.org/10.1016/j.ijpharm.2018.07.013

[89] Kim, H.M. and S.J. Woo, Ocular drug delivery to the retina: current innovations and future perspectives. Pharmaceutics, 2021. 13(1): p. 108. https://doi.org/10.3390/pharmaceutics13010108

[90] Bourges, J., et al., Intraocular implants for extended drug delivery: therapeutic applications. Advanced drug delivery reviews, 2006. 58(11): p. 1182-1202. https://doi.org/10.1016/j.addr.2006.07.026

[91] Lynch, C., et al., Advances in biodegradable nano-sized polymer-based ocular drug delivery. Polymers, 2019. 11(8): p. 1371. https://doi.org/10.3390/polym11081371

[92] Gupta, B., et al., Cellulosic polymers for enhancing drug bioavailability in ocular drug delivery systems. Pharmaceuticals, 2021. 14(11): p. 1201. https://doi.org/10.3390/ph14111201

[93] Mahlumba, P., et al., Stimuli-responsive polymeric systems for controlled protein and peptide delivery: future implications for ocular delivery. Molecules, 2016. 21(8): p. 1002. https://doi.org/10.3390/molecules21081002

[94] Witika, B.A., et al., Nano co-crystal embedded stimuli-responsive hydrogels: A potential approach to treat HIV/AIDS. Pharmaceutics, 2021. 13(2): p. 127. https://doi.org/10.3390/pharmaceutics13020127

[95] Medeiros, F.A., et al., Phase 3, randomized, 20-month study of bimatoprost implant in open-angle glaucoma and ocular hypertension (ARTEMIS 1). Ophthalmology, 2020. 127(12): p. 1627-1641. https://doi.org/10.1016/j.ophtha.2020.06.018

[96] Allyn, M.M., et al., Considerations for polymers used in ocular drug delivery. Frontiers in Medicine, 2022. 8: p. 787644. https://doi.org/10.3389/fmed.2021.787644

[97] Duraccio, D., F. Mussano, and M.G. Faga, Biomaterials for dental implants: current and future trends. Journal of Materials Science, 2015. 50: p. 4779-4812. https://doi.org/10.1007/s10853-015-9056-3

[98] Liu, X., P.K. Chu, and C. Ding, Surface modification of titanium, titanium alloys, and related materials for biomedical applications. Materials Science and Engineering: R: Reports, 2004. 47(3-4): p. 49-121. https://doi.org/10.1016/j.mser.2004.11.001

[99] Dhatrak, P., et al., Influence of cutting flutes on stress distribution for selected dental implants: numerical studies. Materials Today: Proceedings, 2021. 38: p. 2680-2686. https://doi.org/10.1016/j.matpr.2020.08.476

[100] Samieirad, S., et al., Tapered versus cylindrical implant: Which shape inflicts less pain after dental implant surgery? A clinical trial. Journal of Oral and Maxillofacial Surgery, 2019. 77(7): p. 1381-1388. https://doi.org/10.1016/j.joms.2019.02.027

[101] Fang, K., et al., Immobilization of chitosan film containing semaphorin 3A onto a microarc oxidized titanium implant surface via silane reaction to improve MG63 osteogenic differentiation. International Journal of Nanomedicine, 2014: p. 4649-4657. https://doi.org/10.2147/IJN.S68895

Applications of Polymers in Surgery II
Materials Research Forum LLC
Materials Research Foundations 172 (2025) 87-112
https://doi.org/10.21741/9781644903353-4

Chapter 4

Polymers in Cardiovascular Surgery

Sithara Gopinath[1], P. Radhakrishnan Nair[1], Suresh Mathew[1,2*]

[1]Advanced Molecular Materials Research Centre (AMMRC), Mahatma Gandhi University, Kottayam 686560, Kerala, India

[2]School of Chemical Sciences, Mahatma Gandhi University, Kottayam 686560, Kerala, India

[*] gopinathsithara@gmail.com

Abstract

Polymers are very useful in developing cardiovascular surgery, with flexibility, biocompatibility, and possible adjustable mechanical strength. This chapter looks into different forms of polymers, such as natural and synthetic, used in cardiovascular devices, for instance, stents, heart valves, and vascular grafts. Some key advancements have been discussed in polymer synthesis, surface modifications for enhanced hemocompatibility, and their clinical performance. Areas of challenge include the degradation of polymers and future developments that involve biohybrid materials, smart polymers, and so on. The chapter includes discussions on regulatory and ethical considerations so that eventually, polymers shall play a central role in the future of cardiovascular interventions.

Keywords

Cardiovascular Polymers, Biocompatible Materials, Polymeric Biomaterials, Tissue Engineering, Drug-Eluting Polymers, Medical Devices, Surgical Implants

Contents

Materials Research Foundations 172 (2025) 87-112
https://doi.org/10.21741/9781644903353-4

1. Introduction

Cardiovascular diseases (CVDs) are a major cause of death, accounting for over one-third of fatalities globally. Examples of CVDs include heart failure, arrhythmias, peripheral artery disease, congenital heart defects, and coronary artery disease. These conditions may be influenced by genetic predispositions, lifestyle decisions (such as smoking, eating badly, and not exercising), and other medical conditions including diabetes and hypertension [1-4]. When drug treatments and lifestyle changes don't work, surgery becomes required. Cardiovascular surgery is performed to correct anatomical issues in the heart and blood vessels, such as weakened or clogged blood vessels, heart valve malfunctions, or blockages in the arteries [5-7]. Among the frequent surgical treatments are heart valve replacement, angioplasty, coronary artery bypass grafting (CABG), and heart transplants. These surgeries restore blood flow, repair damaged tissues, or replace faulty cardiovascular components.

Cardiovascular procedures now routinely involve the use of medical devices such as pacemakers, vascular grafts, heart valves, and stents. These gadgets either complement or take the place of the heart's and vessels' natural functions. Due to their special qualities, polymers have become more effective and versatile substitutes for the metals and ceramics used in previous electronics. Because they come into direct touch with blood and tissues during cardiovascular surgery, biomaterials including polymers play a critical function [8-10]. These substances must function well without triggering negative reactions or immunological responses. Strict requirements must be met by the biomaterials used in cardiovascular surgery. They must, first and foremost, be biocompatible, which means they can't cause immunological rejection or inflammation [11-13]. Furthermore, they need to have exceptional mechanical qualities, such as strength and flexibility, to endure the continuous movement of cardiovascular tissues. Additionally, these substances must not encourage the formation of clots, which poses a serious risk to blood-contacting equipment. Considerable research has been conducted in the quest for the best biomaterials for cardiovascular applications, and polymers have emerged as a top contender because of their versatility [14]. Unmatched flexibility is provided by polymers, which is essential for applications where materials must replicate the characteristics of biological tissues like heart valves or blood veins. They are perfect for creating gadgets that work in harmony with the body's natural processes since they can be precisely built to match specific needs with their mechanical properties. Additionally, polymers can be altered to have bioactive characteristics that improve their therapeutic efficacy, such as encouraging tissue regeneration or avoiding bleeding. Figure 1 shows the overview of polymers in cardiovascular surgery.

 Materials Research Forum LLC
https://doi.org/10.21741/9781644903353-4

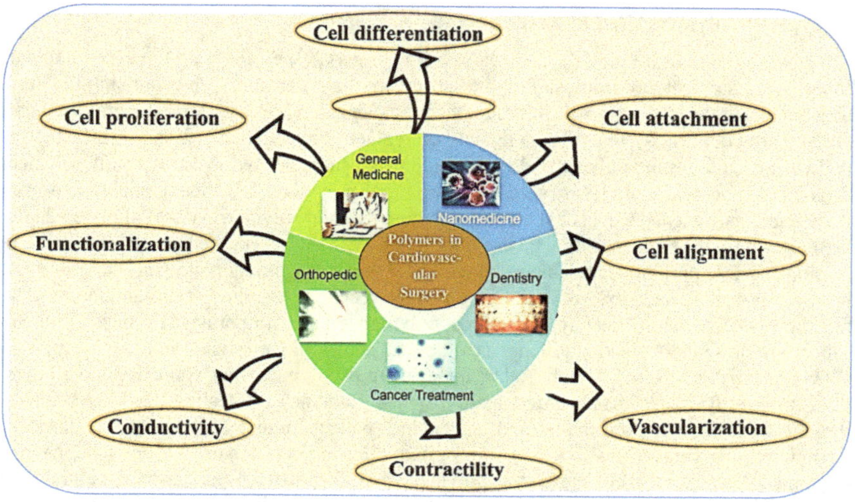

Figure 1. Overview of Polymers in Cardiovascular Surgery.

In cardiovascular surgery, polymers have played a crucial role in developing stent technology. Stents are tiny, tube-shaped devices that maintain arterial patency and improve blood flow. While early metal-only stents worked well, they also came with risks, such as restenosis a condition in which scar tissue causes the arteries to shrink again. The discovery of polymer-coated, drug-eluting stents (DES), which release medication gradually to inhibit the growth of scar tissue and lower the risk of restenosis, completely changed the field of stent technology. The polymer covering guarantees that the stent stays biocompatible and stable inside the artery [15].

Polymers have also advanced heart valve technology. Conventional heart valves have drawbacks since they are constructed either of biological tissues or mechanical parts. Biological valves deteriorate over time, but mechanical valves need to be kept open at all times with anticoagulants [16]. Heart valves made of polymers offer a possible substitute by fusing the enhanced biocompatibility of biological valves with the robustness of mechanical ones. This makes younger patients or those who are more susceptible to clotting issues particularly interested in polymer-based valves. Furthermore, vascular grafts which replace or avoid damaged blood vessels frequently employ polymers. Synthetic grafts composed of polymers like expanded polytetrafluoroethylene (ePTFE) or polyethene terephthalate (PET) offer a flexible and long-lasting substitute for autologous grafts, which are obtained from the patient's own body [17]. The grafts are engineered to emulate the characteristics of native vessels, guaranteeing their ability to endure the strains of blood circulation without breaking down or leading to issues such as clotting. Polymers are therefore now necessary in heart surgery because of their adaptability, biocompatibility, and capacity to be tailored for a wide range of medical uses. The significance of polymers in cardiovascular therapies will only grow as scientific advancements in the field continue, providing patients all around the world with better, more efficient treatment alternatives [18].

Materials Research Foundations 172 (2025) 87-112
https://doi.org/10.21741/9781644903353-4

2. Types of Polymers Used in Cardiovascular Surgery

Significant progress has been made in cardiovascular surgery, especially with the introduction of polymers into surgical procedures. Polymers, due to their versatility, biocompatibility, and mechanical qualities, play a key role in different applications, ranging from vascular grafts to heart valves and drug delivery systems. The longevity of implants, patient recovery, and surgical results can all be impacted by the polymer selection. In general, there are two basic types of polymers utilized in cardiovascular surgery: synthetic and natural polymers, and biodegradable and non-biodegradable polymers [19]. The benefit of biodegradable polymers is that they gradually decompose within the body, negating the need for subsequent surgical procedures to remove them. On the other hand, non-biodegradable polymers offer long-term functionality and durability but come with potential dangers such as foreign body reactions or chronic inflammation. The difference between natural and manufactured polymers is also important. Because synthetic polymers are designed to have particular qualities, attributes like elasticity and tensile strength may be precisely controlled. On the other hand, natural polymers that come from biological sources frequently show better biocompatibility and can aid in healing [20]. Optimizing cardiovascular surgical procedures and enhancing patient outcomes requires a thorough understanding of the various polymer kinds and their unique features. The subtleties of biodegradable versus non-biodegradable polymers and the differences between natural and manufactured polymers in cardiovascular applications will be examined in this investigation.

2.1 Biodegradable vs. Non-biodegradable Polymers

The choice of materials for implants and other devices is crucial in the field of cardiovascular surgery, and polymers are leading the way in these developments. The two main classes of polymers, biodegradable and non-biodegradable, are crucial in deciding how well surgery works out. Every kind has distinct qualities, benefits, and drawbacks that affect how each is used in the field of cardiovascular medicine [21-23].

a). Biodegradable Polymers

Biodegradable polymers are made to gradually disintegrate within the body, usually by enzymatic or hydrolysis processes. Due to this characteristic, they are very useful in a variety of cardiovascular applications, including grafts, stents, and sutures. Polylactic acid (PLA), polyglycolic acid (PGA), and their copolymers (PLGA) are common biodegradable polymers that are preferred because of their biocompatibility and adjustable rates of decomposition.

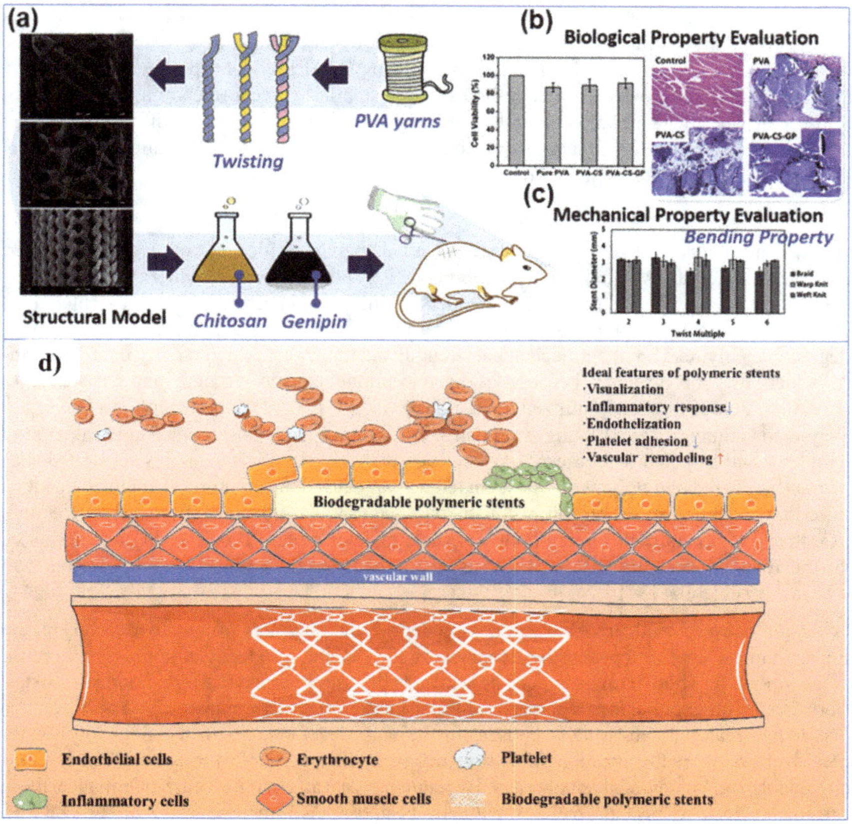

Figure 2. Diagrammatic representation of the development and assessment of a PVA stent. (a) The PVA stent's processing and design. (b) The assessment of the PVA stent's biological properties. (c) The PVA stent's mechanical property assessment. (d) Biodegradable polymeric stents' ideal characteristics. It is important to take into account the characteristics of stents, such as visualization, inflammatory response, endothelization, platelet adhesion, and vascular remodeling (the promotion impact is indicated by a red arrow, and the inhibition effect is indicated by a blue arrow). Reproduced with permission from [25].

Reducing the need for additional surgical procedures is one of the main advantages of biodegradable polymers [24]. Fabricating biodegradable polymeric stents, including injection molding, weaving, 3D printing, and laser cutting, as shown in Figure 2 [25]. Besides, the functional modification of biodegradable polymeric stents is also introduced, including visualization, anti-thrombus, endothelialization, and anti-inflammation For example, biodegradable stents can support a channel temporarily before dissolving, removing the possibility of long-term issues like

 Materials Research Forum LLC
https://doi.org/10.21741/9781644903353-4

restenosis or thrombosis that come with permanent implants [27-29]. Additionally, these materials can act as drug delivery vehicles, enabling targeted therapy and mitigating systemic side effects. Nevertheless, there are several difficulties in using biodegradable polymers. Premature deterioration can cause device failure, while extended stability can cause chronic inflammation, therefore the pace of degradation must be carefully regulated to coincide with tissue healing timeframes. Furthermore, these materials' mechanical qualities can change dramatically over time, calling for a thorough assessment at the design stage [30,31].

b). Non-biodegradable Polymers

In contrast, non-biodegradable polymers are intended to stay inside the body for an extended period. These materials, which include polyurethane, expanded polytetrafluoroethylene (ePTFE), and polyethene terephthalate (PET), are frequently utilized in heart valve and vascular graft applications that call for long-term structural integrity [32]. The main benefits of non-biodegradable polymers are their mechanical strength and longevity, which qualify them for high-stress situations. For instance, because of its exceptional tensile strength and resistance to deterioration, PET is frequently utilized in vascular grafts [33]. These polymers offer an enduring resolution for ailments necessitating extended involvement, permitting uninterrupted operation without the hazards linked with resorption. However, there are significant dangers associated with using non-biodegradable polymers. These materials can cause long-term inflammatory reactions because they decompose, which can result in consequences like infection, fibrosis, or rejection. The existence of a persistent foreign body may also make it more difficult to do more surgeries down the road if necessary. Thus, while using these materials, proper patient selection and monitoring are essential [34].

The decision between biodegradable and non-biodegradable polymers frequently depends on the particular clinical application and the demands of the patient. Biodegradable polymers are an attractive option for short-term uses, such as scaffolds in tissue engineering or temporary vascular support. On the other hand, in circumstances where long-term structural integrity is needed, non-biodegradable materials are usually chosen. In the end, developments in the field of polymer science will result in the creation of hybrid materials that combine the advantages of non-biodegradable and biodegradable polymers. By reducing risks and maximizing performance, these advancements seek to provide patients with the best results from cardiovascular procedures. Cardiovascular implants will be safer and more effective as research develops a better knowledge of how these materials interact with biological systems.

2.2 Synthetic vs. Natural Polymers in Cardiovascular Surgery

Choosing the right materials for cardiovascular surgery is essential to getting the best possible results for the patients. Among these materials, synthetic and natural polymers are essential to the construction and operation of medical devices, including drug delivery systems, heart valves, and vascular grafts [35]. Different polymer types have unique properties, benefits, and drawbacks that influence how they are used in healthcare settings. Figure 3 shows the overview of biomaterials currently used in cardiovascular applications.

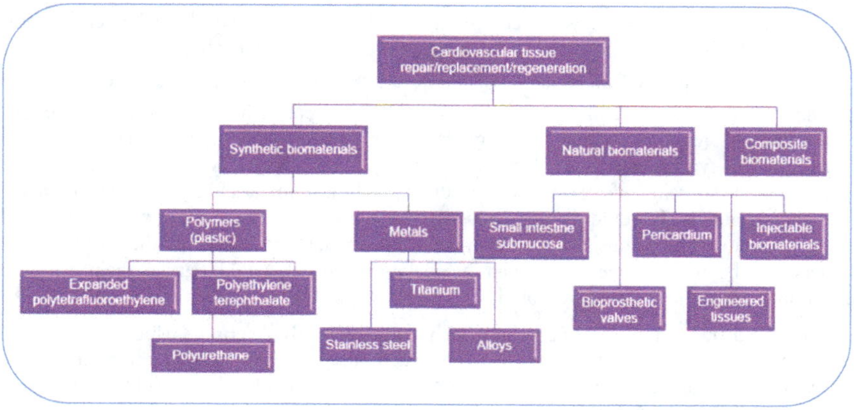

Figure 3. Overview of biomaterials currently used in cardiovascular applications

a). Synthetic Polymers

Man-made materials called synthetic polymers are designed to have certain qualities that are needed for a range of medicinal purposes. Polyethylene (PE), Polypropylene (PP), Polytetrafluoroethylene (PTFE), and Polyurethanes are common examples used in cardiovascular surgery [36]. One of the biggest benefits of synthetic polymers is the capacity to modify their chemical and physical characteristics. To meet specific surgical needs, the mechanical strength, elasticity, and degradation rates all be precisely adjusted.

Applications requiring mechanical stability and long-term durability benefit greatly from synthetic polymers. Because of its low friction coefficient and thrombogenicity, PTFE is frequently utilized in vascular grafts, offering patients undergoing bypass treatments a dependable option. The endurance of synthetic polymers in the body is also facilitated by their resistance to biological breakdown, which makes them perfect for long-term implants.

Synthetic polymers do have disadvantages, though. Even though they might not be as reactive as natural materials, they can nevertheless cause long-term inflammatory reactions that can result in infection or fibrosis. Furthermore, the lack of bioactivity may restrict their capacity to encourage cellular fusion and restoration [37]. Because of this, choosing synthetic polymers for cardiovascular applications requires considerable thought, especially when taking the host tissue's biological response into account.

b). Natural Polymers

Biobased materials such as collagen, gelatin, chitosan, and alginate are examples of natural polymers. These polymers are well known for having outstanding bioactivity and biocompatibility, which makes them appropriate for a range of cardiovascular applications [38]. Collagen-based materials, for example, function similarly to the extracellular matrix and facilitate tissue regeneration, cellular adhesion, and proliferation. Natural polymers have intrinsic benefits because they promote tissue integration and repair. Compared to their synthetic equivalents, they frequently evoke a more positive immunological response since they are derived from biological sources.

Because of this characteristic, they are very useful in applications like medication delivery systems and tissue scaffolds where it is crucial to encourage the body's natural healing processes [39].

Since most natural polymers are biodegradable, it is possible to create temporary implants that the body can eventually absorb and replace with its tissue. This feature helps to lessen the need for follow-up procedures to remove non-biodegradable components, which is a common issue with implants made of synthetic materials. However, natural polymers are not without their restrictions. Disparities in the sourcing and extraction procedures might lead to variability in composition and mechanical qualities. It could be difficult to guarantee consistent performance in clinical applications due to this inconsistency, which could hinder standardization. Furthermore, natural polymers can be more prone to enzymatic breakdown, which could restrict their usage in applications that call for extended durability.

The particular therapeutic needs of a surgical treatment frequently determine which polymers are best, synthetic or natural. Synthetic polymers are usually used for applications needing long-term stability and mechanical strength. On the other hand, because of their biocompatibility and capacity for healing, natural polymers are favoured. To maximize the benefits of both, hybrid materials which blend synthetic and natural polymers are the subject of growing amounts of emerging research [40]. Researchers hope to develop gadgets that improve biological activity and provide mechanical stability by combining these materials. These developments could enhance the functionality and security of heart implants.

Polymers, both natural and synthetic, are essential to cardiovascular surgery. To maximize their use and enhance patient outcomes, a thorough grasp of their characteristics, benefits, and limitations is necessary. The development of innovative hybrid materials holds great promise for safer and more effective cardiovascular therapies in the future as technology progresses. Table 1 discusses the overview of different properties of different types of polymers and their features.

Table 1. Overview of different properties of different types of polymers and their features.

Type of Polymer	Material	Material Combi-nations	Advantages	Applications	Biocomp-atibility	Degrad-ability	Mechanical Properties
Biodegradable Polymers	Poly(lactic acid) (PLA)	PLA with collagen or gelatin	Biocompatible, promotes healing	Vascular grafts, stents	High	Yes	Moderate strength and flexibility
	Polycaprolactone (PCL)	PCL with hydroxyapatite	Good elasticity and slow degradation	Drug delivery systems	High	Yes	Good tensile strength
	Poly(glycolic acid) (PGA)	PGA with PLA	Rapid degradation, excellent tissue integration	Sutures, tissue scaffolds	High	Yes	High tensile strength
	Polyethylene glycol (PEG)	PEG with chitosan	Highly biocompatible and hydrophilic	Hydrogels for drug delivery	Very High	Yes	Varies
	Polyurethane (PU)	PU with silk fibroin	Good mechanical properties	Heart valves, flexible conduits	High	Yes	Excellent elasticity

Materials Research Foundations 172 (2025) 87-112
https://doi.org/10.21741/9781644903353-4

			and biocompatibility				
	Polyvinyl alcohol (PVA)	PVA with alginate	Biocompatible, non-toxic	Hydrogel applications	High	Yes	Moderate flexibility
	Polylactic-co-glycolic acid (PLGA)	PLGA with ceramic composites	Tunable degradation rates, excellent drug release	Stents, scaffolds	High	Yes	Moderate to high strength
	Polycaprolactone-trimethylene carbonate (PCL-TMC)	Blends with PLA	Biocompatibility, controlled degradation	Vascular grafts	High	Yes	Good elasticity
	Chitosan	Chitosan with alginate	Antimicrobial properties and biocompatibility	Drug delivery and wound healing	Very High	Yes	Moderate strength
	Gelatin	Gelatin with PCL or PLA	Promotes cell adhesion and proliferation	Scaffolds for tissue engineering	High	Yes	Moderate flexibility
Non-biodegradable Polymers	Polyethylene (PE)	PE with silicone	Excellent chemical resistance and durability	Catheters, tubing	Moderate	No	High tensile strength
	Polyvinyl chloride (PVC)	PVC with polyurethane	Cost-effective and versatile	Vascular grafts, tubing	Moderate	No	Good mechanical properties
	Polystyrene (PS)	PS with acrylics	Lightweight and easy to mould	Various devices, support structures	Moderate	No	Moderate strength
	Nylon	Nylon with TPU	High strength and flexibility	Sutures, surgical meshes	High	No	High tensile strength
	Polycarbonate (PC)	PC with glass fibres	High impact resistance and durability	Heart valve components	High	No	Excellent mechanical properties
	Polypropylene (PP)	PP with elastomers	Good chemical resistance and low-density	Surgical meshes, sutures	Moderate	No	Moderate strength
	Teflon (PTFE)	PTFE with silicone coatings	Excellent biocompatibility and low friction	Vascular grafts	High	No	Excellent chemical resistance

Materials Research Forum LLC
https://doi.org/10.21741/9781644903353-4

	Polyester (PET)	PET with elastomers	High strength and durability	Vascular grafts, stents	Moderate	No	High tensile strength
	Polyurethane (PU)	PU with various additives	Flexible and durable	Heart valves, artificial heart components	High	No	Excellent elasticity
Synthetic Polymers	Polyethylene glycol (PEG)	PEG with various biopolymers	Highly hydrophilic and biocompatible	Drug delivery systems	Very High	No	Varies
	Polymethyl methacrylate (PMMA)	PMMA with fillers	Optical clarity and good mechanical strength	Bone cement, cardiac implants	Moderate	No	Moderate strength
	Polyimide	Polyimide with carbon composites	Excellent thermal stability and chemical resistance	Flexible circuits in devices	High	No	Excellent mechanical properties
	Silicone	Silicone with various fillers	Good biocompatibility and flexibility	Catheters, heart valves	High	No	Excellent elasticity
Natural Polymers	Collagen	Collagen with elastin	Biocompatible and promotes cell adhesion	Tissue scaffolds, wound dressings	Very High	Yes	Moderate strength
	Fibrin	Fibrin with PEG	Promotes healing and tissue integration	Surgical adhesives, scaffolds	Very High	Yes	Moderate flexibility
	Alginate	Alginate with gelatin	Biocompatible and hydrophilic	Drug delivery and scaffolds	High	Yes	Moderate flexibility
	Chitosan	Chitosan with other biopolymers	Antimicrobial properties	Wound dressings, drug delivery	Very High	Yes	Moderate strength
	Silk fibroin	Silk with PEG	High tensile strength and biocompatibility	Scaffolds for tissue engineering	High	Yes	High tensile strength
	Hyaluronic acid	HA with other polymers	Promotes cell proliferation and healing	Vascular grafts, drug delivery	Very High	Yes	Moderate flexibility
	Keratin	Keratin with PEG	Biocompatible and promotes tissue healing	Scaffolds for tissue engineering	High	Yes	Moderate strength

Materials Research Foundations 172 (2025) 87-112 https://doi.org/10.21741/9781644903353-4

	Cellulose	Cellulose with chitosan	Biodegradable and promotes cell adhesion	Scaffolds, drug delivery	High	Yes	Moderate strength
	Gelatin	Gelatin with alginate	Biocompatible and promotes healing	Tissue scaffolds	High	Yes	Moderate flexibility

4. Polymer Properties Relevant to Cardiovascular Applications

Cardiovascular surgery relies heavily on polymers because of their many characteristics, which have an impact on the functionality and design of devices including heart valves, stents, and vascular grafts. For these devices to be able to survive the physiological forces they come into contact with, two critical mechanical properties elasticity and strength must be present. Durability demanding applications frequently use strong polymers like polytetrafluoroethylene (PTFE) and polyethene terephthalate (PET), which preserve structural integrity in the face of high pressure and pulsatile blood flow [41]. Figure 4 shows the mechanism of polymer working in cardiovascular applications.

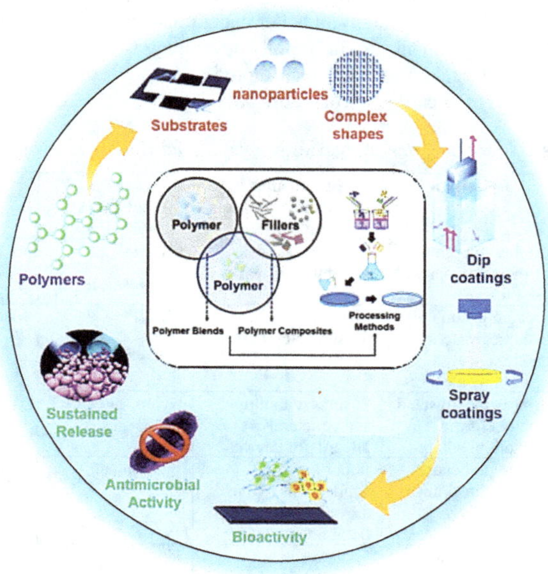

Figure 4. Mechanism of polymer working in cardiovascular applications.

On the other hand, materials that have to allow for dynamic movements like stents, which have to expand and contract in sync with blood vessels' natural pulsation need to have elasticity. A variety of cardiovascular applications can benefit from the unique strength and elasticity balances that can be achieved by engineering synthetic polymers, such as polyurethanes. Apart from its mechanical attributes, biocompatibility a polymer's capacity to properly interact with biological tissues without causing unfavourable reactions is another important consideration when choosing a material [42]. Low reactivity is a common feature of synthetic polymers, which lowers the possibility of problems and inflammation. Because they are more biocompatible and closely resemble the extracellular matrix, natural polymers like collagen and chitosan are preferred for enhancing cellular integration. This characteristic is essential to the durability and efficacy of implants because it promotes better healing results when host tissues are successfully integrated. Moreover, bioactivity a material's capacity to actively encourage advantageous cellular interactions adds still another level of functioning. Tissue integration and regeneration can be aided by bioactive polymers that promote cellular adhesion, proliferation, and differentiation. The bioactivity of natural polymers usually facilitates these processes, but bioactive coatings on artificial devices can improve endothelialization and lower thrombosis risk [43]. With the advancement of research, it is becoming more and more possible to create hybrid materials that combine the best qualities of natural and synthetic polymers, enabling customized mechanical and biological features that maximize device performance. Gaining an understanding of these polymer characteristics is essential to solving the problems in cardiovascular applications, which will ultimately result in safer, more efficient medical procedures and better patient outcomes. Cardiovascular devices will function better and last longer thanks to this rapidly developing field of polymer science, meeting the high standards set by patients and healthcare professionals. Table 2 discusses the different polymer properties relevant to cardiovascular applications.

Table 2. Different polymer properties relevant to cardiovascular applications.

Property	Definition	Significance in Cardiovascular Applications	Examples of Polymers	Measurement Techniques
Biocompatibility	Ability to perform with an appropriate host response	Reduces risk of inflammation and rejection	PLA, PCL, PU	ISO 10993-1
Biodegradability	Decomposition of polymer materials in biological environments	Eliminates the need for surgical removal	PLA, PGA, PCL	Mass loss, weight loss
Mechanical Strength	Ability to withstand applied forces without breaking	Ensures durability and structural integrity of devices	Nylon, TPU, PET	Tensile testing
Flexibility	Ability to bend without breaking	Important for applications requiring movement or flexibility	PU, silicone, PCL	Dynamic mechanical analysis
Elasticity	Ability to return to original shape after deformation	Enhances comfort and adaptability in implants	Silicone, elastomers	Stress-strain curves
Hydrophilicity	Affinity for water	Affects drug delivery and tissue integration	PEG, PVA, HA	Water contact angle measurements

Materials Research Forum LLC
https://doi.org/10.21741/9781644903353-4

Thrombogenicity	Tendency to promote blood clot formation	Critical for vascular applications	Heparin-coated polymers	Thromboelastography
Permeability	Ability to allow substances to pass through	Influences drug release rates	PEG, PVA, PLGA	Gas permeability tests
Surface Roughness	Texture of the polymer surface	Impacts cell adhesion and integration	Various polymers	Atomic force microscopy
Thermal Stability	Resistance to thermal degradation	Ensures performance under body temperature	PTFE, PC, PMMA	Thermogravimetric analysis (TGA)
Chemical Resistance	Ability to withstand chemical exposure	Important for long-term durability in bodily fluids	PTFE, PEEK, silicone	Chemical exposure tests
Bioactivity	Ability to promote biological activity	Enhances healing and integration with tissues	Collagen, gelatin	Cell culture assays
Surface Energy	Measure of the energy at the surface of a polymer	Influences cell adhesion and protein adsorption	Various polymers	Contact angle measurements
Ductility	Ability to deform without breaking	Important for the mechanical behavior of devices	PU, PET	Tensile testing
Fatigue Resistance	Ability to withstand repeated loading	Important for devices subjected to cyclic loading	Nylon, PEEK	Fatigue testing
Optical Clarity	Transparency of the polymer	Essential for applications requiring visibility	PMMA, PC	Light transmittance measurements
Drug Release Kinetics	Rate and manner in which drugs are released	Critical for drug-eluting devices	PLGA, PCL	Release rate studies
Adhesion Properties	Ability to bond with surfaces or tissues	Important for device stability and integration	Polyurethanes, silicones	Peel testing
Solubility	Ability to dissolve in solvents	Impacts on drug formulation and delivery	PEG, PVA, gelatin	Solubility testing
Density	Mass per unit volume of the polymer	Affects the overall weight and performance of devices	Various polymers	Density measurements
Ageing Resistance	Ability to resist degradation over time	Important for long-term implantation	PEEK, PTFE	Accelerated aging tests
Pore Structure	Distribution and size of pores in polymer	Affects cellular infiltration and nutrient transport	Hydrogels, porous scaffolds	Scanning electron microscopy
Antimicrobial Properties	Ability to inhibit microbial growth	Reduces infection risk in devices	Chitosan, silver-infused polymers	Microbial culture methods
Formability	Ease of shaping the polymer into desired forms	Affects the manufacturability of devices	Thermoplastics, elastomers	Molding and extrusion processes

Cost-effectiveness	The economic viability of the polymer	Affects overall device cost and accessibility	Various polymers	Market analysis
Biostability	Resistance to biological degradation	Ensures longevity of implants	PEEK, PTFE	In vivo testing
Specific Heat Capacity	Amount of heat required to change the polymer's temperature	Influences thermal management in devices	Various polymers	Differential scanning calorimetry (DSC)

5. Applications of Polymers in Cardiovascular Devices

While bioactive coatings on synthetic devices can improve endothelialization and lower the risk of thrombosis, natural polymers usually have inherent bioactivity that supports these processes [44]. The creation of hybrid materials which mix the advantages of natural and synthetic polymers is becoming more and more possible as research progresses. These materials offer customized mechanical and biological features that enhance device performance. To solve the difficulties in cardiovascular applications and eventually produce safer, more efficient medical therapies as well as better patient outcomes, it is important to comprehend these polymer features.

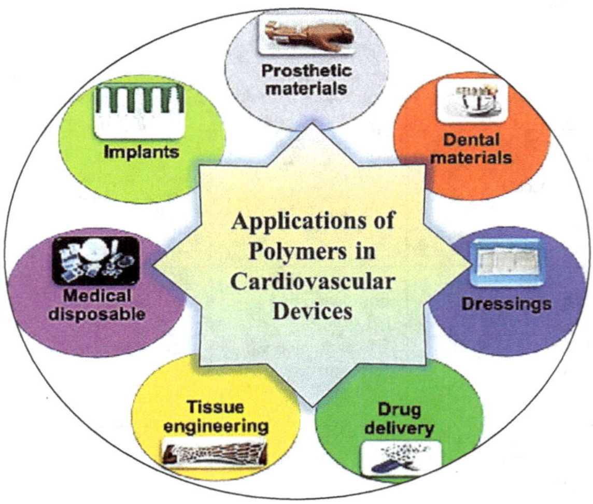

Figure 5. Applications of Polymers in Cardiovascular Devices.

Cardiovascular devices will be able to fulfil the high standards of patients and healthcare providers thanks to the advancements in polymer science, which should improve their longevity and performance. Figure 5 shows the various applications of polymers in cardiovascular devices.

Vascular grafts, which replace damaged vessels or are utilized in bypass surgeries, have progressed from conventional materials to sophisticated polymer compositions. Because of their strength and ease of handling, biocompatible polymers like Dacron (polyester) and expanded polytetrafluoroethylene (ePTFE) are frequently utilized [45]. Biodegradable grafts composed of polymers such as polylactic acid (PLA) and polyglycolic acid (PGA) have been developed recently. These grafts progressively resorb and are replaced by native tissue, eliminating the need for subsequent surgical procedures. These grafts minimize long-term problems by encouraging spontaneous tissue regeneration and supporting blood flow during the healing process. Another important use of polymers in cardiovascular medicine is in heart valves. Heart valves are made of mechanical materials or animal tissue; however, polymers are being used more and more to improve their biocompatibility and longevity. Advanced polymer compositions have replaced conventional materials in vascular grafts, which are utilized in bypass procedures and to repair damaged vessels. Because of their ease of handling and durability, biocompatible polymers such as expanded polytetrafluoroethylene (ePTFE) and polyester Dacron are frequently utilized. The introduction of biodegradable grafts, such as polylactic acid (PLA) and polyglycolic acid (PGA), is one of the more recent advances. These grafts progressively resorb and are replaced by native tissue, eliminating the need for subsequent surgical procedures. In addition to supporting blood flow during the healing process, these grafts encourage spontaneous tissue regeneration, which reduces the likelihood of long-term problems. Heart valves represent yet another essential use of polymers in cardiovascular therapy. Heart valves are being made using more and more polymers to improve their biocompatibility and longevity. Previously, they were made of mechanical materials or animal tissue [46]. Next-generation cardiovascular devices that offer customized therapy and continuous monitoring are made possible by the integration of sensors and bioelectronics with polymer materials. This results in improved management of cardiovascular illnesses. All things considered, the flexibility and versatility of polymers are propelling advancements in cardiovascular devices, tackling the intricate problems of mechanical performance, biocompatibility, and long-term functionality, and greatly improving the standard of care for patients suffering from cardiovascular disorders [47]. The area of cardiovascular medicine is expected to undergo a further revolution as research into innovative polymer compositions and designs advances. This will ultimately lead to improved patient outcomes and a wider range of therapeutic choices for cardiovascular disorders. Table 3 discusses different applications of Polymers in Cardiovascular Devices.

Table 3. Applications of Polymers in Cardiovascular Devices.

Application	Working Mechanism	Advantages Of Conventional Methods	Efficiency Percentage
Drug-Eluting Stents	Polymer coating releases medication to prevent restenosis	Reduces the need for multiple procedures	80-90% effective
Vascular Grafts	Synthetic grafts provide a pathway for blood flow	Reduced risk of infection and improved biocompatibility	75-85% patency rate
Heart Valves	Polymer-based valves mimic natural function	Less calcification and improved durability	90% effective
Balloon Catheters	Expandable polymer balloons open narrowed vessels	Minimally invasive with lower recovery time	85-95% successful
Pacemaker Leads	Flexible polymer insulations protect leads from bodily fluids	Improved longevity and reduced lead fractures	90% reliability
Artificial Hearts	Polymer materials create durable heart chambers	Better biocompatibility and reduced thrombogenicity	80-90% performance

Materials Research Foundations 172 (2025) 87-112 https://doi.org/10.21741/9781644903353-4

Biodegradable Stents	Polymers that dissolve over time release drugs and support vessel healing	Eliminates the need for a second procedure	75-85% effective
Coronary Bypass Grafts	Polymers used in synthetic grafts for bypass surgery	More flexible and easier to implant	80-85% patency rate
Endovascular Aneurysm Repair	Polymer stents used to reinforce weakened artery walls	Less invasive than traditional surgery	85-90% effectiveness
Cardiac Patch	Polymer patches used for myocardial repair after infarction	Promotes healing and reduces scarring	75-80% effectiveness
Left Ventricular Assist Devices	Polymer-based components provide mechanical support for heart function	Enhanced durability and reduced weight	80-85% performance
Hemodialysis Catheters	Biocompatible polymers minimize clotting in dialysis	Improved blood flow and reduced complications	85-90% effectiveness
Antimicrobial Coatings	Polymers infused with antimicrobial agents to reduce infection risk	Longer-lasting protection compared to conventional coatings	80-90% reduction in infection
Bioactive Scaffolds	Polymers that promote tissue regeneration	Enhanced healing compared to inert materials	70-80% effectiveness
Drug Delivery Systems	Polymers control the release of therapeutics directly to affected areas	Targeted delivery reduces side effects	75-85% efficiency
Transcatheter Aortic Valves	Polymer valves delivered via catheter for minimally invasive surgery	Reduced recovery time compared to traditional surgery	85-90% Success
Suture Materials	Biodegradable polymers used in vascular sutures	Eliminates the need for suture removal	80-85% effectiveness
Stent Grafts	Combination of stents and grafts using polymers for support	Improved healing and reduced complications	80-90% effectiveness
Microcatheters	Flexible polymer microcatheters for precise delivery	Reduced risk of damage to surrounding tissues	85-90% effectiveness
Flexible Electrode Arrays	Polymer-based electrodes for cardiac monitoring	Improved comfort and reduced infection risk	90-95% reliability
Coated Balloons	Balloons coated with polymers for targeted drug delivery	Enhanced localized drug delivery and reduced systemic exposure	80-85% effective
Vascular Implants	Polymers used in implants to repair blood vessels	Improved integration and reduced rejection rates	75-80% effectiveness
Biosensors	Polymer-based sensors for real-time monitoring of cardiovascular parameters	Continuous monitoring for early detection	85-90% accuracy
Angioplasty Devices	Polymer balloons used to open blocked arteries	Less trauma and shorter recovery time compared to surgery	80-90% Success
Left Atrial Appendage Occlusion	Polymer devices to close off the left atrial appendage	Reduced risk of stroke in atrial fibrillation patients	85-90% efficacy
Biodegradable Sutures	Polymers that dissolve after wound healing	No need for removal and reduced risk of infection	80-85% effectiveness
Tissue Engineering	Polymer scaffolds for regenerating vascular tissues	Better support for cell attachment and growth	70-80% Success
3D Printed Vascular Models	Polymers used to create patient-specific vascular models	Improved surgical planning and simulation	90% accuracy
Endoscopic Devices	Polymers in endoscopic tools for minimally invasive procedures	Reduced patient trauma and faster recovery	80-90% Success

Materials Research Foundations 172 (2025) 87-112
https://doi.org/10.21741/9781644903353-4

Heart Pump Components	Polymer components in ventricular assist devices for heart support	Lightweight and durable, reducing overall device weight	80-85% reliability
Cardiac Resynchronization Therapy	Polymer leads to optimal electrical stimulation of the heart	Improved patient outcomes compared to conventional pacing	85-90% effectiveness
Custom Implants	3D-printed polymer implants tailored to patient anatomy	Enhanced fit and integration with the body	90-95% Success

6. Coatings and Surface Modifications in Cardiovascular Devices

Surface alterations and coatings are important tactics for improving the functionality of cardiovascular devices [48]. These methods offer extra features like antibacterial qualities, lessen thrombogenicity, and increase biocompatibility. It is crucial to alter the surfaces of materials used in devices like stents, vascular grafts, heart valves, and implantable devices as cardiovascular therapies continue to advance [49].

6.1 Enhancing Biocompatibility and Reducing Thrombogenicity

The capacity of a material to safely interact with biological systems without causing an undesirable reaction is known as biocompatibility. In cardiovascular applications, improving biocompatibility is critical for decreasing problems such as inflammation, thrombosis, and tissue rejection. Thrombogenicity, or a material's propensity to encourage the formation of blood clots, is especially important when it comes to implants that come into direct contact with blood. Cardiovascular devices' biocompatibility can be greatly improved by surface modifications such as the application of bioactive coatings [50]. Anticoagulant heparin is frequently immobilized on device surfaces to prevent platelet activation and lessen the occurrence of clots. This method has been demonstrated to lower the risk of thrombosis in stents and grafts, which will ultimately increase the devices' long-term viability.

Surface alterations can be made to encourage endothelialization, the process by which endothelial cells cover an implant's surface, in addition to anticoagulant coatings. To sustain blood flow and prevent thrombosis, a healthy endothelium layer is essential for device integration and function, which makes it imperative to accelerate endothelialization. This process can be aided by biomimetic coatings, which imitate the characteristics of real tissues. Examples of such coatings include those that include peptides or proteins that stimulate cell adhesion and growth. For example, the introduction of RGD (arginine-glycine-aspartic acid) peptides into polymeric coatings can increase cellular connections, enabling quicker healing and integration of the device with the host tissue [52].

Materials Research Foundations 172 (2025) 87-112
https://doi.org/10.21741/9781644903353-4

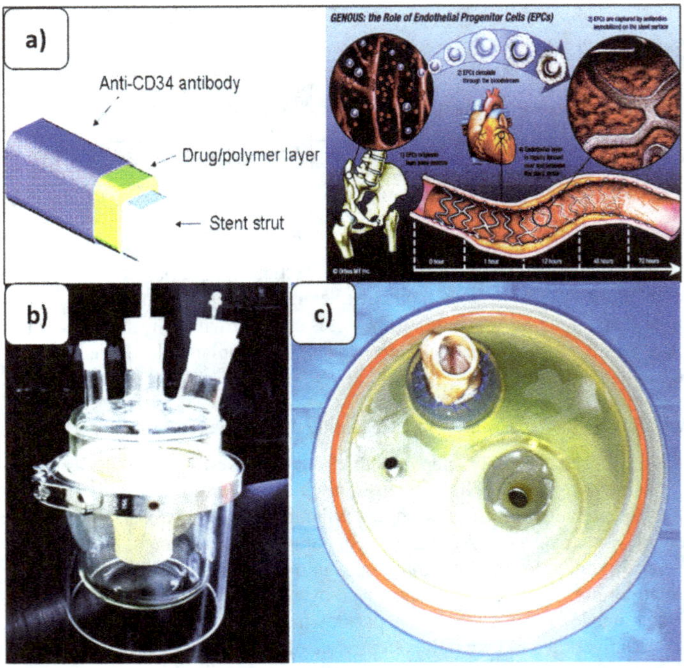

Figure 6. depicts the Sirolimus+EPC capture DES system, including surface modification and drug/polymer coating. On the right, a schematic representation of the Endothelial Progenitor Cell (EPC) capture mechanism, (b) antimicrobial study setup, and (c) an antimicrobial cell. Reproduced with permission from [58]

Furthermore, precise control over surface chemistry and morphology is made possible by sophisticated surface modification techniques such as chemical vapour deposition, layer-by-layer assembly, and plasma treatment [53]. By generating micro- and nanoscale characteristics that affect protein adsorption and cell behaviour, these techniques can improve biocompatibility and decrease thrombogenicity. For example, it has been demonstrated that nanostructured surfaces enhance protein adsorption and encourage advantageous biological responses, which improves the results of healing.

6.2 Antimicrobial Coatings

Cardiovascular procedures are still plagued by infections, especially with implanted devices that can become breeding grounds for bacteria. Antimicrobial coatings are increasingly being developed to overcome this difficulty by reducing biofilm formation and bacterial adherence. These coatings fall into a few categories: those that integrate antimicrobial elements directly into the coating and those that release antimicrobial agents. Adding antibiotics or antimicrobial peptides (AMPs) to polymeric coatings is one popular method [54]. These substances can be released gradually to stop the growth of bacteria and stop infections. For example, because of the well-established antibacterial characteristics of silver, coatings that emit silver ions have become more and more common. Silver ions are useful in avoiding infection because they can damage bacterial cell membranes, impair the activity of enzymes, and obstruct DNA replication [55]. Silver-coated devices have been shown in clinical investigations to improve overall cardiovascular intervention safety and considerably lower the frequency of catheter-associated infections.

Using hydrophilic polymers, like polyethene glycol (PEG), to make surfaces slick and less bacterially sticky is another tactic [56]. These coatings reduce the chance of illness by keeping bacteria from adhering to the surface. By lowering protein adsorption, PEG-based coatings can also improve the biocompatibility of devices and further reduce the risk of thrombosis. Furthermore, the development of intelligent antimicrobial coatings that react to environmental cues is gaining traction. For example, coatings that react to pH changes or bacterial colonization by releasing antimicrobial compounds can provide targeted infection management, reducing the chance of resistance developing [57]. By activating these intelligent systems solely upon detecting germs, antimicrobial agents can be released less frequently overall, maintaining their efficacy over time. Future research has an intriguing opportunity because a single layer possesses both antibacterial and anticoagulant characteristics. Devices that not only resist infection but also encourage healing and integration with host tissues can be made by combining several functions into a single covering. This two-pronged strategy has the potential to greatly enhance the safety and functionality of cardiovascular implants, as shown in Figure 6 [58].

Developments in surface modifications and coatings are an important part of the evolution of cardiovascular devices. Implant long-term success is largely dependent on surface treatments that improve biocompatibility and decrease thrombogenicity. Advanced surface chemistry changes, bioactive coatings, and nanostructuring are some of the techniques that reduce inflammation and thrombosis problems and promote endothelialization. Furthermore, the creation of antimicrobial coatings presents a preventative approach to infections, which continue to be a problem in cardiovascular therapies. The possibility of developing multifunctional coatings that meet biocompatibility and antimicrobial requirements will help to further improve the safety and effectiveness of cardiovascular devices as research into new materials, technologies, and approaches continues. By incorporating cutting-edge coating technologies, cardiovascular medicine may be able to advance and improve patient outcomes.

7. Challenges, Innovations, Future Directions and Limitations of Polymers

Although polymers are essential to the development of medical technology, especially in the field of cardiovascular surgery, they confront many difficulties concerning long-term durability and deterioration. Even though novel polymers have been created to enhance mechanical and biocompatible qualities, problems like biofouling, chemical deterioration, and wear still pose serious problems, particularly in dynamic environments like the cardiovascular system. An additional degree of complexity arises from patient-specific considerations since individual differences in anatomy and physiology might impact the way polymers interact with biological tissues, requiring customized techniques in both design and application. Novel polymer technologies, including drug-eluting implants and bioresorbable stents, have the potential to alleviate some of these issues and improve

Materials Research Foundations 172 (2025) 87-112
https://doi.org/10.21741/9781644903353-4

healing while lowering the risk of consequences. Furthermore, the ability to create bespoke implants that precisely match patient anatomy is transforming the industry because of advancements in 3D printing. In addition to enhancing functionality, this customisation helps to lessen compatibility and integration problems. Notwithstanding this progress, many constraints still exist, such as legal barriers, the requirement for comprehensive testing to guarantee safety and effectiveness, and the intricacy of extending production procedures for clinical use. Looking to the future, interdisciplinary collaboration will be important, merging materials science, engineering, and medicine to develop breakthroughs that enhance outcomes in cardiovascular surgery. With an emphasis on customized treatment and ongoing research into innovative polymer compositions and production processes, it is possible to go beyond present constraints and revolutionize the way that patients are treated. Table 4 discusses different challenges, innovations, future Directions and limitations of polymers.

Table 4. Challenges, Innovations, Future Directions and Limitations of Polymers

Challenges	Innovations	Future Directions	Limitations
Limited long-term durability	Development of biodegradable polymers	Increased focus on personalized medicine	Some polymers may not fully mimic natural tissues
Degradation products may cause inflammation.	Use of nanotechnology to enhance properties	Research into regenerative medicine approaches	Limitations in scalability for manufacturing
Mechanical failure under stress or fatigue	3D printing of custom polymer devices	Development of self-healing polymers	Cost barriers for innovative polymer solutions
Difficulty in achieving optimal biocompatibility	Enhanced drug-eluting stents	Greater integration of AI in design and testing	Regulatory challenges associated with novel materials
Variability in patient-specific responses to materials	Smart polymers that respond to physiological conditions	Innovations in minimally invasive surgical techniques	Difficulty in achieving uniformity in polymer fabrication
Thrombogenicity of some polymer surfaces	Antimicrobial coatings to reduce infection risks	Improved understanding of long-term polymer interactions	Limited clinical data on long-term outcomes
Challenges in sterilization and maintaining integrity	Hybrid materials combining polymers with metals	Collaborative efforts between material scientists and surgeons	Potential for adverse immunological responses
Regulatory hurdles and lengthy approval processes	Advances in polymer chemistry for tailored properties	Focus on sustainability and eco-friendly materials	Dependence on multidisciplinary collaboration
Balancing flexibility and strength in device design	Incorporation of bioactive agents for healing	Exploration of hybrid biomaterials for enhanced performance	Challenges in achieving optimal surface characteristics
Cost implications associated with advanced technologies	Improved imaging techniques for assessing polymer devices	Research into patient-specific polymer designs	Limited availability of high-quality materials
Long-term thrombogenic risk	Development of bioresorbable stents	Integration of smart monitoring technologies	Potential for material degradation over time
Issues with mechanical bonding in hybrid devices	Use of shape-memory polymers	Exploration of nanofibers for vascular applications	Variation in polymer performance based on environment

Materials Research Foundations 172 (2025) 87-112 — https://doi.org/10.21741/9781644903353-4

Limited understanding of polymer-tissue interactions	Coating technologies for enhanced functionality	Advancements in telemedicine for post-surgical monitoring	Potential ethical concerns with novel materials
Variability in manufacturing processes	Use of polymer scaffolds in tissue engineering	Increased focus on patient-centred design	Difficulty in replicating complex shapes and structures
Uncertainty in long-term biocompatibility	Development of flexible and stretchable materials	Focus on real-time monitoring capabilities	High costs of R&D for new polymer formulations
Mechanical properties under physiological conditions	Use of hybrid organic-inorganic polymers	Integration of machine learning for predictive analytics	Limitations in polymer resistance to bodily fluids
Biomechanical mismatch with native tissues	Novel polymer blends for enhanced properties	Increased collaboration with industry partners	Challenges in recycling and disposal of polymer devices
Lack of standardized testing protocols	Development of multi-functional polymers	Research into non-invasive surgical approaches	Risk of device failure leading to serious complications

Conclusion and Future Perspectives

Polymers have become vital in the field of cardiovascular surgery, providing unique qualities that improve the performance and durability of medical devices. Polymers' adaptability allows for the development of a wide range of devices, including stents and grafts, heart valves, and medication delivery systems, all adapted to patients' individual needs. As the desire for new solutions to complicated cardiovascular conditions grows, the future of polymers in this arena appears promising, thanks to breakthroughs in materials science and engineering. Prospects for polymers in cardiovascular surgery are particularly promising, with the possibility of advancements in bioresorbable materials that integrate seamlessly with the body and lower the risk of long-term issues.

Ongoing research focuses on creating polymers that not only meet mechanical and biocompatibility norms, but also have unique capabilities like antibacterial characteristics or controlled medication release. Furthermore, the use of smart materials that respond to physiological inputs may result in devices that adapt to changing conditions within the body, improving patient outcomes. The use of 3D printing technology opens up new possibilities, allowing for the customisation of implants to meet each patient's unique anatomy. This tailored strategy promises to reduce complications and shorten healing times. However, comprehensive testing and compliance with regulatory criteria are required to ensure safety and efficacy. Finally, collaboration among researchers, doctors, and industry stakeholders will be crucial for overcoming current limits, such as durability and degradation concerns, as well as resolving patient-specific variability. By embracing novel ideas and promoting interdisciplinary collaborations, the future of polymers in cardiovascular surgery has the potential to not only improve surgical outcomes but also redefine patient care standards in this critical sector.

References

[1] B. Gaye, G.S. Tajeu, R.S. Vasan, C. Lassale, N.B. Allen, A. Singh-Manoux, X. Jouven, Association of changes in cardiovascular health metrics and risk of subsequent cardiovascular disease and mortality, J. Am. Heart Assoc. 9 (2020) e017458. https://doi.org/10.1161/JAHA.120.017458

 Materials Research Forum LLC
https://doi.org/10.21741/9781644903353-4

[2] L.L. Boland, A.R. Folsom, P.D. Sorlie, H.A. Taylor, W.D. Rosamond, L.E. Chambless, L.S. Cooper, Occurrence of unrecognized myocardial infarction in subjects aged 45 to 65 years (the ARIC study), Am. J. Cardiol. 90 (2002) 927–931. https://doi.org/10.1016/S0002-9149(02)02691-0

[3] V. Fuster, R.A. O'Rourke, R. Walsh, P. Poole-Wilson, *Hurst's the Heart*, McGraw Hill Professional, New York, NY, 2007.

[4] A.S. Go, D. Mozaffarian, V.L. Roger, E.J. Benjamin, J.D. Berry, M.J. Blaha, S. Dai, E.S. Ford, C.S. Fox, S. Franco, et al., Executive summary: Heart disease and stroke statistics—2014 update: A report from the American Heart Association, Circulation 129 (2014) 399–410. https://doi.org/10.1161/01.CIR.0000442015.53336.12

[5] ARIC Study, Community surveillance event rates. Atherosclerosis Risk in Communities (ARIC) Study Website, https://sites.cscc.unc.edu/aric/ (accessed on 30 September 2020).

[6] T. Thom, W.B. Kannel, H. Silbershatz, R.B. D'Agostino, Cardiovascular diseases in the United States and prevention approaches, in: V. Fuster, R.W. Alexander, R.A. O'Rourke, R. Roberts, S.B. King, H.J.J. Wellens (Eds.), *Hurst's the Heart*, McGraw-Hill, New York, NY, 2001, pp. 3–18.

[7] C. Spadaccio, C. Antoniades, A. Nenna, C. Chung, R. Will, M. Chello, M.F.L. Gaudino, Preventing treatment failures in coronary artery disease: What can we learn from the biology of in-stent restenosis, vein graft failure, and internal thoracic arteries?, Cardiovasc. Res. 116 (2019) 505–519. https://doi.org/10.1093/cvr/cvz199

[8] L. Räber, S. Brugaletta, K. Yamaji, C.J. O'Sullivan, S. Otsuki, T. Koppara, M. Taniwaki, Y. Onuma, X. Freixa, F.R. Eberli, et al., Very late scaffold thrombosis: Intracoronary imaging and histopathological and spectroscopic findings, J. Am. Coll. Cardiol. 66 (2015) 1901–1914. https://doi.org/10.1016/j.jacc.2015.08.859

[9] A.V. Finn, G. Nakazawa, M. Joner, F.D. Kolodgie, E.K. Mont, H.K. Gold, R. Virmani, Vascular responses to drug eluting stents: Importance of delayed healing, Arterioscler. Thromb. Vasc. Biol. 27 (2007) 1500–1510. https://doi.org/10.1161/ATVBAHA.106.139154

[10] M. Joner, A.V. Finn, A. Farb, E.K. Mont, F.D. Kolodgie, E. Ladich, R. Kutys, K. Skorija, H.K. Gold, R. Virmani, Pathology of drug-eluting stents in humans: Delayed healing and late thrombotic risk, J. Am. Coll. Cardiol. 48 (2006) 193–202. https://doi.org/10.1016/j.jacc.2006.03.042

[11] G. Nakazawa, A.V. Finn, M. Joner, E. Ladich, R. Kutys, E.K. Mont, H.K. Gold, A.P. Burke, F.D. Kolodgie, R. Virmani, Delayed arterial healing and increased late stent thrombosis at culprit sites after drug-eluting stent placement for acute myocardial infarction patients, Circulation 118 (2008) 1138–1145. https://doi.org/10.1161/CIRCULATIONAHA.108.769331

[12] T. Tada, R.A. Byrne, I. Simunovic, L.A. King, S. Cassese, M. Joner, M. Fusaro, S. Schneider, S. Schulz, T. Ibrahim, et al., Risk of stent thrombosis among bare-metal stents, first-generation drug-eluting stents, and second-generation drug-eluting stents: Results from a registry of 18,334 patients, JACC Cardiovasc. Interv. 6 (2013) 1267–1274. https://doi.org/10.1016/j.jcin.2013.07.008

[13] M.S. Chen, J.M. John, D.P. Chew, D.S. Lee, S.G. Ellis, D.L. Bhatt, Bare metal stent restenosis is not a benign clinical entity, Am. Heart J. 151 (2006) 1260–1264. https://doi.org/10.1016/j.ahj.2006.01.021

[14] D.R. Holmes, B.G. Firth, D.L. Wood, Paradigm shifts in cardiovascular medicine, J. Am. Coll. Cardiol. 43 (2004) 507–512. https://doi.org/10.1016/j.jacc.2003.11.011

[15] F. Bozsak, D. Gonzalez-Rodríguez, Z. Sternberger, P. Belitz, T. Bewley, J.-M. Chomaz, A.I. Barakat, Optimization of drug delivery by drug-eluting stents, PLoS ONE 10 (2015) e0130182. https://doi.org/10.1371/journal.pone.0130182

[16] G. Acharya, K. Park, Mechanisms of controlled drug release from drug-eluting stents, Adv. Drug Deliv. Rev. 58 (2006) 387–401. https://doi.org/10.1016/j.addr.2006.01.016

[17] T. Seo, A. Lafont, S.-Y. Choi, A.I. Barakat, Drug-eluting stent design is a determinant of drug concentration at the endothelial cell surface, Ann. Biomed. Eng. 44 (2016) 302–314. https://doi.org/10.1007/s10439-015-1510-5

[18] C.-W. Hwang, A.D. Levin, M. Jonas, P.H. Li, E.R. Edelman, Thrombosis modulates arterial drug distribution for drug-eluting stents, Circulation 111 (2005) 1619–1626. https://doi.org/10.1161/01.CIR.0000158471.92910.59

[19] B. Balakrishnan, J.F. Dooley, G. Kopia, E.R. Edelman, Intravascular drug release kinetics dictate arterial drug deposition, retention, and distribution, J. Control. Release 123 (2007) 100–108. https://doi.org/10.1016/j.jconrel.2007.07.015

[20] J.E. Sousa, M.A. Costa, A. Abizaid, B.J. Rensing, A.S. Abizaid, L.F. Tanajura, K. Kozuma, G. Van Langenhove, A.G. Sousa, R. Falotico, et al., Sustained suppression of neointimal proliferation by sirolimus-eluting stents: One-year angiographic and intravascular ultrasound follow-up, Circulation 104 (2001) 2007–2011. https://doi.org/10.1161/hc440

[21] B. Nasseri, N. Soleimani, N. Rabiee, et al., Point-of-care microfluidic devices for pathogen detection, Biosens. Bioelectron. 117 (2018) 112–128. https://doi.org/10.1016/j.bios.2018.05.050

[22] S. Bahrami, N. Baheiraei, M. Mohseni, et al., Three-dimensional graphene foam as a conductive scaffold for cardiac tissue engineering, J. Biomater. Appl. 34 (2019) 74–85. https://doi.org/10.1177/0885328219839037

[23] M.G. Toudeshkchoui, N. Rabiee, M. Rabiee, et al., Microfluidic devices with gold thin film channels for chemical and biomedical applications: a review, Biomed. Microdevices 21 (2019) 93. https://doi.org/10.1007/s10544-019-0439-0

[24] C. Bearzi, C. Gargioli, D. Baci, et al., PlGF–MMP9-engineered iPS cells supported on a PEG–fibrinogen hydrogel scaffold possess an enhanced capacity to repair damaged myocardium, Cell Death Dis. 5 (2014) e1053. https://doi.org/10.1038/cddis.2014.12

[25] Shen, Yihong, Xiao Yu, Jie Cui, Fan Yu, Mingyue Liu, Yujie Chen, Jinglei Wu, Binbin Sun, and Xiumei Mo. "Development of biodegradable polymeric stents for the treatment of cardiovascular diseases." Biomolecules 12, no. 9 (2022): 1245.

[26] S. Nour, et al., A review of accelerated wound healing approaches: biomaterial-assisted tissue remodeling, J. Mater. Sci. Mater. Med. 30 (2019) 120.

[27] W.-H. Zimmermann, T. Eschenhagen, Cardiac tissue engineering for replacement therapy, Heart Fail. Rev. 8 (2003) 259–269. https://doi.org/10.1023/A:1024725818835.

[28] G. Vunjak-Novakovic, N. Tandon, A. Godier, et al., Challenges in cardiac tissue engineering, Tissue Eng. Part B Rev. 16 (2009) 169–187. https://doi.org/10.1089/ten.teb.2009.0352

[29] G. Eng, et al., Cardiac tissue engineering, in: Principles of Tissue Engineering, Elsevier, 2014, pp. 771–792.

[30] M. Shachar, S. Cohen, Cardiac tissue engineering, ex-vivo: design principles in biomaterials and bioreactors, Heart Fail. Rev. 8 (2003) 271–276. https://doi.org/10.1023/A:1024729919743

[31] S. Huang, Y. Yang, Q. Yang, et al., Engineered circulatory scaffolds for building cardiac tissue, J. Thorac. Dis. 10 (2018) S2312. https://doi.org/10.21037/jtd.2017.12.92

[32] Ambreen, Jaweria, Thasleema Parveen Malick, Jia Fu Tan, Harith Syahmie Zulfikree, Rathosivan Gopal, Yong Kim Hak, Sivakumar Sivalingam, Hirowati Ali, and Syafiqah Saidin. "Scientific evaluation on biodegradable and biofunctional polyurethane incorporated chitosan/elastin electrospun membranes in compared with synthetic polymeric expanded polytetrafluoroethylene and polyethylene terephthalate vascular membranes." *Journal of Drug Delivery Science and Technology* (2024): 106221. https://doi.org/10.1016/j.jddst.2024.106221

[33] Slart, Riemer HJA, Andor WJM Glaudemans, Olivier Gheysens, Mark Lubberink, Tanja Kero, Marc R. Dweck, Gilbert Habib et al. "Procedural recommendations of cardiac PET/CT imaging: standardization in inflammatory-, infective-, infiltrative-, and innervation (4Is)-related cardiovascular diseases: a joint collaboration of the EACVI and the EANM." *European journal of nuclear medicine and molecular imaging* 48 (2021): 1016-1039. https://doi.org/10.1007/s00259-020-05066-5

[34] Lauri, Chiara, Roberto Iezzi, Michele Rossi, Giovanni Tinelli, Simona Sica, Alberto Signore, Alessandro Posa et al. "Imaging modalities for the diagnosis of vascular graft infections: a consensus paper amongst different specialists." *Journal of Clinical Medicine* 9, no. 5 (2020): 1510. https://doi.org/10.3390/jcm9051510

[35] Ciolacu, Diana Elena, Raluca Nicu, and Florin Ciolacu. "Natural polymers in heart valve tissue engineering: strategies, advances and challenges." *Biomedicines* 10, no. 5 (2022): 1095. https://doi.org/10.3390/biomedicines10051095

[36] Rezvova, Maria A., Kirill Y. Klyshnikov, Aleksander A. Gritskevich, and Evgeny A. Ovcharenko. "Polymeric heart valves will displace mechanical and tissue heart valves: a new era for the medical devices." *International Journal of Molecular Sciences* 24, no. 4 (2023): 3963. https://doi.org/10.3390/ijms24043963

[37] Mu, Lei, Ruonan Dong, and Baolin Guo. "Biomaterials-based cell therapy for myocardial tissue regeneration." *Advanced Healthcare Materials* 12, no. 10 (2023): 2202699. https://doi.org/10.1002/adhm.202202699

[38] Tamimi, Maryam, Sarah Rajabi, and Mohamad Pezeshki-Modaress. "Cardiac ECM/chitosan/alginate ternary scaffolds for cardiac tissue engineering application." *International Journal of Biological Macromolecules* 164 (2020): 389-402. https://doi.org/10.1016/j.ijbiomac.2020.07.134

[39] Beltran-Vargas, Nohra E., Eduardo Peña-Mercado, Concepción Sánchez-Gómez, Mario Garcia-Lorenzana, Juan-Carlos Ruiz, Izlia Arroyo-Maya, Sara Huerta-Yepez, and José Campos-Terán. "Sodium alginate/chitosan scaffolds for cardiac tissue engineering: The influence of its three-dimensional material preparation and the use of gold nanoparticles." *Polymers* 14, no. 16 (2022): 3233. https://doi.org/10.3390/polym14163233

[40] Bharadwaz, Angshuman, and Ambalangodage C. Jayasuriya. "Recent trends in the application of widely used natural and synthetic polymer nanocomposites in bone tissue regeneration." *Materials Science and Engineering: C* 110 (2020): 110698. https://doi.org/10.1016/j.msec.2020.110698

[41] Suhag, Deepa. "Biomaterials for Cardiovascular Applications." In *Handbook of Biomaterials for Medical Applications, Volume 2: Applications*, pp. 105-139. Singapore: Springer Nature Singapore, 2024. https://doi.org/10.1007/978-981-97-5906-4_4

[42] Major, Roman, Maciej Gawlikowski, Marek Sanak, Juergen M. Lackner, and Artur Kapis. "Design, manufacturing technology and in-vitro evaluation of original, polyurethane, petal valves for application in pulsating ventricular assist devices." *Polymers* 12, no. 12 (2020): 2986. https://doi.org/10.3390/polym12122986

[43] Patel, Raj, and Dhruvi Patel. "Injectable Hydrogels in Cardiovascular Tissue Engineering." *Polymers* 16, no. 13 (2024): 1878. https://doi.org/10.3390/polym16131878

[44] Zhao, Jing, and Yakai Feng. "Surface engineering of cardiovascular devices for improved hemocompatibility and rapid endothelialization." *Advanced Healthcare Materials* 9, no. 18 (2020): 2000920. https://doi.org/10.1002/adhm.202000920

[45] Desai, Mital, and George Hamilton. "Graft Materials: Present and Future." *Mechanisms of Vascular Disease: A Textbook for Vascular Specialists* (2020): 621-651. https://doi.org/10.1007/978-3-030-43683-4_28

[46] Şensu, Berna. "Design of electrospun cardiovascular bypass graft using derivative of poly (Alkylene terephthalate)." PhD diss., Institute of Science And Technology, 2020.

[47] Singh, Sameer K., Mateusz Kachel, Estibaliz Castillero, Yingfei Xue, David Kalfa, Giovanni Ferrari, and Isaac George. "Polymeric prosthetic heart valves: A review of current technologies and future directions." *Frontiers in Cardiovascular Medicine* 10 (2023): 1137827. https://doi.org/10.3389/fcvm.2023.1137827

[48] Zhao, Jing, and Yakai Feng. "Surface engineering of cardiovascular devices for improved hemocompatibility and rapid endothelialization." *Advanced Healthcare Materials* 9, no. 18 (2020): 2000920. https://doi.org/10.1002/adhm.202000920

[49] Badv, Maryam, Fereshteh Bayat, Jeffrey I. Weitz, and Tohid F. Didar. "Single and multi-functional coating strategies for enhancing the biocompatibility and tissue integration of blood-contacting medical implants." *Biomaterials* 258 (2020): 120291. https://doi.org/10.1016/j.biomaterials.2020.120291

[50] Hong, Jun Ki. "Liquid-Infused Surfaces for Anti–Thrombogenic Cardiovascular Medical Devices." PhD diss., 2023.

[51] Kumar, Vijay Bhooshan, Om Shanker Tiwari, Gal Finkelstein-Zuta, Sigal Rencus-Lazar, and Ehud Gazit. "Design of functional RGD peptide-based biomaterials for tissue engineering." *Pharmaceutics* 15, no. 2 (2023): 345. https://doi.org/10.3390/pharmaceutics15020345

[52] Luu, Cuong Hung, Nam-Trung Nguyen, and Hang Thu Ta. "Unravelling Surface Modification Strategies for Preventing Medical Device-Induced Thrombosis." *Advanced Healthcare Materials* 13, no. 1 (2024): 2301039. https://doi.org/10.1002/adhm.202301039

[53] Escobar, Ane, Nicolas Muzzio, and Sergio Enrique Moya. "Antibacterial layer-by-layer coatings for medical implants." *Pharmaceutics* 13, no. 1 (2021): 16. https://doi.org/10.3390/pharmaceutics13010016

[54] Negut, Irina, Bogdan Bita, and Andreea Groza. "Polymeric Coatings and Antimicrobial Peptides as Efficient Systems for Treating Implantable Medical Devices Associated-Infections." *Polymers* 14, no. 8 (2022): 1611. https://doi.org/10.3390/polym14081611

[55] Fooladi, Saba, Mohammad Hadi Nematollahi, Navid Rabiee, and Siavash Iravani. "Bacterial cellulose-based materials: A perspective on cardiovascular tissue engineering applications." *ACS Biomaterials Science & Engineering* 9, no. 6 (2023): 2949-2969. https://doi.org/10.1021/acsbiomaterials.3c00300

[56] Patel, Unnati, and Emily C. Hunt. "Recent advances in combating bacterial infections by using hybrid Nano-Systems." *Journal of Nanotheranostics* 4, no. 3 (2023): 429-462. https://doi.org/10.3390/jnt4030019

[57] Cassa, Maria Antonia, Piergiorgio Gentile, Joel Girón-Hernández, Gianluca Ciardelli, and Irene Carmagnola. "Smart self-defensive coatings with bacteria-triggered antimicrobial response for medical devices." *Biomaterials Science* (2024). https://doi.org/10.1039/D4BM00936C

[58] Abizaid, Alexandre, and J. Ribamar Costa Jr. "New drug-eluting stents: an overview on biodegradable and polymer-free next-generation stent systems." *Circulation: Cardiovascular Interventions* 3, no. 4 (2010): 384-393. https://doi.org/10.1161/CIRCINTERVENTIONS.109.891192

Applications of Polymers in Surgery II
Materials Research Foundations 172 (2025) 113-162
Materials Research Forum LLC
https://doi.org/10.21741/9781644903353-5

Chapter 5

Polymers in Neurosurgery

Joy Hoskeri H.[1,*], Preeti Gadyal[1], Preeti Shinde[1], Vidyashree Suryavanshi[1], Nivedita Pujari S[2], Arun K. Shettar[3]

[1]Department of Bioinformatics and Biotechnology, Karnataka State Akkamahadevi Women University, Vijayapura-586109, Karnataka, India

[2]Albert Einstein College of Medicine, Bronx-10461, New York USA

[3]Cytxon Biosolutions Pvt Ltd, Huballi, 580031, Karnataka, India

joybioinformatics@gmail.com, joybioinfo@gmail.com

ORCID ID: 0000-0001-5763-7115

Abstract

Neurosurgery, a field that operates at the intricate intersection of medicine and engineering, has long sought innovative solutions to address the complicated hindrances faced in the treatment of neurological disorders and nerve injuries. One such promising avenue is exploration of polymers in neurosurgical practice. Polymers can be employed as conjugated, doped polymers, or nanocarriers, fabricated hydrogels, scaffolds, or drug delivery vehicles, to treat brain tumors, craniotomies, spine surgery, and aneurysms. As the field of neurosurgery persists in pushing the horizon towards incorporating polymers in neurosurgery holds immense promise. This chapter covers polymer synthesis, polymer fabrication, and polymer applications in neurosurgery.

Keywords

Polymer, Neurosurgery, Nano-Polymers, Craniotomy, Spine Surgery

Contents

Materials Research Forum LLC
https://doi.org/10.21741/9781644903353-5

 Materials Research Forum LLC
https://doi.org/10.21741/9781644903353-5

1. Introduction

Most people contemplate that neurosurgery is brain surgery. Even today, it is a prophylactic word that pertains to diagnosing and treating individuals with disorders or injuries to the brain, spinal cord, spinal column, and peripheral nerves in any region of the body. It is a preventive measure, but neurosurgery comprises both adult and pediatric patients. The sensorium plays a crucial role in an individual's physiology; if the nerves are damaged, the typical motor and sensory processes would be impaired [1]. Nerves possess the capacity for self-regeneration, and certain biodegradable materials have been identified to promote this process. Consequently, the tissue engineering of guided nerve regeneration has emerged as a focal area in molecular biology, clinical medicine, and material science [1]. Neurosurgery provides neurosurgical care and also resuscitative, practically ensuring procedures. On a global basis, more than 9000 surgeries are achieved by neurosurgeons every year.

Neurosurgery involves diagnosing, examining, and treating conditions that affect the central nervous system. Neurosurgery confers distinctive surgical exceptions due to restricted accessibility to cranial and spinal structures, risks to cerebrospinal fluid (CSF) leakage, and the need to handle exquisite nervous tissue. Traditional methods such as suturing and stapling are also not reliable in the nervous system due to their less solidarity strength, inadequate dural repair risks, CSF leakage, and meningitis. The use of polymers in neurosurgery is a promising solution for

Materials Research Foundations 172 (2025) 113-162 https://doi.org/10.21741/9781644903353-5

neurosurgeons because it helps to prevail over constraints such as the above-mentioned traditional methods [2].

Diagnosis can involve a variety of techniques including neurological exams, imaging studies, and laboratory tests, once a diagnosis has been made a treatment plan can be developed which may involve medication, surgery, or other interventions. Neurosurgery is a technique in which so many innovations are evolved/developed specifically in this area of study. Nowadays, robotics-assisted surgeries have become ubiquitous in neurosurgeries allowing for greater precision and less invasive procedures, advanced imaging technology has also greatly improved surgical outcomes. In condition with that, there is ongoing research into new treatments for neurological conditions such as gene therapy and stem cell therapy and the use of polymers in neurosurgery. However, these new ideas and transformations also come with ethical and regulatory thoughts and deliberations that must be carefully identified or addressed. As technology continues to advance, the field of neurosurgery will undoubtedly continue to evolve and improve.

2. Types of Neurosurgeries

2.1 Vascular or Endovascular Neurosurgery

Vascular/Endovascular neurosurgery is a medical/surgical procedure to treat injuries such as strokes or hemorrhages in the brain or spinal cord. Vascular neurosurgery addresses medical disorders such as aneurysms, which are characterized by a weakened or thin area on the cerebral artery that bulges and fills with blood. Vascular malformation (a condition in which an abnormal connection occurs between more than one artery and veins), and Dural arterio-venous fistulas (is abnormal connection arises between arteries and veins). Shape memory polymer coils are good working polymers used to treat aneurysms and they have no side effects over the 12 months after the procedure [3].

2.2 Neurosurgical Oncology

The term brain tumor is concerned with a collection of neoplasms and all the neoplasms with their physiology prognosis and treatment [4]. A brain tumor is an aggregation of unregulated cellular proliferation inside the cerebral system. Neurosurgical oncology encompasses the surgical management of both malignant and benign tumors located in the brain, skull, skull base, spine, spinal cord, and peripheral nerves. Brain tumors are treated by traditional methods such as chemotherapy and radiotherapy, etc., but these methods often cause price toxicity in the body. So, the use of polymers to treat tumors is an attractive and safe method. The material like poly- [bis (p-carboxyphenoxy) propane sebasic acid] is a considerable polymer used to treat brain tumors [5].

2.3 Spine Neurosurgery

Spine surgery is a prevalent surgical intervention for assessing, diagnosing, and managing neurological disorders associated with the spine and spinal cord. Spine surgery is preferred for medical concerns like spinal stenosis, degenerative spondylolisthesis, herniated disks, scoliosis, and spinal deformities. Spinal functions can be rebuilt or restored with the use of graft materials in a variety of forms, including metals, polymers, ceramics, and composites [6]. Although naturally degradable graft implanting is a relatively new technique in the field of spinal neurosurgery. It is

Materials Research Foundations 172 (2025) 113-162 https://doi.org/10.21741/9781644903353-5

a highly anticipated and well-recognized method that is only starting to take shape [7]. In addition, alpha polyesters such as polylactides and polyglycolides are substances that the bone tissue can fully absorb [8-9].

2.4 Pediatric Neurosurgery

A subspecialty of neurosurgery called pediatric neurosurgery treats children from birth to sixteen years old with spine, brain, and spinal cord. In pediatric patients, the optimal skull reconstruction or scientific cranioplasty remains challenging for neurosurgeons. Due to this reason many xenografts, allografts, and autografts are used to treat cranioplasty, but biocompatibility and immune reaction are considerable things when polymers are used for surgical methods [10]. The most popular and recognized polymer for pediatric cranial restoration is polymethyl methacrylate, a clear, transparent, thermostable plastic with impressive outcomes [11]. Especially in pediatric spinal surgery, lumbar myelomeningocele is the most common complication. Usually, myelomeningocele is asymptomatic and gets cured by bed rest and medications, but when it becomes severe it needs surgical care. During pediatric neurosurgery, myelomeningocele has been treated with fast-resorbing polymer mesh [11].

2.5 Peripheral Nerve Neurosurgery

Peripheral nerve neurosurgery is used to treat injuries or pain caused due to trauma in the peripheral nervous system. Polymeric materials are usually preferred for artificial nerve repair or regeneration, the synthetic polymers such as polyesters are highly considered for nerve repair over natural polymers because of their biocompatibility, cost, and fabrication process. Polyethylene glycol and polyethylene lactides are recommended for nerve regeneration approach [12]. To initiate nerve regeneration after peripheral nerve injury, polymer-based nerve guide conduits join the proximal and distal ends of damaged nerve guide conduits, offering both biochemical and physical guidance for the regeneration process [13].

2.6 Neuro-Endoscopy Surgery

Neuroendoscopic surgery is used to treat disorders such as intracranial cysts, intraventricular tumors, skull base tumors, craniosynostosis, and degenerative spine. Endoscopy, which is a tube-like structure or material that enters the skull and helps treat these disorders, can perform complex operations through a small incision [14]. The polymethyl siloxane is polymer molded on microchannels to act as sensors and these micro channels were wrapped around the endoscope operating sheath. Polymethyl siloxane was used to design pressure-sensing skin technology to control the collateral tissue damage during endoscopy [15].

2.7 Laser-based Neurosurgery

Carbon- and carbon-based polymers were used in the treatment of many medicinal conditions, in that carbon-based lasers play a considerable role. A laser is a tiny device that produces light through stimulated radiation emission. An atom has its electrons in its orbit at different energy levels. When the unit of light (photon) strikes an electron in a low energy state then it will accelerate on a high energy level. In this technique carbon dioxide-reinforced polymeric material can be used to treat brain tumors, the emitted laser light targets the brain tumor cells and destroys them [16].

Materials Research Foundations 172 (2025) 113-162
https://doi.org/10.21741/9781644903353-5

2.8 Craniotomy

To view the brain and carry out intracranial surgical treatments, a small portion of the skull is temporarily removed during a craniotomy, a neurosurgical procedure. Metal-based materials are prioritized by many neurosurgeons due to their availability and efficiency, but they also have many disadvantages like lower biocompatibility biodegradability and flexibility. Polymers have become the choice of materials for craniotomy due to their flexibility, high biocompatibility, and biodegradability properties. One of the remarkable polymers is poly-methyl methacrylate act is a good implant material and aids in cranial reconstruction with good results over titanium products [17].

3. Polymers

Polymers are a class of synthetic or natural substances composed of very large molecules and these are also known as macromolecules. Biopolymers have huge potential for use in various biomedical fields, with ongoing research and development opportunities. Research and development in the field of biopolymers has significantly increased globally over the last 20 years. The international scientific and engineering community recognizes the commercial potential of biomaterials and their potential to benefit society and biomedical science more than previously thought. This is a big and positive realization for the health and well-being of all people.

Polymers are present all around us, natural polymers like cellulose, starch, lignin, and protein are used in our daily lives, and synthetic polymers like plastics, and synthetic fibers are also familiar to us. Meanwhile, synthesized or fabricated polymers are used in neurosurgery. During the time of surgery intentionally or unintentionally many injuries occur by mechanical contact of surgical devices or by surgeons' gloves to patients' internal organs, to overcome these problems polymers are the most suitable materials for their functionality and biological properties [18]. For use in medical applications, these polymers can be produced into hydrogels, spun into fibers, and created into fibrous scaffolds.

Polymers trump up many other different forms of materials in plants, animals, or microorganisms and are the basis of many minerals and natural deposits. It is possible to create polymers from both synthetic and natural sources. Polymers have developed and acquired acceptance from the scientific community in the last twenty years. As polymer materials applications showed rapid development, the field of polymer science has observed evolution in the field of regenerative medicine [19]. The awareness we have regarding the nervous system is escalating, at the same time availability of polymers is also increasing.

Natural and synthetic polymers are two of the classes of polymers that are known, according to research in polymer science. Those polymers that are not hurting tissue or other parts of the body can only be used in surgery techniques. These polymeric materials could be grafted into biological systems or introduced through surgical methods. Given the technology's demonstrated effectiveness, more studies on biopolymer materials and their nanocomposites will undoubtedly be conducted for regenerative (replacing damaged tissues or organs due to aging, illness, trauma, or other conditions) and reconstructive (restoring the body parts after an injury) medicine.

The polymers which are used in neurosurgeries should have specific properties like,

- It should be biocompatible

Materials Research Foundations 172 (2025) 113-162
https://doi.org/10.21741/9781644903353-5

- It should be easily fabricated into the desired shape or structure
- Polymer can be easily sterilized, in addition with that there are no changes should occur in its properties due to physical factors
- For functionality, they should become an auspicious contender for different surgical applications [20].
- Polymers should have the capacity to integrate with surrounding tissue or environment.
- It should have cell adhesion properties.
- Polymers should have wound healing and anti-inflammation properties.
- Potential ability to initiate an immunological reaction that could result in inflammation or rejection.

4. Types of Polymer Matrix in Neurosurgery

The polymer matrix serves as a medium for binding and holding the reinforcements together into solids. The polymer matrix protects the reinforcement from degradation and helps to transfer materials from one insert to another (fiber, flake, particles). Polymer matrix which provides finishing, texture, durability, and other functional properties. The polymer matrix can be divided into 3 types such as.

- Synthetic polymer matrix
- Natural polymer matrix
- Semisynthetic polymer matrix

4.1 Synthetic Polymer Matrix in Neurosurgery

Synthetic polymers are types of polymers that can be synthesized with high-quality primeval purity with no immune reaction, and they can be fabricated into many forms with desired morphology. Since these medical devices made of synthetic polymeric materials might disappear once they accomplish their intended function, biodegradation and more advanced properties of certain polymers find use in an expanding range of industries, including suture materials used in neurosurgery [21]. Applications requiring controlled medication release are also made possible by polymers that degrade responsively to specified triggers. These ideas represent the most active research areas at the moment, and items should hit the medical device market soon. The following list of synthetic polymers that are used in neurosurgery includes alpha-polyesters, poly-vinyl alcohol, poly-acrylates, poly-ethylene, poly-ether ketone, poly-urethane, poly(N-isopropyl acrylamide), poly-L-Lactic acid, polypyrole, carbon tubes, poly- lactic acid, polycaprolactone, and poly-L-lactic acid.

4.1.1 Poly Lactic Acid (PLA)

Polylactic acid or lactides is a biodegradable, biocompatible synthetic biopolymer, generated from sugar cane and starch, with the hydrolytic enzyme, or by ring-opening polymerization of lactides [22]. Polylactides are thermoplastic polymers, due to their low melting point and good layer adhesion properties used in many surgical procedures. Because of their high strength and no

Materials Research Foundations 172 (2025) 113-162 https://doi.org/10.21741/9781644903353-5

immune reaction property polylactides can be used in 3D printing. PLA-based nanofibers are fabricated, and they are biomimetic as Dura mater substitute. Due to its low cost and flexibility used to manufacture scaffolds in neurosurgery [23].

4.1.2 Poly Glycolic Acid (PGA)

Polyglycolic acid or polyglycolide is a thermo polymer that has been accepted and used in medicinal surgeries since the decade. Due to its biocompatibility and flexibility used in the medical field to synthesize absorbable sutures and for drug delivery. Polyglycolic acid serves as a membrane to prevent skull deformity, which shows that PGA can regenerate the skull membrane by interpositions in craniosynostosis [24]. PGA can replace suture and biological tissues for closing and knotting wounds and act as a dural substitute without causing incompatibility [25]. The polyglycolic acid-based suturing threads were soft, smooth, and safe for knotting and easily hydrolyzed by bodily enzymes. The collagen sponge-containing tubes were used instead of conventional auto grafts to repair peripheral nerve defects, PGA performs excellent and potential polymer for sensory recovery. According to the findings, using a polyglycolic acid-collagen tube could become a feasible alternative to traditional autografting for the healing of peripheral nerve lesions. It is the first accurate clinical study of sensory recovery utilizing a nerve tube, performed under objective evaluation criteria [26].

4.1.3 Poly L Glycolic acid (PLGA)

As a copolymer of PLA and PGA, poly-L-glycolic acid (PLGA) is a well-liked biodegradable polymer because of its many therapeutic uses, including medication delivery. PLGA is physically strong and has favorable degradation characteristics, due to this, which can be used as a drug delivery vehicle and for nanofibers manufacturing for brain tissues. Poly glycolic acid-based scaffolds are involved to facilitate axonal nerve regeneration and are also involved in anti-inflammatory drug-releasing absorbable surgical sutures [27-28].

4.1.4 Polycaprolactone (PCL)

Polycaprolactone is crystalline, synthetic, aliphatic hydroxyl acidic polyester. However, it is a biodegradable, biocompatible, bioresorbable polymer. The Carothers group produced PCL in early 1930; it is made by polymerizing e-caprolactone in a ring-opening manner. PCL lost its popularity due to the booming of highly biodegradable and biocompatible PLA (polylactides) and PGA (polyglycolide). After twenty years of gap the PCL trend has been changed and used in tissue engineering and slow drug delivery systems and used in cranium regeneration [29].

4.1.5 Carbon Nanotubes (CNTbs)

Carbon nanotubes are nanomaterials, made up of carbon, and their size ranges from 0-100nm in length and diameter. The nanotubes are covered with graphene (a graphite sheet) into a seamless cylinder. Iijima in 1991 was first described about carbon nanotubes. Carbon nanotubes have been used in neurosurgery because of their small size and adaptable external modifications. These properties allow them to function as delivery vehicles for the target brain by overcoming the blood-brain barrier through several targeting methods. [30-31].

Carbon nanotubes have several applications in the nervous system such as electrical and chemical signal monitoring and modulation and also in building neural networks. CBNs are highly valued in the realm of biomedical materials, and efforts are being made to incorporate them into existing materials and technologies, including cellular sensors, tissue scaffolds reinforcements, and drug delivery systems [32].

4.1.6 Polyethylene Glycol (PEG)

Polyethylene glycol is a petroleum-derived bioresorbable polyester and non-toxicbiopolymer, PEG has many applications in medicine and surgery, also it is a USDAFA-approved polymer of choice in drug delivery systems. Polyethylene glycol-based dural sealants were used to control incisional cerebrospinal fluid leakage [33]. Polyethylene glycol is also used in nerve repair and nerve regeneration mechanisms because it acts as a neural cell membrane sealant [34].

4.1.7 Poly(ether-sulfone)

Polyether-sulfone generally called PES is an amorphous, aromatic performance thermoplastic polymer. PES is sulfur sulfur-containing highly physically and mechanically stable biopolymer. This has applications in medicine and surgery. Polyether sulfone plays a role in fabricating electro-spun nanofibers to examine malignant brain tumor migration [35].

4.1.8 Polyvinyl Chloride

Polyvinyl chloride has found the most varied uses in medical products. It provides the best possible balance of low-temperature ductility, high-temperature strength, changeable flexibility, engineering stiffness and strength, and clarity. Because of its exceptional performance-to-cost ratio, PVC will remain a popular choice for packaging and disposable gadgets [36]. The adaptable polymer utilized in a variety of surgical procedures is polyvinyl chloride. Due to its biological characteristics, which include serializability, stability, biodegradability, and biocompatibility. It has been utilized as a cranial implant in cranioplasty, as a dural closure to stop CSF leaking, and as a catheter in neurosurgery.

Different kinds of fabricated synthetic polymeric materials are used in all fields of medicine. The structural alteration and tailoring of polymer structures help to aid their physical and biological properties also engineering-related properties like stiffness, and tensile strength properties may lead to their uses in neurosurgery. Many other synthetic polymers were used in neurosurgery, they are briefly discussed in Table 1.

4.2 Natural Polymer Matrix in Neurosurgery

Natural polymers generally originated from plants and animal sources, natural polymers containing carbohydrates and repeating units of monosaccharides. Compared to synthetic polymers, natural polymers offer several benefits, including safety and chemical-free nature. These are biodegradable, biocompatible, and economically and environmentally friendly. However, apart from applications these are having side effects also, microbial contamination, and metal contamination are the major concern of natural polymers [49-50]. In addition, due to the availability and renewability of natural polymers, these were used in surgical procedures. A few of the commonly used natural polymers are listed below such as polysaccharides (cellulose, hemicellulose, starch, inulin, pectin, carrageenan, chondroitin sulfate), guar gum, chitosan, hyaluronic acid, collagen, gelatin, albumin, fibrin, silk, fibroin, soy protein, alginate, textran, polyamides, poly hydrides, resin (shellac)) are utilized extensively.

Materials Research Foundations 172 (2025) 113-162
https://doi.org/10.21741/9781644903353-5

4.2.1 Collagen

Collagen is a protein, made up of amino acids, and is present abundantly in akin, bones, etc. Selecting material for a nervous disorder is not an easy task, but when the discussion is on collagen, it has having biggest biocompatibility and low cytotoxicity response. Due to this collagen has a wide range of interventions in the development of many neurosurgical scaffolds, acts as a hemostatic agent, and involved in the regeneration of nerves in axons, and provides excellent structural proteins for injured spinal cord [52]. Collagens are components of blood, either they are involved in the replacement of nerve cells or they strengthen them. If collagen is altered by its physical properties, then it's easy to fabricate them and use them in neural tissue engineering, and 3D printing scaffolds are produced. However, collagen has rigidity properties, so it's difficult to fabricate the collagen-based materials. Also, sometimes it shows humoral immune reactions in some people [53].

4.2.2 Hyaluronic Acid

Hyaluronic acid, sometimes referred to as hyaluronan, is a polysaccharide composed of N-acetyl-Dglucosamine and glucuronic acid. A naturally occurring polymer found in numerous body tissues, including skin, cartilage, and vitreous humor, is hyaluronic acid.This can build polymeric cross-linkage and exhibit as 3D printing hydrogels. The unique properties of hydrogels-such as their high water content, remarkable biocompatibility, and capacity to decompose into safe products attracted a lot of attention and have been extensively employed in a range of biomedical applications, such as drug delivery, regenerative medicine, and cosmetic medicine [54].

Hyaluronic acid is the best polymer among the entire natural polymers because of its low immunogenicity, and high biocompatibility. Hyaluronic acid-based 3D printing hydrogels are watery and swollen polymers that promote nerve regeneration. By their soft tissue nature, these polymers do replace the injured neurons. Hyaluronic acid has good porosity, and granular nature due to that hyaluronic acid-based hydrogels have injectability and cell migration properties [55].

4.2.3 Gelatin

A colorless, collagen-derived gelatin is a biopolymer present in animals. Gelatin is utilized to create many scaffolds for the regeneration of peripheral nerves. Gelatin and poly-vinyl-pyrrolidone composite matrix is porous, cost-effective, and has good mechanical properties, and biocompatibility, and also supports cell adhesion and proliferation and helps to differentiate nerve cells. A gelatin-based 3D sponge scaffolds are used as drug delivery vehicles for spinal cord injury [56-57].

Materials Research Foundations 172 (2025) 113-162
https://doi.org/10.21741/9781644903353-5

Table 1. Synthetic polymeric materials used in neurosurgery.

Polymer	Modification	Application	References
Polyethylene glycol	Hydrogel based Sealant Hydrogel sealant Injectable hydrogel	• Dural closure augmentation in cranial surgery • Used as effective closure/suture material in cranial surgery • Used as a dural sealant in a wet environment	37 38 39
Polyethylene terephthalate	Modified as coils Coating material	• The coils were seeded with fibroblast, which helps to aid aneurysm • Coated on nitinol for the treatment of arteriovenous fistulas	40 41
Polyvinyl chloride	3D skull model	• A cranioplasty flap can be created	42
Poly-vinyl alcohol	Modified as sponges Modified as core-coil	• PVA is used to facilitate cerebral bypass surgery • Used for drug delivery system to treat aneurysm	43 44
Poly-propylene polyester	Modified as polymeric knit	Used in cranioplasty as cranioplasty material	45
Poly-uratheneurea	Coating material	Covered on metal coils, used in the treatment of aneurysm	46
Oxidized dextran	Hydrogel	Helps to improve functional motor recovery	47
Carbon dioxide	CO_2 laser	Targeting tumor treatment	48

4.2.4 Silk

Silk is one of the primordial and natural polymeric materials, but today also it will lead the industries. Silk is employed as a medication delivery system in medicine, surgery, and tissue engineering. Two different types can be found in silk: silk fibroin (SF) and silk sericin (SC). Silk sericin has low immunogenicity compared to silk fibroin, and also silk sericin shows good biocompatibility and water-soluble polymeric material. However, both proteins extend with great neuroprotective properties, such as nerve regeneration, and are used as sutures in neurosurgery [58]. The manufactured materials are based on modified silk fibroin and are effective in treating neurodegenerative diseases that cause Parkinson's and Alzheimer's disease [59]. Silk fibroin is also used as suturing material in peripheral nerve injury surgery treatment.

Materials Research Foundations 172 (2025) 113-162
https://doi.org/10.21741/9781644903353-5

4.2.5 Starch

Starch is a polysaccharide (bio-polymer) made up of monomers or monosaccharides called glucose, these monosaccharides are joined together by glycosidic-bond and form polymer (starch). Starch is a colorless and odorless component present in most of the plants, vegetables, and fruits. Starch is a carbohydrate that gives energy to the body, in addition, to having scope and application in neurosurgery. Hemostasis is a very considerable thing in neurosurgery, traditional hemostatic agents are available in the market but they also have many side-effects like low-degradability, high-costs, and safety issues. To overcome these problems, the potato starch-based microporous polysaccharides hemispheres are looking more advantageous than traditional approaches. MPHs are good in the reduction of time to achieve hemostasis and help to reduce excess blood during the time of surgery. However, Microporous polysaccharide hemispheres are also considered an important material in managing arterial bleeding. So, biopolymer starch has many advantages in the study area of neurosurgery [60].

4.2.6 Chitosan

Chitosan is a deacetylated chitin-derived biopolymer. Chitosan is a renewable polymer that can be found in large quantities in the cell walls of numerous fungi, crustaceans, insects, and algae. Numerous biological characteristics of chitosan include itsantibacterial,antioxidant, and anti-inflammatory qualities. Meanwhile, because of its widespread use and several health benefits, chitosan has a variety of uses in neurosurgery. An FDA (Food and Drug Administration)- approved biopolymer called chitosan is used to make laser-activated film surgical adhesives, which are utilized to replace microsurgery in peripheral nerve regeneration procedures [61]. Chitosan has good biocompatibility and host-immune responsibility and is also flexible to do 3D (three-dimensional) scaffolds for nerve tissue regeneration. However, chitosan has many valuable properties but the problem with chitosan is low solubility and poor biodegradability [62].

Natural polymers are available in nature abundantly and these are renewable sources. Due to their good biodegradability and immunogenicity properties, they have many biological applications in neurosurgery. These natural polymers were produced from many of the plants, animals, algae, fungi, and also from aquatic environments. They have synthesized during their life cycle. This content helps discuss a few of the common natural polymers and other excipient natural biopolymers and their modification, applications are given below in Table2.

Table 2. Natural polymeric materials used in neurosurgery.

Polymer	Modification	Application	References
Pectin	Sprayable Hydrogels	For the treatment of malignant Brain tumor	63
Heparin + alginate gel	Artificial nerve graft	Used for digital nerve repair	64
Calcium alginate gel	Gel	To treat endovascular aneurysm	65
Alginate	Hydrogel	• Used for the treatment of spinal cord injury • Brain drug (citicoline) delivery vehicle	66
Sterculia gum	Hydrogel	Brain drug (citicoline) delivery vehicle	67
Hyaluronic acid	Hydrogel based scaffolds	Used as a neuroprotective agent on spinal cord injury	68
Dextran	Coating material	Coated on neural probes (neural probes used for diagnosis of neural disease)	69
Chitosan	film	Used for the treatment of traumatic brain injury to replace injured tissues	70

4.3 Semi-Synthetic Polymer Matrix in Neurosurgery

By combining, grafting, cross-linking, and structurally altering natural and synthetic polymers, semi-synthetic polymers are created by combining both natural and synthetic polymeric components. Purely natural carriers and synthetic polymers with a chemical basis are surpassed by semi-synthetic polymers. Similar to natural polymers, semi-synthetic polymers possess biological attributes such as biocompatibility, biodegradability, flexibility, immunogenicity, and solubility [71]. Semi-synthetic polymers offer a variety of uses in neurosurgery, including scaffolds, drug delivery vehicles, and suturing materials.

4.3.1 Cellulose acetate

Cellulose acetate is a biodegradable, biocompatible, semi-synthetic polymercompound derived from acetylation of cellulose (natural polymer). Cellulose acetate is used in controlled drug delivery systems as film coating on tablets. Cellulose acetate propionate is a derived form of cellulose and combined with cellulose nano whiskers fully biodegradable and biocompatible polymer scaffolds were used in tissue engineering [72]. Cellulose acetate polymers also act as hemostatic agents in neurosurgery and they are used in emergency treatment.

4.3.2 Rayon

Rayon is a semisynthetic biopolymer derived from cotton (Gossypium). Rayon is one of the oldest cotton-based biopolymers. Which has excellent biological properties like biocompatibility, biodegradability, flexibility, and ease of fabricating rayon-based materials. Rayon composite

Materials Research Foundations 172 (2025) 113-162 https://doi.org/10.21741/9781644903353-5

polymeric materials are used as hemostatic agents in neurosurgery and also in the treatment of intracranial tumors [73].

4.3.3 Methylcellulose

Methyl cellulose is a cellulose-derived semi-synthetic biopolymer that is available in a whitish powdery form. Hyaluronans comprised of methylcellulose were fabricated as hydrogels. These fabricated hydrogels were used to manage post-operative adhesion in dura-mater [74].

Various types of natural, synthetic, and semisynthetic polymer materials are available for use in medicine and surgery fields. Their chemical, physical, structural, and biological applications help to assemble them and suitable for use in different neurosurgical procedures. Different polymers have different properties, they may be altered or reconstructed by their structure, composition, physical properties, polymerization, crystallization, and composition, are boosting up the polymer functions such as nerve regeneration, dural repair, and sutures. These kinds of applications are avail or aid them to use in neurosurgery [19].

5. Synthesis of Polymers for Neurosurgery

The synthesis of polymers is also known as polymerization. Polymerization is a chemical reaction process in which monomers act as building blocks of polymers. The monomers are joined by the influence of physical factors like pressure, temperature, etc. Also, these monomers are covalently bonded and monomers can join the particles to form polymers in polymerization [75].

In this chapter, we discuss different synthetic approaches that are regularly used for the preparation of natural and synthetic polymer-based materials which are used in neurosurgery by tailoring their structure, crystallography, etc. However, synthesis of polymer is not an easy task, so different synthetic methods should be followed for synthesizing of polymer are discussed in the below subsection.

5.1 Condensation Polymerization

Condensation polymerization is the polymer preparation method in which monomers are joined together to make a polymer. Furthermore, during condensation polymerization, tiny molecules including water, methanol, and hydrochloric acid are emitted as byproducts. [75]. Several natural polymers are available and those that are used in neurosurgery are synthesized by the condensation polymerization approach. Polyesters are a class of biopolymers that have a diverse range of applications in neurosurgery because of their biocompatibility and biodegradability. Polymerization occurs between either carboxylic acids and diols or hydroxylic acids or diols in the presence of chemical or enzymatic impetus [76].

Many polymers can be synthesized by using condensation polymerization. For example:

Polylactic acidis a biodegradable, biocompatible polymeric material. Many poly-lactic acid-based scaffolds and hydrogels are used in neurosurgery. Synthesis of PLA involves tailoring of lactic acid. Meanwhile, lactic acid acts as a monomer or precursor molecule used for the synthesis of poly-lactic acid. The direct Condensation reaction of PLA is given in Fig. 1. [76].

 Materials Research Forum LLC
https://doi.org/10.21741/9781644903353-5

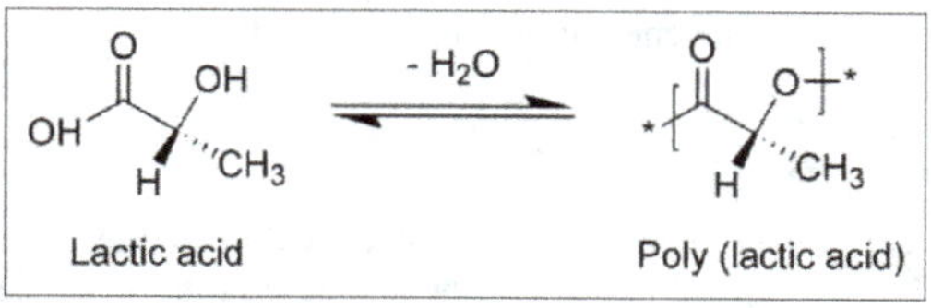

Figure 1. The condensation reaction of Poly-Lactic acid

Here water molecules should be in a lower level or no water molecule also provides a good environment for synthesizing poly-lactic acid because the equilibrium reaction is always transferred to the right. So, water molecules should be evaporated by increasing the temperature above the boiling point. The use of SnO_2 and $SnCl_2$ helps to increase the rate of reaction and form the poly-lactic acid while eliminating the water [76].

5.2 Ring-opening polymerization

Ring-opening polymerization is the chemical reaction in which, ROP uses only the same type of or a single type of ionic precursor molecule. The initiation, propagation, and termination of the chain are all involved in ring-opening polymerization. Here monomers add to the chains only, but not to the branching. In the terminal end, one polymer has its active center and it will react with another monomer to open its ring system to form a long polymer chain. Ring-opening polymerization is a synthetic method used to create polymers with a broad range of industrial applications.

Caprolactone and poly-L-lactides are biopolymers with biological properties like biodegradability and compatibility and are synthesized by ring-opening polymerization. (ring-opening polymerization of caprolactone and polylactic acid is given in Fig.2 [77].

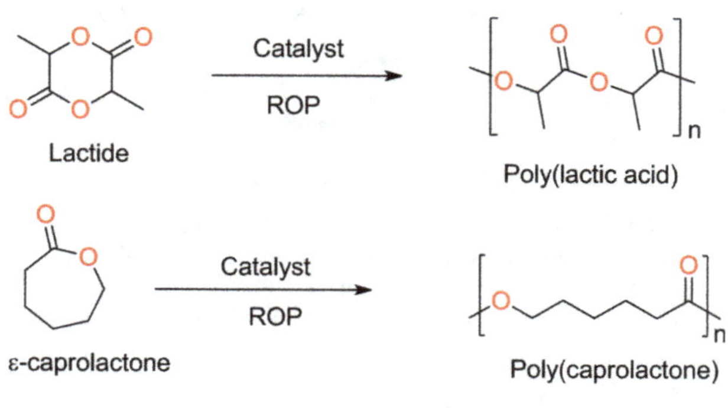

Figure 2. Ring-opening polymerization of Poly-caprolactone

These polymers can be synthesized by tailoring their stereochemistry and molar masses [77].Poly-caprolactone is a biodegradable, biocompatible, and flexible biopolymer, that often has many applications in neurosurgery. Many PLAS and polycaprolactone-based nanomaterials, hydrogels, and scaffolds were used in the regeneration of nerves and for the treatment of traumatic brain injury and aneurysm.

5.3 Addition polymerization

Addition polymerization is the polymerization method used to synthesize the polymers, here an intra-molecular reaction occurs between atoms hence there is the formation of covalent bonds between atoms without elimination of any by-products. The series of events known as addition polymerization occurs when monomers are incorporated into an expanding chain of polymers to create double bonds. Poly-addition reaction is facilitated by a catalyst that will speed up the reaction. Some most common polymers like polyethylene, Teflon, and polyvinyl chloride are synthesized by additional polymerization reactions. Addition polymerization provides stiffness, flexibility, durability, and biological properties to the polymers. Due to these properties, many polymers synthesized based on additional polymerization are used in neurosurgical processes.

Polyurethane is a biopolymer which is either polyester or polyether-based polymer synthesized from the poly-addition method (poly addition reaction of poly-urethane is given in Fig.3 [78].

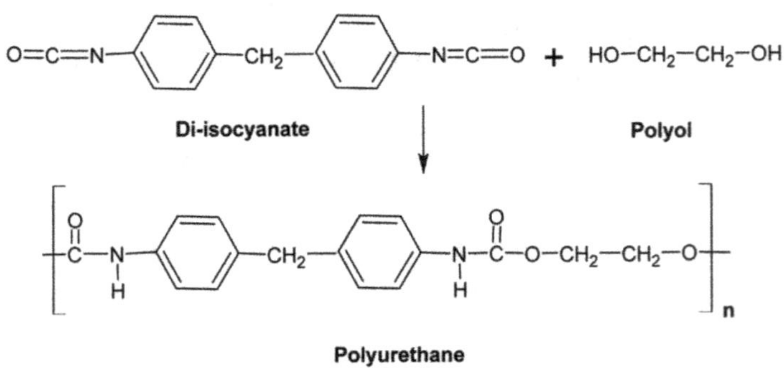

Figure 3. Addition polymerization of Polyurethane

Polyurethane can be used as a coating material, Adhesives, hydrogels, and sealants (polyurethane has excellent sealing properties) in neurosurgery. Poly-urethane involves polyol (two or more hydroxyl groups in alcohol) di-isocyanate acts as an additive and dimethyl hexylamine is a stimulus to inflation the rate of response and reaction in poly addition reaction of polyurethane synthesis [78].

6. Conjugated/Doped Polymers in Neurosurgery

Repeating units of macromolecules (poly-thiophene, poly-aniline, poly-pyrrole, and poly-vinylene, poly-acetylene) in the backbone and side chain of an artificially produced polymer, as well as conjugated polymers with delocalized π electrons in the main chain and SP2 hybridization[79]. Our specific focus is on conjugated polymeric materials that have demonstrated

Materials Research Foundations 172 (2025) 113-162
https://doi.org/10.21741/9781644903353-5

clinical viability in research involving humans or animals and have the potential to be adopted in clinical settings. For therapeutic purposes, conjugated polymers are crucial because they sense electrophysiological signals, control the release of biomolecules at the interfaced tissues, and move charges to alter electrogenic tissues.

The chapter is organized based on several clinical and biological applications that illustrate and call attention to one or more special characteristics of conjugated polymers. We will first discuss some of the special benefits of conjugated polymers in neurosurgery. The content is discussed according to physiological applications that are clinically significant and linked to diseases, such as intracranial, peripheral nerve, biomolecular, and newly developed distributed interfaces [80].

For electrical communication and function, biochemical tissues and cells depend on the mobility of hydrated ions in their intra-and extracellular environment. Conjugated polymers have mixed conductivity, which combines both ionic and electronic properties. A perfect match between conjugated polymer-based electronic-mediated device circuits and ionic-mediated biological systems, providing an increased signal-to-noise ratio, higher current density, and less electrochemical impedance [81]. The common conductive polymers such as poly-pyrrole, poly-(3,4 ethylenedioxythiophene), and poly-aniline are conjugated polymers with having bioelectronics (multidisciplinary field convergence of both biology and electronics)properties like mixed conductors, biomechanical compatibility, chemical tunability made polymers utility high in neurosurgery. Conjugated polymer-made intracranial electrodes are favorable and promising materials for memory, sensory functions, and regaining motor functions and for improving speech ability.

Solution casting, electrospinning, in-situ polymerization, and melt mixing, batch mixing procedures are used in the production of conjugated polymers in neurosurgery. Conjugated polymers like gelatin, hyaluronic acid, and collagen have better biocompatible properties with cell adhesion, cell-binding properties, tissue regenerative properties, etc. [81-82]. To effectively support tissue regeneration, any scaffold biomaterial must possess the key attribute of biocompatibility. Conjugated polymers with strong biocompatibility, such as poly-aniline and poly-pyrrole, have comparatively low cytotoxicity. However, adding natural biomaterials to these artificial polymers can improve the biocompatibility of the final composites much further [83-84].

The relationship between structure and property in macromolecules, particularly conjugated polymers, is significant. Structurally perfect materials are difficult to synthesize, and structural flaws can seriously impair a material's photophysical, charge transport, and stabilizing capabilities [85].

6.1 Conjugated Polymers in Neural Regeneration

Neural regeneration has attracted a lot of attention in tissue engineering and regenerative medicine as a potentially effective treatment option for damaged or diseased nerves. Conjugated polymer nanocomposites have emerged as a treatment [86]. These conductive polymer-based nanocomposites are perfect for brain regeneration because of their unique mechanical and electrical characteristics. When included in tissue engineering constructions, they provide a conducive environment that mimics native tissue, promoting nerve repair [87].

Conjugated polymers, like poly-aniline and poly-pyrrole-based nanofilms, are useful in the treatment and management of epilepsy, Parkinson's disease, and Alzheimer's disease, as well as in

Materials Research Forum LLC
https://doi.org/10.21741/9781644903353-5

drug delivery systems to inject nerve growth factors and neural implants to enhance motor function.

Table 3. Other conjugated polymers and their applications.

Sl. No.	Conjugated polymers	Applications
1	Diketopolypyrole + hyaluronic acid	Targets CD44+ receptors on tumors to control glioblastoma [86]
2	Poly-LacticCo-glycolic acid + temozolomide	Used asa chemotherapy agentforthe Treatment of brain tumors [87]
3	Poly-(9,9-dioctyflurenealt-benzothiadiazole) + Poly-(styrene-co-maleic anhydride)	Promising tumor growth inhibition materials [88]
4	Poly-9,9-dioctyflurene-alt-benzothiadiazole + platinum octaethylporphyrin	Induce oxidative stress in glioblastoma and cause apoptosis [89]

6.2 Doped polymers in neurosurgery

Doping is the process of introducing a negative or positive charge to the polymer backbone through oxidation or reduction, which increases the conductivity of the polymers. Ion implantation, chemical doping, and electrochemical doping are examples of doping methods. Applications for doped polymers include electrochemical sensors used to enhance neural communication and recover motor functions. Polymers like poly-ethylene glycol, poly-aniline, poly-acetylene, and poly-pyrrole act as well as doped polymers used in neurosurgery. Among these polymers, poly-pyrrole is a less-used doped polymer due to its high rigidity and insolubility properties. Acid-base chemistry, protonation techniques, and charge transfer redox chemistry are the main techniques used to dope conjugated polymers. Because doping engineering makes it feasible for conjugated polymers to have customizable properties, these smart materials are sensitive to a wide range of stimuli and changes in reaction to their environment expected to offer a great deal of potential for sensor applications, including biosensors [90].

Using pyrrole and plasma polymerization, non-biodegradable semiconductor biomaterial was created and inserted into spinal cord tissue that had been cut. Tissue sparing and cell proliferation were encouraged by these substances. Potential new biomaterials for spinal cord repair include polypyrrole co-polymerized with PEG and polypyrrole doped with iodine [91].

7. Polymers in Cranial Reconstruction/Craniotomy

A decompressive craniectomy is a type of surgery used to treat severe and unresponsive intracranial hypertension brought on by conditions such as intracranial bleeding, ischemic stroke, traumatic brain injury, and subarachnoid hemorrhage. [92]. In addition to lowering mortality, this procedure may have psychological, social, and neurological effects because the cranial bone is missing. Cranioplasty is recommended when the patient's clinical circumstances are good and the cerebral edema has subsided [93]. The objectives of cranial reconstruction are to restore the cranial contour, restore cerebral shielding against damage, and relieve neurological symptoms by reestablishing intracranial physiological pressure [94]. The anatomic barrier that divides intracranial structures from the surrounding environment is restored when cerebral blood flow

Materials Research Foundations 172 (2025) 113-162
https://doi.org/10.21741/9781644903353-5

dynamics and cerebrospinal fluid return to normal. The term syndrome of the trephined refers to the collection of signs and symptoms connected to partial cranial bone loss.

Regeneration is a very complicated process that cannot always be achieved during healing. Rather, fibrous tissue replaces the functioning tissue due to reparation [95]. Numerous factors can affect both regeneration and reparation and deducing ultimately producing a subpar healing and curing result [95]. Many polymeric materials have been used in cranial reconstruction and a few are discussed below.

7.1 Poly (methyl methacrylate) [PMMA]

Poly (methyl methacrylate) [PMMA] is comparable to bone strength, good compression and torsion test inexpensive cost, and readily available led to its acceptance in medicine. In addition to its strength heat resistance, radiolucency characteristic can be beneficial for detecting the cerebral vasculature adopting angiography, but it became more challenging to identify the plate fractures [96]. Barium is introduced or injected within the plate, detectable by radiography to resolve this problem. Compared to HA, PMMA demonstrates superior stress resistance. When titanium was employed to support wire mesh to implant massive cranioplasties, thereby was a decrease in breakage and an improved aesthetic conclusion [97].

Despite these advantages, PMMA carries a high (5%) chance of extrusion, deterioration, and infection. Patients with an initial infection in the reconstruction region were observed with a high rate of infection (23%). [98-99]. Another method to combine liquid and powder is a mixing pack which includes two different chambers; one for the liquid and one for the powder. After that, the pack is subsequently cut and the resin is squeezed or pressed onto the application site, without off odors, combatted, and allowed to solidify. PMMA commonly adheres to bone but perhaps it may be necessary to employ tiny material plates or screws for skull anchoring [100-101].

During the one-stage operation, poly-methyl-methacrylate is an appropriate material that allows for quick intraoperative rebuilding with great brain protection and a cosmetic result. These goods are easily accessible. Their simple, time-saving application produces the best esthetic and healing results by providing good dural coverage and brain protection. Since some of these goods are more expensive and more readily available than others, the possibilities of infections and host versus graft immunological reactions, just like with any artificial material. The results of our investigation provide credence to the reconstruction of cranial bone abnormalities using poly-methyl-methacrylate [102].

Compared to cranial implants made using comparable RP (rapid prototyping) procedures, PMMA (Poly-methyl-methacrylate) implants made in-house via direct casting utilizing a stereolithographic bio-model of the patient's head fit correctly, are quickly produced, and are less expensive [103].

By employing non-titanium materials for cranial implants and employing finite element analysis to assess their mechanical performance, this work aims to improve cranial reconstruction. Alumina, zirconia, zirconia-reinforced PMMA, hydroxyapatite, and PMMA were used in stress and strain experiments on the brain and skull under various loading scenarios, and a 3D model of the cranial implant was made [104].

Materials Research Foundations 172 (2025) 113-162 https://doi.org/10.21741/9781644903353-5

7.2 Polyetheretherketone [PEEK]

PMMA and PEEK comprise the most fascinating materials used throughout the medical field. These polymers are employed in cardiac, orthopedic, prosthodontic, maxilla-facial, spine, and more recently, cranioplasty surgeries. [105-108]. With the aid of 3D printing technology, these aromatic and semicrystalline thermoplastic polymer-which are chemically inert-can be properly inserted during cranioplasty and merely sculpted with a smooth surface. Another new technique or approach that surgeons are employing to mitigate surgery time and blood loss during PEEK head restoration, is computer-aided design and production [CAM][CAD][109-110]. PEEK performed more effectively when compared to widely used components for cranioplasty, comprising autologous bone, titanium, and PMMA. It additionally showed superior compatibility with the skull and high strength and toughness. Despite this, PEEK implants are quite expensive and have limited osteo-integrative qualities [107].

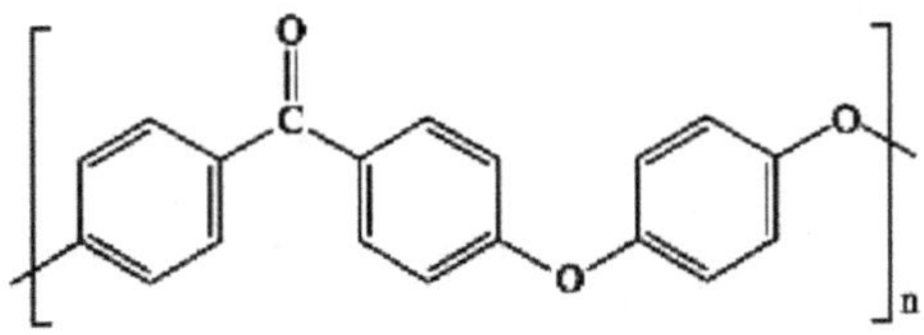

Figure 4. PEEK chemical formula

7.3 Polyethylene [PE]

In the cranioplasty field, PE is an interesting material. It is quite easy to implant requires minimal time and tends to facilitate early vascularization [111-113]. It fails to promote tissue regeneration and in-growth as it is a bioinert material, it could promote a steadier bioactive role. Porous mesh is the most prevalent form of PE utilized for cranioplasty techniques [114-116]. Porous PE demonstrates soft tissue and vascular growth through its pores or small holes in seven days, in addition to fixing surrounding bone and soft tissue in-growth. Through 3D printing, it can be generated and molded. For large-format cranial implants, a comparable polymer, namely polypropylene [PP] polyester(knitwear) [117].

The studies that are being provided show how to duplicate the autogenous bone flap to treat cranial deformity. Cranioplasty is a frequent procedure used to repair the calvaria's shape and structure. Fabricating the cranial prosthesis was a process that involved both prosthetic and surgical phases. Additionally, the characteristics of PMMA and PEEK with their biocompatibility with the calvarial site are highlighted in this work. When treating cranial abnormalities, PMMA is the most often used material for cranial reconstruction [104].

8. Polymers in Spine Surgery

The majority of spinal implants are utilized in conjunction with bone grafts to offer immediate structural stability while awaiting bone fusion i.e., a procedure that fuses two or more bones, so they grow into one solid bone. The contrary drawbacks or complications with metallic spinal

Materials Research Foundations 172 (2025) 113-162
https://doi.org/10.21741/9781644903353-5

implants comprise implant migration, deterioration of imaging quality, and fusion stress shielding [118]. Due to certain complications, the removal of spinal implants is carried out during the perioperative period. Novel devices for internal fixation are being developed to mitigate such constraints in an attempt to address these challenges. These implants ought to be ideally radiolucent [refers to that portion of a processed imaging technique by using X-ray gamma rays which appear as dark or black that occurs due to the passage of maximum photons through the objects], to avoid deterioration of postoperative imaging. For the purpose, of promoting fusion maturation, they should be able to resorb progressively and deliberately under axial loading conditions. An array of polymer implants both bioabsorbable and non-bioabsorbable have been developed according to demand.

Bioabsorbable implants undergo homogenous degradation in the biological milieu or the environment, the breakdown of byproducts integrates into regular physiological, biochemical, and cellular functions. Throughout this period the materials progressively resorb as biological fixation takes place. The load is progressively transferred to the healing bone as they deteriorate or are degraded [120]. In spinal surgery, resorbable implants might offer an array of advantageous characteristics to over rigid metal devices, especially a modulus of elasticity that is more comparable to the instinctual vertebral bone, no radiographic imaging disruptor or interference, anticipated absorption through bulk hydrolysis into carbon dioxide and water, and the elimination of long-term residual metal hardware, which decreases the risk of complications such as migration or dislodgement. Although it is still in its beginning stages, the application of bioresorbable implants in spine surgery is desirable and anticipated development. The overwhelming majority of human clinical applications have been reported so far including anterior column interbody repair, posterior spinal graft containment, arterial cervical and lumbar spine tension band plating, and reestablishing of the harvest site for bone grafts.

8.1 Alpha-polyesters

The majority of research has been concentrated on a class of bioresorbable compounds or materials known as alpha-polyesters, or poly (α-hydroxyl acid). Polyglycolides [PGA], are generated or developed from glycolic acid monomers, and polylactides [PLA], derived from lactic acid monomers are classified as belonging to this group of compounds. These compounds have proven to be fully absorbed by bone tissue. Moreover, PGA wouldn't be an appropriate option to be the main material in a cage device considering it is unstable and may lose its strength and structural integrity in a month. It could merely be considered as only a tiny component of a copolymer [120-121]. PLA can be engineered to possess or access better mechanical properties and they are resistant to hydrolytic degradation compared to PGA [121]. There are two optical isomers of lactic acids the D[right-handed] isomer and the naturally occurring L [left-handed] isomer. Common names of these polymers are PDLA and PLLA, respectively. Although both homopolymers are isotactic (stereochemical regularity of structure in the repeating units of polymer) with their other side groups which are aligned on the same side of the backbone, they satisfy the fundamental condition for crystal formation and are therefore considered semi-crystalline. Although polymers are never completely crystalline; for example, PLLA which occurs naturally, is only about 37% crystalline [122-123]. A racemic polylactide is formed when L and D-isomers are co-polymerized in equal proportion. Crystallinity and average molecular weight or inherent viscosity, are the most important parameters that characterize polymers like PLA. The glass transition temperature is an additional significant metric. A polymer with higher crystallinity will be stiffer, stronger, and typically degrade at a slower rate. The higher degree of crystallinity is directly proportional to the higher glass transition. For spinal cages, which have to withstand strong dynamic loads and operate at 37^0 C for a minimum of six months, these parameters are critical. Because of its symmetric repeat unit lengthy chains, and flexibility, PLLA particular crystallizes easily. Amorphous areas trapped between the crystallites form as a result of crystallization at various places along the chain or polymer impurities. The polymer is thus generally semi-crystalline. The crystalline regions break down more slowly and prevent water penetration because they have stronger secondary connections between the polymer chains. On the other hand, high crystalline PLLA can take more than four years to disintegrate, whereas racemic polylactide, which is completely amorphous, breaks down in a matter of months [122].

8.2 Polyvinyl Alcohol

Polyvinyl alcohol [PVA] is the largest synthetic resin produced all over the world and a soluble hydrophilic polymer [123]. Numerous commercial products based on PVA have recently been developed as a result of their outstanding chemical resistance, physical qualities, and biodegradability. PVA is a polymer that can degrade completely naturally by using carbon dioxide and water. Therefore, it is utilized in numerous biological and pharmaceutical purposes, as a result of its benefits, which include non-toxicity, features that are bio-adhesive, non-carcinogenic, and relatively easy to process [124]. Furthermore, PVA is a polymer with a high degree of crystallinity and strong hydrogen bonding. Typically, melting point ranges from 228-240^0 C for atactic structure i.e., a polymer structure that is amorphous and lacks a regularly ordered structure, 212 to 235^0C for isotactic structure i.e., all of the groups are located on the same side of the hydrocarbon chain, and 230-270^0C for syndiotactic structures [124]. For profoundly hydrolyzed PVA, the glass transition temperature is 85^0C, while for 87-90% hydrolyzation it is 58^0C [125]. PVA has usually several applications, especially for biomedical and pharmaceuticals like blood contact, artificial kidney, and drug delivery. Due to the small size of the hydroxyl group PVA monomers can fold up easily and efficiently although samples of PVA are generally atactic. Therefore, PVA constitutes among the few polymers that can attain high crystallinity within the narrow range of 22-55% [124]. PVA implants maintain more than 70% of their moisture content during physiological loading conditions exhibiting both physiological and mechanical attributes with the nucleus. Moreover, PVA is a biocompatible polymer that endures mechanical loading comparable to that observed in the intervertebral disc nucleus region and assimilates physiological fluid. PVA, however, is not completely stable in the body's physiological environment and gradually decreases as a result of smaller crystallites melting down. This might give rise to a decrease in mechanical qualities and the leaching of molecules into the physiological environment [126].

In-vitro dissolution of hydrogels composed of PVA and PVP blends is lower than that of semi-crystalline PVA gels. Hydrogen bonds between the PVA and PVP chains serve gels their improved stability. The hydrogen bonds in controversy, referred to as interpolymer complexes, perform as secondary physical crosslinks that strengthen the networks against PVA dissolution in the gel. Since PVA/PVP gels are typically elastic, they have shape-memory attributes, which allow them to show consistent, repeatable changes in structure throughout the transition from hydrated to dehydrated states [127]. The hydrogel implant revealed nonlinear stress-strain conduct or behavior typically of soft tissue in two types the confined and unconfined compression tests. Additionally, after the 10 million cycles of compression-compression fatigue to 15% displacement no noticeable modification in the mechanical properties of the hydrogel was detected or observed. Samples

Materials Research Foundations 172 (2025) 113-162
https://doi.org/10.21741/9781644903353-5

subjected to confined compression demonstrated a higher modulus than samples tested in an unconfined condition. Mass stability and recoverable changes were observed. Invitro research employing PVA/PVP hydrogel implant in place of the human lumbar nucleus strengthened the in-situ investigation [128].

8.3 Chitosan

Deacetylated chitin derivative chitosan is a naturally occurring high molecular weight biopolymer that is found in fungal shells and cell walls of marine crustaceans. Chitosan is the second most abundant biopolymer in nature. The amount of glucosamine to N-acetyl glucosamine, which is recognized as the degree of deacetylation, comprises the linear polysaccharide known as chitosan, molecular weight might range from 300 to more than 1000kDa, with a deacetylation degree ranging from 30% to 95%, depending on the source and manufacturing method. When it is in crystalline state chitosan is typically insoluble in aqueous solutions above pH 7 whereas in diluted acids ($pH \leq 6.0$) the glucosamines protonated, free amino groups help the molecule dissolve [129-130]. N-acetylglucosamine is the moiety on chitosan that raises the possibility of associated bioactivities. Indeed, chitosan oligosaccharides activate macrophages, and chitosan, and chitin are the neutrophil chemo-attractants. The extent degree of crystallinity and deacetylation, have an increased relationship with the rate of chitosan degradation.

Thus, highly deacetylated forms may persist for several months, with chitosan oligosaccharides of varying lengths serving as the final byproduct of breakdown [131]. The method of lyophilization, essentially eliminates ice, generating the porous material whose pore size and orientation are controllable by manipulation of the ice crystal size, the freezing rate, and the geometry of the thermal gradient incline during the freezing phase. The resulting material can then be molded into beads, blocks, tubes, and porous membranes. It has recently been demonstrated that pore size and orientation affect the mechanical and structural properties of chitosan scaffolds. Hydrated sample tensile tests revealed that porous membranes have considerably smaller elastic moduli [0.1-0.5 MPa]. The function of both orientation and pore size and the extensibility of porous membrane ranged from values comparable to non-porous chitosan [about 30%] to greater than 100% [131-132]. Additionally, chitosan is a preferred drug delivery vehicle that combines the antibacterial qualities of the connected antibiotic with its intrinsic properties. For potential applications used in orthopedics', chitosan has been combined with a range of delivery materials, including hyaluronic acid, alginate, hydroxyapatite, calcium phosphate, PMMA, Poly-l-lactic acid [PLLA], and growth factors [133-135].

8.4 Polyacrylates

Acrylic acid, methacrylic acid, as well as associated esters, constitute the foundation for polyacrylates. The biologically active materials polymethylmethacrylate [PMMA] and polyhydroxy ethyl methacrylate [PHEMA] are currently and extensively utilized. According to normal circumstances, PMMA is a transparent, unbending substance. The main characteristic determiner is its molecular weight. Bulk, emulsion, and suspension polymerization are examples of free radical polymerization strategies, which may generate high molecular weight PMMA. Cast semi-finished products are produced by bulk polymerization; these products are distinguished for processing outstanding mechanical properties and high surface area [136]. The melting Tg is around 106^0C, whereas the average molecular weight cast material is approximately 106 whilst. Its melting point, which is approximately 107^0C, is prevented from falling below due to its

Materials Research Foundations 172 (2025) 113-162
https://doi.org/10.21741/9781644903353-5

substantial molecular entanglement. PMMA is a polymer with an amorphous structure that dissolves in a variety of liquids with solubility values that are comparable to 18.8 MPa1/2. Ethylene dichloride, ethyl acetate, trichloroethylene, chloroform, and toluene are examples of solvents. Minerals acids may break down the polymer, although it is capable of withstanding the attacks from alkalis, water, and the majority of aqueous inorganic salt solutions. The main properties of PMMA are:

- Strength, stiffness, and high hardness
- Copolymers are tough and homopolymers are brittle
- Capable of being polished, high glass surface scratch-resistant
- High heat torturing temperature
- Machinability and good processability and
- Due to high chain stiffness, the high melt viscosity occurred by restricted rotation about C-bond as the backbone chains [136].

PMMA is a thermoplastic or thermosoftening polymeric material that is extremely versatile. It can be sculpted or molded into an implant with a fixed pattern or maybe polymerized during surgery to create custom implants, like filler material. The emancipation of the monomer, MMA into the bloodstream is another concern with the usage of in-situ polymerization of PMMA; this might result in direct harmful effects of a sharp drop in blood pressure. Multiple experimental studies have demonstrated that the MMA has significant impacts on the peripheral and central neurological systems[137].

9. Polymers in Sealants

After intradural surgery or an unintended dural tear, postoperative cerebrospinal fluid [CSF] leaking is associated with serious challenges such as meningitis and delayed wound healing [138]. Watertight dural closure is an obvious step in preventing CSF leakage. Although it became controversial aphorism is generally acknowledged [139]. The two sides of the incised dura are attached as a component of the traditional dural closure procedure. Consequently, it is challenging to achieve impermeable or waterproof dural closure considering the biomechanical properties of the dura. Even if the needle is brittle and has no flexibility in the dura, the suture causes pinholes to leak. Furthermore, the dura is genuinely open to shrink during surgery as a result of dehydration. The soothing light of the microscope quickens this process. These limitations have resulted in the development, of sealants aimed to stop CSF from leaking. The majority of the existing sealants were employed for gastrointestinal, thoracic, and trauma surgery [140-142]. Sealants developed specifically in the dural region have been developed in the past few decades. In the next years, breakthroughs in biotechnology will provide dural sealants with higher caliber.

9.1 Dural Sealant Biomaterial

The two types of biomaterials that are currently and potentially available for dural sealing include biopolymers and artificial polymers. Biopolymers are naturally occurring polymers that originate or emanate from plants, animals, bacteria, and humans. Biopolymers may be further classified into biomimetic polymers, polysaccharide-based biopolymers, including protein-based biopolymers. The most significant or prominent synthetic polymer groups comprise polycyanoacrylates,

Materials Research Forum LLC
https://doi.org/10.21741/9781644903353-5

polyethylene glycol [PEG], aliphatic polyesters, and polyurethanes [PU] considering their large range of synthetic polymer adhesives available [143-145].

9.1.1 Protein-based Adhesive

Presently, the overwhelming majority of protein-based polymers are generated from humans and animals, which differentiate themselves by their striking similarity to human tissue. The potential of protein-based polymers to trigger desired biological responses, such as clot formation, is an essential feature of hemostasis. Typically, they are biocompatible and biodegradable. Protein-based adhesives interact directly with the underlying tissue to produce a clot, thereby providing them with sealing qualities. The primary drawbacks of protein-based adhesives are composed of materials derived from animals or humans, thereby increasing the possibility of the spread of viruses or prion diseases. Consequently, there is a chance that biomaterials derived from animals will cause an immunological response. One main drawback of some protein-based adhesives is that to achieve sealing attributes, they frequently require to be crosslinked with hazardous substances like acrylates or aldehydes. These groups are related to neurotoxicity [146].

9.1.2 Polysaccharide Adhesives

Sugar-built polymers referred to as polysaccharides are ubiquitous because they are obtained from natural sources likewise plants, animals, humans, and micro-organisms. Polysaccharides such as chitin, chitosan, dextran, and chondroitin sulfate are readily accessible for use in tissue adhesive compounds or molecules. Excellent biodegradable, biocompatible, and non-immunogenic properties characterize polysaccharides. As they can often be modified using physical or chemical techniques. Research that sealed rabbit corneal lacerations revealed the efficacy of a polysaccharide-based glue called methacrylate functionalized hyaluronic acid. 97% of the samples were sealed as a result of photopolymerization [147].

9.1.3 Biomimetic Adhesive

Sealants or Adhesives generated by geckoes, Endo parasitic worms, and mussel sticky protein are prominent instances of biomimetic adhesives that are now gaining popularity. Although the last category generates mechanical attachment to the underlying tissue, these adhesives are either naturally inspired or natural products with adhesive abilities. Initial consequences regarding the mussel adhesive protein require 12-24 hours to bind fully. As of right now, there are no clinically exploitable biomimetic biopolymers for dural sealing. [145-148].

9.2 Synthetic Polymer

9.2.1 Polycyanoacrylates

Cyanoacrylates have exceptional tenacious qualities. Whenever blood and water exist, the cyanoacrylate monomers polymerize rapidly and generate a covalent bond with the rudimentary tissue proteins. Cyanoacrylate doesn't require functionalization, preparation, or any additional ingredients like heat, pressure, and catalyst to seal [142-143]. Modifications can be made to their adhesive strength and biodegradability [147]. Furthermore, by forming a barrier, polycyanoacrylates offer antibacterial properties [148]. The main limitation of polycyanoacrylates is that they break down into formaldehyde and cyanoacrylate, which are the two harmful degradation products [149].

Materials Research Forum LLC
https://doi.org/10.21741/9781644903353-5

9.2.2 Poly (ethylene glycol) [PEG] Polymers

Is a synthetic polymer that is non-immunogenic, water-soluble, and biocompatible. Although photopolymerization was necessary for the adhesive property of the first generation of PEG adhesive, N-hydroxy succinimide [NHS] ester is utilized to functionalize modern PEG-based sealants. These sealants are composed of two parts that combine to generate clinging hydrogel through the covalent relationship of the N-hydroxy succinimide [NHS] ester with the amines in both the mixture and the humid tissue underneath [150]. Furthermore, the hydrogels create mechanical interlocking or blocking by filling the spaces left by the underlying tissues' surface voids [143]. One of the main disadvantages of PEG polymers is their tendency to absorb water, which causes the sealants to size small between 87-95% [151].

9.2.3 Aliphatic Polyesters

The synthetic polymer that has been studied the most is aliphatic polyester. Since the initial synthetic structure based on polyester-based materials has been effectively developed, polyesters have found wider use in medicine. Glycolic acid, lactic acid, and caprolactone are the three main bases for the most commonly utilized biomedical polyesters. A method used to accelerate the biodegradation of certain polyesters, including poly(caprolactone), is co-polymerizing lactide with caprolactone. Co-polymerization of glycoside with lactide can reduce the rigidity of poly (glycolic acid) although it is exceptionally challenging and infrequent [152].

10. Nano polymers in Neurosurgery

The term 'nano' implies to the fabrication of materials and objects with distinct features on the nanometer scale (10^{-9}), generally narrower than 100nm. Microelectronics, molecular biology, surface science, molecular imaging, and tissue engineering are among the many scientific fields or scientific disciplines that require collaboration to achieve the current state of nanotechnology. Nanotechnology holds the potential/capacity to transform medicine by facilitating subcellular manipulation enhancing control over intracellular networks and paving the possibilities of targeted therapies [153-154]. Nanoparticles and nanotools may minimize interference while enhancing the regulation and accuracy of therapeutic intervention by their small stature. The aforementioned features are anticipated to mitigate the elevated rates of complications associated with conventional pharmaceutical treatments by avoiding harm to close-by healthy tissue, an issue associated with most cytotoxic therapies.

The organization makes it simpler to treat lesions profoundly or automatically exacerbated areas of the body [155]. The advantages of nanotechnology might prove particularly noteworthy in the area of neurosurgery, where it is frequently required to target intervention within individual neural networks or neural pathways, such as in functional neurosurgery or within a particular cluster of wayward or aberrant cells, as in neuro-oncology [156-157]. Fortunately, the central nervous system (CNS) poses particular obstacles regarding applying revolutionary nanotechnology-based treatment approaches.

The barrier between the blood and the brain, varying molecular environments throughout various CNS domains, and the distinct cytoarchitectural (systematic study of the arrangement of neural cell bodies within the cortex) features associated with various processing centers [158-159]. In the series, the researchers conduct a comprehensive review or analysis of the most recent advancements in nanotechnology research, particularly those technologies that are anticipated to

Materials Research Foundations 172 (2025) 113-162
https://doi.org/10.21741/9781644903353-5

exert a significant impact on the enhancement of neurosurgical therapeutics in the years that follow. The initial section of this review glances at how nanotechnology has been utilized to abode neurosurgical complications or difficulties at the macro-scale (more effective biological and physical variable evaluation) and micro-scale (spontaneous neuronal regeneration scaffolds quicker and more precise anastomosis (the surgical connection between two structures). At the molecular and subcellular levels, finally, explore how nanotechnology could potentially be used for developing novel therapeutic approaches. The implementation of nanotechnology in targeted molecular medications (focuses on enhancing the CNS drug delivery field of theragnostic medicine (a two-pronged approach to diagnosing and treating cancer through the use of radiotracers) [160].

10.1 Nanoelectromechanical Systems (NEMS)

Devices fabricated at the micrometer scales and the nanoscale are called nanoelectromechanical systems [NEMS] and microelectromechanical systems [MEMS], respectively. Various professionally designed or developed, minuscule components, comprising actuators, sensors, beams, pumps, motors, and resonators, contribute to these systems in overwhelming circumstances [161-163]. The potential of NEMS and MEMS to record and analyze biomechanical physiological variables including intracranial pressure (pressure exerted by fluids such as cerebrospinal fluid inside the skull and on the brain tissue), plasticity (beating rhythmically) of CSF fluid (cerebrospinal fluid) weight load and strain is particularly noteworthy to neurosurgery. Along with the contribution to rendering it simpler to assess these crucial mechanical attributes, scientists have developed nanowires and microelectrode arrays that can determine electrical field potentials [164-165]. Likewise, there is still a biocompatibility consideration that requires to be addressed to monitor brain variables employing NEMS and MEMS directly. These challenges include the formation of scar tissue, probable long-term toxicity, and subsequent inflammatory response.

Numerous techniques or approaches especially integrating biopolymers into these devices are currently being investigated to alleviate these challenges [166-168]. A pressure sensor with integrated valves represents one of the first practical uses of NEMS/MEMS technology in neurosurgery. This type of sensor helps continuous monitoring of intracranial pressure [ICP] and cerebrospinal fluid [CSF] flow within CSF shunts [169]. In addition, considering that it has been established to render it more straightforward to investigate the biochemical implications of novel prosthetic devices [devices designed to replace a missing part of the bodywork better] *in-vivo* NEMS/MEMS technology has been regarded as a significant accomplishment in spinal biomechanics [163].

Furthermore, recent advancements in NEMS technology demonstrated possibilities in physiological monitoring additionally they have also been utilized for mechanical treatments at the cellular and subcellular levels [163-165]. As an illustration, it has been proven that a recently developed nano knife can precisely slash particular axons in peripheral nerves in an in-vivo mouse model [166]. This approach seems extremely intriguing for numerous subspecialties in neurosurgery. For instance, it might constitute an advantageous method in the context of epilepsy surgery permitting for the exceptionally precise disconnection from particular white matter bundles. This enhanced precision might significantly improve surgical outcomes condition attributed to present surgical techniques [163-165].

Materials Research Foundations 172 (2025) 113-162 https://doi.org/10.21741/9781644903353-5

10.2 Nano-scaffolds for Neuroregeneration

Since the peripheral nervous system possesses an inherent capacity to regenerate neurons through axonal development, surgical treatment for injuries to the peripheral nerves generally tends to comply with the expectations of patients and physicians. Complications underlying existing surgical anastomosis treatments, including fibrosis, adhesions [a band of scar tissue that joins two internal body surfaces that are not usually connected], scar formation, and disordered axonal advancement are essentially partially responsible for these less-than-ideal results [170]. Investigations have demonstrated the carbon nanotubes [carbon atoms covalently linked to each other with a hexagonal structure and nanoscale scaffolds may provide the organized framework that most effectively promotes axonal growth by approximating the structure of peripheral nerve myelin sheaths [171-172].

Moreover, the accurate fabrication of perpetual nanofiber scaffolds is currently rendered accessible through revolutionary fabrication techniques including electrospinning. These approaches or techniques generate intricate nanostructures that closely replicate the extracellular matrix that is encountered in the environment [173]. The aforementioned nanofiber matrices. e, is formed by self-assembly of peptide amphiphiles consisting of a poly(lysine) NO donor endothelial cell adhesive ligand, and enzyme-mediated degradable sites have distinctive characteristics, which closely mimic the extracellular matrix's microenvironment have been demonstrated to affect cellular behavior dramatically. These characteristics encompass their high surface volume ratio porosity and different pore-diameter distribution [174].

10.3 Nanoparticulate-Based Technologies for Brain Targeting

Technologies involving a nanoparticulate for the brain targeting the emerging field of nanotechnology, particularly residing at the convergence of biology, biotechnology, engineering, chemistry, and physics continue to transform the field of medicine. Nanotechnology and nanomedicine constitute a leading field in 21st-century medical advancement owing its applications in cancer therapy and diagnosis as well as in screening, diagnosis, and treatment [175]. Research in biomedical and pharmaceutical disciplines has increased tremendously through the potential uses of nanomedicine, which has led to improvements in targeted drug delivery imaging for diagnosis, tissue engineering, and biomaterial development, among other areas of study [176].

Drug immersion inside the brain is potentially increased by the surface-decorated nanoparticles as these particles are capable of passing through the blood barrier [BBB] thereby preventing reticuloendothelial phagocytic activity. Additionally, mucoadhesive nanocapsules [formulated with chitosan by ionotropic gelatine with tripolyphosphate which is negatively charged oligomers of phosphate ions] made of proteins and peptides that are loaded with small-molecules medication/pharmaceuticals and delivered through intranasal that pass through the blood-brain barrier [BBB] and access the cerebrospinal fluid directly through the olfactory mucosa and connective tissues [177].

Due to constraints on the amount of medication dosage in addition to its physical and chemical properties, it frequently stops reaching therapeutic drug concentrations in several brain regions. Customized reagents such as mucoadhesive chemical substances or permeation [process of a liquid, gas, etc. spreading to every part of an object or enhancement elements, are capable of improving the efficiency of drug transportation to the brain to alleviate this particular issue. Mucoadhesive chemical compounds can enhance the therapeutic effect of medication

Materials Research Foundations 172 (2025) 113-162
https://doi.org/10.21741/9781644903353-5

administration on the brain [178]. An additional popular approach or strategy to enhance drug delivery to the brain is to discombobulate the blood-brain barrier [BBB] that enables medications to pass through the brain directly from the circulatory system. It is certainly feasible to discombobulate the BBB through an osmotic ultrasonic or magnetic approach. High osmatic pressure is commonly attributed to substances such as mannitol [diuretic medication], arabinose, urea, fructose, milk amide, and glycerol, which can cause the blood-brain barrier to open. Nevertheless, permitting potentially hazardous or poisonous compounds to enter the brain disrupts the normal function of the CNS [179].

The efficient method for drug delivery across the BBB is convection-enhanced delivery [CED]. By exerting positive pressure and continuous administration of a medical from hours to days, the approach improves the diffusion of macromolecules, protein, and other medications. This convection-enhanced delivery enables a more reliable and monitored supply of drugs by enhancing their osmotic effect [180]. The transport of magnetic drug-coated nanoparticles to the brain can be done via passive and active techniques. Utilizing this passive and active technique combined with focused ultrasound [FUS] targeting enhances the delivery or transport of the nanoparticles through the blood-brain barrier. However, nanoparticles have overcome the biological barriers, and their potential to treat CNS- central nervous system disorders [181].

Cell-penetrating peptides [CPPs] i.e., peptides that are primarily consisting of positively charged amino acid residues which can interact with negatively charged cell-surface molecules [lysine, arginine], therefore helping the peptides across the cell membrane, are well-known for their capacity to enter with minimum cytotoxicity. Proteins, plasmids, siRNA, nanoparticles, nucleic acids, and liposomes are a few of the large and small compounds that CPPs can deliver into cells; by using energy-dependent mechanisms the peptides can enter the cell directly like endocytosis. There are mainly three endocytosis processes involved i.e., micropinocytosis, caveolin-mediated endocytosis, and clathrinid-mediated endocytosis [endocytic pathways that do not use clathrinid can either use the scission GTPase dynamin or be independent [182].

By reducing the toxicity to non-targeted areas and improving the drug efficacy of targeted drug delivery at specific sites thereby intensifying the therapeutic index of the drug. Active targeting for biological specificities such as ligand-receptor or antigen-antibody interaction to deliver the drug, by utilizing the conveyor or transporter like transferrin or antibodies. On the other hand, passive targeting depends on the preferential accumulation of drugs i.e., the difference between the amount of the drug that enters the body and the amount of the drug that is eliminated from the body at a specific time in targeted diseased tissues by its unique physiological condition or parameter. For example, liposomes, nanoparticles, and nanoemulsions can selectively enhance supplements in the liver and spleen through the reticuloendothelial system they can accumulate in the tumors and inflammatory tissues through enhanced permeability and retention (EPR) by effecting intravenous administration [183-184].

11. Polymers in stereotactic neurological procedure

In preliminary research on neurological and psychiatric disorders, stereotactic neurosurgery is used in lab rat or rodent models to implant an electrode or syringe at a predefined site in the brain. Misleading targeting may distort the conclusions of these research investigations. Contrary to clinical practice, imperfections in rodent targeting are usually not detected until they have been assessed using ex-vivo endpoint histology.

Materials Research Foundations 172 (2025) 113-162
https://doi.org/10.21741/9781644903353-5

To treat patient neurological diseases symptoms stereotactic neurosurgery includes accurately inserting a needle or electrode into the brain. Some instances of these techniques includes deep brain stimulation (DBS), lesioning, or cell-based therapy. Experimental research on rat models has an advantage for the clinical use of stereotactic neurosurgery for instance in pinpointing putative locations. Similar research has been conducted on small animals using DBS for neuromodulation [185-186]. Tackling the confrontation in assessing the target precision of stereotactic neurosurgery in presymptomatic research is gaining momentum with the introduction of customized small animal imaging technology or equipment. Numerous investigations have previously documented the significance and advantages of image-based pre-operative surgical planning in small lab animals [187-189]. This requires

- Accretion of multi-modal
- Integrating 3D images generated within an animal along with their spatial normalization to produce a stereotaxic atlas
- 3D reconstruction
- Localization of an electrode tip concerning the target
- Quantification of analyzing targeting accuracy
- Introducing image-based proof regarding possible surgical side effects

In stereotactic neurologic processes, polymers play a crucial role in the context of medical equipment and techniques which require precise targeting and minimal invasiveness. Steps involved in stereotactic neurological procedure are

Biocompatible polymers

Implants as well as other devices that need to interact with neural tissue without creating adverse effects are frequently made up of polymers. For instance, bioresorbable and biodegradable polymers can be developed or generated such as silicone,[PLGA] poly [lactic-co-glycolic acid], and [PCL] polycaprolactone. These elements are utilized in devices like brain implants, drug delivery systems, and neurostimulators.

Customizable Stereotactic Frames

Accurate striving is essential in stereotactic neurosurgery. It is feasible to generate personalized stereotactic frames and guides employing polymers. High-strength polymers [Polyetheretherketone] [PEEK] or acrylic are typical materials utilized to make these frames, that can be customized to fit the patient's specific physique or anatomy. Accuracy of operations like lesioning, deep brain stimulation, and biopsy are improved by this customization.

Microsphere Drug Delivery System

Nanoparticles or microspheres for targeted drug delivery or smart drug delivery are composed of different types of polymers. Therapeutic agents can be delivered directly to specific brain regions through the utilization of a device that may be customized to release drugs in a controlled manner. Although systemic adverse effects are decreased therapeutic efficacy is enhanced this approach is beneficial in the treatment of neurological conditions.

Neuroprotection and repair:

Strategies involving neuroprotection and rehabilitation may require the polymers. Hydrogels and other materials based on polymers, for instance, are being investigated as scaffolds for nerve regeneration and repair. These compounds or polymers possess the capacity to promote brain cell proliferation and promote tissue healing.

Imaging and tracking

Polymers can be utilized as tracking markers or imaging agents that are utilized during the stereotactic process. For instance, polymer-based contrast agents may assist with the precise targeting of therapeutic interventions by improving the precision of imaging techniques like 3D, 2D imaging, CT, or MRI scanning.

12. New Horizons in Polymer Science and Neurosurgery

Furthermore, 3D printing has achieved widespread acceptance as a potentially groundbreaking and revolutionary medical advancement. This technical advancement holds the capacity to drastically change the manufacturing or production processes associated with prosthetics and prosthetics, which holds immense promise for the future of this medical field [190]. 3D printing is an approach that enables the immediate epitome of virtually any regardless of structure or its complexity, in a wide range of materials [191]. Among the medical demands that potentially profit from this technology are patient-specific medical prostheses, anatomical models for surgical planning and exercise, and customized medical delivery systems. Tissue engineering and regenerative medicine have greatly assisted in the discovery of safe and bioresorbable materials for 3D printing [192]. Moreover, the development of 3D printing holds the potential or capacity to transform the arena of customized healthcare interventions that have been meticulously created with their distinctive anatomical and physiological features [193]. In addition to minimizing healthcare costs, this could enhance patient results. The specific obstacles confronting the medical industry preclude 3D printing from being frequently used. Designing improved materials and strategy and promoting testing and gauging protocols are scarcely any challenges that require attention. An alternative obstacle specifically before the appropriate regulatory organizations can authorize 3D printed or additive manufacturing medical equipment and prostheses, they are still imperative or necessary to undergo extensive testing and verification. For instance, 3D-printed renal models can be utilized for pre- and intra-operative planning for the treatment of kidney tumors as well as for understanding patients' renal architecture in the backdrop of orthopedics, craniofacial reconstruction, and cardiovascular and liver illness [194-198]. Despite there are innumerable challenges that need to be overcome before 3D printing's entire potential in the field of medicine is capable of being realized, attempts at research and development are being made with great diligence to address these challenges. This technology could be advantageous in several areas, particularly patient outcomes, medical expenses, and the dart at which medical equipment or devices and artificial equipment can be developed and produced [199]. Throughout an object's existence, three-dimensional printing and additive manufacturing may additionally amend the material's composition and structure [200]. Narrower-diameter lattices and lines, patterned dots, layers, porous membranes, or permeable, and depth-changing motifs may all be designed into complex structures like nacre and bone as well as functionally graded systems [201-202]. It is feasible to modify surface roughness, foam density, form optimization, and new topology by using changeable materials [203].

Materials Research Forum LLC
https://doi.org/10.21741/9781644903353-5

Polymer particles are employed as building constituents in a sort of 3D printing called polymer particle polymerization (PPP) that produces complex three-dimensional structures [204]. In polymer particle polymerization (PPP), a monomer is polymerized throughout a draw breath polymer molecule to generate or bring about a new structural layer. As a consequence of its capacity to create intricate structures with a high degree of control over and resolution of the final product, this strategy has become more and more popular recently for example in the 3D printing of carbon fiber [205]. High-resolution structures are possible to be built in PPP by employing minuscule particles of polymers as elementary units. The technique provides explicit control over the culminating products, size, dimensions, shape, and mechanical attributes PPP may generate on an extensive array of complex structures, incorporating porous structures, hollowed frameworks, and structures with sophisticated geometry [206-208]. 3DP is frequently employed for the production of nanoparticles, epoxy resins, polymeric materials, and assorted materials. The properties of polymer nanocomposites can be influenced by an array of variables such as stress, heat, moisture, the rate of photo polymerization, crosslinking phenomena, erosive compounds, and UV light. A liquid monomer solution is solidified by complex physical and chemical processes. A photoinitator in the solution decomposes or breaks down into active radicals when exposed to light, which results in polymerization. Polymer nanocomposites are generated when monomers attack unsaturated double bonds, higher molecular weight, and polymer chains as a result thick polymer solutions modify the characteristics of the resulting polymer [209]. Natural polymers incorporating rigid body systems may be employed to enhance the frameworks, biological applications, thermal transmission, electromechanical micro-machines, optic electronics, and surface modifications [210].

The predominantly impenetrable blood-brain barrier appears to be a significant obstacle to the effectual delivery of payload or draft into the central nervous system, regardless of the quiescent for nanoscale applications to transform the disciplines of neuro-oncology, neuroradiology, and neurosurgery. Researchers have utilized both intrathecal and intranasal administration methods to go directly into the [CNS] central nervous system and avoid the blood-brain barrier. Through the cribriform plate of the skull base, the mucous membrane of the nasal cavity provides the olfactory epithelium a beeline into the brain. For conditions like allergic rhinitis, the majority of intranasal drugs have been licensed for local or systemic administration to treat Alzheimer's disease, insulin has been using micellar, liposomal, and chitosan nanoparticles, providing the foundation for the eventual use of insulin as a backbone conjugate linker for transferring extra minim mite molecules into the central nervous system [CNS] for the treatment of multiple neurological conditions [211]. Intrathecal delivery of therapeutics has demonstrated both the clinical and preclinical disciplines and areas for the treatment of leptomeningeal disease and intraventricular tumors. Subsequently, this can additionally prevent systemic barriers to central nervous system drug delivery to achieve engrossment of drugs in cerebrospinal fluid without exposing the patient to the system [212-214].

Following the recent research investigation describing the biodistribution or bioaccumulation of 100nm fluorescent-conjugated PEGylated NPs administered into the cisterna of healthy mice, the NPs were merely absorbed into the brain parenchyma but evenly dispersed within the brain and spinal cord in the subarachnoid space, remaining privy to leptomeninges for a maximum of three weeks [215]. Finally, theragnostic the amalgamation of restorative and diagnostic competence into a single NP consignment or delivery system has become increasingly advanced in the frontline of personalized or customized medicine, offering the potential to reduce costs and conserve time while enhancing patient outcomes [216].

 Materials Research Forum LLC
https://doi.org/10.21741/9781644903353-5

Conclusion

The utility of polymers in neurosurgery has revolutionized the field, providing a multitude of advantages that have significantly improved patient outcomes and the overall efficacy of surgical procedures. These versatile materials have a wide range of applications, serving as biocompatible components for implants and scaffolds that can be used to repair or replace damaged neural tissues. The use of polymers in neurosurgery has revolutionized the field, providing a multitude of advantages that have significantly improved patient outcomes and the overall efficacy of surgical procedures. These versatile materials have found a wide range of applications, serving as biocompatible components for implants and scaffolds that can be used to repair or replace damaged neural tissues. One of the key advantages of polymers in neurosurgery is their ability to mimic the natural extracellular matrix, facilitating the integration of implanted materials with the surrounding tissue. This property is particularly important in the [CNS] central nervous system, where the delicate balance of the microenvironment is crucial for successful regeneration and functional recovery. Furthermore, the tunable mechanical and degradation characteristics of polymers allow for the creation of customized scaffolds that can guide the growth and differentiation of neural cells, promoting the restoration of neurological function. The current book chapter covered all aspects of polymers and the types and uses of polymers in neurosurgery. It has given a detailed insight into the information and application of polymers.

References

[1] Wang, S., Hou, J., Bei, J., & Zhao, Y. (2001). Tissue engineering and peripheral nerve regeneration (III) -Sciatic nerve regeneration with PDLLA nerve guide. Science in China Series B: Chemistry, 44, 419-426. https://doi.org/10.1007/BF02879817

[2] Qiu, L., See, A. A. Q., Steele, T. W., & King, N. K. K. (2019). Bioadhesives in neurosurgery: a review. Journal of neurosurgery, 133(6), 1928-1938. https://doi.org/10.3171/2019.8.JNS191592

[3] Echeverría, D., Rivera, R., Giacaman, P., Sordo, J. G., Einersen, M., & Badilla, L. (2023). A novel self-expanding shape memory polymer coil for intracranial aneurysm embolization: 1 year follow-up in Chile. Journal of NeuroInterventional Surgery, 15(8), 781-786. https://doi.org/10.1136/jnis-2022-018996

[4] DeAngelis, L. M. (2001). Brain tumors. New England journal of medicine, 344(2), 114-123. https://doi.org/10.1056/NEJM200101113440207

[5] Leong, K. W., D'Amore, P., Marletta, M., & Langer, R. (1986). Bioerodible polyanhydrides as drug-carrier matrices. II. Biocompatibility and chemical reactivity. Journal of biomedical materials research, 20(1), 51-64. https://doi.org/10.1002/jbm.820200106

[6] Goel, V. K., Martz, E. O., & Park, J. B. (1998). Materials in Spine Surgery. Journal of Musculoskeletal Research, 2(02), 73-88. https://doi.org/10.1142/S021895779800010X

[7] Raucci, M. G., Gloria, A., De Santis, R., Ambrosio, L., & Tanner, K. E. (2012). Introduction to biomaterials for spinal surgery. In Biomaterials for Spinal Surgery (pp. 1-38). Woodhead Publishing. https://doi.org/10.1533/9780857096197.1

[8] Therin, M., Christel, P., Li, S., Garreau, H., & Vert, M. (1992). In vivo degradation of massive poly (α-hydroxy acids): validation of in vitro findings. Biomaterials, 13(9), 594-600. https://doi.org/10.1016/0142-9612(92)90027-L

[9] Middleton, J. C., & Tipton, A. J. (2000). Synthetic biodegradable polymers as orthopedic devices. Biomaterials, 21(23), 2335-2346. https://doi.org/10.1016/S0142-9612(00)00101-0

[10] Fiaschi, P., Pavanello, M., Imperato, A., Dallolio, V., Accogli, A., Capra, V., ... &Piatelli, G. (2016). Surgical results of cranioplasty with a polymethylmethacrylate customized cranial implant in pediatric patients: a single-center experience. Journal of Neurosurgery: Pediatrics, 17(6), 705-710. https://doi.org/10.3171/2015.10.PEDS15489

[11] Ruggiero, C., Barbato, M., Spennato, P., Russo, C., Cicala, D., & Cinalli, G. (2020). Repair of giant lumbosacral pseudomeningocele with fast-resorbing polymer mesh in a pediatric patient operated for posterior dysraphism. Child's Nervous System, 36, 1777-1780. https://doi.org/10.1007/s00381-020-04613-7

[12] Gregory, H., & Phillips, J. B. (2021). Materials for peripheral nerve repair constructs: Natural proteins or synthetic polymers?. Neurochemistry International, 143, 104953. https://doi.org/10.1016/j.neuint.2020.104953

[13] Zhang, X., Qu, W., Li, D., Shi, K., Li, R., Han, Y., ... & Chen, X. (2020). Functional polymer-based nerve guide conduits to promote peripheral nerve regeneration. Advanced Materials Interfaces, 7(14), 2000225. https://doi.org/10.1002/admi.202000225

[14] Shim, K. W., Park, E. K., Kim, D. S., & Choi, J. U. (2017). Neuroendoscopy: current and future perspectives. Journal of Korean Neurosurgical Society, 60(3), 322-326. https://doi.org/10.3340/jkns.2017.0202.006

[15] Codd, P. J., Veaceslav, A., Gosline, A. H., & Dupont, P. E. (2014). Novel pressure-sensing skin for detecting impending tissue damage during neuroendoscopy. Journal of Neurosurgery: Pediatrics, 13(1), 114-121. https://doi.org/10.3171/2013.9.PEDS12595

[16] Jain, K. K. (1983). Lasers in neurosurgery: a review. Lasers in Surgery and Medicine, 2(3), 217-230. https://doi.org/10.1002/lsm.1900020305

[17] Okishev, D. N., Cherebylo, S. A., Konovalov, A. N., Chelushkin, D. M., Shekhtman, O. D., Konovalov, N. A., ... & Eliava, S. S. (2022). Features of modeling a polymer implant for closing a defect after decompressive craniotomy. ZhurnalVoprosyNeirokhirurgiiImeni NN Burdenko, 86(1), 17-27. https://doi.org/10.17116/neiro20228601117

[18] Suzuki, S., &Ikada, Y. (2015). Polymers for Surgery. Advanced Polymers in Medicine, 219-264. https://doi.org/10.1007/978-3-319-12478-0_8

[19] Naureen, Z., Zubair, M., Rasul, F., Nadeem, H., Siddique, M. H., Afzal, M., & Rasul, I. (2022). Applications of Polymers in Neurosurgery. Applications of Polymers in Surgery, 123, 92-122. https://doi.org/10.21741/9781644901892-4

[20] Velnar, T., Bosnjak, R., &Gradisnik, L. (2022). Clinical applications of poly-methyl-methacrylate in neurosurgery: The in vivo cranial bone reconstruction. Journal of functional biomaterials, 13(3), 156. https://doi.org/10.3390/jfb13030156

[21] Maitz, M. F. (2015). Applications of synthetic polymers in clinical medicine. Biosurface and Biotribology, 1(3), 161-176. https://doi.org/10.1016/j.bsbt.2015.08.002

[22] Mussi, E., Mussa, F., Santarelli, C., Scagnet, M., Uccheddu, F., Furferi, R., ... &Genitori, L. (2020). Current practice in preoperative virtual and physical simulation in neurosurgery. Bioengineering, 7(1), 7. https://doi.org/10.3390/bioengineering7010007

[23] Chuan, D., Wang, Y., Fan, R., Zhou, L., Chen, H., Xu, J., & Guo, G. (2020). Fabrication and properties of a biomimetic dura matter substitute based on stereocomplex poly (Lactic Acid) nanofibers. International journal of nanomedicine, 3729-3740. https://doi.org/10.2147/IJN.S248998

[24] Antikainen, T., Kallìoinen, M., Pohjonen, T., Törmälä, P., Waris, T., & Serlo, W. (1994). Polyglycolic acid membrane interpositioning for the prevention of skull deformity following experimental craniosynostosis. Pediatric neurosurgery, 21(1), 77-82. https://doi.org/10.1159/000120819

[25] Terasaka, S., Iwasaki, Y., Shinya, N., & Uchida, T. (2006). Fibrin glue and polyglycolic acid nonwoven fabric as a biocompatible dural substitute. Operative Neurosurgery, 58(1), ONS-134. https://doi.org/10.1227/01.NEU.0000193515.95039.49

[26] Inada, Y., Morimoto, S., Takakura, Y., & Nakamura, T. (2004). Regeneration of peripheral nerve gaps with a polyglycolic acid-collagen tube. Neurosurgery, 55(3), 640-648. https://doi.org/10.1227/01.NEU.0000134388.86603.11

[27] Tseng, Y. Y., Kao, Y. C., Liao, J. Y., Chen, W. A., & Liu, S. J. (2013). Biodegradable drug-eluting poly [lactic-co-glycol acid] nanofibers for the sustainable delivery of vancomycin to brain tissue: in vitro and in vivo studies. ACS chemical neuroscience, 4(9), 1314-1321. https://doi.org/10.1021/cn400108q

[28] Makadia, H. K., & Siegel, S. J. (2011). Poly lactic-co-glycolic acid (PLGA) as biodegradable controlled drug delivery carrier. Polymers, 3(3), 1377-1397. https://doi.org/10.3390/polym3031377

[29] Mondal, D., Griffith, M., & Venkatraman, S. S. (2016). Polycaprolactone-based biomaterials for tissue engineering and drug delivery: Current scenario and challenges. International Journal of Polymeric Materials and Polymeric Biomaterials, 65(5), 255-265. https://doi.org/10.1080/00914037.2015.1103241

[30] Yang, Z., Zhang, Y., Yang, Y., Sun, L., Han, D., Li, H., & Wang, C. (2010). Pharmacological and toxicological target organelles and safe use of single-walled carbon nanotubes as drug carriers in treating Alzheimer disease. Nanomedicine: Nanotechnology, Biology and Medicine, 6(3), 427-441. https://doi.org/10.1016/j.nano.2009.11.007

[31] Cha, C., Shin, S. R., Annabi, N., Dokmeci, M. R., &Khademhosseini, A. (2013). Carbon-based nanomaterials: multifunctional materials for biomedical engineering. ACS nano, 7(4), 2891-2897. https://doi.org/10.1021/nn401196a

[32] Schantz, J. T., Lim, T. C., Ning, C., Teoh, S. H., Tan, K. C., Wang, S. C., & Hutmacher, D. W. (2006). Cranioplasty after trephination using a novel biodegradable burr hole cover: technical case report. Operative Neurosurgery, 58(1), ONS-E176. https://doi.org/10.1227/01.NEU.0000193533.54580.3F

 Materials Research Forum LLC
https://doi.org/10.21741/9781644903353-5

[33] Iijima, S. (1999). Helical microtubules of graphite carbon. Nature, 56, 354. https://doi.org/10.1038/354056a0

[34] J.Li, A.Cassell, L.Delzeit, J.Han, and M.Meyyappan, Novel three-dimensional electrodes: electro-chemical properties of carbon nanotube ensembles. J. Phys. Chem. B 106, 9299 (2002) https://doi.org/10.1021/jp021201n

[35] Hong, K. Z. (1996). Poly (vinyl chloride) in medical device and packaging applications. Journal of Vinyl and Additive Technology, 2(3), 193-197. https://doi.org/10.1002/vnl.10123

[36] Osbun, J. W., Ellenbogen, R. G., Chesnut, R. M., Chin, L. S., Connolly, P. J., Cosgrove, G. R., ... & Wilberger, J. E. (2012). A multicenter, single-blind, prospective randomized trial to evaluate the safety of a polyethylene glycol hydrogel (Duraseal Dural Sealant System) as a dural sealant in cranial surgery. World neurosurgery, 78(5), 498-504 https://doi.org/10.1016/j.wneu.2011.12.011

[37] Cosgrove, G. R., Delashaw, J. B., Grotenhuis, J. A., Tew, J. M., Van Loveren, H., Spetzler, R. F., ... &Norbash, A. (2007). Safety and efficacy of a novel polyethylene glycol hydrogel sealant for watertight dural repair. Journal of neurosurgery, 106(1), 52-58. https://doi.org/10.3171/jns.2007.106.1.52

[38] Zhu, T., Wang, H., Jing, Z., Fan, D., Liu, Z., Wang, X., & Tian, Y. (2022). High efficacy of tetra-PEG hydrogel sealants for sutureless dural closure. Bioactive materials, 8, 12-19. https://doi.org/10.1016/j.bioactmat.2021.06.022

[39] Kawakami, Osamu M.D.; Miyamoto, Susumu M.D.; Hatano, Taketo M.D.; Yamada, Keisuke M.D.; Hashimoto, Nobuo M.D.; Tabata, Yasuhiko Ph.D., D.Med.Sci., D.Pharm.. Accelerated Embolization Healing of Aneurysms by Polyethylene Terephthalate Coils Seeded with Autologous Fibroblasts. Neurosurgery 56(5):p 1075-1081, May 2005.

[40] Schellhammer, F., Walter, M., Berlis, A., Bloss, H. G., Wellens, E., & Schumacher, M. (1999). Polyethylene terephthalate and polyurethane coatings for endovascular stents: Preliminary results in canine experimental arteriovenous fistulas. Radiology, 211(1), 169-175. https://doi.org/10.1148/radiology.211.1.r99ap14169

[41] Prajapati, H. P., & Singh, D. K. (2024). A Single Standard Polyvinyl Chloride 3D Skull Model to Create the Polymethyl Methacrylate Cranioplasty Flap: A Novel and Low-Cost Technique. Journal of Neurological Surgery Part A: Central European Neurosurgery. https://doi.org/10.1055/s-0044-1785648

[42] Verbraeken, B., Lavrysen, E., Aboukais, R., &Menovsky, T. (2021). Polyvinyl Alcohol Sponges to Facilitate Cerebral Bypass Surgery. World neurosurgery, 156, 53-55. https://doi.org/10.1016/j.wneu.2021.09.007

[43] Matsumoto, H., Terada, T., Tsuura, M., Itakura, T., & Ogawa, A. (2003). Experimental polyvinyl alcohol core coil for a drug delivery system. Interventional Neuroradiology, 9(1_suppl), 107-111. https://doi.org/10.1177/15910199030090S114

[44] Andrzejak, S., Fortuniak, J., Wróbel-Wiśniewska, G., &Zawirski, M. (2005). Clinical evaluation of the polypropylene-polyester knit used as a cranioplasty material. Acta Neurochirurgica, 147, 973-976. https://doi.org/10.1007/s00701-005-0581-8

[45] Kim, S., Nowicki, K. W., Ye, S., Jang, K., Elsisy, M., Ibrahim, M., ... & Wagner, W. R. (2022). Bioabsorbable, elastomer-coated magnesium alloy coils for treating saccular cerebrovascular aneurysms. Biomaterials, 290, 121857. https://doi.org/10.1016/j.biomaterials.2022.121857

[46] Qi, Z., Zhang, T., Kong, W., Fu, C., Chang, Y., Li, H., ... & Pan, S. (2022). A dual-drug enhanced injectable hydrogel incorporated with neural stem cells for combination therapy in spinal cord injury. Chemical Engineering Journal, 427, 130906. https://doi.org/10.1016/j.cej.2021.130906

[47] Jain, K. K. (1983). Lasers in neurosurgery: a review. Lasers in Surgery and Medicine, 2(3), 217-230. https://doi.org/10.1002/lsm.1900020305

[48] Iqbal, J., Courville, E., Kazim, S. F., Kogan, M., Schmidt, M. H., & Bowers, C. A. (2024). Role of nanotechnology in neurosurgery: A review of recent advances and their applications. World Neurosurgery: X, 100298. https://doi.org/10.1016/j.wnsx.2024.100298

[49] Than, K. D., Baird, C. J., & Olivi, A. (2008). Polyethylene glycol hydrogel dural sealant may reduce incisional cerebrospinal fluid leak after posterior fossa surgery. Operative neurosurgery, 63(1), ONS182-ONS187 https://doi.org/10.1227/01.NEU.0000313116.28200.67

[50] Paskal, A. M., Paskal, W., Pietruski, P., & Wlodarski, P. K. (2019). Polyethylene glycol: the future of posttraumatic nerve repair? Systemic review. International journal of molecular sciences, 20(6), 1478. https://doi.org/10.3390/ijms20061478

[51] Jia, X., Yeo, Y., Clifton, R. J., Jiao, T., Kohane, D. S., Kobler, J. B., ... & Langer, R. (2006). Hyaluronic acid-based microgels and microgel networks for vocal fold regeneration. Biomacromolecules, 7(12), 3336-3344. https://doi.org/10.1021/bm0604956

[52] Rao, S. S., Nelson, M. T., Xue, R., DeJesus, J. K., Viapiano, M. S., Lannutti, J. J., ... & Winter, J. O. (2013). Mimicking white matter tract topography using core-shell electrospun nanofibers to examine migration of malignant brain tumors. Biomaterials, 34(21), 5181-5190. https://doi.org/10.1016/j.biomaterials.2013.03.069

[53] Kaushik, K., Sharma, R. B., & Agarwal, S. (2016). Natural polymers and their applications. International Journal of Pharmaceutical Sciences Review and Research, 37(2), 30-36.

[54] Yeh, J. Z., Wang, D. H., Cherng, J. H., Wang, Y. W., Fan, G. Y., Liou, N. H., ... & Chou, C. H. (2020). A collagen-based scaffold for promoting neural plasticity in a rat model of spinal cord injury. Polymers, 12(10), 2245. https://doi.org/10.3390/polym12102245

[55] Mealy, J.E.; Chung, J.J.; Jeong, H.-H.; Issadore, D.; Lee, D.; Atluri, P.; Burdick, J.A. Injectable Granular Hydrogels with Multifunctional Properties for Biomedical Applications. Adv. Mater. 2018, 30, 1705912. https://doi.org/10.1002/adma.201705912

[56] Jensen, G., Holloway, J. L., &Stabenfeldt, S. E. (2020). Hyaluronic acid biomaterials for central nervous system regenerative medicine. Cells, 9(9), 2113. https://doi.org/10.3390/cells9092113

[57] Zeng, X., Wei, Q. S., Ye, J. C., Rao, J. H., Zheng, M. G., Ma, Y. H., ... & Zeng, Y. S. (2023). A biocompatible gelatin sponge scaffold confers robust tissue remodeling after spinal

cord injury in a non-human primate model. Biomaterials, 299, 122161. https://doi.org/10.1016/j.biomaterials.2023.122161

[58] McCrorie, P., Mistry, J., Taresco, V., Lovato, T., Fay, M., Ward, I., ... & Rahman, R. (2020). Etoposide and olaparib polymer-coated nanoparticles within a bioadhesive sprayable hydrogel for post-surgical localised delivery to brain tumours. European Journal of Pharmaceutics and Biopharmaceutics, 157, 108-120. https://doi.org/10.1016/j.ejpb.2020.10.005

[59] Amr, S. M., & Ekram, B. (2017). Speculations on the use of marine polysaccharides as scaffolds for artificial nerve 'side-'grafts. In Biological activities and application of marine polysaccharides (pp. 146-179). IntechOpen. https://doi.org/10.5772/66460

[60] Becker, T. A., Preul, M. C., Bichard, W. D., Kipke, D. R., & McDougall, C. G. (2007). Preliminary investigation of calcium alginate gel as a biocompatible material for endovascular aneurysm embolization in vivo. Neurosurgery, 60(6), 1119-1128. https://doi.org/10.1227/01.NEU.0000255447.90106.12

[61] Jarrah, R., El Sammak, S., Onyedimma, C., Ghaith, A. K., Moinuddin, F. M., Bhandarkar, A. R., ... &Bydon, M. (2022). The role of alginate hydrogels as a potential treatment modality for spinal cord injury: a comprehensive review of the literature. Neurospine, 19(2), 272. https://doi.org/10.14245/ns.2244186.093

[62] Singh, B., & Kumar, A. (2020). Synthesis and characterization of alginate and sterculia gum based hydrogel for brain drug delivery applications. International journal of biological macromolecules, 148, 248-257. https://doi.org/10.1016/j.ijbiomac.2020.01.147

[63] Kushchayev, S. V., Giers, M. B., Eng, D. H., Martirosyan, N. L., Eschbacher, J. M., Mortazavi, M. M., ... &Preul, M. C. (2016). Hyaluronic acid scaffold has a neuroprotective effect in hemisection spinal cord injury. Journal of Neurosurgery: Spine, 25(1), 114-124.S https://doi.org/10.3171/2015.9.SPINE15628

[64] Kil, D., Bovet Carmona, M., Ceyssens, F., Deprez, M., Brancato, L., Nuttin, B., ... & Puers, R. (2019). Dextran as a resorbable coating material for flexible neural probes. Micromachines, 10(1), 61. https://doi.org/10.3390/mi10010061

[65] Panteleichuk, A., Kadzhaya, M., Biloschytsky, V., Shmeleva, A., Petriv, T., Gnatyuk, O., ... &Tyortyh, V. (2020). Composite chitosan/polyethylene oxide film for duraplasty in traumatic brain injury model in rats. Cell and Organ Transplantology, 8(1). https://doi.org/10.22494/cot.v8i1.105

[66] Martin, C. A., Radhakrishnan, S., Nagarajan, S., Muthukoori, S., Dueñas, J. M., Ribelles, J. L. G., ... &Subbaraya, N. K. (2019). An innovative bioresorbable gelatin based 3D scaffold that maintains the stemness of adipose tissue derived stem cells and the plasticity of differentiated neurons. RSC advances, 9(25), 14452-14464. https://doi.org/10.1039/C8RA09688K

[67] Vepari, C.; Kaplan, D.L. Silk as a Biomaterial. Prog. Polym. Sci. 2007, 32, 991-1007 https://doi.org/10.1016/j.progpolymsci.2007.05.013

[68] Chon, J.-W.; Kim, H.; Jeon, H.-N.; Park, K.; Lee, K.-G.; Yeo, J.-H.; Kweon, H.; Lee, H.-S.; Jo, Y.-Y.; Park, Y.K. Silk fibroin hydrolysate inhibits osteoclastogenesis and induces

apoptosis of osteoclasts derived from RAW 264.7 cells. Int. J. Mol. Med. 2012, 30, 1203-1210. https://doi.org/10.3892/ijmm.2012.1120

[69] Sirajudheen, P., Poovathumkuzhi, N. C., Vigneshwaran, S., Chelaveettil, B. M., & Meenakshi, S. (2021). Applications of chitin and chitosan based biomaterials for the adsorptive removal of textile dyes from water-A comprehensive review. Carbohydrate polymers, 273, 118604. https://doi.org/10.1016/j.carbpol.2021.118604

[70] Bayer, O. (1947). Das di-isocyanat-polyadditionsverfahren (polyurethane). AngewandteChemie, 59(9), 257-272. https://doi.org/10.1002/ange.19470590901

[71] Södergård, A. and M. Stolt, (2010). Industrial Production of High Molecular Weight Poly(Lactic Acid), in Poly(Lactic Acid), John Wiley & Sons, Inc. p. 27-41. https://doi.org/10.1002/9780470649848.ch3

[72] Yang, Yadie& Zhang, Minglonghai& Ju, Zixin & Tam, Po & Hua, Tao & Waseem Younas, Muhammad & Hasan, Kamrul & Hu, Hong. (2020). Poly(lactic acid) fibers, yarns and fabrics: Manufacturing, properties and applications. Textile Research Journal. 91. 004051752098410. https://doi.org/10.1177/004051752098410l

[73] Munzeiwa, W.A., Omondi, B.O. &Nyamori, V.O. A perspective into ring-opening polymerization of ε-caprolactone and lactides: effect of, ligand, catalyst structure and system dynamics, on catalytic activity and polymer properties. Polym. Bull. 81, 9419-9464 (2024). https://doi.org/10.1007/s00289-024-05149-5

[74] Akindoyo, J. O., Beg, M., Ghazali, S., Islam, M. R., Jeyaratnam, N., & Yuvaraj, A. R. (2016). Polyurethane types, synthesis and applications-a review. Rsc Advances, 6(115), 114453-114482. https://doi.org/10.1039/C6RA14525F

[75] Huang, Y., Sohail, H. M., & Lu, J. (2020). Fabrication, assembly, and optoelectric properties of layered double hydroxide/conjugated polymer nanocomposites. In Layered Double Hydroxide Polymer Nanocomposites (pp. 497-529). Woodhead Publishing. https://doi.org/10.1016/B978-0-08-101903-0.00012-X

[76] C. Deslouis, T. El Moustafid, M. M. Musiani, B. Tribollet, Electrochim. Acta 1996, 41, 1343. https://doi.org/10.1016/0013-4686(95)00455-6

[77]] J. Rivnay, S. Inal, B. A. Collins, M. Sessolo, E. Stavrinidou, X. Strakosas, C. Tassone, D. M. Delongchamp, G. G. Malliaras, Nat. Commun. 2016, 7 https://doi.org/10.1038/ncomms11287

[78] Kausar, A. (2021). Conducting polymer-based nanocomposites: fundamentals and applications. Elsevier.

[79] P. Moutsatsou, K. Coopman, S. Georgiadou, Biocompatibility assessment of conducting pani/chitosan nanofibers for wound healing applications, Polymers 9 (2017) 687, https://doi.org/10.3390/polym9120687

[80] P.T. Bertuoli, J. Ordono, E. Armelin, S. P'erez-Amodio, A.F. Baldissera, C. A. Ferreira, J. Puiggalí, E. Engel, L.J. Del Valle, C. Alem' an, Electrospun conducting and biocompatible uniaxial and core-shell fibers having poly(lactic acid), poly(ethylene glycol), and polyaniline for cardiac tissue engineering, ACS Omega 4 (2019) 3660-3672, https://doi.org/10.1021/acsomega.8b03411

[81] Qiu, Z., Hammer, B. A., &Müllen, K. (2020). Conjugated polymers-Problems and promises. Progress in Polymer Science, 100, 101179. https://doi.org/10.1016/j.progpolymsci.2019.101179

[82] G. Jia, A. Zheng, X. Wang, L. Zhang, L. Li, C. Li, Y. Zhang, L. Cao, Flexible, biocompatible and highly conductive MXene-graphene oxide film for smart actuator and humidity sensor, Sens. Actuators B Chem. 346 (2021), 130507, https://doi.org/10.1016/j.snb.2021.130507

[83] Lubanska, D., Alrashed, S., Mason, G. T., Nadeem, F., Awada, A., DiPasquale, M., ... & Rondeau-Gagné, S. (2022). Impairing proliferation of glioblastoma multiforme with CD44+ selective conjugated polymer nanoparticles. Scientific Reports, 12(1), 12078. https://doi.org/10.1038/s41598-022-15244-0

[84] Hicks, J., Platt, R. S., Holmes, P. S., Howerth, E., Haley, A., Kaplan, M. J., & Kaplan, J. E. (2016). Clinical, imaging and pathological characteristics of brain implanted polylactic co-glycolic acid polymers conjugated with temozolomide. https://doi.org/10.4172/2157-7579.1000325

[85] Caverzán, M. D., Oliveda, P. M., Beaugé, L., Palacios, R. E., Chesta, C. A., & Ibarra, L. E. (2023). Metronomic photodynamic therapy with conjugated polymer nanoparticles in glioblastoma tumor microenvironment. Cells, 12(11), 1541. https://doi.org/10.3390/cells12111541

[86] Ibarra, L. E., Porcal, G. V., Macor, L. P., Ponzio, R. A., Spada, R. M., Lorente, C., … Palacios, R. E. (2018). Metallated porphyrin-doped Conjugated Polymer Nanoparticles for Efficient Photodynamic Therapy of Brain and Colorectal Tumor Cells. Nanomedicine, 13(6), 605-624. https://doi.org/10.2217/nnm-2017-0292

[87] Zhai, D., Liu, B., Shi, Y., Pan, L., Wang, Y., Li, W., ... & Yu, G. (2013). Highly sensitive glucose sensor based on Pt nanoparticle/polyaniline hydrogel heterostructures. ACS nano, 7(4), 3540-3546. https://doi.org/10.1021/nn400482d

[88] Olayo, R., Ríos, C., Salgado-Ceballos, H., Cruz, G. J., Morales, J., Olayo, M. G., ... & Diaz-Ruiz, A. (2008). Tissue spinal cord response in rats after implants of polypyrrole and polyethylene glycol obtained by plasma. Journal of materials science: materials in medicine, 19, 817-826 https://doi.org/10.1007/s10856-007-3080-z

[89] Aarabi B, Hesdorffer DC, Ahn ES, Aresco C, Scalea TM, Eisenberg HM. Outcome following decompressive craniectomy for malignant swelling due to severe head injury. J Neurosurg2006;104:469 79. https://doi.org/10.3171/jns.2006.104.4.469

[90] Guerra WK, Gaab MR, Dietz H, Mueller JU, Piek J, Fritsch MJ. Surgical decompression for traumatic brain swelling: Indications and results. J Neurosurg1999;90:187 96 https://doi.org/10.3171/jns.1999.90.2.0187

[91] Baumeister S, Peek A, Friedman A, Levin LS, Marcus JR. Management of postneurosurgical bone flap loss caused by infection. PlastReconstr Surg 2008;122:195e 208e. https://doi.org/10.1097/PRS.0b013e3181858eee

Materials Research Forum LLC
https://doi.org/10.21741/9781644903353-5

[92] Binyamin, G., Shafi, B. M., & Mery, C. M. (2006, November). Biomaterials: a primer for surgeons. In Seminars in pediatric surgery (Vol. 15, No. 4, pp. 276-283). WB Saunders. https://doi.org/10.1053/j.sempedsurg.2006.07.007

[93] Charnley, J. Total hip replacement by low-friction arthroplasty. Clin. Orthop. Relat. Res. 1970, 72, 7-21. https://doi.org/10.1097/00003086-197009000-00003

[94] Galicich, J.H.; Hovind, K.H. Stainless Steel Mesh-Acrylic Cranioplasty. J. Neurosurg. 1967, 27, 376-378. https://doi.org/10.3171/jns.1967.27.4.0376

[95] Kwarcinski, J.; Boughton, P.; Ruys, A.; Doolan, A.; Van Gelder, J. Cranioplasty and Craniofacial Reconstruction: A Review of Implant Material, Manufacturing Method and Infection Risk. Appl. Sci. 2017, 7, 276. https://doi.org/10.3390/app7030276

[96] Gosain, A.K. Biomaterials in facial reconstruction. Oper. Tech. Plast. Reconstr. Surg. 2002, 9, 23-30. https://doi.org/10.1016/S1071-0949(03)90005-9

[97] Antonelli, V.; Maimone, G.; D'Andrea, M.; Tomassini, A.; Bassi, M.; Tosatto, L. "Single-step" resection and cranio-orbital reconstruction for spheno-orbital metastasis with custom made implant. A case report and review of the literature. Int. J. Surg. Case Rep. 2021, 81, 105755. https://doi.org/10.1016/j.ijscr.2021.105755

[98] Eppley, B.L. Alloplastic cranioplasty. Oper. Tech. Plast. Reconstr. Surg. 2002, 9, 16-22. https://doi.org/10.1016/S1071-0949(03)90004-7

[99] Velnar, T.; Bosnjak, R.; Gradisnik, L. Clinical Applications of Poly-Methyl-Methacrylate in Neurosurgery: The In Vivo Cranial Bone Reconstruction. J. Funct. Biomater. 2022, 13, 156 https://doi.org/10.3390/jfb13030156

[100] Unterhofer, C., Wipplinger, C., Verius, M., Recheis, W., Thomé, C., & Ortler, M. (2017). Reconstruction of large cranial defects with poly-methyl-methacrylate (PMMA) using a rapid prototyping model and a new technique for intraoperative implant modeling. Neurologiaineurochirurgiapolska, 51(3), 214-220. https://doi.org/10.1016/j.pjnns.2017.02.007

[101] Shash, Y. H. (2024). Assessment of cranial reconstruction utilizing various implant materials: finite element study. Journal of Materials Science: Materials in Medicine, 35(1), 50. https://doi.org/10.1007/s10856-024-06816-9

[102] Alkhaibary, A.; Alharbi, A.; Alnefaie, N.; Almubarak, A.O.; Aloraidi, A.; Khairy, S. Cranioplasty: A Comprehensive Re-view of the History, Materials, Surgical Aspects, and Complications. World Neuros. 2020, 139, 445-452. https://doi.org/10.1016/j.wneu.2020.04.211

[103] Morselli, C.; Zaed, I.; Tropeano, M.P.; Cataletti, G.; Iaccarino, C.; Rossini, Z.; Servadei, F. Comparison between the different types of heterologous materials used in cranioplasty: A systematic review of the literature. J. Neurosurg. Sci. 2020, 63, 723-736. https://doi.org/10.23736/S0390-5616.19.04779-9

[104] Zhang, J.; Tian, W.; Chen, J.; Yu, J.; Zhang, J.; Chen, J. The application of polyetheretherketone (PEEK) implants in cranioplasty. Brain Res. Bull. 2019, 153, 143-149. https://doi.org/10.1016/j.brainresbull.2019.08.010

[105] Najeeb, S.; Zafar, M.S.; Khurshid, Z.; Siddiqui, F. Applications of polyetheretherketone (PEEK) in oral implantology and prosthodontics. J. Prosthodont. Res. 2016, 60, 12-19. https://doi.org/10.1016/j.jpor.2015.10.001

[106] Kwarcinski, J.; Boughton, P.; Ruys, A.; Doolan, A.; Van Gelder, J. Cranioplasty and Craniofacial Reconstruction: A Review of Implant Material, Manufacturing Method and Infection Risk. Appl. Sci. 2017, 7, 276. https://doi.org/10.3390/app7030276

[107] DeBarros, A.; Brauge, D.; Quehan, R.; Cavallier, Z.; Roux, F.E.; Moyse, E. One-step customized peek cranioplasty after 3D printed resection template assisted surgery for a frontal intraosseous meningioma: A case report. Turk. Neurosurg. 2020, 31, 142-147. https://doi.org/10.5137/1019-5149.JTN.30192-20.2

[108] Abuzayed, B.; Tuzgen, S.; Canbaz, B.; Yuksel, O.; Tutunculer, B.; Sanus, G.Z. Reconstruction of Growing Skull Fracture within Situ Galeal Graft Duraplasty and Porous Polyethylene Sheet. J. Craniofacial Surg. 2009, 20, 1245-1249. https://doi.org/10.1097/SCS.0b013e3181acdfaf

[109] Kucukyuruk, B.; Biceroglu, H.; Abuzayed, B.; Ulu, M.O.; Sanus, G.Z. Intraosseous meningioma: A rare tumor reconstructed with porous polyethylene. J. Craniofacial Surg. 2010, 21, 936-939. https://doi.org/10.1097/SCS.0b013e3181d84050

[110] Liu, J.K.; Gottfried, O.N.; Cole, C.D.; Dougherty, W.R.; Couldwell, W.T. Porous polyethylene implant for cranioplasty and skull base reconstruction. Neurosurg. Focus 2004, 16, 1-5. https://doi.org/10.3171/foc.2004.16.3.14

[111] Janecka, I.P. New reconstructive technologies in skull base surgery: Role of titanium mesh and porous polyethylene. Arch. Otolaryngol. Head Neck Surg. 2000, 126, 396-401. https://doi.org/10.1001/archotol.126.3.396

[112] Ridwan-Pramana, A.; Wolff, J.; Raziei, A.; Ashton-James, C.E.; Forouzanfar, T. Porous polyethylene implants in facial reconstruction: Outcome and complications. J. Cranio-Maxillofac. Surg. 2015, 43, 1330-1334. https://doi.org/10.1016/j.jcms.2015.06.022

[113] Wang, J.-C.; Wei, L.; Xu, J.; Liu, J.-F.; Gui, L. Clinical Outcome of Cranioplasty With High-Density Porous Polyethylene. J. Craniofacial Surg. 2012, 23, 1404-1406. https://doi.org/10.1097/SCS.0b013e31825e3aeb

[114] Gosain, A.K. Biomaterials in facial reconstruction. Oper. Tech. Plast. Reconstr. Surg. 2002, 9, 23-30. https://doi.org/10.1016/S1071-0949(03)90005-9

[115] Pathak, A., Dhamande, M. M., Pisulkar, S. G., Dubey, S. A., Bhoyar, A., Beri, A., & Sonar, P. R. (2024). The Reconstruction of Post-traumatic Cranial Bone Depression by Duplicating an Autogenous Bone Flap With a Prosthetic Polymethyl Methacrylate Cranial Stent. Cureus, 16(7). https://doi.org/10.7759/cureus.65353

[116] Laakman, R. W., Kaufman, B., Han, J. S., Nelson, A. D., Clampitt, M., O'Block, A. M., Haaga, J. R. and Alfidi, R. J. (1985) 'MR imaging in patients with metallic implants', Radiol J, 157, 711-714. https://doi.org/10.1148/radiology.157.3.4059558

[117] Ciccone, W. J., Motz, C., Bentley, C. and Tasto, J. P. (2001) 'Bioabsorbable implants in orthopaedics: new developments and clinical applications', J Am AcadOrthop Surg, 9, 280-288. https://doi.org/10.5435/00124635-200109000-00001

[118] therin, M., christel, P., li, S., Garreau, H. and Vert, M. (1992) 'In vivo degradation of massive poly(alpha hydroxy acids): validation of in vitro findings', Biomater, 13, 594-600. https://doi.org/10.1016/0142-9612(92)90027-L

[119] Middleton, J. c. and tipton, a. J. (2000) 'Synthetic biodegradable polymers as orthopaedic devices', Biomater, 21, 2335-2346. https://doi.org/10.1016/S0142-9612(00)00101-0

[120] Vert, M. (2004) 'Poly (lactic acid) s', in Wnek, G.E. and Bowlin, G.L., Encyclopedia of Biomaterials and Biomedical Engineering, New York: Marcel Dekker, pp. 1254-1264.

[121] Wright, D. D. (2004) 'Degradable polymer composites', in Wnek, G. E. and Bowlin, G. l., Encyclopedia of Biomaterials and Biomedical Engineering, New York: Marcel Dekker, pp. 423-432.

[122] Bostman, o. M. and Pihlajamaki, H. K. (2000) 'adverse tissue reactions to bioabsorbable fixation devices', Clin Orthop, 371, 216-227. https://doi.org/10.1097/00003086-200002000-00026

[123] Briscoe, B., Luckham, P. and Zhu, S. (2000) 'The effects of hydrogen bonding upon the viscosity of aqueous poly(vinyl alcohol) solutions', Polymer, 41, 3851-3860. https://doi.org/10.1016/S0032-3861(99)00550-9

[124] Cai, W. and Gupta, R. B. (2002) 'Hydrogels', in Kirk-Othmer Encyclopedia of Chemical Technology, New York: John Wiley & Sons. https://doi.org/10.1002/0471238961.0825041807211620.a01

[125] Molyneux, P. (1983) Water-Soluble Synthetic Polymers, Boca Raton, FL: CRC Press

[126] Marten, F. L. (2002) 'Vinyl alcohol polymers', in Kirk-Othmer Encyclopedia of Chemical Technology, New York: John wiley& Sons. https://doi.org/10.1002/0471238961.2209142513011820.a01.pub2

[127] Mallapragada, S. K. and Peppas, N. A. (1996) 'Dissolution mechanism of semicrystalline poly(vinyl alcohol) in water', J Polym Sci Polym Phys Ed, 34, 1339-1346. https://doi.org/10.1002/(SICI)1099-0488(199605)34:7<1339::AID-POLB15>3.0.CO;2-B

[128] Thomas, J., Lowman, A. and Marcolongo, M. (2003) 'Novel associated hydrogels for nucleus pulposus replacement', J Biomed Mater Res, 67A, 1329-1337. https://doi.org/10.1002/jbm.a.10119

[129] Joshi, A., Mehta, S., Vresilovic, E., Karduna, A. and Marcolongo, M. (2005) 'Nucleus implant parameters significantly change the compressive stiffness of the human lumbar intervertebral disc', J Biomech Eng, 127, 536-540. https://doi.org/10.1115/1.1894369

[130] Dornish, M., Kaplan, D. and Skaugrud, O. (2001) 'Standards and guidelines for biopolymers in tissue-engineered medical products', Ann NY Acad Sci, 944, 388-397 https://doi.org/10.1111/j.1749-6632.2001.tb03850.x

[131] Athanasiou, K. A., Shah, A. R., Hernandez, R. J. and LeBaron, R. G. (2001) 'Basic science of articular cartilage repair', Clin Sports Med, 20, 223-247 https://doi.org/10.1016/S0278-5919(05)70304-5

[132] Madihally, S. V. and Matthew, H. W. T. (1999) 'Porous chitosan scaffolds for tissue engineering', Biomater, 20, 1133-1142. https://doi.org/10.1016/S0142-9612(99)00011-3

Materials Research Forum LLC
https://doi.org/10.21741/9781644903353-5

[133] Suh, J. K. F. and Matthew, H. W. T. (2000) 'Application of chitosan-based polysaccharide biomaterials in cartilage tissue engineering', Biomater, 21, 2589-2598. https://doi.org/10.1016/S0142-9612(00)00126-5

[134] Aimin, C., Chunlin, H., Juliang, B., Tinyin, Z. and Zhichao, D. (1999) 'Antibiotic loaded chitosan bar. An in vitro, in vivo study of a possible treatment for osteomyelitis', Clin Orthop, 366, 239-247. https://doi.org/10.1097/00003086-199909000-00031

[135] Domininghaus, H. (1993) Plastics for Engineers, Materials Properties, Applications, Munich, Carl-Hanser Verlag.

[136] Pohler, O. (1983) 'Degradation of metallic orthopaedic implants in biomaterials', in Rubin, L., Biomaterials in Reconstructive Surgery, St Louis, MO: Mosby, pp. 158-228.

[137] Horowitz G, Fliss DM, Margalit N, et al. Association between cerebrospinal fluid leak and meningitis after skull base surgery. Otolaryngol Head Neck Surg. 2011;145(4):689-693. https://doi.org/10.1177/0194599811411534

[138] Barth M, Tuettenberg J, Thome C, et al. Watertight dural closure: is it necessary? A prospective randomized trial in patients with supratentorial craniotomies. Neurosurgery. 2008;63(4 Suppl 2): 352-358. discussion 8 https://doi.org/10.1227/01.NEU.0000310696.52302.99

[139] Epstein NE. Dural repair with four spinal sealants: focused review of the manufacturers' inserts and the current literature. Spine J. 2010;10(12):1065-1068 https://doi.org/10.1016/j.spinee.2010.09.017

[140] Kumar A, Maartens NF, Kaye AH. Evaluation of the use of BioGlue® in neurosurgical procedures. J Clin Neurosci. 2003;10(6):661-664 https://doi.org/10.1016/S0967-5868(03)00163-2

[141] Hutter G, von Felten S, Sailer MH, et al. Risk factors for postoperative CSF leakage after elective craniotomy and the efficacy of fleece-bound tissue sealing against dural suturing alone: a randomized controlled trial. J Neurosurg. 2014;121(3):735-744. • One of the few tandomized control trials testing the effectiveness of a dural sea https://doi.org/10.3171/2014.6.JNS131917

[142] Bhagat V, Becker ML. Degradable Adhesives for Surgery and Tissue Engineering. Biomacromolecules. 2017;18(10):3009-3039. https://doi.org/10.1021/acs.biomac.7b00969

[143] Bouten PJM, Zonjee M, Bender J, et al. The chemistry of tissue adhesive materials. Prog Polym Sci. 2014;39:1375-1405. A great overview of the current adhesive biomaterials that could potentially used for dural sealant development https://doi.org/10.1016/j.progpolymsci.2014.02.001

[144] Nair LS, Laurencin CT. Biodegradable polymers as biomaterials. Prog Polym Sci. 2007;32(8-9): 762-798. %@ 0079-6700 https://doi.org/10.1016/j.progpolymsci.2007.05.017

[145] Miki D, Dastgheib K, Kim T, et al. A photopolymerized sealant for corneal lacerations. Cornea. 2002;21(4):393-399 https://doi.org/10.1097/00003226-200205000-00012

[146] Annabi N, Yue K, Tamayol A, et al. Elastic sealants for surgical applications. Eur J Pharm Biopharm. 2015;95(Pt A):27-39. https://doi.org/10.1016/j.ejpb.2015.05.022

[147] Singer AJ, Quinn JV, Hollander JE. The cyanoacrylate topical skin adhesives. Am J Emerg Med. 2008;26(4):490-496 https://doi.org/10.1016/j.ajem.2007.05.015

[148] Mertz PM, Davis SC, Cazzaniga AL, et al. Barrier and antibacterial properties of 2-octyl cyanoacrylate-derived wound treatment films. J Cutan Med Surg. 2003;7(1):1-6 https://doi.org/10.1177/120347540300700101

[149] Leggat PA, Smith DR, Kedjarune U. Surgical applications of cyanoacrylate adhesives: a review of toxicity. ANZ J Surg. 2007;77 (4):209-213 https://doi.org/10.1111/j.1445-2197.2007.04020.x

[150] Preul MC, Bichard WD, Spetzler RF. Toward optimal tissue sealants for neurosurgery: use of a novel hydrogel sealant in a canine durotomy repair model. Neurosurgery. 2003;53(5): 1189-1198. discussion 98-9 https://doi.org/10.1227/01.NEU.0000089481.87226.F7

[151] Campbell, P. K., Bennett, S. L., Driscoll, A., & Sawhney, A. S. (2005). Evaluation of absorbable surgical sealants: in-vitro testing. In-vitro testing, 1170576081-1559295193..

[152] Miller RA, Brady JM, Cutright DE. Degradation rates of oral resorbable implants (polylactates and polyglycolates): rate modification with changes in PLA/PGA copolymer ratios. J Biomed Mater Res. 1977;11(5):711-719 https://doi.org/10.1002/jbm.820110507

[153] Ko WH, Meyrick CW, Rekate HL (1988) Cerebrospinal fluid control system. Proc IEEE 76:1226-1235 https://doi.org/10.1109/5.9669

[154] Silva GA (2004) Introduction to nanotechnology and its applications to medicine. Surg Neurol 61(3):216-220 https://doi.org/10.1016/j.surneu.2003.09.036

[155] Liu CY, Wang MY, Apuzzo ML (2004) The evolution and future of minimalism in neurological surgery. Childs Nerv Syst 20(11-12): 783-789 https://doi.org/10.1007/s00381-004-0931-5

[156] Andrews RJ (2009) Nanotechnology and neurosurgery. J NanosciNanotechnol 9(8):5008-5013 https://doi.org/10.1166/jnn.2009.GR03

[157] Saini R, Saini S, Sharma S (2010) Nanotechnology: the future of medicine. J CutanAesthet Surg 3(1):32-33 https://doi.org/10.4103/0974-2077.63301

[158] Ott B, Constantinescu MA, Erni D, Banic A, Schaffner T, Frenz M (2004) Intramural laser light source and external solder: in vivo evaluation of a new technique for microvascular anastomosis. Lasers Surg Med 35(4):312-316 https://doi.org/10.1002/lsm.20096

[159] Stewart RB, Benbrahim A, LaMuraglia GM et al (1996) Laser assisted vascular welding with real time temperature control. Lasers Surg Med 19(1):9-16 https://doi.org/10.1002/(SICI)1096-9101(1996)19:1<9::AID-LSM2>3.0.CO;2-W

[160] Chang WC, Sretaven DW (2007) Microtechnology in medicine: the emergence of surgical microdevices. Clin Neurosurg 54:137-147

[161] Melosh NA, Boukai A, Diana F, Gerardot B, Badolato A, Petroff PM, Heath JR (2003) Ultrahigh-density nanowire lattices and circuits. Science 300(5616):112-115 https://doi.org/10.1126/science.1081940

[162] Roy S, Ferrara LA, Fleischman AJ, Benzel EC (2001) Microelectromechanical systems and neurosurgery: a new era in a new millennium. Neurosurgery 49(4):779-797 https://doi.org/10.1227/00006123-200110000-00003

[163] Hollenberg BA, Richards CD, Richards R, Bahr DF, Rector DM (2006) A MEMS fabricated flexible electrode array for recording surface field potentials. J Neurosci Methods 153(1):147-153 https://doi.org/10.1016/j.jneumeth.2005.10.016

[164] Nabavi A, Thurm H, Zountsas B, Pietsch T, Lanfermann H, Pichlmeier U, Mehdorn M (2009) 5-ALA Recurrent Glioma Study Group. Five-aminolevulinic acid for fluorescence-guided resection of recurrent malignant gliomas: a phase II study. Neurosurgery 65(6):1070-1076 https://doi.org/10.1227/01.NEU.0000360128.03597.C7

[165] He W, Bellamkonda RV (2005) Nanoscale neuro-integrative coatings for neural implants. Biomaterials 26(16):2983-2990 https://doi.org/10.1016/j.biomaterials.2004.08.021

[166] He W, McConnell GC, Bellamkonda RV (2006) Nanoscale laminin coating modulates cortical scarring response around implanted silicon microelectrode arrays. J Neural Eng 3(4): 316-326 https://doi.org/10.1088/1741-2560/3/4/009

[167] Ignatius MJ, Sawhney N, Gupta A, Thibadeau BM, Monteiro OR, Brown IG (1998) Bioactive surface coatings for nanoscale instruments: effects on CNS neurons. J Biomed Mater Res 40(2):264-274 https://doi.org/10.1002/(SICI)1097-4636(199805)40:2<264::AID-JBM11>3.0.CO;2-M

[168] Ko WH, Meyrick CW, Rekate HL (1988) Cerebrospinal fluid control system. Proc IEEE 76:1226-1235 https://doi.org/10.1109/5.9669

[169] Elder JB, Liu CY, Apuzzo ML (2008) Neurosurgery in the realm of 10(−9), part 1: stardust and nanotechnology in neuroscience. Neurosurgery 62(1):1-20 https://doi.org/10.1227/01.NEU.0000311058.80249.6B

[170] Chang WC, Hawkes EA, Kliot M, Sretaven DW (2007) In vivo use of a nanoknife for axon microsurgery. Neurosurgery 61(4):683-691 https://doi.org/10.1227/01.NEU.0000298896.31355.80

[171] Lundborg G (1990) Nerve regeneration problems in a clinical perspective. Restor Neurol Neurosci 1(3):297-302 https://doi.org/10.3233/RNN-1990-13418

[172] Chang WC, Kilot M, Sretavan DW (2008) Microtechnology and nanotechnology in nerve repair. Neurol Res 30(10):1053-1062 https://doi.org/10.1179/174313208X362532

[173] Jiang X, Mi R, Hoke A, Chew SY (2012) Nanofibrous nerve conduit-enhanced peripheral nerve regeneration. J Tissue Eng Regen Med. doi:10.1002/term.1531 https://doi.org/10.1002/term.1531

[174] Ko WH, Meyrick CW, Rekate HL (1988) Cerebrospinal fluid control system. Proc IEEE 76:1226-1235 https://doi.org/10.1109/5.9669

[175] Kumbar SG, James R, Nukavarapu SP, Laurencin CT (2008) Electrospun nanofiber scaffolds: engineering soft tissues. Biomed Mater 3(3):034002 https://doi.org/10.1088/1748-6041/3/3/034002

[176] Mousa SA, Bharali DJ. Nanotechnology-based detection and targeted therapy in cancer: nano-bio paradigms and applications. Cancers 2011; 3(3): 2888-903. https://doi.org/10.3390/cancers3032888

[177] Antunes AM, Alencar MS, da Silva CH, Nunes J, Mendes FM. Trends in nanotechnology patents applied to the health sector. Recent Pat Nanotechnol 2012; 6(1): 29-43. https://doi.org/10.2174/187221012798109309

[178] Pardeshi CV, Belgamwar VS. Direct nose to brain drug delivery via integrated nerve pathways bypassing the blood-brain barrier: an excellent platform for brain targeting. Expert Opin Drug Deliv 2013; 10(7): 957-72 https://doi.org/10.1517/17425247.2013.790887

[179] Kozlovskaya L, Abou-Kaoud M, Stepensky D. Quantitative analysis of drug delivery to the brain via nasal route. J Control Release 2014; 189: 133-40. https://doi.org/10.1016/j.jconrel.2014.06.053

[180] Kozler P, Pokorny J. Effect of methylprednisolone on the axonal impairment accompanying cellular brain oedema induced by water intoxication in rats. Neuroendocrinol Lett 2012; 33(8): 782-6.

[181] Foley CP, Nishimura N, Neeves KB, Schaffer CB, Olbricht WL. Real-time imaging of perivascular transport of nanoparticles during convection-enhanced delivery in the rat cortex. Ann Biomed Eng 2012; 40(2): 292-303 https://doi.org/10.1007/s10439-011-0440-0

[182] Diaz RJ, McVeigh PZ, O'Reilly MA, et al. Focused ultrasound delivery of Raman nanoparticles across the blood-brain barrier: potential for targeting experimental brain tumors. Nanomedicine 2014; 10(5): 1075-87. https://doi.org/10.1016/j.nano.2013.12.006

[183] Eugenin EA, Clements JE, Zink MC, Berman JW. Human immunodeficiency virus infection of human astrocytes disrupts bloodbrain barrier integrity by a gap junction-dependent mechanism. J Neurosci 2011; 31(26): 9456-65 https://doi.org/10.1523/JNEUROSCI.1460-11.2011

[184] Balducci A, Wen Y, Zhang Y, et al. A novel probe for the noninvasive detection of tumor-associated inflammation. OncoImmunology 2013; 2(2): e23034. https://doi.org/10.4161/onci.23034

[185] Kuo YC, Ko HF. Targeting delivery of saquinavir to the brain using 83-14 monoclonal antibody-grafted solid lipid nanoparticles. Biomaterials 2013; 34(20): 4818-30 https://doi.org/10.1016/j.biomaterials.2013.03.013

[186] van Kuyck, K. et al. Effects of electrical stimulation or lesion in nucleus accumbens on the behaviour of rats in a T-maze after administration of 8-OH-DPAT or vehicle. Behav Brain Res 140, 165-173 (2003). https://doi.org/10.1016/S0166-4328(02)00295-4

[187] van Kuyck, K., Brak, K., Das, J., Rizopoulos, D. &Nuttin, B. Comparative study of the effects of electrical stimulation in the nucleus accumbens, the mediodorsal thalamic nucleus and the bed nucleus of the stria terminalis in rats with schedule-induced polydipsia. Brain Res 1201, 93-99 (2008). https://doi.org/10.1016/j.brainres.2008.01.043

[188] Aggarwal, M., Zhang, J., Miller, M. I., Sidman, R. L. & Mori, S. Magnetic resonance imaging and micro-computed tomography combined atlas of developing and adult mouse

brains for stereotaxic surgery. Neuroscience 162, 1339-1350 (2009). https://doi.org/10.1016/j.neuroscience.2009.05.070

[189] Li, X., Aggarwal, M., Hsu, J., Jiang, H. & Mori, S. AtlasGuide: Software for stereotaxic guidance using 3D CT/MRI hybrid atlases of developing mouse brains. J. Neurosci. Methods 220, 75-84 (2013). https://doi.org/10.1016/j.jneumeth.2013.08.017

[190] Chan, E., Kovacevíc, N., Ho, S. K. Y., Henkelman, R. M. & Henderson, J. T. Development of a high resolution three-dimensional surgical atlas of the murine head for strains 129S1/SvImJ and C57Bl/6J using magnetic resonance imaging and micro-computed tomography. Neuroscience 144, 604-615 (2007). https://doi.org/10.1016/j.neuroscience.2006.08.080

[191] Ramrath, L., Hofmann, U. G. & Schweikard, A. A robotic assistant for stereotactic neurosurgery on small animals,. Int J Med Robot Comp 4(4), 295-303 (2008). https://doi.org/10.1002/rcs.218

[192] Waspe, A. C. et al. Integration and evaluation of a needle-positioning robot with volumetric microcomputed tomography image guidance for small animal stereotactic interventions. Med. Phys. 37, 1647-1659 (2010). https://doi.org/10.1118/1.3312520

[193] Tabaczynski, J.R.; Stoll, T.; Shepard, L.M.; Siddiqui, M.I.G.; Karkhanis, N.V.; Sommer, K. Use of Patient Specific 3D Printed (3DP) Neurovascular Phantoms for Mechanical Assessment of Devices Used in Image Guided Minimally Invasive Procedures. Available online: https://lens.org/035-591-417-289-741 (accessed on 27 September 2021).

[194] Udelson, J.E.; Stevenson, L.W. The Future of Heart Failure Diagnosis, Therapy, and Management. Circulation 2016, 133, 2671-2686. https://doi.org/10.1161/CIRCULATIONAHA.116.023518

[195] Mandelbaum, B.R.; Browne, J.E.; Fu, F.; Micheli, L.; Mosely, J.B.; Erggelet, C.; Minas, T.; Peterson, L. Articular Cartilage Lesions of the Knee. Am. J. Sports Med. 1998, 26, 853-861. https://doi.org/10.1177/03635465980260062201

[196] Miriyev, A.; Xia, B.; Joseph, J.C.; Lipson, H. Additive Manufacturing of Silicone Composites for Soft Actuation. 3D Print. Addit. Manuf. 2019, 6, 309-318. https://doi.org/10.1089/3dp.2019.0116

[197] Michler, R.E. Stem Cell Therapy for Heart Failure. Cardiol. Rev. 2014, 22, 105-116. https://doi.org/10.1097/CRD.0000000000000018

[198] Punyaratabandhu, T.; Liacouras, P.C.; Pairojboriboon, S. Using 3D models in orthopedic oncology: Presenting personalized advantages in surgical planning and intraoperative outcomes. 3D Print. Med. 2018, 4, 12. https://doi.org/10.1186/s41205-018-0035-6

[199] Moldovanu, C.-G.; Lebovici, A.; Buruian, M.M. A systematic review of the clinical value and applications of three-dimensional virtual reconstructions in renal tumors. Med. Pharm. Rep. 2021, 95, 11-23. https://doi.org/10.15386/mpr-2129

[200] Lupulescu, C.; Sun, Z. A Systematic Review of the Clinical Value and Applications of Three-Dimensional Printing in Renal Surgery. J. Clin. Med. 2019, 8, 990. https://doi.org/10.3390/jcm8070990

[201] Lee, H.; Nguyen, N.H.; Hwang, S.I.; Lee, H.J.; Hong, S.K.; Byun, S.S. Personalized 3D kidney model produced by rapid prototyping method and its usefulness in clinical applications. Int. Braz. J. Urol. 2018, 44, 952-957. https://doi.org/10.1590/s1677-5538.ibju.2018.0162

[202] Dai, S.; Wang, Q.; Jiang, Z.; Liu, C.; Teng, X.; Yan, S.; Xia, D.; Tuo, Z.; Bi, L. Application of Three-Dimensional Printing Technology in Renal Diseases. Front. Med. 2022, 9, 1088592. https://doi.org/10.3389/fmed.2022.1088592

[203] Hoon Kang, S. 3D Printing and Characterization of a Soft and Biostable Elastomer with High Flexibility and Strength for Biomedical Applications. Morressier. 18 September 2020. https://doi.org/10.26226/morressier.5f5f8e69aa777f8ba5bd6150

[204] Bandyopadhyay, A.; Traxel, K.D.; Bose, S. Nature-inspired materials and structures using 3D Printing. Mater. Sci. Eng. R Rep. 2021, 145, 100609. https://doi.org/10.1016/j.mser.2021.100609

[205] Piao, Y.; You, H.; Xu, T.; Bei, H.P.; Piwko, I.Z.; Kwan, Y.Y.; Zhao, X. Biomedical applications of gelatin methacryloyl hydrogels. Eng. Regen. 2021, 2, 47-56. https://doi.org/10.1016/j.engreg.2021.03.002

[206] Kreider, P.B.; Cardew-Hall, A.; Sommacal, S.; Chadwick, A.; Hümbert, S.; Nowotny, S.; Nisbet, D.; Tricoli, A.; Compston, P. The effect of a superhydrophobic coating on moisture absorption and tensile strength of 3D-printed carbon-fibre/polyamide. Compos. Part A Appl. Sci. Manuf. 2021, 145, 106380. https://doi.org/10.1016/j.compositesa.2021.106380

[207] Wu, G.-H.; Hsu, S. Review: Polymeric-Based 3D Printing for Tissue Engineering. J. Med. Biol. Eng. 2015, 35, 285-292. [PubMed] https://doi.org/10.1007/s40846-015-0038-3

[208] Morteza, Z.; James, W.J.; Mostafa, Y. 3D Printing of Short-Carbon-Fiber-Reinforced Thermoset Polymer Composites via Frontal Polymerization. ACS Appl. Mater. Interfaces 2022, 14, 16694-16702. https://doi.org/10.1021/acsami.2c02076

[209] Yang, W.; Liu, L.; Zhou, Z.; Liu, H.; Xie, B.; Xu, W. Rational preparation of dibenzothiophene-imprinted polymers by surface imprinting technique combined with atom transfer radical polymerization. Appl. Surf. Sci. 2013, 282, 809-819. https://doi.org/10.1016/j.apsusc.2013.06.063

[210] Okubo, M.; Yonehara, H.; Kurino, T. Influence of viscosity within polymerizing particle on the morphology of micron-sized, monodisperse composite polymer particles produced by seeded polymerization for the dispersion of highly monomer-swollen polymer particles. Colloid Polym. Sci. 2003, 281, 1002-1005. https://doi.org/10.1007/s00396-003-0875-4

[211] Coppola, A.M.; Huelskamp, S.R.; Tanner, C.; Rapking, D.; Ricchi, R.D. Application of tailored fiber placement to fabricate automotive composite components with complex geometries. Compos. Struct. 2023, 313, 116855. https://doi.org/10.1016/j.compstruct.2023.116855

[212] Khan, S.B.; Li, N.; Liang, J.; Xiao, C.; Sun, X.; Chen, S. Influence of Exposure Period and Angle Alteration on the Flexural Resilience and Mechanical Attributes of Photosensitive Resin. Nanomaterials 2022, 12, 2566. https://doi.org/10.3390/nano12152566

[213] Tashima, T. (2020). Shortcut Approaches to Substance Delivery into the Brain Based on Intranasal Administration Using Nanodelivery Strategies for Insulin. Molecules 25 (21). https://doi.org/10.3390/molecules25215188

[214] Sandberg, D. I., Peet, M. M., Johnson, M. D., Cole, P., Koru-Sengul, T., and Luqman, A. W. (2012). Chemotherapy Administration Directly into the Fourth Ventricle in a Nonhuman Primate Model. Ped 9 (5), 530-541. https://doi.org/10.3171/2012.1.PEDS11410

[215] Sandberg, D. I., Rytting, M., Zaky, W., Kerr, M., Ketonen, L., Kundu, U., et al. (2015). Methotrexate Administration Directly into the Fourth Ventricle in Children with Malignant Fourth Ventricular Brain Tumors: a Pilot Clinical Trial. J. Neurooncol. 125 (1), 133-141. https://doi.org/10.1007/s11060-015-1878-y

[216] Bottros, M. M., and Christo, P. J. (2014). Current Perspectives on Intrathecal Drug Delivery. J. Pain Res. 7, 615-626. https://doi.org/10.2147/JPR.S37591

[217] Householder, K. T., Dharmaraj, S., Sandberg, D. I., Wechsler-Reya, R. J., and Sirianni, R. W. (2019). Fate of Nanoparticles in the central Nervous System after Intrathecal Injection in Healthy Mice. Sci. Rep. 9 (1), 12587. https://doi.org/10.1038/s41598-019-49028-w

[218] Lammers, T., Kiessling, F., Hennink, W. E., and Storm, G. (2010). Nanotheranostics and Image-Guided Drug Delivery: Current Concepts and Future Directions. Mol. Pharmaceutics 7 (6), 1899-1912. https://doi.org/10.1021/mp100228v

Applications of Polymers in Surgery II
Materials Research Foundations 172 (2025) 163-186
Materials Research Forum LLC
https://doi.org/10.21741/9781644903353-6

Chapter 6

Polymers in Orthopedic Surgery

Muhammad Shahbaz[1], Nadia Akram[1]*, Tanveer Hussain Bokhari[1], Muhammad Saeed[1], Khalid Mahmood Zia[1]

[1]Department of Chemistry, Government College University Faisalabad, Faisalabad-38000, Pakistan

*nadiaakram@gcuf.edu.pk

Abstract

There is advancement in the field of orthopedic surgery with the innovation in polymers having diverse range of applications by providing variety of materials, having tailored properties which can be used in replacement of joint and in bone replacement. This chapter gives detailed overview of polymers applications in orthopedic surgery applications. Also, polymer-based scaffolds derived from polymers are explained which can be used in orthopedic applications. Limitations of using polymers in orthopedic surgery and advanced polymer materials in surgery are explained.

Keywords

Orthopedic Implants, Mechanical and Biological Properties, Acrylic Polymers, Polyurethanes, Fabrication Methods

Contents

Materials Research Foundations 172 (2025) 163-186
https://doi.org/10.21741/9781644903353-6

1. Introduction to orthopedic surgery

There are almost 206 bones in the skeleton of humans. The skeleton is divided into appendicular and axial with 126 and 80 bones respectively. There are lower and upper limbs in the appendicular skeleton including pelvic and shoulder girdle as well while the axial skeleton contains vertebral column, skull as well as thoracic cage. Along with that, pelvis, scapulae and clavicle belong to appendicular skeleton while hyoid bone and middle ear ossicles are in the axial skeleton.

Nicholas Andry, professor from Paris university wrote a book in 1741 and used the word “Orthopedia” as book title. Orthopedia word derived by combining two Greek words which are (orthos) and (padios) mean straight and child respectively [1]. However, orthopedic surgery is not only limited to children but also applied to older people as well. Other name of orthopedics is orthopedic surgery in which defected or injured as well as abnormalities of muscular skeleton are fixed. The defected or abnormal parts may be bones, joints, muscles, nerves, tendon or ligaments which perform specific functions in body. The term has different subspecialties like elbow and shoulder surgery, lower limb surgery, foot and knee surgery, ankle surgery, trauma surgery, spinal surgery as well as hand surgery. It is an efficient step toward pointing out defects in patients and restoring their function in an efficient way. For instance, osteoarthritis is a condition in which there is pain in the hip joint and restricts them to do work in normal way. So surgical treatment is a way in which that problem is fixed in patients by effective physiotherapy resulting in excellent mobility and pain relief[2].

Overall, Orthopedic surgery is associated with the condition of musculoskeletal system. Musculoskeletal disorders of muscular skeleton are basic reason of disability in all areas of body and overall considered as half of the disease (chronic)[3].

Orthopedic surgery in other words is the surgery of appendicular and axial skeleton and structures related to it. There are different subdivisions which include arthritis, fractures, tumor, fractures,

Materials Research Foundations 172 (2025) 163-186
https://doi.org/10.21741/9781644903353-6

soft tissue processes, conditions of metabolism and acquired and congenital conditions[4]. The aim of orthopedic surgery is to do alignment of bones and deformation correction, spasticity control on joints of individual, dislocation and contraction correction as well as scoliosis for the purpose of function maximization in patients[5].

Some of the issues which are commonly found in orthopedics are given below:

- Carpal tunnel syndrome Fibromyalgia
- Bursitis
- Osteoarthritis
- Back and neck pain
- Torn meniscus
- Hip fracture
- Osteoporosis

Some of the types of orthopedics are as follows.

1.1 Arthroscopy

Arthroscopy word is derived from two words "arthro" and "skopein" mean "joint look" Simply called as looking into joint.

In the procedure of arthroscopy, the doctor investigates inside, and that procedure gives a clear overlook into the joint. By that way, they are able to diagnose the problem and to resolve them. Overall, arthroscopy is a surgical way to point out joint problems and treat them.

The working mechanism of arthroscopy is incorporation of thin and small instrument (arthroscope) into joint. That instrument contains a small camera along with a light source. Pictures are displayed on screen of television which gives clear overview to doctor. The problems assessed during this procedure are cured by the doctor. As instruments like arthroscope and other instruments are small. A small incision can be used for open surgery as compared to large incision. Commonly investigated joints are elbow joint, shoulder, knee, hip and wrist [6].

1.2 Anterior cruciate ligament reconstruction

Rupturing of anterior cruciate ligament is mostly found as injury of knee. Complete anterior cruciate ligament tears are usually handled by orthopedic surgeons and physicians of sports medicine with restoration surgery of cruciate ligament, in which tissue graft is placed on place of torn ligament in order to mimic the function of natural anterior cruciate ligament and tissue graft replaces torn ligaments.

The lifestyle of patients and anterior cruciate ligament tear severity tells us about the need whether to do surgery or not. Anterior cruciate ligament does not heal completely by itself anterior cruciate ligament. It is confirmed that, if there is partially tear problems of anterior cruciate ligament in patients, ligament healing occurs without surgery. On the other hand, partial tear in anterior cruciate ligament may heal without surgery [7].

Materials Research Foundations 172 (2025) 163-186
https://doi.org/10.21741/9781644903353-6

1.3 Spinal fusion

Spinal fusion is a method used in surgery that induces stability as well as reinforcement in back structure by joining together two vertebrae. Joining vertebrae provides support to each other which results in reducing pain as well as nerves and spinal cord easing symptoms. As low back pain is increasing all over the world, which leads to productivity loss. Lower back pain is caused by degenerative changes (non- specific) affect soft tissues and bone of spine along with infection, acquired and congenital deformity, trauma and malignancy comprising much smaller cohort [8].

1.4 Hip replacement

Modern types of total hip replacement results in better life quality of patients as compared to other procedures of elective surgery. The major indication which tells abouts the need for total hip replacement is arthritis at end stage [9]. This procedure is efficient and less costly than other types of surgery in which defected part of the hip joints are replaced with new and artificial parts which mimic the function same as joint of normal hip [10].

1.5 Knee replacement

In 1970 and 1980, replacement of knee was done first time on a wide area, and it is now considered as less costly and effective method for treatment of knee replacement when the arthritis is at end stage. The design of surgical implants along with other devices was done for the first time and were delivered by surgical developers with collaboration of industry. Identification of patients in which joint is being replaced tells us about success rate of surgery. Arthritis at the end stages as well as severe pain confirm the knee replacement. Partial knee replacement is also possible or can possibly be done in medical, lateral or patella femoral knee compartments. In partial replacement of knee, the parts which are replaced are severely affected as compared to complete joint to total knee replacement [11].

2. Requirement of material for orthopedic implants

2.1 Mechanical Requirements

Mechanical property requirements of any implants depend on its use like shear stress elasticity and loading strength. In orthopedics, the implant device in orthopedic should work in repeated download-upload cycles under the influence of different force types such as twisting, shearing stress and bending. As devices are being implanted face conditions which are corrosive for a longer time, and that affects their properties. In order to overcome that problem, the device should have property to overcome/ maintain its reduction from fracture.

Mechanical property of material is defined as deformation which occurs in material with respect to force applied (stress to strain ratio). Stress is mostly generated as a result of bending, loading, compression, shear action or torsion. So, to investigate material mechanical property, it is subjected to stress-strain process and diagram is generated for deformation of material under force applied. Diagrams may show two regions i-e elastic and plastic where there is no deformation and where deformation is permanent respectively. Curve slope in elastic region is called rigidity which is material's property to show resistance against deformation. Increasing stress gradually will cause accumulation of micro fractures and then plastic deformation and break up of material at the end. Rigidity is not constant in specific cases of bone as the case for devices (implanted). It is

Materials Research Foundations 172 (2025) 163-186
https://doi.org/10.21741/9781644903353-6

mentioned that the nature of bone tissue is anisotropic i-e., it shows different mechanical properties depending on their determination direction. So, the mechanical property of bone depends on applied force direction. For instance, tissue of bones shows more rigidity when force applied from external way act in longitudinal direction of bone as compared to force which is applied on its surface.

2.2 Biological requirements

It is the most important parameter as the body shows certain reactivity against medical device being used in orthopedic surgery. In that case there will be limited degradation in implanted devices. For the device to be implanted, the process should not damage mechanical property and should not produce byproducts which are harmful. Side by side, it is also investigated that to make healing process faster, there should be limited reactivity between biological surroundings and implanted devices.

The incorporation of biomaterials has provided the way of creating materials with specific properties that help in adjusting the implant to the morphological as well as mechanical characteristics of the acceptor structure. It is mentioned by Dolcimascolo et al. that there are three groups of generations in which biomaterials (being used from the last six years) can be placed which are

- Bio-inert materials of first generation
- Biodegradable/bioactive materials (2nd generation)
- In the 3rd generation, the design material to be implanted triggers molecular response to do healing process in fast way[12].

3. Polymers for Orthopedic Implant Materials

3.1 Ultrahigh MW Polyethylene

The Ultra-high MW polyethylene is a specific polymer having excellent physical as well as mechanical properties. It is a chemically inert polymer having abrasion and impact resistance and have lubricity. It is considered a good option for the replacement of joints in articulating bearing surfaces.

High density polyethylene is having molecular weight up to 200000 gram per mole and has a linear structure. When there is a comparison of high-density polyethylene, ultra-high MW polyethylene has MW up to 6000000 gram per mole with ethylene units up to 200000 gram per mole. The packing of linear chains in better way increases the crystalline nature in ultra-high molecular weight polyethylene results in excellent mechanical properties which are required for use in orthopedics even of decrease in fracture toughness and ductility [13].

Ultra-high MW polyethylene (medical grade) is used in orthopedics that have crystallinity up to 55% and have molecular weight within the range of 3.5 to 6 million gram per mole.

From the introduction, that was started in 1962, ultra-high MW polyethylene used as bearing material for all the time in joint arthroplasties as of its higher wear resistance and excellent mechanical properties. However sometimes in literature, polyethylene of high density and ultra-high MW are used alternatively but the main difference between these two polymers lies is in

Materials Research Foundations 172 (2025) 163-186 https://doi.org/10.21741/9781644903353-6

terms of their properties like impact strength, molecular weight, abrasive wear and density as well. No doubt, the difference between density between these polymers is not too much (0.07 to 0.02) per cubic centimeter, but it is considered as noticeable difference between polyethylene materials. Overall, there is crystalline as well as amorphous phase in ultra-high MW polyethylene. Crystalline phase gives stiffness or modulus while amorphous rubber like phase gives toughness as well as ductility to material. Rates of high wear in commonly used ultrahigh molecular weight polyethylene being sterilized by using gamma radiation in air, decreases with ethylene oxide, stable environment, and sterilization by gas plasma. Also, it is confirmed from chemistry that irradiation can also cause cross linking and have the ability to increase the wear resistance of ultra-high MW polyethylene. That step of cross-linking phenomenon in end of 19th century enabled the researchers to synthesize cross linked ultra-high MW polyethylene for joint arthroplasty[14].

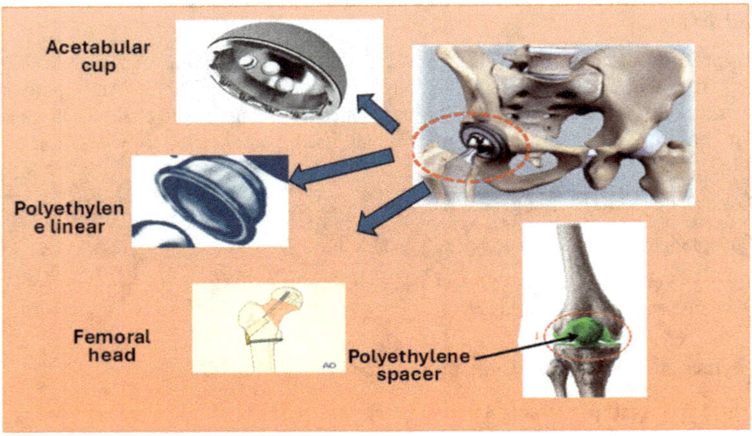

Figure 1. Applications of Ultrahigh Molecular Weight Polyethylene in orthopedic surgery

3.2 Acrylic polymer

The acrylic polymers or polyacrylates are compounds that have high MW. Their physiochemical properties are dependent on the chain substituents of polymer. They have good transmittance of light and as a result are the best materials for contact lens and intraocular lens (hard and soft). Poly (methyl methacrylate) is one of the most used polymers for repairing bones. It is also known as "Plexiglas". It is water resistant with linear chain polymer having transparency with glassy and amorphous nature at room temperature.

Poly(methyl methacrylate) is synthesized by type of addition polymerization reaction of methyl methacrylate with radical polymerization along with activator for generation of radicals. It is applied in joint arthroplasty as bone cement. The bone cement's function is in the artificial joint fixation. Available Space between bone and prosthesis is filled by the cement and impart very significant zone. As optimal rigidity, cements have ability to evenly buffer the forces which are in against the bone. Because of the close relationship between bone and cement as well as between

 Materials Research Forum LLC
https://doi.org/10.21741/9781644903353-6

prosthesis and cement enables equal and optimum distribution of interface strain energy and stresses.

Poly(methyl methacrylate) based cement has two components:

- Powder components which are mainly Poly(methyl methacrylate)
- Liquid components which are monomers of methyl methacrylate

Both components were introduced by Charnley[15].

As mentioned above, powder components are usually poly (methyl methacrylate) , The polymer powder components also have benzoyl peroxide which works as initiator in radical polymerization, including in the polymer beads or admixed to powder simply. Along with that, power may also have antibiotic or radio pacifier as well. So, the resulting bone cement can perform carrier matrix function. The function of radiopacified is to investigate and to monitor faults/defects in bone cement.

On the other hand, in liquid components, methyl methacrylate monomer is the major ingredient. N,N-dimethyl-p-toluidine (an aromatic amine) is usually added for the formation of free radicals as an activator. Additionally, it also has an inhibitor in order to avoid polymerization which is premature during the storage and have coloring agent as optional material. When both components (powder or liquid) are mixed together then reaction starts between initiator and activator. The N,N-dimethyl-p-toluidine from liquid component causes decomposition of benzoyl peroxide in powder component in a process of oxidation reduction results in radical generation of benzoyl by electron transfer. These are also chemical moieties which have short lives and are reactive, may start the process of polymerization by reacting themselves in carbon-to-carbon double bond of methyl methacrylate. In case of these reactive species and radicals, polymer chains continue to grow rapidly having molecular weight from 100000 gram per mole to 1000000 gram per mole at the end. As the viscosity of mixed material (dough) increases, the monomer mobility decreases and with two radicals chains recombination, radicals depletion occurs which results in end of polymerization process. This process of polymerization is exothermic with heat of polymerization of 57 KJ mol^{-1}[16].

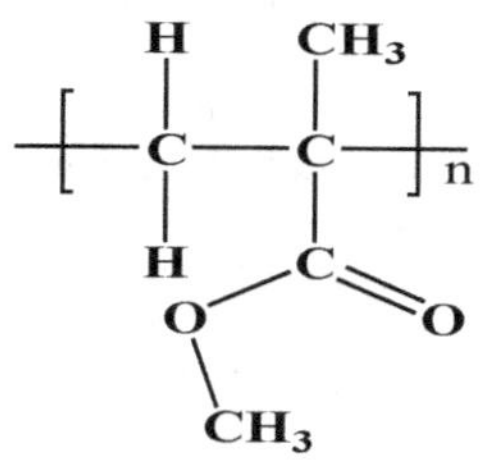

Figure 2. Structure of Polymethyl methacrylate

3.3 Biodegradable polymers in orthopedics

Since last decades, the use of polymers in the field of orthopedic surgery is increasing exponentially. These types of polymers are used in the fabrication of rods for fracture fixation, suture anchors, clips, staples, plates, hooks, screws sutures, arrows and more importantly to synthesize scaffolds for tissue engineering of muscular skeleton. Biodegradable polymers usually degrade into their respective elements by the process of enzymes or hydrolysis. That is an efficient phenomenon which eliminates the need to remove unnecessary implants from the body by doing second surgery. That step is not only cost effective and beneficial but also provides ease to patients. On top of that, the effect of stress shielding can be reduced by the system potentially by using a fixation device which is biodegradable.

A bone (living tissue) which when faces or experiences any load, it remodels, and that act is called Wolff's law. In stiff metal implants, the reduction of load on the surrounding bones occur, and hence, over the longer period bone will have a tendency for osteopenia which is a phenomenon termed as stress-shielding. Biodegradable polymers which are used for the fabrication of fixation devices counterbalance that problem as these devices lose their mechanical properties and degrades at the end. Thus, it can only face decreasing load level, which results in moderate reloading of the repaired or supported bone up to the time when full load bearing is restored [17]

3.4 Polylactide (PLA)

Polylactide is also called polylactic acid and is applied on larger scale in surgical sutures which are biodegradable, for controlled drug delivery and for making implants for bone fracture fixation. Polylactic acid has been used for many years and have ability to cure bone fracture in effective way. The polymer was first used as biodegradable implant for the synthesis of scaffold in tissue engineering of cartilage and bone.

There are several methods for the synthesis of polylactic acids. The suitable and reliable method is synthesis of polylactide by bulk polymerization (ring opening) of lactide. There are different compounds like organic, inorganic, organic and tin compounds and metallic compounds which are used as initiators. Among the initiators, stannous chloride, stannous octoate and tetraphenyl tin are mostly use and are very effective. Usually, tetraphenyl tin is preferred as it is passed and said to be safe as food additive by Food and Drug Administration, Washington, D.C. Monomers and reaction of polymerization are very reactive and sensitive towards OH group which works as chain transfer agent during reaction of polymerization. That high nature of sensitivity toward impurities of hydroxyl like lactic acid and moisture makes rection of polymerization very difficult. These impurities must be avoided in polymerization process and care should be taken during the process of synthesis. The mostly used method for the process of polymerization is described by Kulkarni and coworkers in which method is done containing monomer mixture in flask along with catalyst sealed under vacuum. Polymerization process starts when temperature exceeds from melting point of monomer. The reaction of polymerization is studied in a broader way. It is recently investigated that the main factor for handling polymerization reaction and in order to obtain high MW polylactide, monomer recrystallization is done by sealing the reaction containing flask in high vacuum [18].

Poly lactic acid (PLA) is well known polymer having biocompatibility and bioresorbable use from a decade in the applications of biomedical. Implants which are absorbable have more benefits as compared to metals when there is temporary application required only since no operation for removal is needed and complications of long term are substantially decreased. Other common applications of polylactic acid along with their copolymers are screw for fixation of bone as well as in bone plates. Degradation of polylactic acid occurs by chain scission and it is converted into

Materials Research Foundations 172 (2025) 163-186
https://doi.org/10.21741/9781644903353-6

monomers of lactic acid in the body, an intermediate produced in metabolism of carbohydrates. Therefore, it is considered as a suitable polymer for orthopedic and in drug delivery applications. There are two optical isomers of polylactide. Lactide is dimer of lactic acid which has L and D isomers. Lactide of L types of isomers exist naturally whereas DL lactide is produced synthetically by blending L lactide and D lactide. There is similarity between the polymerization of lactide and glycolide. The L-lactide homopolymer is a polymer of semicrystalline nature. Homopolymer along with PGA shows low elongation, high modulus and high tensile strength which enables them to be more applicable than polymer of amorphous nature in application of load bearing like sutures and orthopedic fixation. Poly(DL lactide) is polymer of amorphous nature have random distribution of isomeric form of both lactide isomers which do not set into crystalline form and are unable to make well-ordered structure. Absorbable screws of polylactide as compared with screws of stainless steel, are well tolerated without incidence of sterile effusion and help in medial malleolar fractures healing [19,20].

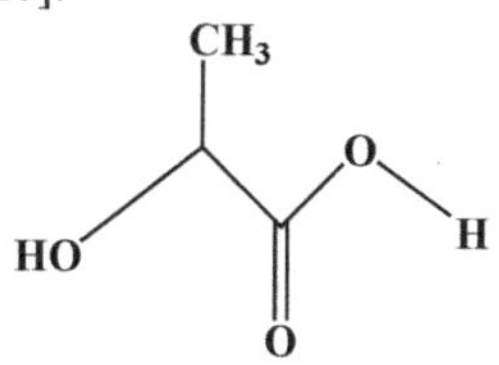

Figure 3. Structure of polylactic acid

3.5 Poly(glycolic acid)

Polyglycolic acid is the first synthetic and biodegradable polymer used on a large scale in orthopedic surgery applications.

Poly(α-hydroxy acids)are specific types of polymers having repeating units of -(O-CO-CHR)- and are prepared from monomers α-hydroxy acids. Class of these types of polymers is under the process of development for the synthesis of osteosynthesis devices from last 65 years.

Polyglycolic acid with high MW is crystalline, hard and tough polymer which melts at 224-228°C and has a glass transition temperature of 36°C.

As for the highly crystalline nature of polyglycolic acid, it is not soluble in many solvents and only show solubility in highly fluorinated solvent like hexafluoro isopropanol with molecular weight of 45000 gram per mole.

Polyglycolic acid preparation at industrial level is very difficult in terms of high molecular weight synthesis because of unstable nature and degradability of polyglycolic acid. In synthesis method of polyglycolic acid, two types of monomers are used which are glycolides and glycolic acid. Synthesis of high molecular weight polyglycolic acid by the process of ring opening polymerization of glycolide is acceptable but that method is costly and gives lower yield of polyglycolic acid. Another method described by Yamane and his coworkers pointed out that synthesis of polyglycolic acid via glycolic acid condensation along with reaction of dehydration is cost-effective[21]. However, the problem with that method is the production of high-molecular

Materials Research Foundations 172 (2025) 163-186
https://doi.org/10.21741/9781644903353-6

weight polyglycolic acid due to equilibrium between chain extension for termination of polyglycolic acid hydroxyl group. Other than that, glycolide ring opening polymerization was found for producing high molecular weight polyglycolic acid. However, there are limitations which are still present on the industrial level synthesis of high molecular weight polyglycolic acid such as attachment of glycolide on the wall of inner side of distillation lines.

Polymer with an average molecular weight from 20000 to 145000 can be spun into fibers. Polyglycolic acid strength increases when it changes its form into fiber because of preferred high orientation of polymer. Self-reinforcement is the production of a composite material made of some polymeric material consisting of reinforcing units in specific orientation like fibers or fibrils with binding matrix, both have same chemical structure. This process of reinforcement increases mechanical properties of partial crystalline polyglycolic acid implants and is used for production of strong devices [22-24]Click or tap here to enter text..

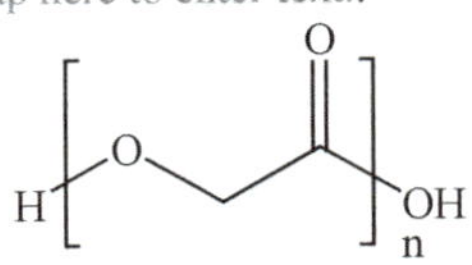

Figure 4. Structure of polyglycolic acid

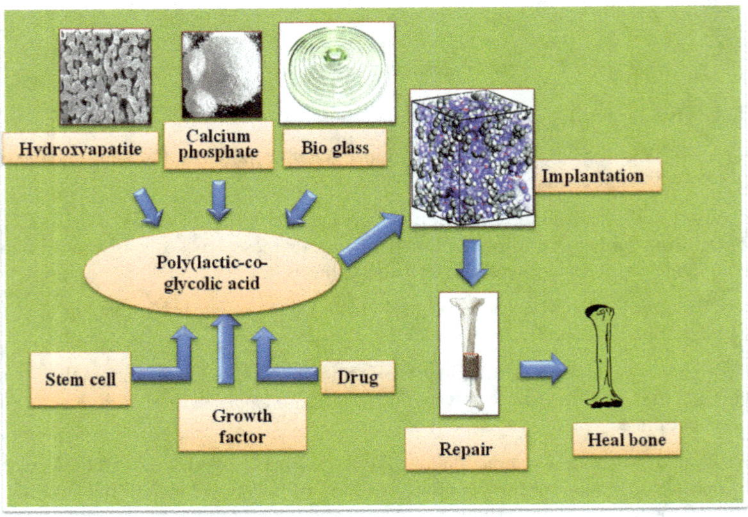

Figure 5. Applications of Polyglycolic acid in orthopedic surgery

3.6 Poly(dioxanone)

Polydioxanone is a type of polyester commonly used for the synthesis of medical devices(biodegradable). It is manmade polymer, having no color, absorbable polyester which was

Materials Research Forum LLC
https://doi.org/10.21741/9781644903353-6

synthesized in the start of 1980. Polydioxanone is generally prepared by doing ring opening polymerization of p- dioxanone with suitable catalyst like organometallic with heat. After various investigations, it is found that polydioxanone product has no adverse reactions with tissues in the process of adsorption after implantation and is considered as nonpyrogenic and nonantigenic. Degradation of Polydioxanone occurs by hydrolysis and metabolize completely in the body by the process. As this polymer is synthetic and the duration of absorption of polydioxanone is up to 238 days which is very long, it is used in medical devices which are undyed or dyed devices used in medical applications and is implanted in area where absorbable material is required which is long lasting. It has semicrystalline nature and almost 55% crystallinity with temperature of glass transition between -10 to 0°C and melting temperature range from 110-115°C. These properties enable it to make sutures synthesized by process of extrusion at temperature, which is lowest possible and prevents depolymerization which is spontaneous.

In the field of medicine, Polydioxanone is usually implanted in varying sizes and shapes. Polydioxanone based medical devices are already being used in surgeries as standard practice named orthopedic pins and PDSTM II sutures Suture made by Polydioxanone have high flexibility because of ether oxygen group existence in the backbone of polymer [25].

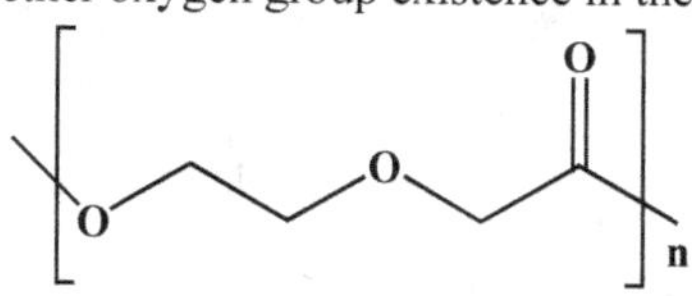

Figure 6. Structure of poly(dioxanone)

3.7 Polyanhydrides

Polyanhydrides are biodegradable polymer types. It can be obtained in various characteristics by changing methods of synthesis. That polymer is biocompatible, non-toxic and excreted from the body at the end[26]. There are various methods of Polyanhydrides synthesis given as

- Activated diacids condensation in bulk melt
- Polymerization by ring opening
- Reaction between diacid chlorides and dibasic acid

Polyanhydrides with highest molecular weight are obtained by using process of melt condensation, under optimized conditions of reaction and polymerization and by minimizing process of depolymerization at the same time. Prepolymers in pure form is prepared individually and then purified by the process of recrystallization. The conditions of reaction, temperature of reaction, time of reaction, presence of coordinative catalysts and vacuum created are significant factors for the synthesis of high MW polymers. Also, these are pure polymers and copolymer composition which is same as that of monomers composition used in reaction [27].

Polyanhydrides from monomers (non physiological) have been synthesized having favorable properties like biocompatibility, exhibiting well-defined degradation properties, and are marked as green by the FDA in 1996 to denote agents of chemotherapy for treatment of patient's brain cancer. Also, these polymers have water resistant backbones and anhydride linkage (hydrolytically

 Materials Research Forum LLC
https://doi.org/10.21741/9781644903353-6

labile) that give the property of surface degradation While the mechanism has been confirmed as significant in the drug delivery.

It is also mentioned that surface-eroding polymers might give additional benefits in some load-bearing applications like in orthopedic surgery. Mechanical properties of linear prepared polyanhydride are lower than that of ultra-high molecular weight polylactic acid which is highly crystalline. Polyanhydrides are already prepared by Staubli & Uhrich et al. which possess enhanced mechanical properties by addition of imide group in the backbone of polymer. Polyimides are highly stable (thermally and mechanically) along with aromatic polyimides which are biocompatible [17]

On the other hand, a new class of anhydride monomer is also developed recently having methacrylate group at the end, photopolymerized readily and generate polyanhydride network which is highly cross linked. Through this approach, the density of crosslinking of the obtained network of polymer can be applied to control the properties of the final obtained material while the degradation scale of time can be handled by network chemical composition. At the top of that, additional advantage of that process is to form cross linked polymer by photopolymerization. The system can be cured by light in vivo and possess required mechanical properties and rate of degradation having application in various fields. For instance, in a bone defect of irregular shape trabecular, a monomer having viscous liquid can be implemented in defected area and light in vivo, induces cross linking in polymer having biodegradable nature and gives dimensions which are required. Fixation plates for flexible fracture are synthesized by partially reacting monomers like post gelation which is incomplete. Surgeons usually introduce plates which are flexible, and these plates become rigid when they are cured by light to give crosslinked network when polymerization reaction is completed.

Degradation of polyanhydrides occurs by hydrolysis but the mechanism of its degradation is different from polylactic acid and mostly degrades by surface erosion. This property of degradation enables it attractive polymer to be used as drug delivery system. It is investigated that poly anhydride-co-imide polymers are same as that of PLCA having biocompatibility and can support cortical and endosteal bone regeneration. So, that polymer is considered as an efficient candidate for manufacturing scaffolds for orthopedic surgery in weight bearing applications [28,29].

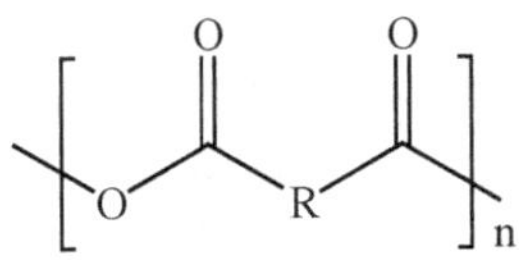

Figure 7. Structure of polyanhydride

3.8 Polyfumarates

Polyfumarates is a copolymer alternating with that fumaric acid and propylene glycol. They have attractive features including the double bond in each structural unit of fumarate, two monomers nature and ester linkage between each structural unit gives alternative unit. Double bond present in polymers makes it able to do cross-linking after injection by radical polymerization. Degradation of the polymer is because of easter linkage present. The degradation rate of composite material is usually handled by changing the initial MW of polyfumarates and cross-linking ratio in formulation of composite.

Two monomers are involved in the synthesis of polyfumarates which are fumaric acid as well as propylene glycol both are biocompatible substances. Their cross linking can be done by using vinyl monomer and system of free radical consist of benzoyl peroxide as initiator along with dimethyl toluidine as accelerator. Mechanical stability of cured polymer depends on the polymer to monomer in the mixture of cross linking. Composite material is synthesized by incorporation of solid phase components. The added components serve as porogen like sodium chloride or can increase the mechanical properties of polymer.

The major benefit of using unsaturated polymers is its tendency to cure in vivo material, hence filling defects of skeleton which are of any size or shape with lowest surgical inventions.

Tubular bones of human can be replaced with composites made of polyfumarates having mechanical properties. Other than that, high molecular weight addition of polyfumarates in formulations increase modulus and strength of composites at the time of start and then mechanical property start to decrease slowly over time. Study of in vivo degradation also tells the same trend as increase and decrease in mechanical properties. These confirmed results are very attractive for human tubular bone replacement specifically at the site of load bearing[30].

Materials based on polyfumarates are used for tissue engineering of bone. Payne et al. also confirmed the study of in vitro and mentioned the polypropylene fumarate (cross linked) is potential substrate for the support of rat osteoblast

From last few years, there has been a lot of work being done by scientists on polyfumarate. For example, it is also investigated by Mikos and coworkers that linear polyester (partially saturated) based on polypropylene fumarate can be used as skeletal defects filling. Those polymers have properties which are the same as those of trabecular bone. These polymer have also the ability to form composites with beta-tricalcium phosphate. Biodegradable bone plates by the combination of polyfumarates (cross linked) with polylactic acid are already prepared by Hasiric et al. [31,32][33]

3.9 Polyurethane

Polyurethane is attractive class of polymers which consist of repeating urethane linkage synthesized by reaction between diisocyanate and polyol. There are hard and soft segments in polyurethane present at alternate position. The soft segment is defined as the part of macrodiol which contributes to the final chain of polymer. On the other hand, hard segments are defined as the part generated by the components of chain extender and diisocyanates. The properties of the final prepared polyurethane depend on molecular weight, structure and composition of monomer[34].

Polyurethane is attractive and cheap material has excellent mechanical properties and hence can be used for biomedical applications. There are various types of polyurethane depending on the type of device being implanted. The properties are flexibility, stiffness, elasticity as well as mechanical strength. On top of that, the specific property which polyurethane shows is to mimic some biological structure of body. Polyurethane can be biodegradable for absorbable scaffolds as regeneration of tissue as well as it can be biocompatible for permanent implants in body.

Materials Research Foundations 172 (2025) 163-186
https://doi.org/10.21741/9781644903353-6

The application of polyurethane in the medical field is semi-permeable membrane which are transparent, vascular catheters, valves for heart, wound dressing, orthopedic implants, ureteral and stents. As the requirement for bone substitutes are biodegradable or bioresorbable polymers, so polyurethane is a suitable candidate for that purpose. This polymer may be biodegradable, biocompatible and calcified. Polyurethane has a wide range of mechanical properties that depend on chemical composition and may be from flexible to rigid. It is used to synthesize a variety of products like coating, films, fibers, foam etc. Polyurethane also supports adhesion of cells and human proliferation of osteoblasts as well as other types of cells. They are important in tissue engineering biomaterials. Various researches describe the possibility of polyurethane application in regeneration of bone [35].

Polyurethanes of third generation are used for scaffolds as tissue reinforcement as biodegradable material with urea) or for peripheral nerves. Also, these are mostly used polymers in devices for orthopedic fixation. By doing chemical modification in polyurethane, they can mimic the function of spongy bone as well as cortical bone compact structure[36].

Research is going on to prepare polyurethane of improved properties and also it is applied as articular surface. It is also considered as attractive candidate for tibial insertion of total knee replacement. Main reason to give priority to polyurethane over polyethylene is its stability over elevated temperature, larger size particles, hydrophilicity and less cytokine response produced by debris and osteolysis.

The rigid foam of polyurethane called ostamer is considered as excellent at this time for orthopedic applications. It has excellent properties like bonding strength to bone range between 170 to 190 pounds per square inch; compressive strength is 20,000 pounds per square inch tensile strength is 800 pound per square inch; flexural strength between 1,000-1,200 pounds per square inch density is between 25-30 pounds per cubic foot strength and shear strength is 500 pounds per square inch. Polyurethane was used to cure defects made by surgically in femurs and dogs radii, and dogs were followed up histologically and radiologically. The polymer was proved to be excellent and had satisfying requirements for fragment fixation without osteogenesis inhibition [37].

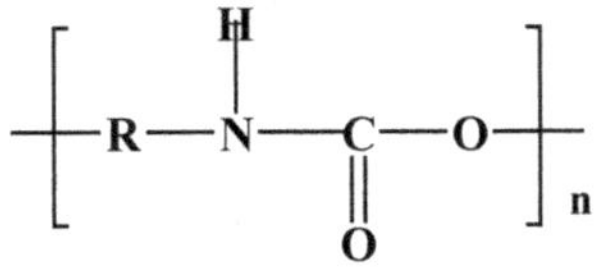

Figure 8. Structure of polyurethane

3.10 Polyetheretherketone (PEEK)

Polyether ether ketone abbreviated as PEEK is very common in orthopedic surgery as of its super biocompatibility, radiolucency, high stability under temperature, efficient sterilizability, resistivity against biological and chemical degradation, extra order fatigue properties and modulus which is similar to cortical human bone. It has tensile strength approximately 90-100Mpa and its elastic modulus is round about 3.6 Gpa and it is semicrystalline thermoplastic polymer. The melting temperature of polyether ether ketone is 343°C. Polyether ether ketone have high rigidity and

Materials Research Foundations 172 (2025) 163-186 https://doi.org/10.21741/9781644903353-6

strength along with mechanical properties like low friction coefficient and excellent impact resistance as compared to other thermoplastic polymers.

Polyether ether ketone is polyaromatic and semicrystalline thermoplastic polymer which belongs to polyaryl ether ketone. It is chemically inert with elastic modulus of up to 4 GPa which is quite close to human cortical bone. It also has excellent resistance to chemicals and radiations, compatibility with magnetic resonance imaging enables it to be applicable for the use in orthopedic surgery. As of its inert nature, cell adhesion promotion and bone contact ability, researchers are trying to enhance bioactivity of polyether ether ketone by the process of surface modification, incorporating materials and other processing techniques to improve its osteointegration tendency and hence to increase the lifespan of polyether ether ketone implant.

One of the attractive uses of polyether ether ketone in orthopedic surgery is surgery of spinal diffusion. In 1998, the first cage of polyether ether ketone was synthesized which was applied in spine in intervertebral body. Comparing polyether ether ketone cage with the cage of titanium i-e prevalent fusion implant in spinal body, the former cage has edge of improving the fusion rate as well as reducing subsidence rate of implant. However, there is still requirement for polyether ether ketone surface modification as its inert nature cause osteointegration hinderance and further create bone nonunion. There are mainly two types of polyether ether ketone cage surface modification

- Polyetheretherketone coated by titanium
- Polyetheretherketone reinforced by carbon fiber

Other various uses of PEEK are in Joint arthroplasty, surgery of sports, Osteosynthesis etc[38]

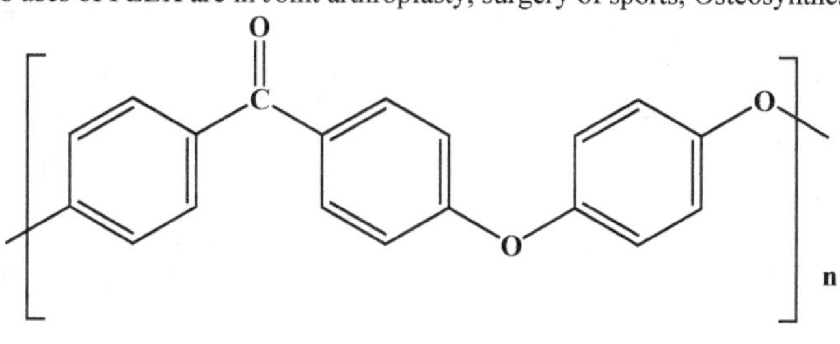

Figure 9. Structure of polyether ether ketone

Materials Research Foundations 172 (2025) 163-186
https://doi.org/10.21741/9781644903353-6

***Table 1**: List of polymers used in orthopedic surgery applications*

Sr No.	Polymer name	Applications
1	Polyglycolic acid	Bone internal fixation devices, sutures[39]
2	Poly(L-lactic acid)	Orthopedic fixation devices, Sutures[40]
3	Poly-L/D-lactide	Bioresorbable implant materials
4	Polydioxanone	Fixation screws for bone
5	Poly(caprolactone-co-trimethylene carbonate) & Poly(glycolide-co-ε-caprolactone)	Orthopedic tacks and screws[41]
6	Poly(3-hydroxybutyrate)3-hydroxyvalerate	Bone plates and pins[42]
7	Lysine diisocyanate based Polyurethane	Bone cement & orthopedic applications[43]
8	Poly(Propylene Fumarate)	spine fusion and intervertebral cages[44]
9	Polyphosphazenes	Skeletal reconstructions[45]
10	Poly(amino acids)	Implantation of muscle and bone model[46]
11	Polyglycolides	Suture, clips
12	Poly(methyl methacrylate)	Bone replacement
13	Polyether ether ketone	parts in joint replacements, implants
14	UHMWPE	orthopedic prostheses, Knee replacement, treatment of humeral fractures
15	Polysulfone	Devices for fixation of bone, total joint arthroplasties
16	Polydimethylsiloxane	Foot and hand joints, implants of bone, implants of plastic surgery and orthopedics[47]

4. Fabrication methods of orthopedic devices

4.1 3D printing

3D printing initially termed as "rapid prototyping" or referred as additive manufacturing. Its primary means typing is to produce industrial design prototypes in a fast and cheap way. Technology is being used in different ways compared to conventional methods (machining and primarily casting) of processing. This method has originally used layer manufacturing and additive

 Materials Research Forum LLC
https://doi.org/10.21741/9781644903353-6

manufacturing, the method in which the material being passed through extruder and resulted layer deposit occur on the top of each other in pre controlled manner which are fused to form required product.

3D printing (rapid prototyping) has three types of technologies for applications in biomedical field;

- Extrusion (Nozzle),
- Laser
- Printer based

[48].

4.2 Vat photopolymerization

The process of vat photopolymerization is an additive manufacturing technique in which controlled radiations are fall on the material which is photosensitive and layers of polymers are obtained. The obtained layers are joined together to form 3D structures. However, the method is only applicable to materials which are light sensitive and show polymerization under light. The process is applicable to resins and photopolymers. Depending upon the curing source variation, vat polymerization can be divided into digital light processing, stereolithography, volumetric 3D printing and two-photon polymerization [49].

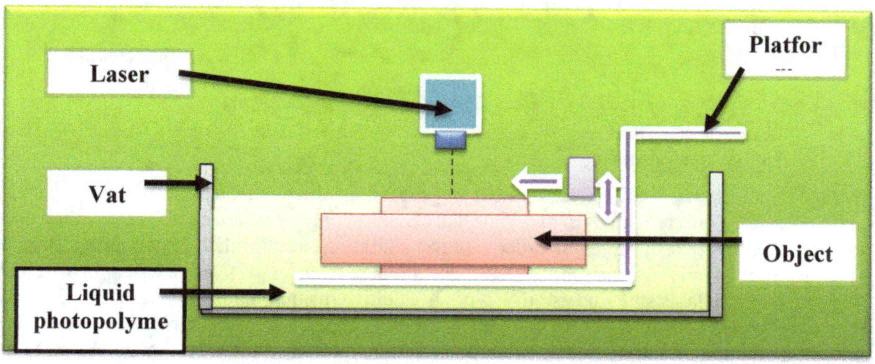

Figure 10. Diagram of Vat photopolymerization

4.3 Fused deposition modeling

It is an additive manufacturing type which is suitable for producing prototypes, products and models. It is simple and fast technology used to prepare personalized things in cost effective way. It is also called extrusion based additive manufacturing. The product is synthesized by extruding tiny beads of material that harden with no time after extrusion to make solid layers. A material filament normally metal or thermoplastic wire is fed into the head of extrusion nozzle which heat filaments, turns off after extruding fibers, and it depends on the given shape as per specific design. However, as the machine is available at commercial level it has a

Materials Research Foundations 172 (2025) 163-186
https://doi.org/10.21741/9781644903353-6

working temperature which is up to 300°C so the polymer having temperature above 300°C cannot be processed by using this technique[50].

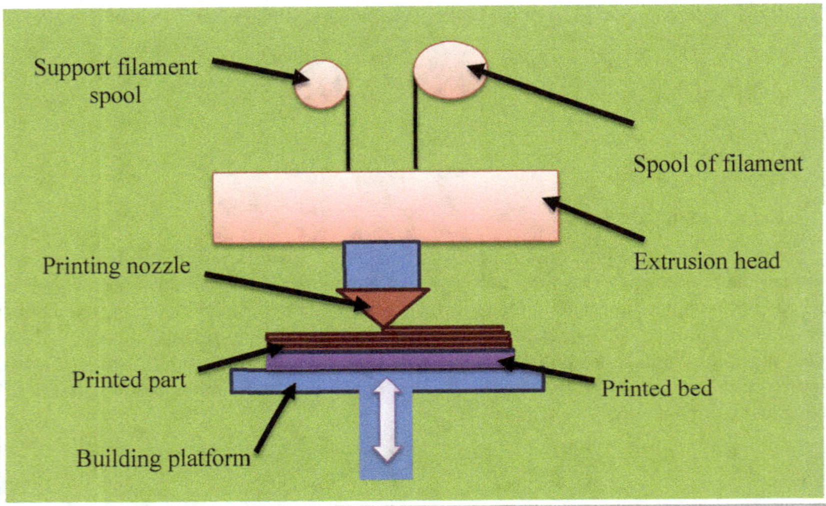

Figure 11. Diagram of fused deposition modeling

4.4 Selective laser sintering

This technique is one of the fastest methods of processing which is increasing rapidly. It has the ability to process different kinds of materials like ceramics, polymers and metals. The method does not depend on any type of support in order to prevent building paths resulted by collapsing. Powder which is not molten is used as support material. Complex types of systems can then be produced by using selective laser sintering as other types of additive manicuring techniques can't be used. The resulting parts are of high strength which can be compared to original material. Other than that, the method does not require any solvent so solvent sensitive molecules can be produced by the method [51].

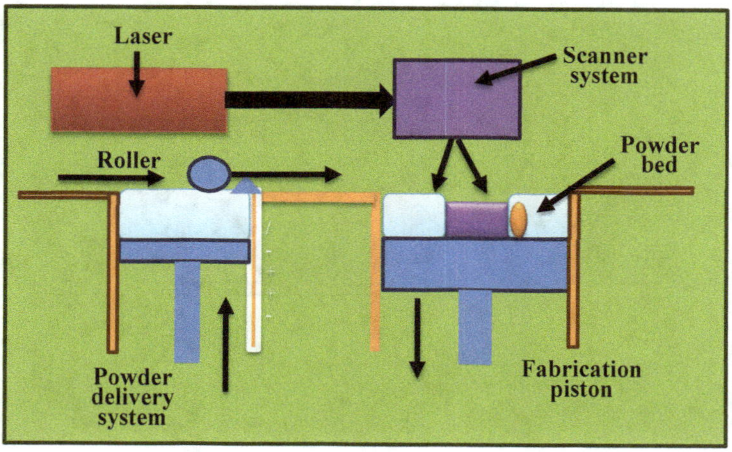

Figure 12. Diagram of Selective laser sintering

4.5 Binder jetting

Binder-jetting is the type of technique which is used with powder binding together to form the green part by selectively jetting polymer binder in layers form. The method gives a lot of benefits as an additive manufacturing. For instance, the method can work with powder of any feedstock and adds materials which are functionally graded. Binder jetting system building rate is very high as compared to other methods of processing like material jetting as the method only requires printing the binder as a total volume fraction. For example, a 100-head nozzle print can make part up to round about 200 cm^3 per minute. To create each layer of part. Powder layer is spread using a counter rotating roller. After that, the head of inkjet print force binding agent in liquid form into powder bed to make 2D pattern for layer. Some systems of powder/binder can use heating systems which help in controlling moisture and curing process but that is not primary requirement for the process. Built platforms goes down which makes room for the synthesis of next layer and the process is repeated each time. As the printed parts are fragile and usually treated further in order to improve the mechanical properties of material. [52]. However, to implement that technique on wide range in industrial area, there is dire need to reduce cost of binder jetting [53].

Materials Research Forum LLC
https://doi.org/10.21741/9781644903353-6

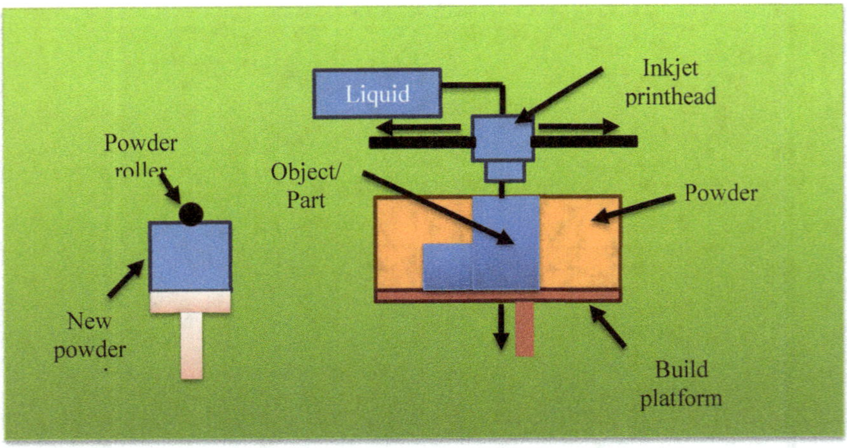

Figure 13. Diagram of Binder jetting

Summary

Polymers have a diverse range of applications in biomedical fields. Their properties can be tailored according to their use. They are used in the manufacturing of various bone implants including screws, pins, plates which are applied as structure stabilizing supporting bone healing. polymers which are biodegradable have the ability of degrade slowly in the body which eliminates the need of second surgery in order to remove the implant. Also, polymers are used as joints and prosthetic limbs having light weight and are compatible as well as durable properties. Other polymers like silicon and polyurethane are employed due to functionality and comfort providers in prosthetic devices. Other than that, these polymers are used as a replacement or repairing damaged cartilage. Polymer based scaffolds like collagen-based hydrogel promote repair and regeneration of cartilage tissues. They may function as drug delivery components which provides therapeutic agents for disease of bone and results in improvement in surgical outcomes.

Overall, polymers have a major role in orthopedic surgery as of their functionality and tailoring ability for specific functions, their biocompatibility as well as outcomes after treatment.

References

[1] C.L. Bennett Wilke, B.M. Dowdle, The Future of Orthopedics: Imagining Orthopedics in the Year 2100, Physician Assist Clin 9 (2024) 149–153.https://doi.org/10.1016/J.CPHA.2023.09.002.

[2] M.H. Mohammad, R. Verma, M.M. Jeyam, An Introduction to Orthopedic Surgery Volume 3 Issue 1 Doctors Academy Publications, 2013. www.wjmer.co.uk.

[3] N. Kose, A. Ayse Kose, Application of Nanomaterials in Prevention of Bone and Joint Infections, in: Nanotechnology in Diagnosis, Treatment and Prophylaxis of Infectious Diseases, Elsevier, 2015: pp. 107–117. https://doi.org/10.1016/B978-0-12-801317-5.00007-4

[4] G.E. Cook, D.C. Markel, W. Ren, L.X. Webb, M.D. McKee, E.H. Schemitsch, Infection in Orthopaedics, J Orthop Trauma 29 (2015) S19–S23. https://doi.org/10.1097/BOT.0000000000000461.

[5] D. Clayton-Krasinski, L. Fieback, Pediatric Nonprogressive Central Nervous System Disorders, in: Physical Rehabilitation, Elsevier, 2007: pp. 367–404. https://doi.org/10.1016/B978-072160361-2.50018-1

[6] D. Buford, B.W. Blatz, N. Hyde, Needle Arthroscopy of the Knee, Shoulder, and Hip, Atlas of Interventional Orthopedics Procedures: Essential Guide for Fluoroscopy and Ultrasound Guided Procedures (2022) 594–598. https://doi.org/10.1016/B978-0-323-75514-6.00036-4.

[7] H.S. Kim, J.K. Seon, A.R. Jo, Current Trends in Anterior Cruciate Ligament Reconstruction, Knee Surg Relat Res 25 (2013) 165–173. https://doi.org/10.5792/ksrr.2013.25.4.165

[8] L. Evans, T. O'Donohoe, A. Morokoff, K. Drummond, The role of spinal surgery in the treatment of low back pain, Medical Journal of Australia 218 (2023) 40–45. https://doi.org/10.5694/mja2.51788

[9] R.J. Ferguson, A.J. Palmer, A. Taylor, M.L. Porter, H. Malchau, S. Glyn-Jones, Hip replacement, The Lancet 392 (2018) 1662–1671. https://doi.org/10.1016/S0140-6736(18)31777-X

[10] R.W. Chang, A Cost-effectiveness Analysis of Total Hip Arthroplasty for Osteoarthritis of the Hip, JAMA: The Journal of the American Medical Association 275 (1996) 858. https://doi.org/10.1001/jama.1996.03530350040032

[11] A.J. Carr, O. Robertsson, S. Graves, A.J. Price, N.K. Arden, A. Judge, D.J. Beard, Knee replacement, The Lancet 379 (2012) 1331–1340. https://doi.org/10.1016/S0140-6736(11)60752-6

[12] N. Filip, I. Radu, B. Veliceasa, C. Filip, M. Pertea, A. Clim, A.C. Pinzariu, I.C. Drochioi, R.L. Hilitanu, I.L. Serban, Biomaterials in Orthopedic Devices: Current Issues and Future Perspectives, Coatings 12 (2022) 1544. https://doi.org/10.3390/coatings12101544

[13] M.A. MATEOS-TIMONEDA, Polymers for bone repair, in: Bone Repair Biomaterials, Elsevier, 2009: pp. 231–251. https://doi.org/10.1533/9781845696610.2.231

[14] A. Bistolfi, F. Giustra, F. Bosco, L. Sabatini, A. Aprato, P. Bracco, A. Bellare, Ultra-high molecular weight polyethylene (UHMWPE) for hip and knee arthroplasty: The present and the future, J Orthop 25 (2021) 98–106. https://doi.org/10.1016/j.jor.2021.04.004

[15] C.P. Reghunadhan Nair, G. Clouet, THERMAL INIFERTERS: THEIR CONCEPT AND APPLICATION IN FREE RADICAL POLYMERIZATION, Journal of Macromolecular Science, Part C: Polymer Reviews 31 (1991) 311–340. https://doi.org/10.1080/15321799108021926

[16] Y.D. K Draenert, Properties of bone cement: the three interfaces, 2005.

[17] C.M. Agrawal, Biodegradable Polymers for Orthopedic Applications, in: Polymer Based Systems on Tissue Engineering, Replacement and Regeneration, Springer Netherlands, Dordrecht, 2002: pp. 25–36. https://doi.org/10.1007/978-94-010-0305-6_3

[18] X. Zhang, Mattheus, F.A. Goosen, S.P. Wyss, D. Pichora, Biodegradable Polymers for Orthopedic Applications, Journal of Macromolecular Science, Part C: Polymer Reviews 33 (1993) 81–102. https://doi.org/10.1080/15321799308021434

[19] H. Altpeter, M.J. Bevis, D.W. Grijpma, J. Feijen, Non-conventional injection molding of poly(lactide) and poly(ε-caprolactone) intended for orthopedic applications, J Mater Sci Mater Med 15 (2004) 175–184. https://doi.org/10.1023/B:JMSM.0000011820.64572.a5

[20] J.C. Middleton, A.J. Tipton, Synthetic biodegradable polymers as orthopedic devices, Biomaterials 21 (2000) 2335–2346. https://doi.org/10.1016/S0142-9612(00)00101-0

[21] K. Yamane, H. Sato, Y. Ichikawa, K. Sunagawa, Y. Shigaki, Development of an industrial production technology for high-molecular-weight polyglycolic acid, Polym J 46 (2014) 769–775. https://doi.org/10.1038/pj.2014.69

[22] N. Ashammakhi, P. Rokkanen, Absorbable polyglycolide devices in trauma and bone surgery, Biomaterials 18 (1997) 3–9. https://doi.org/10.1016/S0142-9612(96)00107-X

[23] O. Böstman, H. Pihlajamäki, Clinical biocompatibility of biodegradable orthopedic implants for internal fixation: a review, Biomaterials 21 (2000) 2615–2621. https://doi.org/10.1016/S0142-9612(00)00129-0

[24] K. Budak, O. Sogut, U. Aydemir Sezer, A review on synthesis and biomedical applications of polyglycolic acid, Journal of Polymer Research 27 (2020) 208. https://doi.org/10.1007/s10965-020-02187-1

[25] J.A. Martins, A.A. Lach, H.L. Morris, A.J. Carr, P.-A. Mouthuy, Polydioxanone implants: A systematic review on safety and performance in patients, J Biomater Appl 34 (2020) 902–916. https://doi.org/10.1177/0885328219888841

[26] P.G. Reddy, A.J. Domb, Polyanhydride Chemistry, Biomacromolecules 23 (2022) 4959–4984. https://doi.org/10.1021/acs.biomac.2c01180

[27] A.J. Domb, S. Amselem, J. Shah, M. Maniar, Polyanhydrides: Synthesis and characterization, in: 1993: pp. 93–141. https://doi.org/10.1007/BFb0027552

[28] Mark. Chasin, R.S. Langer, Biodegradable polymers as drug delivery systems, M. Dekker, 1990.

[29] K.M. Hämäläinen, E. Määttä, H. Piirainen, Marianne Sarkola, A. Väisänen, V.-P. Ranta, A. Urtti, Roles of acid/base nature and molecular weight in drug release from matrices of gelfoam and monoisopropyl ester of poly(vinyl methyl ether–maleic anhydride), Journal of Controlled Release 56 (1998) 273–283. https://doi.org/10.1016/S0168-3659(98)00094-7

[30] S.J. Peter, M.J. Yaszemski, L.J. Suggs, R.G. Payne, R. Langer, W.C. Hayes, M.R. Unroe, L.B. Alemany, P.S. Engel, A.G. Mikos, Characterization of partially saturated poly(propylene fumarate) for orthopedic application, J Biomater Sci Polym Ed 8 (1997) 893–904. https://doi.org/10.1163/156856297X00074

[31] S.J. Peter, M.J. Yaszemski, L.J. Suggs, R.G. Payne, R. Langer, W.C. Hayes, M.R. Unroe, L.B. Alemany, P.S. Engel, A.G. Mikos, Characterization of partially saturated poly(propylene fumarate) for orthopedic application, J Biomater Sci Polym Ed 8 (1997) 893–904. https://doi.org/10.1163/156856297X00074

[32] S.J. Peter, M.J. Yaszemski, L.J. Suggs, R.G. Payne, R. Langer, W.C. Hayes, M.R. Unroe, L.B. Alemany, P.S. Engel, A.G. Mikos, Characterization of partially saturated poly(propylene fumarate) for orthopedic application, J Biomater Sci Polym Ed 8 (1997) 893–904. https://doi.org/10.1163/156856297X00074

[33] M. Martina, D.W. Hutmacher, Biodegradable polymers applied in tissue engineering research: a review, Polym Int 56 (2007) 145–157. https://doi.org/10.1002/pi.2108

[34] N. Akram, M. Shahbaz, K.M. Zia, M. Usman, A. Ali, R. Al-Salahi, H.A. Abuelizz, C. Delattre, Investigation of the *in vitro* biological activities of polyethylene glycol-based thermally stable polyurethane elastomers, RSC Adv 14 (2024) 779–793. https://doi.org/10.1039/D3RA06997D

[35] M. Marzec, J. Kucińska-Lipka, I. Kalaszczyńska, H. Janik, Development of polyurethanes for bone repair, Materials Science and Engineering: C 80 (2017) 736–747. https://doi.org/10.1016/j.msec.2017.07.047

[36] N. Filip, I. Radu, B. Veliceasa, C. Filip, M. Pertea, A. Clim, A.C. Pinzariu, I.C. Drochioi, R.L. Hilitanu, I.L. Serban, Biomaterials in Orthopedic Devices: Current Issues and Future Perspectives, Coatings 12 (2022) 1544. https://doi.org/10.3390/coatings12101544

[37] M.P. Mandarino, J.E. Salvatore, Polyurethane polymer—Its use in fractured and diseased bones, The American Journal of Surgery 97 (1959) 442–446. https://doi.org/10.1016/0002-9610(59)90011-X

[38] Y.-H. Yu, S.-J. Liu, Polyetheretherketone for orthopedic applications: a review, Curr Opin Chem Eng 32 (2021) 100687. https://doi.org/10.1016/j.coche.2021.100687

[39] P.B. Maurus, C.C. Kaeding, Bioabsorbable implant material review, Oper Tech Sports Med 12 (2004) 158–160.

[40] H.H. Lu, J.A. Cooper Jr, S. Manuel, J.W. Freeman, M.A. Attawia, F.K. Ko, C.T. Laurencin, Anterior cruciate ligament regeneration using braided biodegradable scaffolds: in vitro optimization studies, Biomaterials 26 (2005) 4805–4816.

[41] B. Dhandayuthapani, Y. Yoshida, T. Maekawa, D.S. Kumar, Polymeric Scaffolds in Tissue Engineering Application: A Review, Int J Polym Sci 2011 (2011) 1–19. https://doi.org/10.1155/2011/290602

[42] C.W. Pouton, S. Akhtar, Biosynthetic polyhydroxyalkanoates and their potential in drug delivery, Adv Drug Deliv Rev 18 (1996) 133–162.

[43] I.C. Bonzani, R. Adhikari, S. Houshyar, R. Mayadunne, P. Gunatillake, M.M. Stevens, Synthesis of two-component injectable polyurethanes for bone tissue engineering, Biomaterials 28 (2007) 423–433.

[44] Y. Teng, H. Giambini, A. Rezaei, X. Liu, A. Lee Miller, B.E. Waletzki, L. Lu, Poly(Propylene Fumarate)–Hydroxyapatite Nanocomposite Can Be a Suitable Candidate for Cervical Cages, J Biomech Eng 140 (2018). https://doi.org/10.1115/1.4040458

[45] A.L. Baillargeon, K. Mequanint, Biodegradable Polyphosphazene Biomaterials for Tissue Engineering and Delivery of Therapeutics, Biomed Res Int 2014 (2014) 1–16. https://doi.org/10.1155/2014/761373

[46] P. Wang, P. Liu, H. Peng, X. Luo, H. Yuan, J. Zhang, Y. Yan, Biocompatibility evaluation of dicalcium phosphate/calcium sulfate/poly (amino acid) composite for orthopedic tissue engineering *in vitro* and *in vivo*, J Biomater Sci Polym Ed 27 (2016) 1170–1186. https://doi.org/10.1080/09205063.2016.1184123

[47] S.V. Gohil, S. Suhail, J. Rose, T. Vella, L.S. Nair, Polymers and Composites for Orthopedic Applications, in: Materials for Bone Disorders, Elsevier, 2017: pp. 349–403. https://doi.org/10.1016/B978-0-12-802792-9.00008-2

[48] V. Bagaria, R. Bhansali, P. Pawar, 3D printing- creating a blueprint for the future of orthopedics: Current concept review and the road ahead!, J Clin Orthop Trauma 9 (2018) 207–212. https://doi.org/10.1016/j.jcot.2018.07.007

[49] C. Sakib-Uz-Zaman, M.A.H. Khondoker, Polymer-Based Additive Manufacturing for Orthotic and Prosthetic Devices: Industry Outlook in Canada, Polymers (Basel) 15 (2023) 1506. https://doi.org/10.3390/polym15061506

[50] D. Acierno, A. Patti, Fused Deposition Modelling (FDM) of Thermoplastic-Based Filaments: Process and Rheological Properties—An Overview, Materials 16 (2023) 7664. https://doi.org/10.3390/ma16247664

[51] W. Han, L. Kong, M. Xu, Advances in selective laser sintering of polymers, International Journal of Extreme Manufacturing 4 (2022) 042002. https://doi.org/10.1088/2631-7990/ac9096

[52] M. Ziaee, N.B. Crane, Binder jetting: A review of process, materials, and methods, Addit Manuf 28 (2019) 781–801. https://doi.org/10.1016/j.addma.2019.05.031

[53] D.-H. Kim, J. Lee, J. Bae, S. Park, J. Choi, J.H. Lee, E. Kim, Mechanical Analysis of Ceramic/Polymer Composite with Mesh-Type Lightweight Design Using Binder-Jet 3D Printing, Materials 11 (2018) 1941. https://doi.org/10.3390/ma11101941

Applications of Polymers in Surgery II
Materials Research Foundations 172 (2025) 187-198
Materials Research Forum LLC
https://doi.org/10.21741/9781644903537

Chapter 7

Significant Role of Polymers in the Field of Surgery

Niharika Kulshrestha*
Department of Physics, GLA University, Mathura, U.P., India
niharika.kulshrestha@gla.ac.in

Abstract

Polymers have unique features which make their wide application. The unique features of polymers such as high flexibility, durability, mechanical stability, light weight, easy to use, easily available, low cost etc. make them useable in various industrial, homely, electrochemical, electrical, electronics, medical, biomedical, surgery and food packaging. This chapter mainly focused on the application of polymers in the field of surgery. In the older methods, absence of polymers was facing limitations like decreased mechanical stability and low adherence to tissue surface. This chapter also discussed that different shape memory polymers can be used in surgery.

Keywords

Polymers, Surgery, Bio-medical, Shape Memory Polymers, Biopolymers

Contents

1. Introduction

1.1 Polymers

Polymers are chemically repeatedly linked units bonded together. This chemical repeated units also called as monomers. The advantage of using polymers is that they have high flexibility,

Materials Research Foundations 172 (2025) 187-198
https://doi.org/10.21741/9781644903537

durability, mechanical stability, light weight, easy to use, easily available, low cost etc. Polymers have application in various fields such as packaging, transport vehicles, electrochemical devices, fuel cells, solar cells, health care, sensors etc. As they provide adaptable solutions for a wide range of applications, polymers are essential to surgery. In biomedical settings, they are widely used in dental applications, prosthetic limbs, hard and soft tissue healing, and bone fracture repair [1]. Traditional surgical materials have limitations, including decreased mechanical stability and difficulty adhering to tissue surfaces. These issues have been addressed by the use of polymers [2]. Polymers are used as suture materials in surgical applications; a variety of choices are available, such as tissue glues made of bioadhesive polymers and absorbable and nonabsorbable polymers [3]. Polymers are essential for surgical interventions because of their special physical, chemical, and biological characteristics, which also help to enhance regenerative medicine, medical supplies, diagnostics, and therapeutic therapies [4]. All things considered, the use of polymers in surgery is still growing, propelled by constant research and development initiatives aimed at improving surgical results and patient care. Recent developments have produced polymers with characteristics including haemostatic capacity, antibacterial activity, and drug delivery systems, which increase their efficacy in the healing of wounds (Chen et al., 2024 [5]). Research in this area is necessary because, despite the emphasis on using novel polymer architectures to improve surgical outcomes, difficulties persist in striking a balance between mechanical characteristics and biocompatibility and biodegradability ("Applications of Polymers in Surgery", 2022) [6].

1.2 Types of Polymers

The structure of the polymers used in surgical applications is essential in order to maximise healing, reduce the danger of infection, and ensure proper wound closure. To address these needs, a variety of synthetic and natural polymer types have been produced. Natural Polymers are obtained from natural environment and are not hazardous to the environment i.e., natural polymers are biocompatible and biodegradable materials with silk and gut are conventional choices ("Polymers for surgical sutures", 2023). [7]. However, synthetic Polymers are comprised of non-absorbable materials like nylon and polypropylene and absorbable ones like poly(glycolic acid) and poly(lactic acid) ("Polymers for surgical sutures", 2023)(Suzuki & Ikada, 2015) [8]. Biopolymers have gained attention for surgical applications due to their characteristics and can be modified to improve functionality in surgical applications [9]. Chitosan is a promising material for surgical applications and Chitosan-based materials have various surgical uses [10]. According to the research, polymers are employed in medical equipment such as implants and play a significant part in surgery because of their advantages in terms of weight, cost, and performance [11]. Synthetic polymers are widely used in medical devices for benefit whereas Advanced polymers like resin, copolymers, thermoplastics used in surgery [11]. Liquid copolymers serve as biodegradable surgical sealants, enhancing wound closure by controlling mechanical properties and biocompatibility. They offer a promising solution for surgical procedures requiring tissue sealants [12]. Biodegradable surgical sealant is used for wound closure. Neat prepolymers PEG4-PLGA-NHS and PEG4-NH2 yield elastic sealant. Polymeric biomaterials commonly used for surgery include hydrogels and 3D scaffolds. Hydrogels are formed through chemical cross-linking to create permanent junction-type networks, while physical cross-linking leads to transient junction-type networks. 3D scaffolds are designed with properties like suitable surface science, porosity, and interconnectivity systems to support cell communication, migration, and multiplication during tissue recovery processes

Materials Research Foundations 172 (2025) 187-198 https://doi.org/10.21741/9781644903537

2. Surgery

Due to specific physical, chemical, and biomechanical qualities in biomedical applications, polymers are employed in surgery for less intrusive procedures, quicker suturing, and bone fracture repair [13]. In surgery, polymer nanoparticles are used as hemostats, tissue adhesives, wound dressings, adhesion barriers, and sealants to improve patient outcomes and get around material constraints [14]. Polymers are used in surgery to make sutures and tissue glues. They are superior to previous techniques because they promote healing, cause less tissue damage, and can be used in less intrusive procedures [15]. Physical, mechanical, and chemical characteristics of the polymers make them a suitable material in surgery for medical implants and devices, improving patient outcomes and surgical processes [16]. Polymers due to their adaptability and biocomptibility are considered as valuable materials in surgical applications and are widely utilised in bone repair, orthopaedic surgery, and tissue engineering [17].

2.1 Field of surgery

There are various surgical departments e.g. cardiovascular surgery, orthopaedic surgery, tissue engineering, polymer therapeutics etc. Polymers are crucial in cardiovascular surgery, offering diverse applications like filling aneurysms and blocking vessels. They provide essential properties for successful surgical interventions and patient recovery [18]. Vascular prostheses based on PET widely used in practice. Filling aneurysms, emboli for vessel blockage. Polymer therapeutics in surgery offer targeted delivery, smart release, and safety benefits. They enhance biophysical properties and have potential in current surgical practice, attracting clinician-scientists for innovation [19]. Polymers are extensively used in bone repair, orthopaedic surgery, and tissue engineering due to their versatility and biocompatibility, making them valuable materials in surgical applications [20].

2.2 Polymers used in surgery

Polymer nanomaterials are utilized as adjuvant surgical tools, including wound dressings, tissue adhesives, sealants, hemostats, and adhesion barriers, enhancing outcomes in surgical interventions [21]. Polymer options for surgical sutures include absorbable (e.g., poly(glycolic acid), poly(lactide-co-glycolic acid)) and nonabsorbable (e.g., polytetrafluoroethylene, polypropylene) materials, along with emerging bioadhesive polymers for tissue glues [22]. Polymers are utilized in neurosurgery for tissue engineering, drug delivery, and scaffolds due to their ability to mimic the extracellular matrix, aiding in neuronal repair and regenerative medicine advancements [23]. Chitosan is a prominent polymer used in surgical materials due to its bacteriostatic, hemostatic, and tissue regeneration properties, making it valuable for various surgical applications [24]. The chemical structure of Chitosan is shown in Figure 1. Chitosan, a suitable biopolymer can be used for manufacturing wide range of products including scaffolds, sponges, hydrogels, meshes, membranes, sutures, fibers, and nanoparticles [25]. Due to the hemostatic properties of Chitosan, it is used in oral surgery [26]. Polymers like poly(methyl methacrylate), polycaprolactone, and polyacrylic acid are used as carriers of gentamicin in traumatology and orthopaedic surgery for local antibiotic delivery [27]. Polyolefins such as polyethylene and polypropylene can be used for surgical applications. However, polyethylene can recrystallize or can undergo oxidation, making it brittle, therefore polypropylene can be better used for surgical sutures [28].

Polymer materials used in surgery include gloves, sutures, containers, and implants. Additive techniques are employed to create anatomical models, templates, and implants for precise and efficient procedures. [29]. Polymeric materials are utilized for osteofixation surgery in maxillofacial bone surgery as an alternative to titanium, offering comparable physical properties with the aim of avoiding secondary removal surgeries [30]. Polymeric materials are essential for pharyngeal reconstruction in head and neck surgery, requiring stability, tissue regeneration, and minimal systemic inflammation for successful implantation in challenging anatomical sites [31]. Polymer materials like polycaprolactone, polyvinyl alcohol, and Carboxy methyl cellulose are commonly used in surgery to prevent postoperative adhesion, enhancing patient outcomes and reducing complications [32]. Bioabsorbable polymers are utilized in surgery for sealants, anti-adhesion barriers, fixation devices, and sutures, contributing to tissue healing and surgical applications. Polymers like hydrophobic, hydrophilic, acrylic, and silicone are used in intraocular lenses for cataract surgery to enhance patient eyesight and minimize surgical complications, meeting increasing demand for premium lenses [33]. Smart polymers are utilized in minimally invasive surgery, diagnosis, and other medical applications. They respond to environmental stimuli, enabling advancements in biosensors, actuators, and microfluidic systems [34]. The surgical polymer in the research paper includes antibacterial additives like silver, gold, copper, bronze, or ground bamboo fibers for tear-resistant properties in surgically implantable mesh materials [35]. Polymers commonly used in surgery include nylon, polypropylene, polyesters like PET, and polytetrafluoroethylene (PTFE). These materials are utilized in soft tissue surgery for various applications [36]. Albumin polymers, specifically Bioglue®, were utilized during breast cancer surgery to improve postoperative seroma outcomes [37]. The chemical structure of albumin is hown in Figure 2. Polyether-polyamide plastics, specifically Pebax, are utilized for surgical applications such as intraocular lens inserters due to their clean and labile oligomer-free properties [38]. Aqueous polyurethane compositions are suitable for surgeon's gloves, offering toughness, pliability, and some solvent resistance, replacing latex gloves known for allergic reactions [39]. The chemical structure of polyurethrane is shown in Figure 3. Poly(glycerol sebacate)-co-lactic acid (PGS-co-LA) is a synthetic polymer used as a surgical sealant, offering high adhesive strength, improved cytocompatibility, and the ability to solidify at body temperature [40].

S.No.	Polymer Materials	Surgery applications
1.	Polymer Nanomaterials	adjuvant surgical tools, including wound dressings, tissue adhesives, sealants, hemostats, and adhesion barriers
2.	poly(glycolic acid), poly(lactide-co-glycolic acid	Surgical sutures
3.	polytetrafluoroethylene	Tissue glues
4.	poly(methyl methacrylate), polycaprolactone, and polyacrylic acid	carriers of gentamicin in traumatology and orthopedic surgery for local antibiotic delivery
5.	polycaprolactone, polyvinyl alcohol, and Carboxy methyl cellulose	postoperative adhesion, enhancing patient outcomes and reducing complications
6.	Bioabsorbable polymers	sealants, anti-adhesion barriers, fixation devices, and sutures, contributing to tissue healing and surgical applications.
7.	hydrophobic, hydrophilic, acrylic polymers and silicone	intraocular lenses for cataract surgery
8.	nylon, polypropylene, polyethylene terephthalate and polytetrafluoroethylene (PTFE).	Soft tissue surgery
9.	Albumin polymers, specifically Bioglue®,	Breast cancer surgery
10.	Polyether-polyamide plastics, specifically Pebax	intraocular lens inserters due to their clean and labile oligomer-free properties
11.	Aqueous polyurethane	surgeon's gloves
12	Poly(glycerol sebacate)-co-lactic acid (PGS-co-LA)	Surgical sealant
13.	Chitosan	scaffolds, sponges, hydrogels, meshes, membranes, sutures, fibers, and nanoparticles, Oral surgery [52-53]
14	Polypropylene	Suture materials and meshes [54]

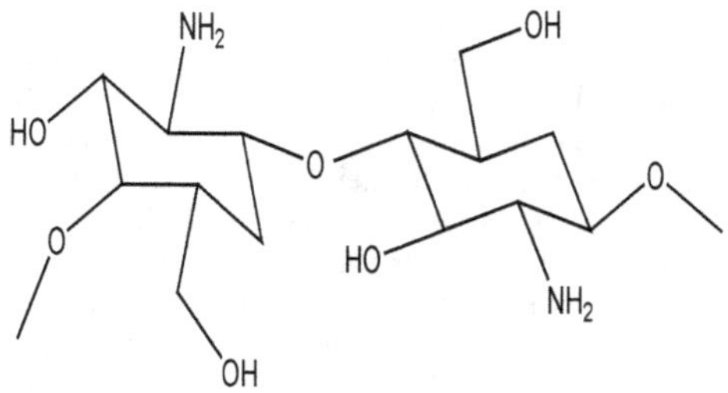

Figure 1. chemical structure of Chitosan

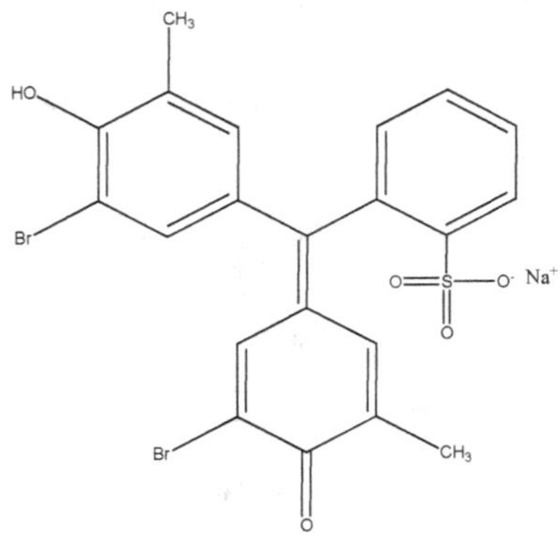

Figure 2. The chemical structure of albumin

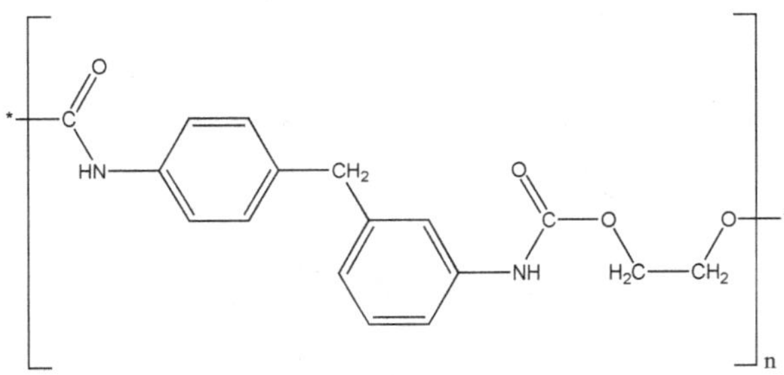

Figure3. The chemical structure of polyurethrane.

3. Shape memory Polymers

Shape-memory polymers (SMPs) are stimuli-responsive materials that, in response to the occurrence of a suitable input, will return to their original form. In recent decades, SMPs have emerged as a class of material that is highly valued. "Wise" materials (SMPs) are gaining a lot of attention in the advanced material areas due to their many, flexible applications in a variety of fields. In addition to others, they require a significant portion of the medical industry. SMPs are flexible enough to facilitate keyhole surgery, which is degradable, delivers steady strength, and applies stabilising forces and rinses therapeutic substances. Biocompatibility, tolerability, disinfection, and non-toxicity are prioritised in basic research on suggested SMP systems for medicine [41]. Shape-memory polymers (SMPs) are utilized in surgery for medical devices, implants, and controlled drug delivery systems due to their tunable properties, biocompatibility, and on-demand shape changes upon stimulation [42]. Chemically cross-linked glassy thermosets, chemically cross-linked semicrystalline, physically cross-linked glassy copolymers, and physically cross-linked block copolymers are the four varieties of SMPs that can be distinguished

Materials Research Forum LLC
https://doi.org/10.21741/9781644903537

by their permanent form, elasticity origin, and temporary shape fixing mechanism. SMPs are widely applicable in the fields of general medicine, drug delivery, regenerative medicine, dentistry, neuromedicine, cancer therapy, orthopedics, and corrosion protection, as demonstrated by the comprehensive summary of current discoveries on SMPs for biomedical applications [43]. Shape memory polymers (SMPs) are a class of programmable materials that, in response to an external input, can change their shape in a predetermined manner. SMPs are perfect for minimally invasive operations since they can expand to cover defects because they can alter shape in response to temperature (Nardo et al., 2006) [44].

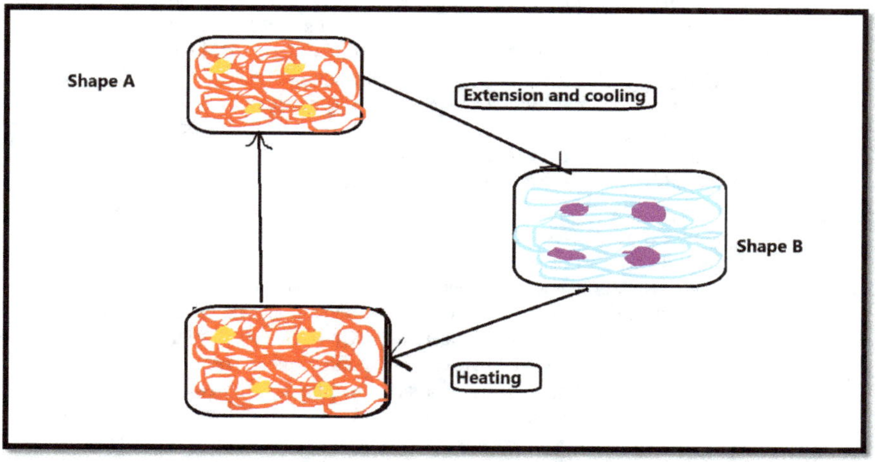

Figure 4. Molecular mechanism of thermally induced effect on Shape Memory polymers [45]

When a bone defect grows to a critical size which can be caused by severe trauma, infection, or bone tumour excision surgery. The bone's natural ability to heal itself is insufficient to fully repair the defect, necessitating bone grafting. Bone tissue engineering now uses synthetic or natural bone grafts, such as bioceramics, but these therapies are expensive and limited by the scarcity of bone. Thus, shape memory polymers (SMPs) have drawn interest in bone tissue engineering recently because they combine biocompatibility, biodegradability, mechanical characteristics, shape tunability, ease of access, and minimally invasive implantation [46]. Polyurethrane and SMPU (Shape Memory Polyrurethrane) have different properties as ordinary polyurethrane does not have restore plastic/inelastic deformation upon heating. Using a flexible crosslinker, a series of shape-memory polyurethanes are created by post crosslinking in hard-segment domains, expanding Cross-linked Polyurethranes (CPUs) with an elastic crosslinking structure as possible implant biomaterials.

Materials Research Foundations 172 (2025) 187-198 https://doi.org/10.21741/9781644903537

S.No	Base polymer	Surgery/Purpose	Conclusions	References
1	SMPU (shape memory polyurethane) used for thrombectomy device prototype. Components include PTFE liner, braided reinforcement, Pebax jacket, copper wire.	A mechanical thrombectomy device for acute ischemic stroke	SMPU tubing recovers from straight to coiled shape when heated. Prototype thrombectomy device successfully demonstrates shape memory polymer application.	[47]
2.	Ultra-stretchable, strong, tough, and highly recoverable shape-memory polymeric string Fabricated solely by tetra-branched poly(ε-caprolactone) (PCL)	Contraction of blood vessels in fetal surgery Resolving difficulties in restricting blood flow in minimally invasive surgeries	Novel ultra-stretchable, strong, tough, and highly recoverable shape-memory polymeric string. Customizable shape-memory contraction force for blood vessel contraction in surger	[48]
3.	Bisphenol A epoxy with E-glass fibers TiO2 nanoparticles for bioinspiration qualities	Non-invasive orthopaedic fractured bone immobilisation Optimising design and geometrical parameters of smart plaster Save	MP smart plaster showed promising results for fractured bone immobilisation. Undercast temperature and pressure were within safe limits for human use.	[49]
4.	Polymer composite based on PLA with shape memory effect. Contains 10% poly (e-caprolactone) to enhance properties.	Temporary self-fitting scaffolds for reconstructive surgery Adaptive medical structures with shape memory effect	Reduced activation initiation temperature for shape memory effect. Decreased activation energy by 85 kJ/mol for safe use.	[50]
5.	Hydrogels synthesized from acrylic acid and N-hydroxyethyl acrylamide. Porous SMO produced from lactose functionalized polyurea-urethranes	Biomedical devices for minimally invasive surgeries. Porous SMPs in Vascular grafts for fluid transfer and tissue growth	Hydrogel coating enhances fluid uptake without hindering shape memory properties. Porous SMPs with hydrogels can be used in medical devices.	[51]
6.	SMPU made of PCL-2OH, IPDI, and GL Molar ratio of 2:1.5:0.33 for best performance	Medical devices Smart textiles	SMPU triggers shape memory at body temperature with 450% recoverable strain.	[52]

Materials Research Foundations 172 (2025) 187-198 https://doi.org/10.21741/9781644903537

Conclusions

In this chapter, the advantages of using polymers over other classes of materials for surgery and surgical applications are seen. It is also observed that polymers have different properties and nearly every kind of polymer has their application in surgery or surgical devices or products. Biopolymers like chitosan have been used now a days for various surgical products such as scaffolds, sutures, meshes etc. Synthetic polymers also have various applications in the field of surgery. There is also one class of polymers called shape memory polymers, that can change their shape in response to the external stimuli. Shape-memory polymers (SMPs) have been used for medical devices such as implants, drug delivery, dentistry, neuromedicnes and Cancer therapy etc. There are various kinds of polymers researched for the applications in the field of surgery have discussed in the above chapter. It is also observed in this chapter that polymers are replacing other class of materials for surgical applications.

References

[1] Heitor, Luiz, Ornaghi., Francisco, Maciel, Monticeli., Lucas, Dall, Agnol. (2023). A Review on Polymers for Biomedical Applications on Hard and Soft Tissues and Prosthetic Limbs. Polymers. https://doi.org/10.3390/polym15194034.

[2] Metecan, Erdi., Anthony, D., Sandler., Peter, Kofinas. (2023). Polymer nanomaterials for use as adjuvant surgical tools.. Wiley Interdisciplinary Reviews-nanomedicine and Nanobiotechnology. https://doi.org/10.1002/wnan.1889.

[3] (2023). Polymers for surgical sutures. https://doi.org/10.1016/b978-0-12-819750-9.00004-8.

[4] Robert, D., Benson., Wei, He. (2024). Polymeric Biomaterials. https://doi.org/10.1016/b978-0-323-88667-3.00022-9.

[5] Yunhao, Chen., Muyuan, Chai., Chen, Xuan., Jiayi, Lin., Huishang, Yang., Chunhui, Li., Manshan, Xie., Serge, Ostrovidov., Xuetao, Shi., Chuanbin, Mao. (2024). 3. Tuning the properties of surgical polymeric materials for improved soft-tissue wound closure and healing. Progress in Materials Science. https://doi.org/10.1016/j.pmatsci.2024.101249

[6] (2022). 2. Applications of Polymers in Surgery. https://doi.org/10.21741/9781644901892

[7] (2023). 1. Polymers for surgical sutures. https://doi.org/10.1016/b978-0-12-819750-9.00004-8

[8] Shuko, Suzuki., Yoshito, Ikada. (2015). 5. Polymers for Surgery. https://doi.org/10.1007/978-3-319-12478-0_8

[9] Tudor, Bibire., Onur, Yilmaz., Cristina, Mihaela, Ghiciuc., Nela, Bibire., Radu, Danila. (2022). Biopolymers for Surgical Applications. Coatings. https://doi.org/10.3390/coatings12020211

[10] Fernando, Notario-Pérez., Araceli, Martín-Illana., Raúl, Cazorla-Luna., Roberto, Ruiz-Caro., María-Dolores, Veiga. (2022). Applications of Chitosan in Surgical and Post-Surgical Materials. Marine Drugs. https://doi.org/10.3390/md20060396

[11] Marios, Koutsoukos. (2022). The Role of Advanced Polymers in Surgery & Medical Devices. https://doi.org/10.21741/9781644901892-2

[12] Neta, Shimony., Alona, Shagan., Bat-hen, Eylon., Abraham, Nyska., Adi, Gross., Boaz, Mizrahi. (2021). Liquid Copolymers as Biodegradable Surgical Sealant.. Advanced Healthcare Materials. https://doi.org/10.1002/ADHM.202100803

[13] Maitz, M. F. (2015). Applications of synthetic polymers in clinical medicine. *Biosurface and Biotribology*, *1*(3), 161-176.

[14] Annabi, N., Tamayol, A., Shin, S. R., Ghaemmaghami, A. M., Peppas, N. A., & Khademhosseini, A. (2014). Surgical materials: current challenges and nano-enabled solutions. *Nano today*, *9*(5), 574-589.

[15] Jain, R., & Wairkar, S. (2019). Recent developments and clinical applications of surgical glues: An overview. *International journal of biological macromolecules*, *137*, 95-106.

[16] Teo, A. J., Mishra, A., Park, I., Kim, Y. J., Park, W. T., & Yoon, Y. J. (2016). Polymeric biomaterials for medical implants and devices. *ACS Biomaterials Science & Engineering*, *2*(4), 454-472.

[17] Guo, L., Liang, Z., Yang, L., Du, W., Yu, T., Tang, H., ... & Qiu, H. (2021). The role of natural polymers in bone tissue engineering. *Journal of Controlled Release*, *338*, 571-582.

[18] Rami, Matarneh., S., Sotnik., Vyacheslav, Lyashenko. (2018). Polymers in cardiovascular surgery. Asian Journal of Pharmaceutical and Clinical Research. https://doi.org/10.22159/AJPCR.2018.V11I5.24576

[19] Ernest, A., Azzopardi., R., Steven, Conlan., Iain, S., Whitaker. (2016). Polymer therapeutics in surgery: the next frontier.. https://doi.org/10.1002/JIN2.6

[20] Sapana, Jadoun., Sampath, Chinnam., Amir, Qureshi. (2023). Polymers in bone and orthopedic surgery. https://doi.org/10.1016/b978-0-12-823797-7.00019-8

[21] Metecan, Erdi., Anthony, D., Sandler., Peter, Kofinas. (2023). Polymer nanomaterials for use as adjuvant surgical tools.. Wiley Interdisciplinary Reviews-nanomedicine and Nanobiotechnology. https://doi.org/10.1002/wnan.1889

[22] (2023). Polymers for surgical sutures. https://doi.org/10.1016/b978-0-12-819750-9.00004-8

[23] İlkay, CİVELEK. (2022). Applications of Polymers in Neurosurgery. https://doi.org/10.21741/9781644901892-4

[24] Fernando, Notario-Pérez., Araceli, Martín-Illana., Raúl, Cazorla-Luna., Roberto, Ruiz-Caro., María-Dolores, Veiga. (2022). Applications of Chitosan in Surgical and Post-Surgical Materials. Marine Drugs. https://doi.org/10.3390/md20060396

[25] Notario-Pérez, F., Martín-Illana, A., Cazorla-Luna, R., Ruiz-Caro, R., & Veiga, M. D. (2022). Applications of chitosan in surgical and post-surgical materials. *Marine drugs*, *20*(6), 396.

[26] Franco, R., Gianfreda, F., Miranda, M., Barlattani, A., & Bollero, P. (2020). The hemostatic properties of chitosan in oral surgery. *Biomedical and Biotechnology Research Journal (BBRJ)*, *4*(3), 186-188.

[27] Shuko, Suzuki., Yoshito, Ikada. (2015). Polymers for Surgery. https://doi.org/10.1007/978-3-319-12478-0_8

[28] Maitz, M. F. (2015). Applications of synthetic polymers in clinical medicine. *Biosurface and Biotribology*, *1*(3), 161-176.

[29] Fernando, Notario-Pérez., Araceli, Martín-Illana., Raúl, Cazorla-Luna., Roberto, Ruiz-Caro., María-Dolores, Veiga. (2022). Applications of Chitosan in Surgical and Post-Surgical Materials. Marine Drugs. https://doi.org/10.3390/md20060396

Applications of Polymers in Surgery II | Materials Research Forum LLC
Materials Research Foundations 172 (2025) 199-230 | https://doi.org/10.21741/9781644903353-8

Chapter 8

Polymers in Tissue Engineering

David Romero-Fierro[1,*], Y. Aylin Esquivel-Lozano[1], Gabriela Casillas-Calzadilla[2], Alexander E. Escobar-Pullas[3], and Emilio Bucio[1,*]

[1]Departamento de Química de Radiaciones y Radioquímica, Instituto de Ciencias Nucleares, Universidad Nacional Autónoma de México, Circuito Exterior, Ciudad Universitaria, C.P. 04510 Ciudad de México, México

[2]Facultad de Química. Universidad Nacional Autónoma de México, Apartado Postal 70-360. Ciudad Universitaria, Coyoacán. C.P 04510, Ciudad De México, México

[3]Departamento de Química, Escuela de Ciencias Química e Ingeniería, Universidad de Investigación de Tecnología Experimental Yachay Tech, Hacienda San José s/n y Proyecto Yachay, Urcuquí, 100650, Imbabura, Ecuador

* david.arf2096@gmail.com (D.Romero-Fierro); ebucio@nucleares.unam.mx (E. Bucio)

Abstract

Polymers display important roles in tissue engineering acting as precursors to synthesize materials that serve as scaffolds for drug delivery and cell growth, or in the creation of implants for regeneration or replacement of damaged organs and tissues. This importance relies on the remarkable properties of these macromolecules and their versatility to obtain scaffolds with desired characteristics and specific applications. This chapter describes the most featured aspects of polymers with application in tissue engineering including desired properties, types of polymers, synthetic methods and future challenges that scientific community must face in this interdisciplinary field.

Keywords

Polymers, Scaffolds, Tissue Engineering, Regeneration, Tissue Replacement

Contents

Materials Research Foundations 172 (2025) 199-230
https://doi.org/10.21741/9781644903353-8

1. Introduction

Every year, millions of patients around the world suffer a deterioration of their organs and tissues, being in the necessity to be replaced. The first human organ successfully transplanted was a kidney in 1954 by Dr. Joseph Murray, who transplanted this organ between two twins. Next, Dr. Murray performed a transplant between non-genetically identical patients, overcoming the immunologic barrier. Latterly, in 1990, he receives Nobel prize by this achievement and his studies about immunosuppression. However, nowadays it is more difficult to meet potential ideal donors due to the increased life expectancy, damage and diseases that people present in their organs and incompatibility of organ to be donated to the patient. Even if the organ is not rejected, the necessary use of immunosuppressive drugs gives several issues in these patients. All these problems led to the scientific community exploring new alternatives for organ transplantation. In 1960, researchers began to combine material science advances with cell biology fundamentals, arising a new scientific field currently known as *tissue engineering* [1,2].

Tissue engineering (TE) is a multidisciplinary field that integrates concepts from basic science, technology and engineering and it refers to the practice to combine scaffolds, cells and molecules with biological activity to create functional tissues of regenerating damaged tissues. The primary goal of TE is to collect ideas and theories of afore mentioned fields to restore, preserve or enhance damaged tissues. Advanced TE involves the development of biomaterials as scaffolds to act as implant and to perform other functions. In this sense, TE deals with the development of scaffolds acting as replacements of natural extracellular matrix (ECM) with the capacity of inducing the correct tissue pattern formation. Natural and synthetic polymers are the idoneous choice of biomaterial for these applications due to their countless properties including availability, versatility to be applied in different contexts, their tunable properties depending on their application, their biodegradability and biocompatibility. In fact, when a polymeric scaffold is obtained from different synthetic routes, they exhibit relevant properties such as good cell adhesiveness, promote cell-biomaterial interactions, display excellent structural and mechanical properties, and possess a microstructure appropriate for TE [1,3].

Typically, in TE, scaffolds are designed using a mixture of cells from the patient and polymeric scaffolds, with the objective of mimic damaged organs or tissues. For that, cells are isolated from the patient and incubated *in vitro*. Then, isolated cells are incorporated into the porous polymer scaffolds and are implanted back to the patient by different surgical or less invasive methods (Fig. 1). Polymer scaffold must agglutinate cells, control the growth of tissue structure, regulate the function of cells, and allow the diffusion of molecules of biological interest. The success of these expected functions depends on the regulation of interactions between cells and scaffolds which depends on the physical properties of polymer scaffolds and the release of soluble species retained in the scaffold. These parameters are difficult to completely control due to the different applications that polymeric scaffolds must be exposed, for this reason there is no polymer to be considered ideal for all types of approaches and tissues and the use of a single polymer scaffold can provide all the requirements for cellular interactions [1,4]. In the latest 20 years, TE scientists have proposed to design and engineer each tissue of the human body, using polymers as scaffolds. This chapter presents the principal fundamentals of polymers used as scaffolds in TE, such as desired properties of polymers for that applications, types of polymers, basic techniques to obtain TE scaffolds and current challenges that scientific community must face in the incoming years.

2. Desired properties of polymers for tissue engineering

The properties of polymers with applications in TE are crucial, thanks to the studies carried out in recent years, it is possible to have control over these factors such as rigidity, mechanical resistance, adaptability, biocompatibility, swelling, low protein adsorption due to low surface tension and biodegradability; which give it a structural similarity with several native tissues, making them suitable for repairing damage in these tissues or replace them.

2.1 Mechanical properties

Mechanical properties, at macroscopic and microscopic scales, play a fundamental role, since they determine the way in which the biomaterial will interact with the biofluid, and therefore, they perform a relevant role in the regulation of cellular behavior. These properties are affected by the

Materials Research Foundations 172 (2025) 199-230
https://doi.org/10.21741/9781644903353-8

scaffold morphology, the degree of swelling, concentration of cross-linking agent, initial concentration of the monomer, and polymerization temperature; these determine parameters such as viscosity, stability, biodegradability, among others [5,6].

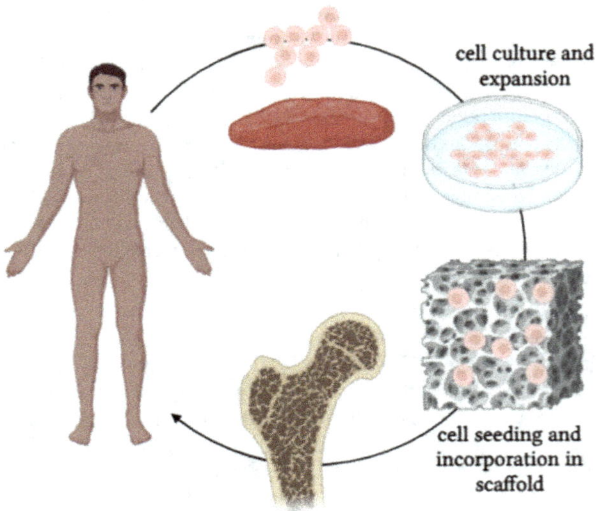

Figure 1. *Basic principle of tissue engineering*

The dynamic bonds present in hydrogels help to effectively dissipate energy, which can be tailored to confer increased stiffness and stretchability that influence cell adhesion. These properties can be tuned by a change in the degree of cross-linking, the change in molecular structure, and the tension of the polymer chains [7].

In recent years, highly rigid natural tissue-like materials have been developed, with elastic moduli with dimensions ranging from MPa to GPa. The elastic modulus (E) of a polymeric scaffold generally comprises the strain in the range of 0-10%, with outstanding cases of maximum values of 15%. For the synthesis of such materials, energy needs to be dissipated along the network. This is achieved by dissipation through the presence of the named sacrificial bonds which convert the kinetic energy concentrated from mechanical impact into chemical energy breaking the sacrificial bonds. There are two classes in which this occurs [8]:

Double networks: In this case, there is a first weakly cross-linked network, while a second, more fragile network is synthesized. When the material is elongated, a part of the bonds in the last one network are broken, this action allows a greater dissipation of the energy from the mechanical impact.

Materials Research Foundations 172 (2025) 199-230
https://doi.org/10.21741/9781644903353-8

Non-covalent sacrificial bonds: These sacrificial bonds can be broken and reformed in both ways and include some chemical interactions including metal-ligand interactions, hydrogen bond interactions, ionic interactions, hydrophobic interactions, polymer-nanomaterial adsorption, supramolecular self-assembly, or host-guest complex formation. Non-covalent bonds confer self-healing and self-recovery capabilities [8].

Due to the aging or manufacturing process, materials may hold or develop cracks in their structure. For the correct performance of scaffold, these cracks must be prevented from propagating during their application, for example, in applications where they are subjected to a multitude of high-impact load cycles [6,8].

In addition, due to the degree of cross-linking of these materials, it is important to measure the fracture stress at break, which is an index of the capability of a material to concentrate or retain energy before its breakage. With this, the resistance of material to the propagation of cracks can be known and whether it can withstand loading conditions [6,8].

2.2 Wetabbility

Polymeric scaffolds in the form have a high absorption capacity, derived from the hydrophobic and hydrophilic groups such as $-NH_2$, -OH, -COOH, $-OSO_3H$ within their structure [9]. The characteristic that determines hydrophilicity or hydrophobicity is wettability, which results from the balance of adhesive forces (responsible for the dispersion of droplets on the surface) and cohesive forces (responsible for preventing droplets from interacting with the surface), which binds or adheres to two materials. These forces are responsible for the dispersion of droplets on the surface, and the cohesion forces of the liquid secure the liquid ball to prevent it from interacting with the surface [6].

2.3 Degradability

The degradation of polymeric scaffolds applied in TE can be carried out in a controlled manner, using factors such as light, pH, temperature, to name a few. In TE, it is important that the polymeric scaffold is removed when it serves the purpose of providing a tissue scaffold in the accumulation, proliferation and dissipation of cells and tissues, thus degradation is a critical parameter [6].

Therefore, cross-linking agents must be biodegradable by chemical processes including hydrolysis and enzymatic degradation so that the polymeric scaffold could act as replacement the extracellular matrix as tissue regeneration progresses. For example, degradable polymeric scaffold applied in bone regeneration provide space for the proliferation of new cells and allow bone to grow [10].

2.4 Swelling properties

Thanks to the hydrophilic fraction present in main polymer chains, they can absorb and retain a substantial amount of water, which makes the polymeric scaffold swell and form a three-dimensional structure; this gives it the ability to absorb body fluids in a biological environment, depending on the functional group, the water state and the density of the cross-linking network in the scaffold [9]. This property is crucial, since it influences tissue repair, regeneration, and function through the gelation mechanism.

Materials Research Foundations 172 (2025) 199-230
https://doi.org/10.21741/9781644903353-8

The equilibrium swelling of polymeric scaffold in the form of hydrogel can be calculated as shown below (Eq. 1):

$$\text{Swelling ratio} = \frac{W_t - W_d}{W_d} \quad (1)$$

where W_t and W_d belong to the weight of scaffold in swollen state and the weight of dried scaffold after freeze-drying process, respectively. In this sense, Xiobin, et al. synthesized an injectable hydrogel to be used as a scaffold in TE by copolymerizing hyaluronic acid and polyglutamic acid, to control protein release. As the concentration of the reagents increases, the degree of swelling is reduced, if there is an increase in cross-linking degree in the structure [11].

2.5 Degree of cross-linking

The degree of cross-linking manages the water uptake and allows us to maintain the network structure in a swollen state [9]. Due to the presence of modifiable functional groups, hydrogels to be used as polymeric scaffolds can be synthesized by two ways: chemical and physical crosslinking, or both, simultaneously [9]. Both methods create a stable three-dimensional structure of the hydrogel, which prevents its dissolution, and allows therapeutic agents (drugs and biomolecules) to be trapped and successfully released [5]. The cross-linking method influences parameters such as mass and molecular weight, improving properties such as mechanical strength, heat resistance, and solvent resistance [5,6].

The functional groups inherent to biopolymers have interactions between the polymer and/or cross-linker chains, which can be chemical or physical. Physical interactions can occur in the presence of the -OH group that enables the formation of intermolecular or intramolecular hydrogen bonds, in biopolymers including starch, cellulose and chitosan. While chemical interactions are more relevant between functional groups containing -COOH, $-CONH_2$, $-OSO_3$ and $-NH_2$, of polymers and crosslinkers [6,8].

It is worth mentioning that physical and chemical cross-linking agents allow hydrogels to be prepared at a precise location in the body, the polymer solution is injected into the body in liquid form and converted into a solid scaffold using cross-linking agents. Materials that are formed *in situ* also have the particularity of being capable of correcting defects in their environment. Thermogels, that are sensitive to temperature, are very useful for generating structures *in situ*. These materials have the ability to change their state passing from a solution to a hydrogel when their critical temperature is oscillating between room and physiological temperature [6,12].

3. Degradable polymers for tissue engineering

The biodegradability of polymers is a mandatory property for the correct performance of the polymeric scaffold, since when eliminated from the body it gives space to regenerative tissues. The degradability of polymers consists of the decomposition of them and, consequently, of polymeric scaffolds into harmless final products by the living organism. In TE, it has been possible to control the synthesis conditions to make them biodegradable, where the design can be based on natural or synthetic polymers.

Synthetic polymers have a greater water absorption capacity and excellent mechanical properties, however, they have low biodegradability and potential toxicity, so natural polymers are a better option, since in addition to their excellent biodegradability, they are found in abundance, are biocompatible and have an innate interaction with the cell surface.

Biodegradability is measured by a change in weight of the material over a specific period in a biological environment. It is achieved by the modification in the composition of the main chain, cross-linking, polymer nature, density, and morphology. In summary, biodegradability depends on the synthetic method, intrinsic properties of reagents, and environmental factors (such as pH and temperature) [6,9,13]

3.1 Biodegradable polymers

By fulfilling their main objective, they can gradually degrade and make room for regenerative tissues. However, due to their low mechanical strength, it is difficult for them to withstand cyclic forces within tissues, such as tendons, bone, cartilage, and muscle, during tissue regeneration. Thus, in recent years, the properties of polymeric scaffolds focused on chitosan, hyaluronic acid, alginate and gelatin have been enhanced and focused on the reparation and regeneration of tendons, bone, cartilage, and muscle [5,14]. Table 1 summarizes the polymers that have been used for TE [9].

Alginate is the second most abundant natural biopolymer. Something peculiar is its ability to physically cross-link at room temperature, through divalent cations such as Ca^{2+}; it can be obtained from oysters, insects, fungi and marine waste. Alginate scaffolds help to the multiplication and diffusion of encapsulated chondrocytes, for these 21-28 days of inoculation are needed, time during which collagen is formed together with the cartilage gene. Likewise, this polymer can develop primary bone cells, for bone regeneration [5].

Biodegradable scaffolds are commonly brittle (0.1–10.0 J/m^2) and weak (typically below 5 kPa) [4]. One way to improve the mechanical properties is the grafting with monomers or synthetic polymers. For example, the synthetic polymer polyvinyl alcohol (PVA) is used as a blending agent to improve the mechanical strength and flexibility of natural polymers [9].

In the design of biodegradable scaffolds, a balance must be struck between degradation time and regeneration or healing time; since slow degradation can lead to inefficient healing, while rapid degradation causes an early loss of the scaffold structure and functions [13]. Thus, it is desirable for degradable scaffolds to maintain their functionalities, particularly mechanical properties, during their useful life [15].

The optimal degradation time is found based on two crucial parameters: the type and size of the tissue. In rats, the ECM requires, at least, 12 weeks to obtain a well-ordered muscle that mimics the normal health tissue [15].

Materials Research Foundations 172 (2025) 199-230 https://doi.org/10.21741/9781644903353-8

Table 1. Biodegradable polymers for tissue engineering [6].

Natural	Synthetics
Chitosan Cellulose Hyaluronic acid Gelatin Chondroitin sulfate Starch Sodium alginate Glucans Pectin Functional proteins Peptides	polylactic acid polyurethanes polycaprolactone polylactic-*co*-glycolic acid

3.2 Degradation mechanisms

Some polymeric scaffolds can undergo different degradation mechanisms, depending on the network topology and the degradation sites. Based on the point where degradation occurs, this can occur as follows (Fig. 2) [15]:

a) Degradation along the main chain.

b) Degradation on cleavable cross-links.

c) Degradation through degradable pendant chains.

Another classification is surface and bulk degradation (Fig. 3). These two processes, experimentally, occur at the same rate [15].

a) *Surface erosion:* During this process, degradation occurs first in the outermost layer of the material, once it has been completely removed, the inner part continues.

b) *Bulk degradation:* It occurs simultaneously throughout the material, from the surface to the core, but mainly in the core.

Degradation processes include hydrolysis and solubilization [9]. Simple hydrolysis occurs in polymeric scaffolds in the form of hydrogel cross-linked by physical interactions. In polymers that do not form hydrogels, but combine with the hydrogel, forming hydrophilic monomers or polymers, it occurs through chemical hydrolysis of the ester bonds. In the case of polysaccharides, proteins and synthetic polypeptides, it occurs by enzymatic hydrolysis [6].

Polymeric scaffolds formed by physical interactions can suffer a reversible transition from gel to sol under exposure to external environmental or induced stimuli including temperature, pH, light, magnetic and electric fields, and ionic groups. These stimuli lead to their dissolution promoting bond breaking and inducing structural rearrangements, this principle is crucial for degradation [15]

Design for degradation also relies on reversible chemical cross-links. Dissociation of these bonds occurs with external triggers. For example, cross-links consisting of epoxide, aromatic or azo functionalities are susceptible to light, while for hydrocarbon-based polymers, the common breaking mechanism is the thermo-oxidative degradation method[15]. Control over the degradation of polymeric scaffolds that are covalently cross-linked has often been reached by the

Materials Research Foundations 172 (2025) 199-230
https://doi.org/10.21741/9781644903353-8

incorporation of degradable functional groups that can undergo ester degradation under a physiological media [16].

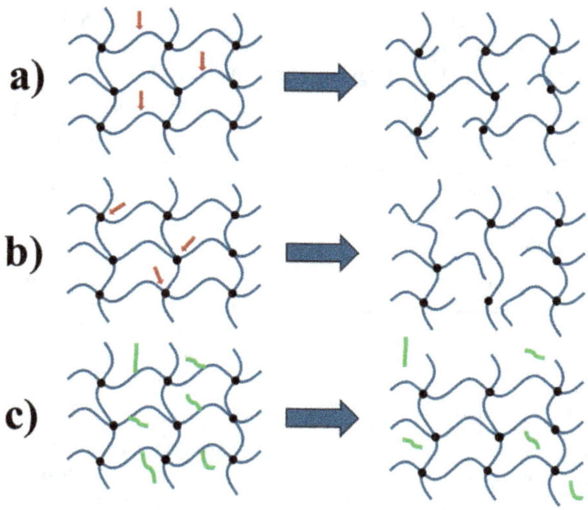

Figure 2. Schematic representation of degradation of degradable polymeric scaffolds

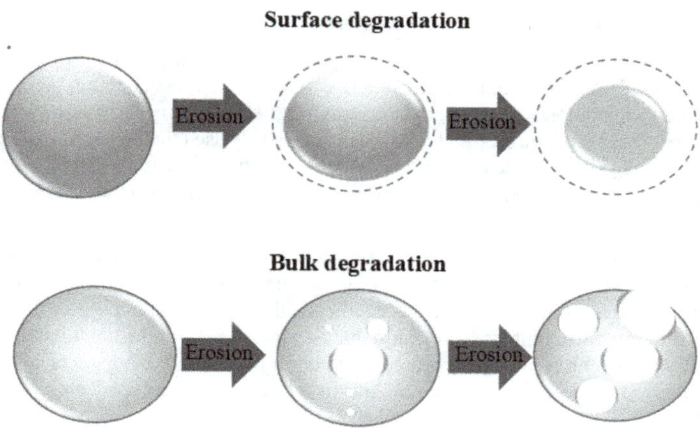

Figure 3. Schematic representation of surface and bulk degradation

3.3 Photodegradation

Bond cleavage is achieved at a certain wavelength, which is absorbed by the photolabile moieties. For example, Raman, Gwyne, et.al. [17], developed a polyacrylamide hydrogel for use as an embedded gastric balloon and an esophageal tube. It was possible to decompose it after one hour at light intensities between 5.1-11.4 $mWcm^{-2}$, which are secure for cellular activity. This irradiation was achieved by ingestion of a light-irradiating tablet [17].

Another way to cleave bonds in functional groups including disulfides and allyl sulfides, is with the use of photoinitiators such as lithium phenyl-2,4,6-trimethylbenzoyl phosphinate (LAP). Compounds containing azo groups can be easily broken by direct irradiation with UV light and with a suitable temperature and even with a mechanical stimulus [17].

3.4 Enzymatic degradation

Enzymatic degradation can occur through chemical reactions such as oxidation-reduction, hydrolysis, because enzymes are a type of proteins functioning as biocatalysts. This method can provide site-specific degradation, since the concentration of enzymes varies depending on the cell or tissue type [9].

3.5 Hydrolytic degradation

In the design of polymeric scaffolds, hydrolysis-sensitive groups are added, either along the chains or at the cross-linking junctions. These groups can be carboxylic acid anhydrides, esters, thioesters. M. Shao, Z. Shi et. al., synthesized a biodegradable polyurethane hydrogel that was applied as a wound dressing [18].

4. Non-degradable polymers for tissue engineering

Non degradable polymers are a kind of polymers with a high stability in environmental conditions and they are in almost any building, instrument or object; the most common members of these polymers are synthetic ones. Non-degradable polymers, known as permanent polymers, involve the non-biodegradable polymers which have been extensively used for TE because of their stability over the time physiological conditions, this characteristic is essential for TE in long term prosthesis, implants, scaffolds, hydrogels or drug delivery devices when their resistant structure is needed [19].

The term non-biodegradable polymer could be misunderstood like synthetic polymers, even when a considerable number of synthetic polymers are permanent some of them do not; the term non-biodegradable is orientated to describe the stability of these materials in physiological conditions and can be referred as bio-integrable or biostable polymers [20]. On the other hand, biopolymers are macromolecules from biological sources but not all of them are biodegradable which means the term biopolymer describe the sources, and in this category, biobased plastics are included [19,21,22].

Over the time the discussion about the sustainability of these materials used for different purposes has increased as a response of greenhouse effect and global warming. The main problem of those materials is their low rate of degradation through the time and the most prominent disposal treatments are landfilling and incineration while recycling has not been enough effective in most countries. In fact, there is a significant amount of research about the degradation of those materials

Materials Research Foundations 172 (2025) 199-230 https://doi.org/10.21741/9781644903353-8

by photodegradation, pyrolysis, chemical catalysis or biodegradation with microorganisms; this field is called degradation technology. Despite this approach, permanent polymers are undisputable materials used for biomedical applications [23].

In addition, other strategies has been taken to produce this nondegradable polymers with a smaller environmental impact, biobased plastics, those materials are plastics but their sources are renewable for example bio-polyethylene (Bio-PE) can be sensitized with ethanol from vegetal biomass, or polyamide 11 (PA11) is derived from castor oil; other examples of body based plastics are polypropylene (PP), polyethylene terephthalate (PET), polyamides (PAs), polythioester (PTE), polytrimethylene terephthalate (PTT), polyurethanes (PURs), among others. By 2018 biobased nonbiodegradable plastics represented 57% of biobased global market [19,20].

The new TE trends are advocating biodegradable polymers as the main material in some synthetic tissues or devices which are designed to interact with the human body in a short term, these polymers are able to be degraded or reabsorbed in physiological conditions, but when a medical device needs to be working inside the body for a long period of time those are not the best option because of their limited stability which can affect directly the function of the biomedical device or synthetic tissues. Both biodegradable and non-biodegradable polymers share a huge capacity of imitating the tissue behavior because of their versatility in the design, as polymers they can be built with different monomers, functionalize the structure or cross-link different kind of polymers biodegradable and non-biodegradable in the same structure [24,25].

Biostable polymers are the scaffold of different technologies we use now for the treatment of different health conditions where an external device is needed to meet the human body and regulate or replace a body part from a patient. This kind of polymer provides durability and chemical stability to the polymers matrix or composite matrix used to fabricate the synthetic tissue or device. Another important highlight of these nonresorbable materials is their facility to adopt different shapes after an easier processing compared with other materials like metals or ceramics [26,27].

Non-degradable polymers which are considered for TE must fulfill those three requirements: stability, biocompatibility and desirable mechanical properties; all those properties of each polymer should be included in the material design according to the tissue host and physiological functional. Biostable polymers are not susceptible to biological degradation and maintain their integrity throughout their whole lifetime of the patient, but the biological body response needs to be evaluated through the time, to prevent undesirable immune reactions [28,29].

Although new trends on research are trying to enhance the usage of natural polymers like collagen, alginate, chitosan, hyaluronic acid, among others; they maintain some disadvantages as their high potential for contamination and poor mechanical properties and for this reason, synthetic biopolymers are a great alternative, even some of them are less immunogenic than the natural ones [30,31].

The most common uses of permanent polymers in TE and medical applications are joint prosthesis, vascular grafts, backbone and neurotransmission devices, bone cement, regenerative scaffolds, pacemakers, eye lenses, mesh implants, dental prosthesis and orthodontic components [19,20]. Moreover, other advanced materials are getting into medical area like composites or smart polymers which can have a specific stimuli response on the body; some of those smart polymers are thermal responsive polymers, pH responsive polymers, conductive polymers, memory shaped polymers, among others [32].

Materials Research Foundations 172 (2025) 199-230 https://doi.org/10.21741/9781644903353-8

The research on these field are not only focused on the material by itself, but the construction method; therefore, new techniques are emerging as alternatives to build more precise devices for instance, in vascular grafts are reported method using 3D printing, electrospinning, molding, decellularization, coculture of cells, among others [33].

Most frequently used polymers for TE are polyethylene, polyurethanes, polyamides, nylon, polyethylene terephthalate, polypropylene, polymethylmethacrylates, among others (Fig. 4), some of them will be described below.

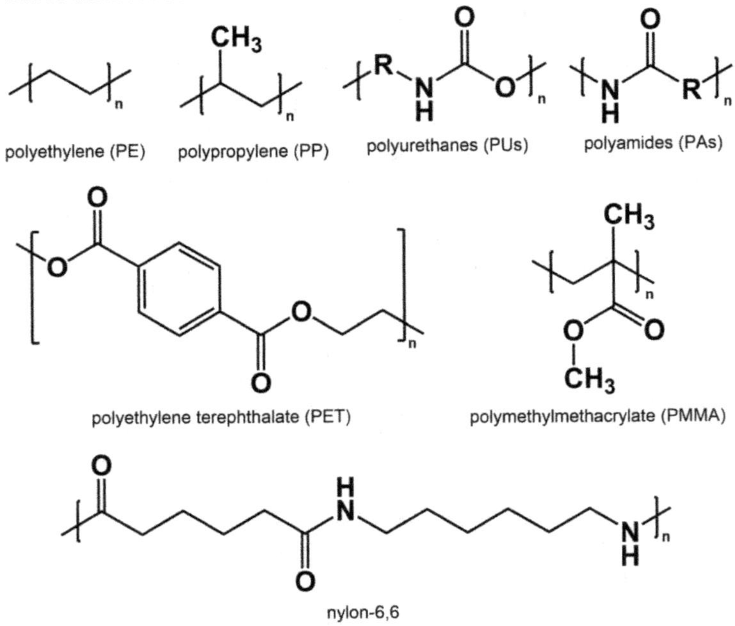

Figure 4. Common non-biodegradable polymers used in tissue engineering

4.1 Polyethylene

Polyethylene is a versatile polymer which can be produced petrochemically or from renewable sources like ethanol; independently of its origin, this polymer maintains quite attractive mechanical properties and biostability. There are different types of polyethylene base on the molecular weight, chemical structure and density such as low-density polyethylene (LDPE), high density polyethylene (HDPE), expanded polyethylene (EPE), reticulated high density polyethylene (XHDPE), ultra-high-density polyethylene (UHDPE), among others. Polyethylene is a thermoplastic therefore its recycling process is suitable for all kinds of PE but expanded variations could need additional steps to reduce their volume [34,35].

The application of each kind of polyethylene depends on the final use of the devices; some polyethylene plastics are more rigid than others or heavier than others; those characteristics will be defined by the chemical interactions in their structure. Although, any polyethylene is a polymer from ethylene, the chemical structure for each one can differ on the kind of connections between each monomer; LDPE present high branching and low crystallinity which produces high flexibility and impact resistance, HDPE is mainly lineal with high crystallinity this therefore it is rigid polymer with a high chemical and mechanical resistance, expanded polyethylene looks like a foam thus it is light than others while reticulated polyethylene has higher thermal and mechanical properties then non reticulated polyethylene [31,36].

There are several research about the applications of the different kinds of polyethylene, some of them are orbital implants, vascular grafts, joint prothesis, craniofacial reconstructive implants, nasal implants, auricular implants, among others. Polyethylene has hydrophobic properties that could interfere with vascularization and integration of implants, but surface modifications could be a great alternative solution (porous polyethylene) or integration of other materials in the polymer matrix such as copolymers and crosslinked copolymers [37]. It exists in the market several materials with polyethylene as part of their composition which are approved for medical applications such as MEDPOR®, Marlex®, Su-Por, among others®.

HDPE has been used for biomedical applications when a light, resistant and porous material is needed, one reported case is its usage as three-dimensional scaffold for reconstruction on bone defects; according of the results from Martinez et al, the *in vivo* study in rat tibia showed polyethylene have an acute inflammatory response after 15 days from implantation, but after 60 day no inflammation response was detected; HDPE let formation of vascular tissue inside pores and on the surface, and quite good osteointegration. HDPE presented suitable osteoconductive properties to be considered as a potential material for intraosseous placements [38].

4.2 Polypropylene

Polypropylene (PP) is the second most produced polymer after polyethylene, PP has attractive mechanical properties for medical applications like polyethylene, but with a lower cost of production. PP is formed by polymerization of propylene; this monomer could be obtained from non-renewable resources or biomass, some of them are metathesis of ethylene and 2-butene from ethanol, reduction of acetone from ABE fermentation (acetone, butanol, and ethanol fermentation) to propylene, dehydration of isopropanol or dehydrogenation of propane [25,30,39].

PP has other attractive characteristics suggest transparency, dimensional stability, recyclability, and resistivity to a wide range of acids, organic solvents and bases, but the most important characteristics for TE are fatigue resistance, low density, tension resistance, low moisture absorption and its sterilization is simple. Those characteristics make it a suitable material for diagnostic fields, fibers, syringes, laboratory material, and biomedical devices including prosthesis [30,39,40].

Although PP is a biocompatible polymer, there are some disadvantages that we can find about this material such as low temperature resistance, for you UV radiation resistance, flammability, difficult to bound with traditional adhesives and PP is degraded by a certain solvents and chemicals. It is relevant to highlight that ignition of polypropylene can generate toxic gases [30,40].

Polypropylene is used in several biomedical fibers such as prosthetics, drug delivery systems, suture threads and permanent meshes; the most common prosthetic made from PP are meshes, the

 Materials Research Forum LLC
https://doi.org/10.21741/9781644903353-8

regular meshes are used to support some tissues as hernia problems. PP meshes have been reported a suitable device in inguinal hernia surgery by its price quality radio, postoperative results and social professional integration time of the patients. The biocompatibility and performance of polypropylene could increase the by surface functional functionalization, different synthetic polymer integration, pores formation or, even inclusion of cells or bioactive molecules on the surface of the material [41,42].

4.3 Polyethylene terephthalate

Polyethylene terephthalate (PET) is also known as Dacron ® and it is a derived polymer from terephthalic acid (TPA) and ethylene glycol (EG). PET is a non-degradable thermoplastic and could be amorphous (transparent) or semi-crystalline (translucent). Although PET is synthetized from non-renewable sources (fossil oil), Bio-PET has been able to catch the interest of industry lately, and for that EG is synthetized from ethanol and TPA from sugar beet paraxylene [43].

PET is more chemically resistant than other thermoplastics, and it could be amorphous or semi-crystalline; PET is resistant to impact, semi-rigid to rigid, impermeable against moisture and carbon dioxide, biocompatible, abrasion resistant and nondegradable [44]. PET is used for different applications in medicine, but in TE the most recurrent form is as fibers, and the most common usages in this field are vascular prothesis, surgical meshes, scaffold, urinary replacement tissue, ligament implants and grafts, vascular grafts, sutures and bloodstream and urinary catheters [20,45].

PET is one of the most used commers for vascular grafts and cardiac support devices. It has been used for aorto-iliac replacements but restricted to medium diameters (6-8 mm) and on those conditions the probability of restenosis and thrombosis is reduced as result of PET inert characteristics; but at lower diameters PET can induce low patency rates and thrombosis [46].

4.4 Polymethylmethacrylate

Polymethyl methacrylate (PMMA) is a polymer from methyl methacrylate, it is an acrylic polymer which main characteristic is its light weight, transparency and toughness. The most common uses in biomedical applications are bone cement, dentures, intraocular lenses or bone fixation. [47,48]

For instance, a decellularized skin dermis and PMMA composite was created by the immersion of dermis in methyl methacrylate monomers followed by polymerization, resulting in a material with elasticity that mimics the dermis, with its compressive modulus value falling between that of the dermis and PMMA. In orthopedic surgeries, PMMA serves as bone cement, providing structural support and stability in joint replacements. Additionally, its use in intraocular lenses for cataract surgeries is well-documented, where its optical clarity and biocompatibility restore vision effectively [25]

In TE, PMMA is combined with patterning techniques, such as solvent-assisted micromolding, to control cellular microenvironments, influencing specific tissue culture responses. Moreover, PMMA is utilized in porous scaffolds that enhance soft tissue regeneration, such as in mandibular defect models, demonstrating potential for clinical translation. Similarly, its role in enhancing osteogenic differentiation on fibrous scaffolds highlights its significance in bone TE. Lastly, PMMA-based devices like TRAFFIC, designed for retinal degeneration and diabetes treatments, exemplify its diverse medical applications.

Materials Research Foundations 172 (2025) 199-230
https://doi.org/10.21741/9781644903353-8

4.5 Polyurethanes

Polyurethanes are synthetic polymers that have shown excellent properties for strength, durability, elasticity and biodegradable[49,50]. These properties of polyurethane made them suitable for biomedical applications for soft and hard tissues that are suitable for applications on blood bags, catheters, artificial blood vessels, wound healing and artificial heart valves [49].

Since the 1960s, polyurethane has been used for cardiovascular devices or allographs [20,49] by its property of been hemo-compatible. The scaffolds used in this area to generate artificial vessels should be degradable, resistant to mechanical fatigue and corrosion [49]. Another biomedical application of the PU is in bone regeneration where modifiable scaffolds are required after injection to the body [51]. Scaffolds of PU exhibit promising application as a temporary extracellular matrix in one healing by their good biocompatibility, bioactivity, injectability, and biodegradability [51]. The characteristic of PU in this application is the biodegradability of scaffold because it allows the formation of natural extracellular matrix. Furthermore, on wound healing applications, PU is a promising material for the development of hydrogels because hydrogels present transparency, adhesion, high mechanical and tensile strength, non-toxic, gaseous permeability and fatigue resistance [50]. The various biomedical applications in different tissues such as bone, skin or veins are just one example of the role that PU plays in the advancement of TE.

4.6 Other non-biodegradable polymers

Moreover, there are more non-degradable polymers which can be used as base materials for biomedical applications like scaffolds, grafts, implants or prosthesis. Some of them are polyamides (nylon-6, nylon-66), polytetrafluoroethylene (PTFE or Teflon®), Polytrimethylene terephthalate (PTT), among others. Ask the other non-degradable polymers, these provide of the resistance and improve the mechanical properties, but the specific final usage must be considering number to select the best option.

Although each permanent polymer was discussed separately, the real applications normally include a combination of them or the inclusion of other kinds of polymers such as biodegradable polymers, in fact copolymers can present better properties. The combination of different kinds of polymers in a material matrix contribute to achieve materials with better biocompatibility, specific mechanical and resistance properties, and thermal and chemical stability. The mixture of non-biodegradable polymers with biodegradable polymers or even with some active molecules or cells could increase the biocompatibility of those materials or enhance a specific property of the material needed for a suitable bio-integration and cell adhesion.

Other important property needed to be considered is the surface of the material, most of the synthetic non-biodegradable polymers have smooth surface and, in addition those materials are hydrophobic which can complicate the bio-integration with the patient body. In many cases the development us porous surface polymers could be a great solution, and it could be achieved by an extra chemical process or integration of biomolecules, nanomaterials, or cells. The non-degradable polymer based biomedical devices are though for a long time and modifications should be considered strategies to reduce the immune response or body rejection of these synthetic tissues.

5. Hydrogels for tissue engineering

Hydrogels are three-dimensional networks of cross-linked polymers that can uptake important amounts of water and retain it which gives them a very important characteristic, like similarity with human tissue. That is, hydrogels have the texture, flexibility and softness of any organ tissue, which is why they have attracted relevance in materials science and in the medical field. Their main application focuses on TE as key pieces to promote cell development and regeneration in damaged tissues.

Synthesis of hydrogels can be carried out using chemical cross-linking agents such as glutaraldehyde, formaldehyde, N-N-methyl bis(acrylamide) (MBAAm), to name a few; physical agents through high-energy radiation such as gamma radiation or ultraviolet (UV) light, and in some cases physical or electrostatic interactions allow cross-linking. The term cross-linking refers to the union at various points of the polymer chains forming a three-dimensional network. Each method offers advantages and disadvantages, both in the ease and speed of synthesis and in the final properties of the material. The main difference lies in the mechanical properties; by chemical method and with gamma radiation it is common to obtain hydrogels with high stability and good mechanical properties, however, those that are cross-linked by means of physical bonds are very unstable, their consistency is not long-lasting, and they are fragile [52].

The characteristics of hydrogels directly influence their application in TE, for example, their biocompatibility, porosity, and size, since adverse reactions in the human body must be avoided when meeting an external material and oxygen and nutrients must also be supplied to the cells in a favorable manner. Therefore, there are two ways in which the hydrogel can intervene in TE, either by depositing cells in the material in question or by placing the material it will provide the optimal conditions for the multiplication and proliferation of cells in the damaged tissue. In both cases, the hydrogels used for these purposes are known as cellular scaffolds [53], some of its main applications are shown in Fig. 5. Furthermore, these hydrogels are not only limited to being cellular scaffolds but also used as vehicles for drug administration.

5.1 Stimuli-responsive hydrogels

Based on the properties that hydrogels have is one way in which they can be classified. Among the most popular are the smart or stimulus-sensitive hydrogels, which are susceptible to an external stimulus such as temperature, light, pH, ionic strength or enzymatic environment. In this case, the materials can have different types of responses to these stimuli, one of the most common being a structural change due to the relaxation and contraction of the polymer chains or a hydrophilic to hydrophobic change of the hydrogel.

 Materials Research Forum LLC
https://doi.org/10.21741/9781644903353-8

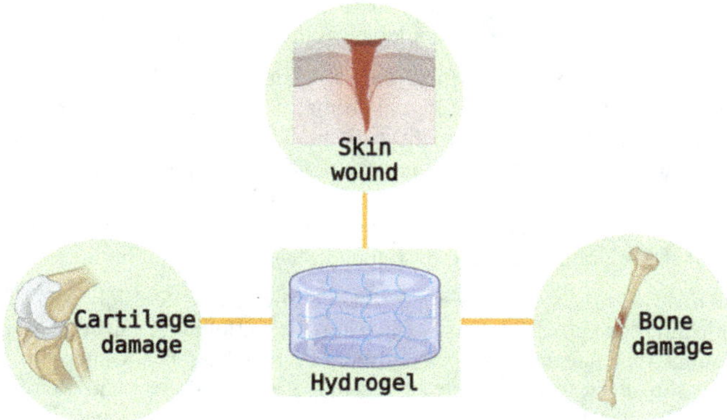

Figure 5. *Diverse applications of hydrogels in tissue engineering*

Recently in 2024, the development of a thermo-responsive injectable hydrogel (HGs) based on calcium glycerophosphate (Ca-GP) was reported, which has the capacity for thermo-response, chitosan (CS), hydroxyapatite (HA), and zein (ZN) which has great potential for regenerative power. The fluid hydrogel becomes a hydrogel when it meets body temperature. Degradation analyses indicate that it degrades up to 21 days later. Its antibacterial effects against *E. coli* and *S. aureus* of samples loaded with metronidazole were also evaluated, demonstrating that they do have activity against these bacteria with a controlled release of the drug. The HGs material presents favorable characteristics that make it ideal for bone regeneration [54].

Another type of stimuli-sensitive hydrogel that has gained great relevance in TE are those sensitive to light. In this case, the hydrogel is required to contain photosensitive molecules called chromophores, which will be able to capture light and carry out reactions such as cleavages or isomerizations, transforming the properties of the material; some proteins have been used for these purposes. These types of reactions are usually fast and clean, that is, no unwanted by-products are produced. This makes it possible to manipulate the structure of the scaffold, i.e. modify the pore size or its geometry, which directly correlates with the variables that influence cell adhesion, growth and development depending on the type of tissue to be repaired. Generally, work is done mainly with ultraviolet-vis light in the range of 200-700 nm and near infrared light from 700 to 1700 nm. This type of hydrogel has been used in tissues such as nerves, muscles and skin [55].

5.2 Natural hydrogels

Natural polymers are one of the favorite candidates for the development of materials for TE and in general with products that are designed to be used in the medical field, due to their relevant properties including a high biocompatibility and excellent degradability. Zahra Barabadi and her research group reported an electroconductive hydrogel of gelatin methacrylate (GelMA) with gold nanoparticles, through a chemical synthesis using curcumin for its antioxidant and redox properties, which allowed generating conductive routes within the polymer network. This research opens the way to the combination of different areas in which electrical activity is involved, for example in cardiac or nervous tissues, managing to reproduce their electrical properties. In addition to having the properties previously described to favor cell growth, it is also important that conductive polymers facilitate electrical communication between cells [56].

Hyaluronic acid (HA) is one of the most encouraging materials around tissue regeneration, specifically for cartilage and bones. Different forms of this material have been used to make hydrogels in TE, where derivatives and different molecular weights of the polymer are used. Commonly, the synthesis of HA hydrogels is carried out through photo-initiators using ultraviolet or visible light. The m-HA hydrogel synthesized from methacrylic anhydride and HA was tested on damaged cartilage in different proportions of minipigs, demonstrating good adhesion, favoring cell proliferation and migration, and demonstrating potential regeneration in cartilage [57,58].

Within natural hydrogels there is a very important classification of decellularized ECM. These are based on eliminating the cells from a tissue, leaving the scaffold that can serve as a substrate for later modifications, but already having the structure. This has an advantage in terms of high biocompatibility, however, there are defects in terms of its insufficient antimicrobial activity. Its application can vary according to the specific matrix of each tissue, since the manufacture of hydrogels can mimic the characteristics of the original tissue [59].

Zheng Zhou et al. reported the synthesis of an ECM hydrogel from rat skin which was treated with a Cu^{2+} solution for the formation of the hydrogel, then exosome-containing factor 2 fibroblasts (exoFDF-2) were encapsulated having the exoFGF-2@ECM/Cu^{2+} hydrogel material. This material presented antimicrobial activity provided by copper ions, said activity is directly related to the copper concentration in the hydrogel, with 100 ng/mL of Cu^{2+} being able to inhibit bacteria without having a cytotoxic effect. In tests with wounds in animals, the hydrogel favored healing, collagen deposition and thus tissue regeneration in these areas [60].

5.3 Composite hydrogels

One of the major disadvantages of hydrogels in general is the deficiency in mechanical properties. Various factors including the degree of cross-linking and the starting material itself are those that directly influence the rigidity and stability of the material. This is why alternatives have been sought that contribute to adequate mechanical stability and that also have additional properties such as antimicrobial activity. The most common way to increase the mechanical properties of hydrogels is through “fillers”, that is, some material in the form of nanoparticles or flakes that are introduced into the matrix and can provide other types of properties to those of the hydrogel. Combining two types of materials and one of them being “nano” in size results in a nanocomposite material [61].

In this sense, both natural and synthetic polymers can be combined with different types of reinforcements. For example, Jae Gu Jung and his research team reported the synthesis of a polyvinyl alcohol (PVA) hydrogel composite with graphene flakes (GF) and graphene oxide (GO), where the reinforcements also act as antimicrobials. Chitosan-gelatin/zinc oxide nanoparticles (CS-GEL/nZnO) hydrogels were also synthesized, which according to what was reported and the tests performed, present excellent properties for skin cell regeneration and have an antimicrobial effect due to the metal oxide nanoparticles. Both cases present potential antimicrobial and tissue regeneration properties for application in TE [61,62].

Another inorganic biomaterial widely used in TE is hydroxyapatite due to its similarity in bone composition. In 2024, a work was reported in which a cellular scaffold for tumor treatment in

Materials Research Foundations 172 (2025) 199-230
https://doi.org/10.21741/9781644903353-8

bones is synthesized by 3D printing. This hydrogel contains hydroxyapatite nanoparticles as a reinforcing agent obtained from a coral that particularly has structural and morphological characteristics very similar to human bones and contains polycaprolactone (PCL)/gelatin (Gel). To give greater use to the material, drugs can be loaded and released. In this case, doxorubicin (DOX) was loaded, which is used in cancer treatment, showing that when included in the cellular scaffold, it has a stable release for 14 h and the cytotoxicity is minimal [63].

6. Synthesis of polymeric materials for tissue engineering

Currently, there are some modern strategies to design and obtain polymeric materials with remarkable properties with application in TE and regenerative medicine. The choosing of one or more synthetic methods to obtain a polymeric scaffold must meet some criteria including cost, efficiency, practicality, versatility, eco-friendliness

6.1 Electrospinning

Electrospinning is a widely used modern and efficient method that uses electric potential to generate micro/nanofibers starting from a polymer solution in a continuous process. One of the most relevant features of this method is that the fiber diameter is adjusted in the range from nanometers to microns, dimensions that, generally, cannot be achieved with other techniques. These fine fibers may be of some types such as metallic, ceramic and, evidently, polymeric, all of these starting from a polymer solution or melt. The properties of electrospinning fibers, including diameter and morphology, can be modulated in dependence on several parameters divided into three categories: properties of solution (concentration, viscosity, electrical conductivity, surface tension, molecular weight), processing (voltage, collector, distance between collector and syringe), and environmental factors (temperature and humidity) [64,65].

A typical electrospinning working system is shown in Fig. 6, where are depicted a syringe pump, grounded collector, and high-voltage DC power as main components. Firstly, polymer is dissolved in a suitable solvent, and it is deposited in the syringe pump. Then, a high electric field is applied to the solution media and an electric charge is induced, causing a repulsive interaction between similar charges in the solution. Next, electrical forces must be balanced by surface tension and when these forces must be greater than this surface tension, a fiber is ejected from the needle towards a grounded collector. Now, in the collector, the synthesized material is present [66,67].

Synthetic and natural polymers are frequently used in the design and manufacturing of electrospinning fibers with application in TE. Synthetic polymers highlight for their high flexibility in the electrospinning process, providing improved mechanical properties to the final biomaterial. However, these polymers possess a high cost of production and display a lack in biocompatibility and bioactivity, needing to be modified to improve biological properties. On the other hand, natural polymers exhibit inherent bioactive properties such as low immune response and an increased interactions of fiber-cell and cell-cell [65]. Their main issue is related to their partial degradation of some natural polymers. An amalgam of both natural and synthetic polymers could complement properties achieving improved materials with remarkable properties.

Materials Research Foundations 172 (2025) 199-230
https://doi.org/10.21741/9781644903353-8

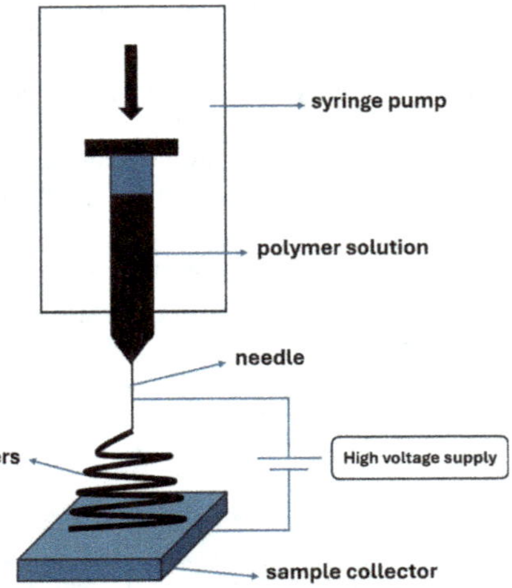

Figure 6. *Basic electrospinning equipment with their main components*

6.2 Additive manufacturing

Additive manufacturing, also known as 3D printing, is a set of techniques including stereolithography, fuse deposition modeling, selective laser sintering, inkjet printing, bioprinting, extrusion, laser beam melting, digital laser printing, electron beam melting, and polyjet [68]. All these methods are based on a common principle which is the layer-by-layer synthesis until the desired material is created. The main advantages of this method are the ability to manufacture prototypes with complex geometries, a variety of materials with multiple properties, and the possibility to be reproducible to an industrial scale with reduced waste and higher energy efficiency compared with conventional methods. In contrast to traditional scaffold fabrication techniques, desired properties can be precisely controlled during additive manufacturing with computer-aided design obtaining scaffolds with different applications.

The main objective of TE is to design and create tissue or organ replacement strategies. The use of polymers oriented to this goal is expanded to many tissues including cardiac tissues and liver kidney, these being the most common transplanted organs. All types of polymers can be used to fabricate scaffolds for TE and in two physical states, solid or liquid. However, polymers to be used in additive manufacturing must be met some properties. Appropriate polymer qualities must fit rheological properties such as viscosity, cross-linkage, Barus effect and non-Newtonian. Generally, polymers with higher viscosity tend to block the press noodle, slowing down the 3D printing or interrupting the process. Also, non-Newtonian fluids tend to display low fluid dilution viscosity, establishing it as an ideal ink given that allow a fluid printing [69].

Scaffolds based on natural polymers are beneficial to assess some necessary biological properties in TE. For example, materials with higher porosity and extended lifetime can be obtained using polymers based on starch, which is fundamental for the engineering of bone tissues. In fact, it has been proved that natural starch polymers used directly with water-based binders in additive

Materials Research Foundations 172 (2025) 199-230
https://doi.org/10.21741/9781644903353-8

manufacturing exhibited positive results, assessing biocompatibility, and mechanical properties can be improved with synthetic polymers [70]. Another relevant polymer for additive manufacturing is gelatin since it is a biocompatible, nontoxic, and nonimmunogenic polymer, additionally to present a biomimetic peptide that prevent cell apoptosis. This polymer has been used to create a 3D-printed scaffold for cartilage TE exhibiting multiporous, uniformly distributed, interconnected along the scaffold. This was implemented into a defective rat cartilage for 12 weeks obtaining excellent repair results [71].

6.3 Phase separation

Phase separation is a scaffold manufacturing technique which main characteristic is to design foams with desired mechanical properties and pore size for TE purposes. Also displays other advantages including being fast, scalable and controllable method. It can be divided into two types: liquid-solid and liquid-liquid phase separation and can be proceed into two ways either thermal or nonsolvent process. It indicates that the porous scaffold can be obtained by the reduction of the solution temperature or by the extraction of liquid phase until obtain a solid matrix after solidification. Both ways include the extraction of obtained materials from the mold, washing with distilled water and drying to remove solvent traces. It is crucial to take control over some experimental conditions such as temperature, rate of cooling, type of solvent, crystallinity and concentration of polymer to manage and ensure the desired pore properties and the interconnectivity between them in the scaffold [72,73].

6.4 Freeze-drying

Freeze drying (FD) technique, also known as lyophilization, is a process based on the principle of sublimation, where a material or solution is freezing below the triple point of solvent and then subjected to a high-pressure vacuum to sublimate water in the form of vapor. Frozen drops of water are sublimed leaving only solid and dried components. This step of evaporation of drop water allows us to create porous scaffolds or templates with pore size in the micrometers range. Specifically in the TE area, scientists establish FD as a promising technology to obtain artificial tissues including bone, muscles, nerves, and skin. These artificial materials must mimic mechanically, biologically, structurally, and functionally in addition to showing other characteristics such as biocompatibility, cell proliferation, adhesion, and differentiation [74,75].

Natural and synthetic polymers are used to synthesize polymeric scaffolds by FD method. In the first case, polysaccharides-based scaffolds were synthesized by FD technique combined with crosslinking obtaining pore size with a mean diameter of 55 ± 4 μm and a porosity of 33 ± 12% for high freeze-drying pressure, whereas low freeze-drying pressure scaffolds showed larger pores with a mean diameter of 243 ± 14 μm and a porosity of 68 ± 3%. These scaffolds exhibited a potential application for biomedical applications such as enhancing vascularization and tissue in-growth *in vivo* [76]. In the case of synthetic polymers, composite scaffolds of poly(hydroxybutyrate-*co*-hydroxyvalerate) (PHBV) and poly(L-lactic acid) (PLLA) with incorporated nano hydroxyapatite (HA) were obtained using the emulsion freezing/freeze-drying technique. The results stated that at -35 °C, synthesized composite exhibited a better pore size distribution and a continuous pore network. This scaffold shows a potential application in bone tissue growth giving an excellent environment for cell attachment, diffusion, and differentiation [77].

Materials Research Foundations 172 (2025) 199-230
https://doi.org/10.21741/9781644903353-8

The morphology of pore, porosity and desired properties can be achieved by a control of the parameters of FD process, including instrumental factors and solution parameters. Among the instrumental factors it is important to mention the rate of decreasing temperature until reach triple point of solvent, chamber pressure, time between each stage, and mold. These parameters are difficult to control because they involve environmental conditions. Freezing rate and temperature are the most important parameters, and it is proved that when decrease the temperature until reach -70 °C, no pore is presented in a PVA/gelatin solution [78]. On the other hand, solution parameters concern several factors including type of polymer, concentration of polymer, and viscosity of solution. These latest parameters are easier to control. In general terms, a slicker solution gives a higher porosity and larger pore diameter on the final product. To obtain a slick solution is highly recommended using a weaker concentration and an adequate molecular weight of polymer [74].

6.5 Gas foaming

Gas foaming is a novelty technique that aims to produce porogenic polymeric biomaterials with desirable pore size. This method requires the use of foaming agents as sieved effervescent salt particles, being ammonium bicarbonate the most common, inducing a blending of polymer through a gas bubble formation and, consequently, pores in the scaffold. Other typical foaming agents used in academia and industry are carbon dioxide, pentane, 1,1,1,3,3-pentafluoropropane, nitrogen, and 1,1,1,2-tetrafluoroethane as well as pressurized gases including argon, nitrogen or air [79]. The process involves the deposition of polymer solution in a cast and the subsequent addition of salt particles or gas. This mixture is dispersed in a cast with hot water forming a polymer gel paste where the solvent is evaporated. Gaseous porogens are produced during the increase of temperature or drop in pressure. This follows a decrease in the solubility of formed carbon dioxide, which tries to escape from the media, increasing the nucleation and giving a porous microstructure as a final product (Fig. 7) [80,81].

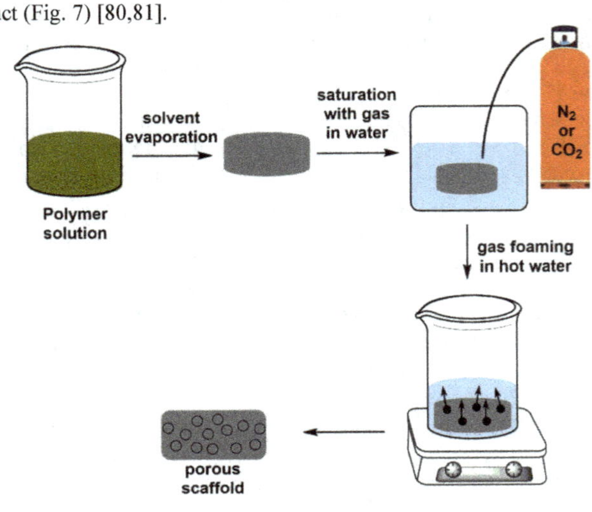

Figure 7. *Gas foaming process obtaining a porous scaffold as final product*

Materials Research Forum LLC
https://doi.org/10.21741/9781644903353-8

Gas foaming show a main advantage which is that allow to create materials with desired porosity which is a key factor for encouraging tissue integration, facilitates nutrient delivery and cell penetration. Also, this technique highlights other techniques due to their versatility allowing to synthesize a wide variety of porous polymeric biomaterials with application in TE, using both natural and synthetic polymers. Gelatin is one of the most common used to this purpose in which a correct combination of conditions and additives gives a porous biomaterial with desired porosity but with a weak mechanical strength, which will be a concern in the field of manufacturing TE scaffolds [82]. As a synthetic polymer, poly (L-lactic acid) is widely used for the manufacturing of TE scaffolds due to their great biocompatibility and their use as bone repair or replacing. These polymeric scaffolds can be manufactured via gas foaming using supercritical CO_2 to induce a heterogeneous nucleation to give a biomaterial with high crystallinity, high strength, and low density with enhanced cellular morphology [83].

7. Challenges and future perspectives.

Although it is undeniable that science and technology have made great progress around polymeric materials aimed at TE and regenerative medicine, there are still many pending issues to be resolved.

A relevant challenge in the TE field is the design of polymeric scaffolds with the ability to mimic the complex architecture of native tissue or organ while maintain their porosity and guarantee properties for cell growth. As described earlier, properties of final synthesized scaffold depend on synthetic method, solvent, type of polymer, and environmental and experimental conditions. Also, a scaffold implanted in a human body suffers several changes in its composition due to mechanical degradation or simply by interaction with their surroundings. This fact can affect their performance and the correct tissue regeneration. Most efforts must be focused on the correct choice of methods to assess the application in a cell environment. The use of new technologies, including modelling and artificial intelligence could be an appropriate tool for this purpose.

The micro/nanostructure and composition of ECM varies significantly between different tissues and organs. In this sense, interactions between scaffold and tissue will depend largely on this fact. Depending on the success or failure of these interactions, the regenerative process will take different routes given that this process promotes distinct cell phenotype changes. Controlling this aspect will be possible if a complete and solid knowledge of the composition of ECM of each target tissue is achieved. If the scaffold does not match correctly with the right cues of tissue, the cell does not attach to the matrix and will suffer apoptosis. Also, a complete characterization of scaffold for short and long-term cell phenotype effects will be imperative.

One of the reasons to use polymers as starting materials to synthesize polymeric scaffolds is that they can be functionalized and change their properties. In this way, a scaffold used in a specific application could be used with other purposes. However, this approach is not widely exploited. Aging is a surface modification technique that uses the plasma effects to induce a change in the conformation of the polymeric matrices. These changes include the grafting of functional monomers and the creation of active sites that will be susceptible to post-reactions. This process gives materials with a wide range of functionalities, but with a poor stability [84]. The use of gamma radiation to induce surface modifications of polymeric scaffolds appears as an alternative. The high energy waves of this type of radiation allow to create actives sites in a polymer both in bulk and surface of matrix. This approach allows to obtain materials with improved properties

Materials Research Foundations 172 (2025) 199-230 | https://doi.org/10.21741/9781644903353-8

through grafting of molecules of interest with high yields and high stability [85–88]. Grafting induced by gamma radiation is a poorly exploited technique with very high potential.

As necessities of patients and the demand to improve the quality-of-life increase, it is evident that there is an increase in the scientific demand and challenges in the TE innovation. Nowadays, TE plays a minor role in the treatment of patients due to that these processes are still experimental and very expensive. Most advanced studies have been carried out in scaffold design have been reached in models like rodents and rabbits. Despite the most demanded organs such as heart, lungs, and liver have been reproduced successfully at laboratory scale, there are some issues with reproducibility, and they cannot yet be implanted in humans. This phase has been very limited to just a few successful results which limits the knowledge about scaffolds performance and interaction with human body, due to the differences between animal and human anatomies. The way is clear, it must exist a synergistic collaboration between different scientists, engineers, medics, physicists, financial entities, and government agencies to accelerate these research steps and to break the academia/industry barrier, to that new innovations in TE scaffolds with a wide range of applications become within everyone's reach.

Conclusions

TE is a multidisciplinary field with opportunities and demands increasingly urgent. This fact has made the search for new materials to be used as implants, as drug delivery systems or as organ or tissue replacement be a relentless challenge. The use of polymers as precursors of TE scaffolds means a significant advance in the area because, with these macromolecules, it is possible to easily obtain sophisticated materials with desirable properties. Scaffolds based on natural, synthetic, degradable, non-degradable polymers or a combination of those structures are idoneous materials for TE due to their availability and versatility to adapt to different contexts. Final attributes of polymeric scaffolds not only will depend on the inherent characteristics of the polymer per se but also on the synthetic method selected to obtain them. A suitable selection of ideal synthetic method will assess easiness, quickness, versatility, and to approach in specific target application. It is important that synthetic methods and final polymeric scaffolds are framed in an awareness of eco-friendliness as the current conditions of our environmental media demand us. Finally, it is a responsibility and a challenge to all TE scientists to bring closer the new advances in the area to the people who really need it, the patients, who desires to overcome their diseases and to improve their quality of life. For that, it will be necessary to break the gap between academy and industry, between microscales and macroscale, between science and human needs to assess better days for our humanity.

Acknowledgements

Thanks to CONAHCyT for the doctoral scholarship provided for David Romero Fierro (1175725) and Yeeimi Aylin Esquivel Lozano (952777). This work was supported by Dirección General de Asuntos del Personal Académico, Universidad Nacional Autónoma de México under Grant IN204223.

References

[1] J.L. Olson, A. Atala, J.J. Yoo, Tissue Engineering: Current Strategies and Future Directions, Chonnam Med J 47 (2011) 1. https://doi.org/10.4068/cmj.2011.47.1.1.

[2] C. Vacanti, The history of tissue engineering, J Cell Mol Med 1 (2006) 569–576. https://doi.org/10.2755/jcmm010.003.20

[3] G.L. Koons, M. Diba, A.G. Mikos, Materials design for bone-tissue engineering, Nat Rev Mater 5 (2020) 584–603. https://doi.org/10.1038/s41578-020-0204-2

[4] K.Y. Lee, Design parameters of polymers for tissue engineering applications, Macromol Res 13 (2005) 277–284. https://doi.org/10.1007/BF03218454

[5] H. Chamkouri, A Review of Hydrogels, Their Properties and Applications in Medicine, Am J Biomed Sci Res 11 (2021) 485–493. https://doi.org/10.34297/AJBSR.2021.11.001682

[6] M.U.A. Khan, G.M. Stojanović, M.F. Bin Abdullah, A. Dolatshahi-Pirouz, H.E. Marei, N. Ashammakhi, A. Hasan, Fundamental properties of smart hydrogels for tissue engineering applications: A review, Int J Biol Macromol 254 (2024) 127882. https://doi.org/10.1016/j.ijbiomac.2023.127882

[7] S. Zhu, W. Dou, X. Zeng, X. Chen, Y. Gao, H. Liu, S. Li, Recent Advances in the Degradability and Applications of Tissue Adhesives Based on Biodegradable Polymers, Int J Mol Sci 25 (2024) 5249. https://doi.org/10.3390/ijms25105249.

[8] N. Petelinšek, S. Mommer, Tough Hydrogels for Load-Bearing Applications, Advanced Science 11 (2024). https://doi.org/10.1002/advs.202307404

[9] S. Bashir, M. Hina, J. Iqbal, A.H. Rajpar, M.A. Mujtaba, N.A. Alghamdi, S. Wageh, K. Ramesh, S. Ramesh, Fundamental Concepts of Hydrogels: Synthesis, Properties, and Their Applications, Polymers (Basel) 12 (2020) 2702. https://doi.org/10.3390/polym12112702

[10] J.L. Guo, Y.S. Kim, V.Y. Xie, B.T. Smith, E. Watson, J. Lam, H.A. Pearce, P.S. Engel, A.G. Mikos, Modular, tissue-specific, and biodegradable hydrogel cross-linkers for tissue engineering, Sci Adv 5 (2019). https://doi.org/10.1126/sciadv.aaw7396

[11] X. Ma, T. Xu, W. Chen, H. Qin, B. Chi, Z. Ye, Injectable hydrogels based on the hyaluronic acid and poly (γ-glutamic acid) for controlled protein delivery, Carbohydr Polym 179 (2018) 100–109. https://doi.org/10.1016/j.carbpol.2017.09.071

[12] B.M. Watson, F.K. Kasper, P.S. Engel, A.G. Mikos, Synthesis and Characterization of Injectable, Biodegradable, Phosphate-Containing, Chemically Cross-Linkable, Thermoresponsive Macromers for Bone Tissue Engineering, Biomacromolecules 15 (2014) 1788–1796. https://doi.org/10.1021/bm500175e

[13] M. Hoque, M. Alam, S. Wang, J.U. Zaman, Md.S. Rahman, M. Johir, L. Tian, J.-G. Choi, M.B. Ahmed, M.-H. Yoon, Interaction chemistry of functional groups for natural biopolymer-based hydrogel design, Materials Science and Engineering: R: Reports 156 (2023) 100758. https://doi.org/10.1016/j.mser.2023.100758

[14]H. Zong, B. Wang, G. Li, S. Yan, K. Zhang, Y. Shou, J. Yin, Biodegradable High-Strength Hydrogels with Injectable Performance Based on Poly(l -Glutamic Acid) and Gellan Gum, ACS Biomater Sci Eng 6 (2020) 4702–4713. https://doi.org/10.1021/ACSBIOMATERIALS.0C00915/ASSET/IMAGES/LARGE/AB0C00915_0009.JPEG

[15]B. Hosseinzadeh, M. Ahmadi, Degradable hydrogels: Design mechanisms and versatile applications, Materials Today Sustainability 23 (2023) 100468. https://doi.org/10.1016/J.MTSUST.2023.100468

[16]P.M. Kharkar, K.L. Kiick, A.M. Kloxin, Design of thiol- and light-sensitive degradable hydrogels using Michael-type addition reactions, Polym Chem 6 (2015) 5565–5574. https://doi.org/10.1039/C5PY00750J

[17]R. Raman, T. Hua, D. Gwynne, J. Collins, S. Tamang, J. Zhou, T. Esfandiary, V. Soares, S. Pajovic, A. Hayward, R. Langer, G. Traverso, Light-degradable hydrogels as dynamic triggers for gastrointestinal applications, Sci Adv 6 (2020). https://doi.org/10.1126/SCIADV.AAY0065/SUPPL_FILE/AAY0065_SM.PDF.

[18]M. Shao, Z. Shi, X. Zhang, B. Zhai, J. Sun, Synthesis and Properties of Biodegradable Hydrogel Based on Polysaccharide Wound Dressing, Materials 2023, Vol. 16, Page 1358 16 (2023) 1358. https://doi.org/10.3390/MA16041358

[19]A. Steinbüchel, Non-biodegradable biopolymers from renewable resources: perspectives and impacts, Curr Opin Biotechnol 16 (2005) 607–613. https://doi.org/10.1016/J.COPBIO.2005.10.011

[20]A. Subramaniam, S. Sethuraman, Biomedical Applications of Nondegradable Polymers, Natural and Synthetic Biomedical Polymers (2014) 301–308. https://doi.org/10.1016/B978-0-12-396983-5.00019-3

[21]U. Kong, N.F. Mohammad Rawi, G.S. Tay, The Potential Applications of Reinforced Bioplastics in Various Industries: A Review, Polymers (Basel) 15 (2023). https://doi.org/10.3390/POLYM15102399

[22]L. Shen, J. Huafe, M.K. Patel, Product overview and market projection of emerging bio-based plastics PRO-BIP 2009 , Utrecht, 2009.

[23]S. Yan, B.K. Biswal, R. Balasubramanian, Insights into interactions of biodegradable and non-biodegradable microplastics with heavy metals, Environ Sci Pollut Res Int 30 (2023) 107419–107434. https://doi.org/10.1007/S11356-023-27906-1

[24]C.J. Demott, M.A. Grunlan, Emerging polymeric material strategies for cartilage repair, J Mater Chem B 10 (2022) 9578–9589. https://doi.org/10.1039/D2TB02005J

[25]C. Andreeßen, A. Steinbüchel, Recent developments in non-biodegradable biopolymers: Precursors, production processes, and future perspectives, Appl Microbiol Biotechnol 103 (2019) 143–157. https://doi.org/10.1007/S00253-018-9483-6/METRICS

[26] A. Steinbüchel, Non-biodegradable biopolymers from renewable resources: perspectives and impacts, Curr Opin Biotechnol 16 (2005) 607–613. https://doi.org/10.1016/J.COPBIO.2005.10.011

[27] Z. Yao, H.J. Seong, Y.S. Jang, Environmental toxicity and decomposition of polyethylene, Ecotoxicol Environ Saf 242 (2022). https://doi.org/10.1016/J.ECOENV.2022.113933

[28] M.A. Rodriguez-Soto, N.S. Vargas, A. Riveros, C.M. Camargo, J.C. Cruz, N. Sandoval, J.C. Briceño, Failure Analysis of TEVG's I: Overcoming the Initial Stages of Blood Material Interaction and Stabilization of the Immune Response, Cells 10 (2021). https://doi.org/10.3390/CELLS10113140

[29] W.G. Ribeiro, A.C.C. Nascimento, L.B. Ferreira, D.D. De Marchi, G.M. Rego, C.T. Maeda, G.E.B. Silva, R.A. Neto, O.J.M. Torres, M.B. Pitombo, Analysis of tissue inflammatory response, fibroplasia, and foreign body reaction between the polyglactin suture of abdominal aponeurosis in rats and the intraperitoneal implant of polypropylene, polypropylene/polyglecaprone and polyester/porcine collagen meshes, Acta Cir Bras 36 (2021). https://doi.org/10.1590/ACB360706

[30] S. S, A.P. R G, G. Bajaj, A.E. John, S. Chandran, V.V. Kumar, S. Ramakrishna, A review on the recent applications of synthetic biopolymers in 3D printing for biomedical applications, J Mater Sci Mater Med 34 (2023). https://doi.org/10.1007/S10856-023-06765-9

[31] M.C. Biswas, B. Jony, P.K. Nandy, R.A. Chowdhury, S. Halder, D. Kumar, S. Ramakrishna, M. Hassan, M.A. Ahsan, M.E. Hoque, M.A. Imam, Recent Advancement of Biopolymers and Their Potential Biomedical Applications, J Polym Environ 30 (2022) 51–74. https://doi.org/10.1007/S10924-021-02199-Y

[32] M.C. Serrano, G.A. Ameer, Recent insights into the biomedical applications of shape-memory polymers, Macromol Biosci 12 (2012) 1156–1171. https://doi.org/10.1002/MABI.201200097

[33] M.X. Li, Q.Q. Wei, H.L. Mo, Y. Ren, W. Zhang, H.J. Lu, Y.K. Joung, Challenges and advances in materials and fabrication technologies of small-diameter vascular grafts, Biomater Res 27 (2023) 1–22. https://doi.org/10.1186/S40824-023-00399-2/FIGURES/9

[34] C.S. Pfeifer, Polymer-Based Direct Filling Materials, Dent Clin North Am 61 (2017) 733–750. https://doi.org/10.1016/J.CDEN.2017.06.002

[35] S. Koltzenburg, M. Maskos, O. Nuyken, Polymer Chemistry, First, Springer , Berlin, 2017.

[36] R.M. Patel, Types and Basics of Polyethylene, Handbook of Industrial Polyethylene Technology (2016) 105–138. https://doi.org/10.1002/9781119159797.CH4

[37] S. Stephan, J. Reinisch, Auricular Reconstruction Using Porous Polyethylene Implant Technique, Facial Plast Surg Clin North Am 26 (2018) 69–85. https://doi.org/10.1016/J.FSC.2017.09.009

[38] J. Martínez Rodríguez, S.J. Renou, M.B. Guglielmotti, D.G. Olmedo, Tissue response to porous high density polyethylene as a three-dimensional scaffold for bone tissue engineering: An experimental study, J Biomater Sci Polym Ed 30 (2019) 486–499. https://doi.org/10.1080/09205063.2019.1582278

[39] P.F.H. Harmsen, M.M. Hackmann, H.L. Bos, Green building blocks for bio-based plastics, Biofuels, Bioproducts and Biorefining 8 (2014) 306–324. https://doi.org/10.1002/BBB.1468

[40] L.K. Zhang, H. Wang, R. Yang, M. Liu, Q. Ban, W. Chen, M. Zhao, R. You, Y. Jin, Y.Q. Guan, Bone marrow stem cells combined with polycaprolactone-polylactic acid-polypropylene amine scaffolds for the treatment of acute liver failure, Chemical Engineering Journal 360 (2019) 1564–1576. https://doi.org/10.1016/J.CEJ.2018.10.230

[41] M. Khaledian, F. Jiroudhashemi, E. Biazar, Chitosan- and polypropylene-oriented surface modification using excimer laser and their biocompatibility study, Artif Cells Nanomed Biotechnol 45 (2017) 135–138. https://doi.org/10.3109/21691401.2016.1138485

[42] F.F. Ai, M. Mao, Y. Zhang, J. Kang, L. Zhu, The in vivo biocompatibility of titanized polypropylene lightweight mesh is superior to that of conventional polypropylene mesh, Neurourol Urodyn 39 (2020) 96–107. https://doi.org/10.1002/NAU.24159

[43] M.A. Rodriguez-Soto, A. Riveros, N.S. Vargas, A.J. Garcia-Brand, C.M. Camargo, J.C. Cruz, N. Sandoval, J.C. Briceño, Failure Analysis of TEVG's II: Late Failure and Entering the Regeneration Pathway, Cells 11 (2022). https://doi.org/10.3390/CELLS11060939

[44] R. Nisticò, Polyethylene terephthalate (PET) in the packaging industry, Polym Test 90 (2020) 106707. https://doi.org/10.1016/J.POLYMERTESTING.2020.106707.

[45] T. Çaykara, M.G. Sande, N. Azoia, L.R. Rodrigues, C.J. Silva, Exploring the potential of polyethylene terephthalate in the design of antibacterial surfaces, Med Microbiol Immunol 209 (2020) 363. https://doi.org/10.1007/S00430-020-00660-8

[46] D. Wang, Y. Xu, Q. Li, L.S. Turng, Artificial Small-Diameter Blood Vessels: Materials, Fabrication, Surface Modification, Mechanical Properties, and Bioactive Functionalities, J Mater Chem B 8 (2020) 1801. https://doi.org/10.1039/C9TB01849B

[47] Q. Chen, G. Thouas, Biomaterials A Basic Introduction, First, CRC Press, 2014.

[48] S.M. Pituru, M. Greabu, A. Totan, M. Imre, M. Pantea, T. Spinu, A.M.C. Tancu, N.O. Popoviciu, I.I. Stanescu, E. Ionescu, A Review on the Biocompatibility of PMMA-Based Dental Materials for Interim Prosthetic Restorations with a Glimpse into their Modern Manufacturing Techniques, Materials (Basel) 13 (2020) 1–14. https://doi.org/10.3390/MA13132894

[49] S. Fathi-Karkan, B. Banimohamad-Shotorbani, S. Saghati, R. Rahbarghazi, S. Davaran, A critical review of fibrous polyurethane-based vascular tissue engineering

scaffolds, Journal of Biological Engineering 2022 16:1 16 (2022) 1–18. https://doi.org/10.1186/S13036-022-00286-9

[50] M. Ribeiro, M. Simões, C. Vitorino, F. Mascarenhas-Melo, Hydrogels in Cutaneous Wound Healing: Insights into Characterization, Properties, Formulation and Therapeutic Potential, Gels 10 (2024). https://doi.org/10.3390/GELS10030188

[51] P. Szczepańczyk, M. Szlachta, N. Złocista-Szewczyk, J. Chłopek, K. Pielichowska, Recent Developments in Polyurethane-Based Materials for Bone Tissue Engineering, Polymers 2021, Vol. 13, Page 946 13 (2021) 946. https://doi.org/10.3390/POLYM13060946

[52] T.-C. Ho, C.-C. Chang, H.-P. Chan, T.-W. Chung, C.-W. Shu, K.-P. Chuang, T.-H. Duh, M.-H. Yang, Y.-C. Tyan, Hydrogels: Properties and Applications in Biomedicine, Molecules 27 (2022) 2902. https://doi.org/10.3390/molecules27092902

[53] H. Zhao, M. Liu, Y. Zhang, J. Yin, R. Pei, Nanocomposite hydrogels for tissue engineering applications, Nanoscale 12 (2020) 14976–14995. https://doi.org/10.1039/D0NR03785K

[54] M.U. din Khan, A. Afzaal, Shahnaz, M.A. Gilani, S. Perveen, F. Sharif, A. Asif, A. Faisal, M.S. Nazir, O. Huck, S. Tabassum, Synergistic utilization of cost-effective glycerophosphate and biologically active zein for innovative minimally invasive smart thermo-responsive hydrogels for potential hard tissue engineering applications, Smart Mater Struct 33 (2024) 085007. https://doi.org/10.1088/1361-665X/ad57a4

[55] R. Luo, X. Xiang, Q. Jiao, H. Hua, Y. Chen, Photoresponsive Hydrogels for Tissue Engineering, ACS Biomater Sci Eng 10 (2024) 3612–3630. https://doi.org/10.1021/acsbiomaterials.4c00314

[56] Z. Barabadi, A. Bahmani, M. Jalalimonfared, M. Ashrafizadeh, M. Rashtbar, E. Sharifi, H. Tian, Design and characterization of electroactive gelatin methacrylate hydrogel incorporated with gold nanoparticles empowered with parahydroxybenzaldehyde and curcumin for advanced tissue engineering applications, J Mater Sci Mater Med 35 (2024) 45. https://doi.org/10.1007/s10856-024-06808-9

[57] F. Ghorbani, B. Ghalandari, M. Khajehmohammadi, N. Bakhtiary, H. Tolabi, M. Sahranavard, S. Fathi-Karkan, V. Nazar, S. Hasan Niari Niar, A. Armoon, M. Ettelaei, M. Tavakoli Banizi, M.N. Collins, Photo-cross-linkable hyaluronic acid bioinks for bone and cartilage tissue engineering applications, International Materials Reviews 68 (2023) 901–942. https://doi.org/10.1080/09506608.2023.2167559

[58] W. Yan, X. Xu, Q. Xu, Z. Sun, Q. Jiang, D. Shi, Platelet-rich plasma combined with injectable hyaluronic acid hydrogel for porcine cartilage regeneration: a 6-month follow-up, Regen Biomater 7 (2020) 77–90. https://doi.org/10.1093/rb/rbz039

[59] M. Gomez-Florit, A. Pardo, R.M.A. Domingues, A.L. Graça, P.S. Babo, R.L. Reis, M.E. Gomes, Natural-Based Hydrogels for Tissue Engineering Applications, Molecules 25 (2020) 5858. https://doi.org/10.3390/molecules25245858

[60] Z. Zhou, Z. Bu, S. Wang, J. Yu, W. Liu, J. Huang, J. Hu, S. Xu, P. Wu, Extracellular matrix hydrogels with fibroblast growth factor 2 containing exosomes for

reconstructing skin microstructures, J Nanobiotechnology 22 (2024) 438. https://doi.org/10.1186/s12951-024-02718-8

[61]R. Rakhshaei, H. Namazi, H. Hamishehkar, H.S. Kafil, R. Salehi, In situ synthesized chitosan–gelatin/ZnO nanocomposite scaffold with drug delivery properties: Higher antibacterial and lower cytotoxicity effects, J Appl Polym Sci 136 (2019). https://doi.org/10.1002/app.47590

[62]J.G. Jung, J.H. Kim, J. Moon, J.H. Kang, Y.J. Kim, H.B. Park, Enhanced antibacterial activity of poly(vinyl alcohol)-graphene composites via graphene oxide surfactancy, J Appl Polym Sci 141 (2024). https://doi.org/10.1002/app.55910

[63]F. Kadi, G. Dini, S.A. Poursamar, F. Ejeian, Fabrication and characterization of 3D-printed composite scaffolds of coral-derived hydroxyapatite nanoparticles/polycaprolactone/gelatin carrying doxorubicin for bone tissue engineering, J Mater Sci Mater Med 35 (2024) 7. https://doi.org/10.1007/s10856-024-06779-x

[64]A. Izadyari Aghmiuni, A. Ghadi, E. Azmoun, N. Kalantari, I. Mohammadi, H. Hemati Kordmahaleh, Electrospun Polymeric Substrates for Tissue Engineering: Viewpoints on Fabrication, Application, and Challenges, in: Electrospinning - Material Technology of the Future, IntechOpen, 2022. https://doi.org/10.5772/intechopen.102596

[65]M.Z.A. Zulkifli, D. Nordin, N. Shaari, S.K. Kamarudin, Overview of Electrospinning for Tissue Engineering Applications, Polymers (Basel) 15 (2023) 2418. https://doi.org/10.3390/polym15112418

[66]R. Khajavi, M. Abbasipour, A. Bahador, Electrospun biodegradable nanofibers scaffolds for bone tissue engineering, J Appl Polym Sci 133 (2016). https://doi.org/10.1002/app.42883

[67]K. Peranidze, T. V. Safronova, N.R. Kildeeva, Electrospun Nanomaterials Based on Cellulose and Its Derivatives for Cell Cultures: Recent Developments and Challenges, Polymers (Basel) 15 (2023) 1174. https://doi.org/10.3390/polym15051174

[68]G.-H. Wu, S. Hsu, Review: Polymeric-Based 3D Printing for Tissue Engineering, J Med Biol Eng 35 (2015) 285–292. https://doi.org/10.1007/s40846-015-0038-3

[69]D. Veeman, M.S. Sai, P. Sureshkumar, T. Jagadeesha, L. Natrayan, M. Ravichandran, W.D. Mammo, Additive Manufacturing of Biopolymers for Tissue Engineering and Regenerative Medicine: An Overview, Potential Applications, Advancements, and Trends, Int J Polym Sci 2021 (2021) 1–20. https://doi.org/10.1155/2021/4907027

[70]K. Tappa, U. Jammalamadaka, Novel Biomaterials Used in Medical 3D Printing Techniques, J Funct Biomater 9 (2018) 17. https://doi.org/10.3390/jfb9010017.

[71]A. Maihemuti, H. Zhang, X. Lin, Y. Wang, Z. Xu, D. Zhang, Q. Jiang, 3D-printed fish gelatin scaffolds for cartilage tissue engineering, Bioact Mater 26 (2023) 77–87. https://doi.org/10.1016/j.bioactmat.2023.02.007

[72] R. Akbarzadeh, A. Yousefi, Effects of processing parameters in thermally induced phase separation technique on porous architecture of scaffolds for bone tissue engineering, J Biomed Mater Res B Appl Biomater 102 (2014) 1304–1315. https://doi.org/10.1002/jbm.b.33101

[73] A. Yadegari, F. Fahimipour, M. Rasoulianboroujeni, E. Dashtimoghadarm, M. Omidi, H. Golzar, M. Tahriri, L. Tayebi, Specific considerations in scaffold design for oral tissue engineering, in: Biomaterials for Oral and Dental Tissue Engineering, Elsevier, 2017: pp. 157–183. https://doi.org/10.1016/B978-0-08-100961-1.00010-4

[74] Z. Fereshteh, Freeze-drying technologies for 3D scaffold engineering, in: Functional 3D Tissue Engineering Scaffolds, Elsevier, 2018: pp. 151–174. https://doi.org/10.1016/B978-0-08-100979-6.00007-0

[75] C.M. Brougham, T.J. Levingstone, N. Shen, G.M. Cooney, S. Jockenhoevel, T.C. Flanagan, F.J. O'Brien, Freeze-Drying as a Novel Biofabrication Method for Achieving a Controlled Microarchitecture within Large, Complex Natural Biomaterial Scaffolds, Adv Healthc Mater 6 (2017). https://doi.org/10.1002/adhm.201700598

[76] A. Autissier, C. Le Visage, C. Pouzet, F. Chaubet, D. Letourneur, Fabrication of porous polysaccharide-based scaffolds using a combined freeze-drying/cross-linking process, Acta Biomater 6 (2010) 3640–3648. https://doi.org/10.1016/j.actbio.2010.03.004

[77] N. Sultana, M. Wang, PHBV/PLLA-based composite scaffolds fabricated using an emulsion freezing/freeze-drying technique for bone tissue engineering: surface modification and *in vitro* biological evaluation, Biofabrication 4 (2012) 015003. https://doi.org/10.1088/1758-5082/4/1/015003

[78] K.S. Vasanthan, A. Subramaniam, U.M. Krishnan, S. Sethuraman, Influence of 3D porous galactose containing PVA/gelatin hydrogel scaffolds on three-dimensional spheroidal morphology of hepatocytes, J Mater Sci Mater Med 26 (2015) 20. https://doi.org/10.1007/s10856-014-5345-7

[79] M. Costantini, A. Barbetta, Gas foaming technologies for 3D scaffold engineering, in: Functional 3D Tissue Engineering Scaffolds, Elsevier, 2018: pp. 127–149. https://doi.org/10.1016/B978-0-08-100979-6.00006-9

[80] F. Dehghani, N. Annabi, Engineering porous scaffolds using gas-based techniques, Curr Opin Biotechnol 22 (2011) 661–666. https://doi.org/10.1016/j.copbio.2011.04.005

[81] Q.Z. Chen, Foaming technology of tissue engineering scaffolds - a review, Bubble Sci Eng Technol 3 (2011) 34–47. https://doi.org/10.1179/1758897911Y.0000000003

[82] S.A. Poursamar, J. Hatami, A.N. Lehner, C.L. da Silva, F.C. Ferreira, A.P.M. Antunes, Gelatin porous scaffolds fabricated using a modified gas foaming technique: Characterisation and cytotoxicity assessment, Materials Science and Engineering: C 48 (2015) 63–70. https://doi.org/10.1016/j.msec.2014.10.074

[83] T. Kuang, F. Chen, L. Chang, Y. Zhao, D. Fu, X. Gong, X. Peng, Facile preparation of open-cellular porous poly (l-lactic acid) scaffold by supercritical carbon dioxide

foaming for potential tissue engineering applications, Chemical Engineering Journal 307 (2017) 1017–1025. https://doi.org/10.1016/j.cej.2016.09.023

[84] F. Habibzadeh, S.M. Sadraei, R. Mansoori, N.P. Singh Chauhan, G. Sargazi, Nanomaterials supported by polymers for tissue engineering applications: A review, Heliyon 8 (2022) e12193. https://doi.org/10.1016/j.heliyon.2022.e12193

[85] M. González-Torres, I.H. Serrano-Aguilar, A. Cabrera-Wrooman, R. Sánchez-Sánchez, R. Pichardo-Bahena, Y. Melgarejo-Ramírez, G. Leyva-Gómez, H. Cortés, M. de los Angeles Moyaho-Bernal, E. Lima, C. Ibarra, C. Velasquillo, Gamma radiation-induced grafting of poly(2-aminoethyl methacrylate) onto chitosan: A comprehensive study of a polyurethane scaffold intended for skin tissue engineering, Carbohydr Polym 270 (2021) 117916. https://doi.org/10.1016/j.carbpol.2021.117916

[86] M.M. Nasef, B. Gupta, K. Shameli, C. Verma, R.R. Ali, T.M. Ting, Engineered Bioactive Polymeric Surfaces by Radiation Induced Graft Copolymerization: Strategies and Applications, Polymers (Basel) 13 (2021) 3102. https://doi.org/10.3390/polym13183102

[87] V.H. Pino-Ramos, H.I. Meléndez-Ortiz, A. Ramos-Ballesteros, E. Bucio, Radiation Grafting of Biopolymers and Synthetic Polymers, in: Biopolymer Grafting: Applications, Elsevier, 2018: pp. 205–250. https://doi.org/10.1016/B978-0-12-810462-0.00006-5

[88] E.I. Ochoa-Segundo, M. González-Torres, A. Cabrera-Wrooman, R. Sánchez-Sánchez, B.M. Huerta-Martínez, Y. Melgarejo-Ramírez, G. Leyva-Gómez, E.M. Rivera-Muñoz, H. Cortés, C. Velasquillo, S. Vargas-Muñoz, R. Rodríguez-Talavera, Gamma radiation-induced grafting of n-hydroxyethyl acrylamide onto poly(3-hydroxybutyrate): A companion study on its polyurethane scaffolds meant for potential skin tissue engineering applications, Materials Science and Engineering: C 116 (2020) 111176. https://doi.org/10.1016/j.msec.2020.111176

Materials Research Foundations 172 (2025) 231-245
https://doi.org/10.21741/9781644903353-9

Chapter 9

Polymer-Based 3D Printed Scaffolds for Tissue Engineering

S. Aslan[1]*, Y. Kaya[2], Y. Öztürk[3]

[1]Department of Chemistry, Faculty of Science, Mugla Sitki Kocman University, Mugla, Turkey

[2]Department of Molecular Biology and Genetics, Faculty of Science, Mugla Sitki Kocman University, Mugla, Turkey

[3]Department of Architecture, Izmir Institute of Technology, İzmir, Turkey

sozcan87@gmail.com*

Abstract

The way of 3D printing technology has diverged into many fields, with tissue engineering standing out as one of the most promising areas of application. Polymer-based 3D printed scaffolds are essential to tissue engineering because they offer structural support and facilitate cell growth and tissue regeneration. This review explores the advances in polymer-based 3D printed scaffolds, including materials used, fabrication techniques, and their applications in tissue engineering. Emphasis is placed on the properties of different polymers, the integration of bioactive factors, and the challenges and potential paths forward in this quickly developing subject.

Keywords

Polymers, 3D Scaffolds, Tissue Engineering, Technology

Contents

Materials Research Foundations 172 (2025) 231-245
https://doi.org/10.21741/9781644903353-9

1. Introduction

Cells, biomaterials, and biological elements to replace or repair damaged tissues and organs can be engineered. In this process, scaffolds play a crucial role as a transient framework supporting cell growth factors. Advances in 3D printing technology have made it possible to precisely fabricate scaffolds with intricate geometries and customized attributes.

Several types of polymers are used for the 3D printed scaffold developments for tissue engineering. Synthetic polymers such as polylactic acid (PLA), polycaprolactone (PCL) or polyethylene glycol (PEG) and natural polymers such as chitosan, collagen, alginate vb. are the most addressed types. Each of those provides unique features in this area [1, 2]. PLA is widely used for their superior mechanical properties and practical usage but, its hydrophobic nature can limit cell adhesion. PCL biodegradates with a slower rate compared to PLA. It provides excellent mechanical strength and is often used in combination with other polymers to enhance scaffold properties. PEG is known for its hydrophilicity and can be modified to incorporate bioactive molecules [3]. Its flexibility and tunable properties make it suitable for creating scaffolds with controlled degradation rates and mechanical strengths. As natural ones, collagen comes forward, and highly biocompatible and supports scaffolding but, challenging to process due to its susceptibility to degradation [4]. Brown seaweed yields the polysaccharide alginate is a kind of hydrogel initiated by calcium ions, which can be advantageous for creating scaffolds with high water content and adjustable mechanical properties. Chitosan is derived from chitin, is known for its antimicrobial properties and biocompatibility. It is often used in combination with other polymers to enhance scaffold performance.

Several 3D printing techniques are employed to fabricate polymer-based scaffolds, each with its advantages and limitations. Fused Deposition Modeling (FDM), stereolithography (SLA) or inkjet printing are the most popular ones that will be defined in the following parts. But the polymer-based 3D printed scaffolds have shown promising scenario. Many obstacles still need to be overcome in the field of polymer-based 3D printed scaffolds, despite tremendous advancements as material limitations, scale up or integration with biological systems.

Tissue engineering has advanced significantly with the use of polymer-based 3D printed scaffolds, which helps to create customized characteristics. Even with the obstacles still present, research and technological developments never stop pushing the limits of tissue regeneration and repair. There is hope for this field's future in creating more potent, clinically useful treatments for a variety of illnesses.

This chapter addresses both existing capabilities and future directions in the field of tissue engineering regarding polymer-based 3D printed scaffolds. It offers a thorough look at the status of technology.

Materials Research Foundations 172 (2025) 231-245 https://doi.org/10.21741/9781644903353-9

2. Polymers

Repeating structural units coupled together to form monomers are the building blocks of large compounds known as polymers by a covalent link to form lengthy chains [1]. These substances possess an extensive array of properties and applications based on chemical structure and the arrangement of the monomers.

2.1 Types of Polymers

2.1.1 Natural Polymers

These polymers are organic structures that exist naturally in living organisms. They play vital roles in biological systems and have been utilized by humans for various applications for centuries [5]. There are some examples of natural polymers and their functions.

Proteins: Peptide bonds provide the structural linkage between the lengthy chains of amino acids that make up proteins. Their functions are serving diverse functions including catalysis (enzymes), being a structural support (collagen), transportation (hemoglobin), and providing immune response (antibodies) [4]. Collagen, for example, gives connective tissues structural strength, while keratin is present in skin, hair, and nails. Moreover, casein is a protein in milk used in cheese-making[5]. As a primary component of the extracellular matrix, collagen is highly biocompatible and supports cell adhesion and growth. However, it can be challenging to process due to its susceptibility to degradation.

Nucleic Acids: Nucleotide's main structure consists of nucleotide monomers. Their functions are storing and transmitting genetic information. DNA transfers genetic blueprints for an organism and RNA plays various roles in translating genetic information into proteins and other functions.

Polysaccharides: Polysaccharides' structure is basically a long chain of mono-sugars. The main function of these polymers is to serve as energy storage or structural components. There are several examples for polysaccharides such as cellulose, starch, glycogen and chitin. Derived from chitin, chitosan is known for its antimicrobial properties and biocompatibility. It is often used in combination with other polymers to enhance scaffold performance.

Cellulose provides structural support in plant cell walls; it's the natural polymer that is most prevalent on Earth. Secondly, starch is a storage form of energy in plants, found in potatoes and grains. Next, animals store glucose in the form of glycogen, which is mainly found in the muscles and liver. Lastly, chitin provides structural support in the exoskeletons of arthropods and insects. Alginate is a polymer obtained from brown algae. If the alginates treated with calcium ions they form hydrogels, which can be advantageous for creating scaffolds with high water content and adjustable mechanical properties.

Natural Rubber: Natural rubbers are composed of polyisoprene, a polymer of the monomer isoprene. They are used for their elasticity and flexibility in general. The most known organic rubber is latex. It is derived from *Hevea brasiliensis* pulp, used in products like gloves and balloons.

Lignin: Lignin is a complex, irregular polymer composed of phenolic compounds. It provides rigidity and resistance to degradation in plant cell walls, especially in wood and bark.

Pectin: Pectin is a polysaccharide found in plant cell walls. Fruit syrups and jellies use pectin as a gelling agent and helps bind cells together in plant tissues.

Gum: Gums are a structure of various polysaccharides or polysaccharide derivatives. Gums are utilized in food and industrial applications as thickening, gelling, and stabilizing agents. Most common variations of gums are gum Arabic and gum guar. Gum Arabic is obtained from the acacia tree and used as a stabilizer in foods and drinks.

Natural polymers are integral to biological processes, and they are often biodegradable and can be more environmentally friendly compared to some synthetic polymers[6].

2.1.2 Synthetic Polymers

These are materials that were produced by humans using chemical reactions with specific properties for various applications. Unlike natural polymers, which are derived from biological sources, synthetic polymers are manufactured through polymerization processes involving chemical reactions [7], [2]. There are some main types of synthetic polymers. Those are plastics, synthetic rubbers, fibers, elastomers, engineering plastics, conductive polymers and biodegradable polymers [8].

PLA is made of renewable resources and is biodegradable. Because of its easy processing and good mechanical qualities, it is frequently used. Its hydrophobic properties, however, may restrict cell attachment [2]. Compared to PLA, PCL is a biodegradable polyester that breaks down more slowly [7]. It provides excellent mechanical strength and is often used in combination with other polymers to enhance scaffold properties [8], [9]. PEG is known for its hydrophilicity and can be modified to incorporate bioactive molecules. Its flexibility and tunable properties make it suitable for creating scaffolds with controlled degradation rates and mechanical strengths[7].

Plastics: In industry, plastics are the most used synthetic polymers, and they have three types, polyethylene (PE), polypropylene (PP) and polyvinyl Chloride (PVC). PEs are classified according to two density variations, low LDPE, and high HDPE. Both are used in packaging materials (bags, films), containers, toys, and pipes. Secondly, polypropylene (PP) is mainly used in automotive parts, textiles, food containers, and laboratory equipment. Lastly, polyvinyl chloride (PVC) can be either rigid (used in pipes) and flexible (used in vinyl flooring). polyvinyl chloride is commonly used on as a construction material for instance pipes, windows, medical devices, and as raw material of variational flooring tiles [2].

Synthetic Rubbers: There are three types of synthetic rubbers as a synthetic polymer. The first one is Styrene-Butadiene Rubber (SBR) which is used in tires, conveyor belts, and footwear. Then, the second is Nitrile Rubber (NBR) which is the main material of fuel and oil-resistant seals, gaskets, and gloves. And the last one is Ethylene Propylene Diene Monomer (EPDM)and it is used in automotive weather striping, roofing membranes, and seals.

Fibers: The garment sector extensively utilizes fibers. Nylon (Polyamide), polyester and acrylic are well-known fabric materials. Nylon (Polyamide) is used for Clothing, carpets, and industrial products like ropes and nets. Besides, Polyester is the fabrics for clothing, home textiles (curtains, bed linens), and PET bottles. Moreover, acrylic is used for knitwear, blankets, and outdoor fabrics.

Elastomers: Two types, silicone rubber and polyurethane, are the elastomers of synthetic polymer goods. Silicone Rubber is one of the main substances of cookware, seals, medical implants, and

Materials Research Foundations 172 (2025) 231-245 https://doi.org/10.21741/9781644903353-9

insulation for electrical cables. Additionally, polyurethane is a regular material for foams (mattresses, cushions), coatings, and adhesives.

Engineering Plastics: Polycarbonate, polyamide and polyetheretherketone are three types of most used engineering plastics. Polycarbonate (PC) is a popular material for Optical discs (CDs, DVDs), eyewear lenses, and impact-resistant materials. Polyamide (Nylon) has a common industrial consumption range in mechanical parts, automotive components, and industrial fibers. Polyetheretherketone (PEEK) is widely used for high-performance engineering applications, including aerospace and medical implants.

Conductive Polymers: Polyaniline (PANI) and polypyrrole (PPy) are main types of this kind. Polyaniline is used in electronics, sensors, and anti-static coatings, polypyrrole (PPy) are utilized in supercapacitors, sensors, and electronics.

Biodegradable Polymers: Biodegradable polymers are polylactic acid (PLA) for compostable food packaging, biodegradable cutlery, and 3D printing, and as a second one polyhydroxyalkanoates (PHA) that is for biodegradable plastics and medical devices.

2.2 Polymerization Processes

They are divided into three categories according to their structures. The linear ones made of long, straight chains (e.g., polyethylene), branched polymers have side chains branching off the main chain (e.g., low-density polyethylene) and cross-linked polymers consist of a lot of chains are interconnected to form a network (e.g., vulcanized rubber) [10].

Besides, polymerization is the chemical process used to create polymers by linking monomers together. There are several key polymerization methods, each producing different types of polymers with unique properties [5], [11], [12]. The addition (chain-growth) and condensation (step-growth) polymerization processes are the two main types of polymerization processes.

2.2.1 Chain-Growth

According to addition polymerization, monomers with double bonds (or other reactive groups) react to form a polymer without producing by-products (e.g., polystyrene from styrene or Polyethylene from ethylene). This process typically involves the following stages.

Initiation: An initiator (e.g., free radicals, cations, or anions), starts polymerization by reacting with the monomer to create a new active site. Free radicals in the polymerization of polyethylene from ethylene is an example of initiation.

Propagation: The active site created during initiation reacts with other monomer molecules, extending the polymer chain. This stage continues until the monomer supply is exhausted or the reaction is otherwise stopped. During the production of polystyrene from styrene, the growing chain reacts with more styrene molecules, and this is a common example of propagation.

Termination: The polymerization ends when two growing polymer chains combine (recombination) or when the active site is deactivated by a terminating agent. two polymer radicals combined to create a stable end group is one of the termination process examples.

There are three types of Addition Polymers, free radical, cationic and anionic polymerizations [11]. To begin, free radical polymerization uses free radicals to initiate the process. Those are common for making polyethylene, polystyrene, and polymethyl methacrylate. Secondly, cationic

Materials Research Foundations 172 (2025) 231-245 https://doi.org/10.21741/9781644903353-9

polymerization uses cations (positively charged ions) to initiate polymerization. Their examples include the production of polyisobutylene. Thirdly, anionic polymerization uses anions (negatively charged ions) to initiate polymerization. Their examples include the production of polybutadiene [8], [9]..

2.2.2 Condensation (Step-Growth) Polymerization

In this kind, the monomers with one or more active groups join together to form polymers, and as a byproduct, a tiny molecule like water is produced (e.g., nylon from diamines and dicarboxylic acids or polyester from diols and dicarboxylic acids.) [13]. This process involves two main stages: the formation of dimers and trimers, and the growth of longer polymer chains.

Formation of Dimers and Trimers: At that phase, monomers react to form dimers (two monomers) and trimers (three monomers), which can further react to form longer chains. For example, the reaction of a diol and a dicarboxylic acid to form a polyester are one of the formation of dimers and trimers.

Growth and Polymer Formation: Dimers and trimers react to form longer chains, continuing until the reaction reaches completion. Small molecules like water or methanol are released as by-products. The formation of nylon or the formation of polyester are examples for growth and polymer formation.

There are three different kinds of condensation polymers that come after these phases. These include polyurethanes, polyesters, and polyamides. Polyesters are created using the process of diols and dicarboxylic acids. Examples include PET (polyethylene terephthalate) and polycarbonate. Polyamides are obtained from the reaction of diamines and dicarboxylic acids. Examples include nylon 6,6 and nylon 6. And polyurethanes include foams and coatings.

2.2.3 Other Polymerization Techniques

There are more polymerization techniques in industry and those are evolving day by day. Most common ones are copolymerization, Ring-Opening polymerization and living Polymerization. The process of copolymerization entails polymerizing two or more distinct monomers to produce copolymers with various characteristics [3], [14]. Styrene-butadiene rubber (SBR), which combines styrene and butadiene are examples for copolymerization. Another one is Ring-Opening Polymerization which requires the opening of a cyclic monomer into create a polymer. Various methods, including catalysis used to achieve this purpose. For instance, the polymerization of lactides to form polylactic acid (PLA). The last one is Living Polymerization, is a good example of this section, where the active chain ends are not terminated and can continue to grow [3], [15]. This allows for precise control over polymer size and architecture. Living anionic polymerization used to create well-defined block copolymers are applications of living polymerization.

The uses for polymers are practically limitless. Plastics like polyethylene and polypropylene are used for containers, bags, and wrap are used in packaging[16]. Fibers like polyester, nylon, and spandex are used in clothing and upholstery are the main components of textile industry. Polymers like PVC and epoxy resins in pipes, sealants and varnishes are used in construction. Lastly, the main topic of this chapter is medical devices, the polymers are used in implants, pharmacy, and tissue engineering very much. Understanding these polymerization processes helps in designing and producing polymers with specific properties for various applications, from everyday items to

Materials Research Foundations 172 (2025) 231-245
https://doi.org/10.21741/9781644903353-9

advanced materials [11], [17]. Every approach has pros and cons of its own that affect which polymerization technique is best for a certain material.

2.3 Properties

Polymers are used based on their chemical susceptibility, mechanical qualities, and thermo features. To begin with, mechanical properties can range from very flexible to highly rigid, depending on the polymer's structure and additives. Thermal properties vary from thermoplastic (can be melted and reformed) to thermosetting (hardens irreversibly) [18]. And chemical resistance is related to synthetic polymers and can be designed to resist various chemicals, making them suitable for different industrial applications.

Understanding polymers involves chemistry, materials science, and engineering, as their design and properties are critical to their functionality in various applications[18]. Synthetic polymers are versatile and tailored to meet specific needs across many industries, from everyday consumer products to advanced technological applications [12]. Their development has revolutionized manufacturing and consumer goods, making modern life more convenient and functional.

3. Polymer-Based 3D Printed Scaffolds for Tissue Engineering

The human body is composed of several tissues that range from macro to nanoscale, providing to perform a wide variety of simple to complex body functions. Each year, many people suffer from tissue damage caused by disease and/or accidents. Because of this reason, patients need tissue/organ transplantation. The challenges of finding suitable organ donors and the risk of transplanted tissues or organs being rejected by the recipient's body have led scientists to create multifunctional materials that can substitute for tissues or organs. At this point, The field of tissue engineering has become a field that requires the contributions of cell therapy, molecular therapy, artificial and bio-organ technology to ensure the regeneration or healing of ill or injured tissues [19] [20].

Tissue engineering combines material science, cell biology, and engineering principles to promote the development or repair of damaged or diseased tissues and organs [21][22]. Tissue engineering studies comprise applications with liver, vascular, kidney tissues and especially non-regenerative tissues such as bone, skin, or nerve cells [22].

As a subfield of regenerative medicine, tissue engineering applications have seen significant development over the past decade. There are three different strategies of tissue engineering to restore the defective tissue/organ: (1) Transplanting replacements to the host, (2) stimulating factors (3) cell positioning into the different scaffolds (Nerem, 1992).

A combination of cells (autologous, allogeneic, xenogeneic, progenitor and stem cells) scaffolds, and growth signals draw attention in tissue engineering applications as the essential components for creating engineered tissues (Fig. 1). A scaffold is a three-dimensional (3D) platform that may control cellular responses by supplying various signals, including mechanical, spatial, and biological ones [23]. There are four different scaffolding strategies in tissue engineering:

- Porous scaffold that is a basement for cell seeding
- Extracellular matrix without cells

Materials Research Foundations 172 (2025) 231-245
https://doi.org/10.21741/9781644903353-9

- Cells are together with extracellular matrix
- Cells encapsulated with 3D hydrogel structure [24].

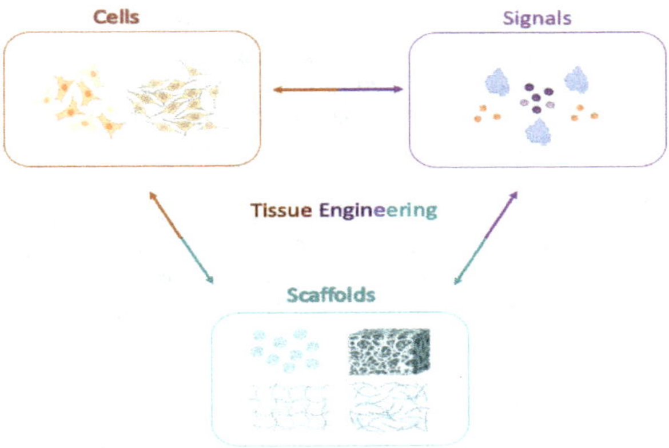

Figure 1. Three main components of scaffolding

The primary aim of a scaffold is to maintain mechanical integrity at the site of damage until regeneration occurs. For this reason, scaffolds need to have some specific features [25]. First, scaffolds must have sufficient space to support cell viability, proliferation, differentiation and vascularization. This allows for the intake of essential nutrients for dividing cells, the removal of cellular waste, proper cell penetration during cell development, and new tissue growing into and integrates with a scaffold [26]. Scaffolds used in tissue engineering applications must be both biocompatible and biodegradable. Following their application, these scaffolds should integrate with the adjacent tissues and endogenous cells, generate minimal inflammation and immune response, and should degrade concurrently with the formation of new tissue and/or cells [27]. In addition to all these properties and functions, scaffolds in tissue engineering need to be compatible with biomolecules such as cytokines, growth factors, adhesion proteins, and should be structurally suitable for the controlled release of these biomolecules [28].

There are a wide range of materials and production techniques used in scaffold fabrication (Fig. 2). Upon examining scaffold fabrication methods, it is observed that conventional methods such as phase separation, electrospinning or solvent casting method provide limited control over scaffold geometry and have a low level of repeatability [29].

Materials Research Foundations 172 (2025) 231-245
https://doi.org/10.21741/9781644903353-9

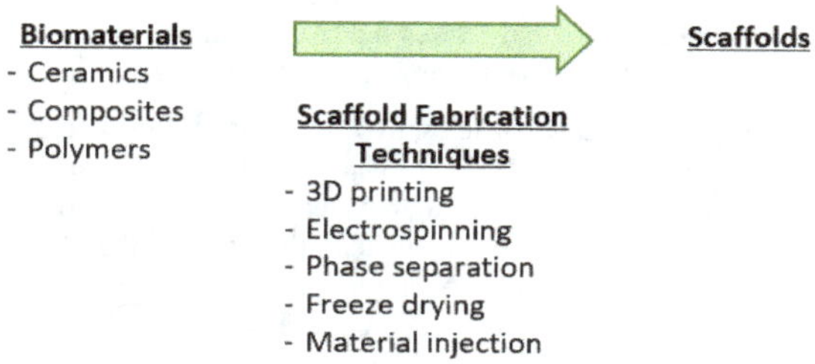

Figure 2. Type of scaffold fabrications

3.1 3D Printing Techniques

Several 3D methods are employed to fabricate polymer-based scaffolds, each with its advantages and limitations such as fused deposition modeling (FDM), Stereolithography (SLA), inkjet printing or electrospinning. Some of them are explained in this part. Beginning with the FDM, in which a thermoplastic polymer is heated and extruded via a nozzle to gradually construct the scaffold. Widely applicable to various polymers, FDM is renowned for its affordability and ease of use. However, it may lack the resolution required for intricate scaffold designs.

SLA solidifies liquid resin into layers by curing it with a laser. This technique offers high resolution and precision, making it suitable for complex scaffold geometries. However, the choice of resins is limited, and post-processing is often required.

Inkjet printing involves depositing droplets of polymer solutions or suspensions onto a build platform. It allows for the creation of scaffolds with high resolution and can incorporate multiple materials. However, the physical features of the printed structures vary onto the ink formulation.

Compared to conventional techniques, 3D printing has several advantages. 3D printing can deliver nearly all components required to build tissues or organs, such as cells, cytoplasmic matrix, and growth factors. In addition, 3D printing enables the creation of flexible models and personalized structures while reducing waste through the use of computer software. Besides these features, 3D printing creates structures that are very similar to natural biological tissues, allowing for the design of particular region of disesase [30] [31].

3.2 Applications for Tissue Engineering

Polymer-based 3D printed scaffolds have shown promising achievements as given below.

Bone Tissue Engineering: Bone tissue scaffolds need to be strength full and bioactive. Polymers like PCL and PLA are often combined with bioactive ceramics or growth factors to enhance bone regeneration. The ability to customize scaffold porosity and architecture supports the formation of new bone tissue.

Cartilage Repair: Cartilage scaffolds must provide flexibility and support while promoting chondrocyte growth. Natural polymers like collagen and alginate are frequently used in cartilage scaffolds due to their resemblance to the native extracellular matrix.

Skin Regeneration: Scaffolds for skin regeneration require both mechanical support and the ability to support vascularization. Incorporated structures made up with natural and synthetic polymers are utilized to imitate the skin's structure and promote wound healing.

Among the products of 3D printing, nanoparticles, hydrogels, nanofibers and organoids are mostly used in tissue engineering applications [32] [33] [34] [35]. Hydrogels are composed of natural, synthetic, or hybrid polymers that swell through water during their formation. Due to their structure, cellular characteristics like adhesion, differentiation, distribution, and migration are supported by hydrogels. In terms of their applications, hydrogels are commonly used in drug delivery, wound healing, contact lenses, and 3D cell cultures [36] [37].

3D hydrogel-based cell culture provides in vitro diseases models that provide to understand mechanisms of various pathologies. Siddiqui and his colleagues have shown that 3D hydrogels composed of Star-PEG and heparin provide a 3D cell culture system that enables the investigation of neural development and the pathology of Alzheimer's Disease. The Star-PEG and heparin-based 3D cell culture system provides a neuron-specific extracellular matrix structure that induces neuronal network formation and the generation of various neuron-specific signals. Following amyloid beta application, the 3D cell culture system revealed Alzheimer's Disease pathologies such as synaptic loss, excessive tau hyperphosphorylation, and the presence of neurofibrillary tangles [38]. Hydrogels have a good potential as tools in wound healing. In this study, MC methylcellulose hydrogels were used to treat the wounded body region of Sprague-Dawley rats (4-week-old). It was reported that MC methylcellulose hydrogels significantly improved the healing of burn wounds compared to the untreated experimental group [39]. Biocompatible hydrogels are also used for controlled drug release. A drug delivery study using chitosan (CS)/polyvinyl alcohol (PVA) hydrogels demonstrated that hydrogels loaded with fluorouracil (5-FU) had a strong ability to retain the drug at pH 7.4, this prevented drugs from seeping into healthy cells and minimized side effects. The drug release capacity is 84.8% at pH 5.0. The current investigation concludes that the hydrogel accomplishes pH-depended drug release to the lesion region in addition to maintaining a greater concentration of 5-FU surrounding tumor cells [40].

Nanoparticles are small entities, from 1 to 100 nm, composed of materials such as carbon, metal, ceramic, lipids, or polymers. They are utilized for therapeutic, medical and biological applications [41]. Gold nanoparticles (AuNPs) employed for drug studies. Findings from research on B16F10 murine melanoma cells reveal that AuNPs, loaded with imatinib mesylate (IM) and STAT3 siRNA decrease cell viability and STAT3 protein expression levels. In addition to in vitro studies, the combination of IM and STAT3 siRNA was applied to mice with melanoma tumors. According to the results, tumor size was significantly reduced after the treatment and STAT3 protein expression levels were reduced. Compared to the control groups, these findings indicate that nanoparticles have target-specific function in drug delivery [42].

Beside drug delivery function, AuNPs induce regeneration of bone tissue. According to a study conducted by Zhang and his colleagues, AuNPs promote addition, cell viability and change of primary osteoblasts. Primary osteoblasts have the potential to form, repair and structurally support bone. After the application of AuNPs to the osteoblasts, alkaline phosphatase levels, the amount of bone nodules and calcium density have been increased. In addition, osteogenic gene (BMP-2, Runx-2, OCN and Col-1) The degrees of communication were significantly elevated. All in all, AuNPs are a possible option for improving bone regrowth [43]

Organoids are three-dimensional cellular models used in tissue and organ regeneration with drug screening applications, which exhibit a significant similarity to in vivo tissues and organs [44] [45]. An organoid structurally consists of a source cell (stem/progenitor cells or induced pluripotent stem cells), various cellular factors (growth factors, small molecules etc.), and a basement similar to extracellular matrix components (collagen, matrigel, polymeric hydrogels etc.) [46]. It has been observed that heart, liver, and kidney organoids have been developed using three-dimensional bioprinting technologies [47] [48] [49]. Organoid models containing tumor stem cells provide preliminary information about the characteristics and heterogeneity of tumors and contribute significantly to personalized cancer treatment. Due to these characteristics, tumor organoid models provide improved knowledge of the structure of tumors compared to animal models [50] [51]. In a current investigation conducted by Chang et. al. has explained that organoid models obtained from various ovarian cancer cells offer a more sustainable approach compared to two-dimensional cell cultures and patient-derived xenografts (PDX). This organoid model facilitates and enables research into the molecular mechanisms, classification, and drug screening of ovarian cancer [52].

Conclusion

Even with great advancements, there are still several obstacles in the field of polymer-based 3D printed scaffolds. There are material limitations, and the features of new polymers and composites should be enhanced, and diverse functionality is needed. Scaling up is another problem, translating 3D printed scaffolds from the lab to clinical applications requires scaling up production methods while maintaining quality and consistency. Also, integration with biological systems still needs some more investigations. Better integration of scaffolds with the surrounding tissue and incorporation of dynamic stimuli to enhance tissue regeneration are ongoing areas of research.

Polymer-based 3D printed scaffolds represent a significant advancement in tissue engineering, providing the capacity to build intricate structures with tailored properties. Even with the obstacles still present, continuous research and technical developments keep pushing the envelope of what is possible in tissue repair and regeneration. This field has a bright future for developing more effective and clinically applicable solutions for a wide range of medical conditions.

References

[1] G. R. Strobl, The physics of polymers, vol. 2. Springer, 1997. https://doi.org/10.1007/978-3-662-03488-0

[2] S. C. Rasmussen, "Revisiting the early history of synthetic polymers: critiques and new insights," Ambix, vol. 65, no. 4, pp. 356-372, 2018. https://doi.org/10.1080/00026980.2018.1512775

[3] G. Wypych, Handbook of polymers. Elsevier, 2022.

[4] O. Olatunji, Natural polymers: industry techniques and applications. Springer, 2015.

[5] K. Kaushik, R. B. Sharma, and S. Agarwal, "Natural polymers and their applications," Int. J. Pharm. Sci. Rev. Res., vol. 37, no. 2, pp. 30-36, 2016.

[6] G. Satchanska, S. Davidova, and P. D. Petrov, "Natural and Synthetic Polymers for Biomedical and Environmental Applications," Polymers (Basel)., vol. 16, no. 8, p. 1159, 2024. https://doi.org/10.3390/polym16081159

[7] M. F. Maitz, "Applications of synthetic polymers in clinical medicine," Biosurface and Biotribology, vol. 1, no. 3, pp. 161-176, 2015. https://doi.org/10.1016/j.bsbt.2015.08.002

[8] M. C. Hacker, J. Krieghoff, and A. G. Mikos, "Synthetic polymers," in Principles of regenerative medicine, Elsevier, 2019, pp. 559-590. https://doi.org/10.1016/B978-0-12-809880-6.00033-3

[9] S. Bhatia and S. Bhatia, "Natural polymers vs synthetic polymer," Nat. Polym. drug Deliv. Syst. nanoparticles, plants, algae, pp. 95-118, 2016. https://doi.org/10.1007/978-3-319-41129-3_3

[10] L. E. Nielsen, "Cross-linking-effect on physical properties of polymers," J. Macromol. Sci. Part C, vol. 3, no. 1, pp. 69-103, 1969. https://doi.org/10.1080/15583726908545897

[11] V. Ambrogi, C. Carfagna, P. Cerruti, and V. Marturano, "Additives in polymers," in Modification of polymer properties, Elsevier, 2017, pp. 87-108. https://doi.org/10.1016/B978-0-323-44353-1.00004-X

[12] R. Geyer, "Production, use, and fate of synthetic polymers," in Plastic waste and recycling, Elsevier, 2020, pp. 13-32. https://doi.org/10.1016/B978-0-12-817880-5.00002-5

[13] S. Ramakrishnan, "Condensation polymerization," Resonance, vol. 22, pp. 355-368, 2017. https://doi.org/10.1007/s12045-017-0475-0

[14] L.-S. Hornberger, P. Weingarten, P. L. Lange, T. Schleid, and F. Adams, "Connecting 16- with 4-Membered Rings: Ring-Opening Polymerization of Bio-Based ω-Pentadecalactone with Amino-Alkoxy-Bis(phenolate) Yttrium Initiators and Copolymerization with β-Butyrolactone," Eur. Polym. J., vol. 199, p. 112449, 2023. https://doi.org/10.1016/j.eurpolymj.2023.112449

[15] A. N. Generalova et al., "Polymers in 3D printing of external maxillofacial prostheses and in their retention systems," Int. J. Pharm., vol. 657, p. 124181, 2024. https://doi.org/10.1016/j.ijpharm.2024.124181

[16] G. W. Ehrenstein and S. Pongratz, "Applications," G. W. Ehrenstein and S. B. T.-R. and S. of P. Pongratz, Eds. Hanser, 2013, pp. 423-884. https://doi.org/10.1007/978-3-446-43709-8_5

[17] V. Mubayi, C. B. Ahern, M. Calusinska, and M. A. O'Malley, "Toward a Circular Bioeconomy: Designing Microbes and Polymers for Biodegradation," ACS Synth. Biol., vol. 13, no. 7, pp. 1978-1993, 2024. https://doi.org/10.1021/acssynbio.4c00077

[18] J. E. Mark, Physical properties of polymers handbook. Newyork: Springer, 2007. https://doi.org/10.1007/978-0-387-69002-5

[19] T. Li, J. Chang, Y. Zhu, and C. Wu, "3D printing of bioinspired biomaterials for tissue regeneration," Adv. Healthc. Mater., vol. 9, no. 23, p. 2000208, 2020. https://doi.org/10.1002/adhm.202000208

Materials Research Forum LLC
https://doi.org/10.21741/9781644903353-9

[20] M. S. Kim, J. H. Kim, B. H. Min, H. J. Chun, D. K. Han, and H. B. Lee, "Polymeric scaffolds for regenerative medicine," Polym. Rev., vol. 51, no. 1, pp. 23-52, 2011. https://doi.org/10.1080/15583724.2010.537800

[21] R. Boni, A. Ali, A. Shavandi, and A. N. Clarkson, "Current and novel polymeric biomaterials for neural tissue engineering," J. Biomed. Sci., vol. 25, pp. 1-21, 2018. https://doi.org/10.1186/s12929-018-0491-8

[22] J. Liu and C. Yan, "3D printing of scaffolds for tissue engineering," by Cvetković D., Intech Open, UK, vol. 7, pp. 137-154, 2018. https://doi.org/10.5772/intechopen.78145

[23] J. A. Hunt, R. Chen, T. van Veen, and N. Bryan, "Hydrogels for tissue engineering and regenerative medicine," J. Mater. Chem. B, vol. 2, no. 33, pp. 5319-5338, 2014. https://doi.org/10.1039/C4TB00775A

[24] B. P. Chan and K. W. Leong, "Scaffolding in tissue engineering: general approaches and tissue-specific considerations," Eur. spine J., vol. 17, pp. 467-479, 2008. https://doi.org/10.1007/s00586-008-0745-3

[25] M. R. Dias, J. M. Guedes, C. L. Flanagan, S. J. Hollister, and P. R. Fernandes, "Optimization of scaffold design for bone tissue engineering: A computational and experimental study," Med. Eng. Phys., vol. 36, no. 4, pp. 448-457, 2014. https://doi.org/10.1016/j.medengphy.2014.02.010

[26] E. Pamuła, Biomateriały dla inżynierii tkankowej: badania nad kształtowaniem struktury i właściwości biologicznych poliestrów alifatycznych. Polskie Stowarzyszenie Biomateriałów, 2008.

[27] H. Patel, M. Bonde, and G. Srinivasan, "Biodegradable polymer scaffold for tissue engineering," Trends Biomater Artif Organs, vol. 25, no. 1, pp. 20-29, 2011.

[28] J. Berger, M. Reist, J. M. Mayer, O. Felt, N. A. Peppas, and R. Gurny, "Structure and interactions in covalently and ionically crosslinked chitosan hydrogels for biomedical applications," Eur. J. Pharm. Biopharm., vol. 57, no. 1, pp. 19-34, 2004. https://doi.org/10.1016/S0939-6411(03)00161-9

[29] A. Zaszczyńska, M. Moczulska-Heljak, A. Gradys, and P. Sajkiewicz, "Advances in 3D printing for tissue engineering," Materials (Basel)., vol. 14, no. 12, p. 3149, 2021. https://doi.org/10.3390/ma14123149

[30] S. Jordahl et al., "Engineered fibrillar fibronectin networks as three-dimensional tissue scaffolds," Adv. Mater., vol. 31, no. 46, p. 1904580, 2019. https://doi.org/10.1002/adma.201904580

[31] H. Huang and D. Dean, "3-D printed porous cellulose acetate tissue scaffolds for additive manufacturing," Addit. Manuf., vol. 31, p. 100927, 2020. https://doi.org/10.1016/j.addma.2019.100927

[32] B. Xu, Y. Li, B. Deng, X. Liu, L. Wang, and Q.-L. Zhu, "Chitosan hydrogel improves mesenchymal stem cell transplant survival and cardiac function following myocardial infarction in rats," Exp. Ther. Med., vol. 13, no. 2, pp. 588-594, 2017. https://doi.org/10.3892/etm.2017.4026

[33] L. Saludas, S. Pascual-Gil, F. Prósper, E. Garbayo, and M. Blanco-Prieto, "Hydrogel based approaches for cardiac tissue engineering," Int. J. Pharm., vol. 523, no. 2, pp. 454-475, 2017. https://doi.org/10.1016/j.ijpharm.2016.10.061

[34] J. Joshi and C. R Kothapalli, "Nanofibers based tissue engineering and drug delivery approaches for myocardial regeneration," Curr. Pharm. Des., vol. 21, no. 15, pp. 2006-2020, 2015. https://doi.org/10.2174/1381612821666150302153138

[35] A. Hasan, M. Morshed, A. Memic, S. Hassan, T. J. Webster, and H. E.-S. Marei, "Nanoparticles in tissue engineering: applications, challenges and prospects," Int. J. Nanomedicine, pp. 5637-5655, 2018. https://doi.org/10.2147/IJN.S153758

[36] Y. Park, K. M. Huh, and S.-W. Kang, "Applications of biomaterials in 3D cell culture and contributions of 3D cell culture to drug development and basic biomedical research," Int. J. Mol. Sci., vol. 22, no. 5, p. 2491, 2021. https://doi.org/10.3390/ijms22052491

[37] E. Hamidi, R. Alhaloush, Ş. Şenol, and E. AKYOL, "An overview on current trends and future outlook of hydrogels in drug delivery.," Sigma J. Eng. Nat. Sci. ve Fen Bilim. Derg., vol. 41, no. 5, 2023. https://doi.org/10.14744/sigma.2023.00071

[38] T. Siddiqui et al., "Three-Dimensional Biohybrid StarPEG-Heparin Hydrogel Cultures for Modeling Human Neuronal Development and Alzheimer's Disease Pathology," in Alzheimer's Disease: Methods and Protocols, Springer, 2022, pp. 159-170. https://doi.org/10.1007/978-1-0716-2655-9_8

[39] M. H. Kim, H. Park, H. C. Nam, S. R. Park, J.-Y. Jung, and W. H. Park, "Injectable methylcellulose hydrogel containing silver oxide nanoparticles for burn wound healing," Carbohydr. Polym., vol. 181, pp. 579-586, 2018. https://doi.org/10.1016/j.carbpol.2017.11.109

[40] G. Chang et al., "Self-healable hydrogel on tumor cell as drug delivery system for localized and effective therapy," Carbohydr. Polym., vol. 122, pp. 336-342, 2015. https://doi.org/10.1016/j.carbpol.2014.12.077

[41] A. Radomska, J. Leszczyszyn, and M. W. Radomski, "The nanopharmacology and nanotoxicology of nanomaterials: new opportunities and challenges," Adv. Clin. Exp. Med., vol. 25, no. 1, pp. 151-162, 2016. https://doi.org/10.17219/acem/60879

[42] S. Labala et al., "Effective melanoma cancer suppression by iontophoretic co-delivery of STAT3 siRNA and imatinib using gold nanoparticles," Int. J. Pharm., vol. 525, no. 2, pp. 407-417, 2017. https://doi.org/10.1016/j.ijpharm.2017.03.087

[43] D. Zhang, D. Liu, J. Zhang, C. Fong, and M. Yang, "Gold nanoparticles stimulate differentiation and mineralization of primary osteoblasts through the ERK/MAPK signaling pathway," Mater. Sci. Eng. C, vol. 42, pp. 70-77, 2014. https://doi.org/10.1016/j.msec.2014.04.042

[44] Y. Ren et al., "Developments and opportunities for 3D bioprinted organoids," Int. J. Bioprinting, vol. 7, no. 3, 2021. https://doi.org/10.18063/ijb.v7i3.364

[45] A. Fatehullah, S. H. Tan, and N. Barker, "Organoids as an in vitro model of human development and disease," Nat. Cell Biol., vol. 18, no. 3, pp. 246-254, 2016. https://doi.org/10.1038/ncb3312

[46] Z. Zhao et al., "Organoids," Nat. Rev. Methods Prim., vol. 2, no. 1, p. 94, 2022. https://doi.org/10.1038/s43586-022-00186-8

[47] M. Takasato et al., "Kidney organoids from human iPS cells contain multiple lineages and model human nephrogenesis," Nature, vol. 536, no. 7615, p. 238, 2016. https://doi.org/10.1038/nature17982

[48] M. E. Kupfer et al., "In situ expansion, differentiation, and electromechanical coupling of human cardiac muscle in a 3D bioprinted, chambered organoid," Circ. Res., vol. 127, no. 2, pp. 207-224, 2020. https://doi.org/10.1161/CIRCRESAHA.119.316155

[49] M. Huch et al., "In vitro expansion of single Lgr5+ liver stem cells induced by Wnt-driven regeneration," Nature, vol. 494, no. 7436, pp. 247-250, 2013. https://doi.org/10.1038/nature11826

[50] L. Ma et al., "The construction of in vitro tumor models based on 3D bioprinting," Bio-Design Manuf., vol. 3, pp. 227-236, 2020. https://doi.org/10.1007/s42242-020-00068-6

[51] A. Dominijanni, A. Mazzocchi, E. Shelkey, S. Forsythe, M. Devarsetty, and S. Soker, "Bioengineered tumor organoids," Curr. Opin. Biomed. Eng., vol. 13, pp. 168-173, 2020. https://doi.org/10.1016/j.cobme.2020.03.005

[52] Y.-H. Chang, K.-C. Wu, T. Harnod, and D.-C. Ding, "The organoid: A research model for ovarian cancer," Tzu Chi Med. J., vol. 34, no. 3, pp. 255-260, 2022. https://doi.org/10.4103/tcmj.tcmj_63_21

Applications of Polymers in Surgery II
Materials Research Foundations 172 (2025) 246-262
Materials Research Forum LLC
https://doi.org/10.21741/9781644903353-10

Chapter 10

Polymers for the Production of Conventional Dosage Forms

Maria Lúcia Mendonça Costa[1], Layla Ribeiro Lucena[1], Brenda Maria Silva Bezerra[2], João Augusto Oshiro-Júnior[2*], Bruna Galdorfini Chiari-Andréo[3*]

[1]Department of Biological and Health Sciences, University of Araraquara – UNIARA, Araraquara, São Paulo, Brazil

[2]Pharmaceutical Sciences Postgraduate Program, Center for Biological and Health Sciences, State University of Paraíba – UEPB, Campina Grande, Paraíba, Brazil

[3]Department of Biological and Health Sciences, University of Araraquara – UNIARA, Araraquara, São Paulo, Brazil

[*]joaooshiro@yahoo.com.br; bgchiari-andreo@uniara.edu.br

Abstract

Performing a surgical procedure aims to improve the patient's health and quality of life. Despite several controls and the advances experienced in this area, there is invariably aggression in organs and tissues in these procedures, subsequently requiring their repair. For this purpose, there are pharmacological resources to guarantee good results, prevent infections, control pain, contribute to the psychological care of patients, and improve the quality of healing. Pharmaceutical dosage forms can carry molecules of interest and contribute to the repair process. In these dosage forms various polymers can be used. Therefore, this chapter discusses the classifications of these different polymers, their particularities, and their use in medicines.

Keywords

Polymers, Pharmaceutical Dosage Forms, Excipients, Drug Delivery Systems

Contents

Materials Research Foundations 172 (2025) 246-262
https://doi.org/10.21741/9781644903353-10

1. Introduction

The hospital context requires expressive care to ensure the health and well-being of patients, and for this, several control systems must be used. These controls must avoid casualties, guaranteeing conditions of maximum security, based on a meticulous process of precautions. Among the controls mentioned, hospital infections can be listed; the rational use of antimicrobials and other medications; the profile of resistance and sensitivity to these substances; hand hygiene; the continuing education of health professionals; standardization, implementation, and monitoring of insulation measures; the education of family members, among several others. Also noteworthy is the control of all surgical procedures.

The performance of a surgical procedure, whatever it is, has a very simple reason to be performed: to improve the patient's health. Therefore, controls and care must occur in the perioperative period, evaluating the necessary controls to reduce the risk of complications related to surgical intervention. The phases that make up the perioperative period are the preoperative - assessment of the patient's initial health status, the intraoperative phase - the moment of the intervention itself and, the postoperative period - in which the monitoring and control of the patient is carried out between the leaving the operating room and until he resumes his daily activities [1].

All this control and planning of actions, when noticed by the patient, generates their confidence in the team that takes care of them and reduces the impact on their emotional health. Understandably, the patient mobilizes a large amount of physical and psychic energy to deal with situational stress [2].

Also, in this sense of ensuring the well-being of the surgical patient, Garcia et al. (2017) [3] suggest strategies to control the impacts in the postoperative period, namely the pre-anesthetic assessment to identify factors for the inclusion or exclusion of a certain technique or pharmaceutical product; analgesia selection; patient education to avoid anxiety and unrealistic expectations regarding postoperative pain; consultation with the anesthesiologist to adapt the analgesia technique to the intraoperative period; postoperative follow-up and periodic evaluation of the success of pain management and the possibility of modifications based on the patient's response [3].

Despite all these precautions and controls mentioned, and also the great advances experienced in surgical procedures, there is invariably aggression to organs and tissues during the procedure, subsequently requiring their repair. The various events that contribute to tissue repair can be organized into a sequence of four recognized stages that define the evolutionary process, namely,

Materials Research Foundations 172 (2025) 246-262
https://doi.org/10.21741/9781644903353-10

(1) hemostasis, (2) inflammation, (3) proliferation, and (4) remodeling. These stages are interrelated events that overlap in time. Each of these phases contributes to the wound healing process and depends largely on the specific characteristics it presents, such as its extension (including area and depth) [4].

It is noted, through this brief contextualization, that allies to guarantee good results must be sought, avoiding infections, controlling pain, contributing to the psychological care of surgical patients and improving the quality of healing, preventing the formation of deforming scars. For this, that is, among these allies, several pharmacological resources and pharmaceutical products in general can be used. In order to carry molecules of interest or contribute to an ideal environment for the repair process, several polymers can be used in conventional pharmaceutical dosage forms.

These polymers, which can be used in the most varied pharmaceutical forms, are expected to be biocompatible and biodegradable, with the possibility of even using natural polymers.

The biodegradability and biocompatibility of a polymer are among the most important properties for pharmaceutical applications. In other words, biocompatibility refers to specific properties of a material that are non-toxic or without harmful effects on biological systems. Non-biocompatible materials can cause tissue damage when placed in contact, including irreversibly, with permanent tissue destruction, necrosis, significant and dystrophic fibrosis, in addition to calcification [5].

Biodegradable and biocompatible materials for pharmaceutical dosage forms have allowed the advancement of pharmaceutical products, providing better therapy and management of the disease state, through the adequate release of drugs, in some situations, immediately and, in others, as controlled release [5].

The use of polymers to carry drugs or active substances increases as researchers have shown their benefits and versatility. Polymers are highly used as excipients in drug formulations. An excipient is any inactive substance, that is, devoid of therapeutic power, but which has various functions, such as ensuring the stability and biopharmaceutical properties of drugs; improving the organoleptic, microbiological, and chemical characteristics of the product; assisting in preparation; and ensure patient acceptability.

Among the various applications of polymers used in medicines are: the control of drug release as previously mentioned, in addition to film-forming properties (coating of dosage forms), thickening (rheology and sensory modifier), gelation (controlled release), adhesion (binding between particles - agglutination), pH-dependent solubility, solubility in organic solvents (taste masking) and barrier properties (protection and packaging) [6].

Based on the above and given the importance of polymers as raw materials to carry substances with a relevant therapeutic effect in the most diverse therapies, including surgical ones, this chapter discusses the classifications of these different polymers, their use in conventional and advanced medicines, present the most used polymers and their particularities.

2. Classification of polymers used in pharmaceutical dosage forms

Different parameters can be adopted for the classification of polymers, including those for pharmaceutical use.

Considering solubility, pharmaceutical polymers can be classified as water-soluble and water-insoluble (in this case, soluble in fatty solvents or organic solvents). Cellulose ethers with methyl

 Materials Research Forum LLC
https://doi.org/10.21741/9781644903353-10

and hydroxypropyl substitutions are soluble in water, while ethylcellulose and a group of cellulose esters such as cellulose acetate, butyrate or phthalate are soluble in organic solvents. Hydrocolloid gums are also used when water solubility is desirable. Water-soluble synthetic polymers have also found extensive applications in the pharmaceutical industries, including polyethylene glycol, vinyl alcohol and polyethylene glycol polymers, polyethylene oxides, polyvinylpyrrolidone, and polyacrylate or polymethacrylate esters [7].

There are also charged polymers used as excipients in pharmaceutical formulations, such as cationic polymers (polyquaternary) with the presence of quaternary ammonium centers, extensively used in hair conditioner formulations and formulations with antimicrobial purposes [7]. According to Canevarolo Jr. (2002) [8], when dealing with polymerizations there is a division into cationic and anionic due to their electron alteration, so that "in the first case a positive charge, carbocation, is generated and the reaction is called cationic polymerization, and, in the second, there is an excess of electrons generating a negative charge, carbanion or anionic polymerization" [8]. Thus, anionic polymers exhibit a negative charge in their chain, attracting predominantly positively charged molecules, for example: carboxymethylcellulose and xanthan gum – widely used in the formulation of syrups, emulsions, suspensions, creams, and tablets to prolong the release time of the active compound.

Considering the classification in terms of charges, it is also necessary to describe amphoteric polymers, those capable of reacting with acidic or basic behavior, such as acrylic acid and methacrylic acid or their esters [6,9,10,11].

The so-called biodegradable polymers are materials that undergo macromolecular degradation in vivo by the action of enzymes from natural microorganisms such as fungi, bacteria, and algae or from cells [12]. Biodegradation corresponds to processes that involve the breaking of the chain and the disintegration of the material caused by biological activity, resulting in structural and chemical changes in the molar mass, color, and solubility of the polymer [12].

Polymers can still be classified according to their properties. Bioadhesive polymers are materials capable of binding to biological substrates and can be used in drugs that undergo rapid absorption since human mucosa is considered relatively permeable – the so-called mucoadhesive polymers. There are also those polymers that interact with the cell plasma membrane – cytoadhesive polymers [13].

Bioadhesion is a property that has been explored as an alternative to promote a more specific targeting and retention of the drug incorporated in the delivery system, using the most varied absorption membranes in the body. This technological platform combined with nanotechnology offers possibilities such as protection and control of the release that can occur in specific sites, which can increase the drug's permanence time in the body. Other advantages are the improved bioavailability of drugs and the achievement of reduced doses and toxicity; therefore, they are excellent strategies for increasing drug bioavailability and therapeutic effectiveness [14].

In addition, hydrophilic swelling polymers can form a three-dimensional (3D) network, called hydrogel, and retain a large amount of water and/or biological fluids while maintaining structure due to chemical or physical crosslinking of individual polymer chains [15]. They can be chemically stable or degrade over time. Thus, hydrogels can be highly used in pharmaceutical formulations and preparations that allow the release of the drug after expansion when in contact with the

Materials Research Foundations 172 (2025) 246-262
https://doi.org/10.21741/9781644903353-10

dissolution medium or when they have physiological stimuli which can be temperature, enzymes, pH, insulin, urea, glucose, oxidation and ionic strength [16,17].

Bioactive and biomimetic polymers are also being researched. Biomimetic materials are defined as those that can selectively interact with the biological microenvironment, mimicking it. Bioactive polymers, on the other hand, are capable of interacting with cell receptors, via biological recognition, providing specific responses. Aliphatic polyesters such as poly(hydroxybutyrate) (PHB), of natural and synthetic origin, have been studied in the preparation of biomimetic materials [18].

3. Properties of polymers in conventional and innovative medicine

Therapeutic responses considered effective and safe are achieved using traditional medicines and conventional pharmaceutical forms. However, the use of an advanced drug delivery system, based on the results of innovative research, can contribute to obtaining even more promising results. Polymers have become essential components for innovative medicines, as they act by exerting a direct action on the release of drugs at their sites of action [19].

Concerning conventional use, polymers are used as excipients in common drug preparations, having several functions, such as providing physicochemical and organoleptic stability; assisting in preparation; maintaining stability; ensuring patient acceptability; and providing security and effectiveness [20].

Conventional solid dosage forms, such as tablets, powders, and capsules, have polymeric excipients performing several functions. Sorbitol, lactose, starch, and microcrystalline cellulose are among the most used pharmaceutical diluents in the preparation of these forms [20]. Polyvinylpyrrolidone (PVP), vinyl acetate, and vinyl pyrrolidone (copovidone) copolymers, polysaccharides, and various cellulose derivatives are used as binding agents (agents used to increase the adhesion of powder particles in the granulation process) and as disintegrating agents (agents facilitating the disintegration of tablets and capsules in biological fluids) [21]. Carboxymethylcellulose (CMC) is an important disintegrant when it comes to formulating fast-disintegrating granules, as well as tablets and capsules [21,22,23].

In liquid medicines for oral use (solutions, syrups, and elixirs) polymers are also used as stabilizing agents and solubilizing agents. Some polymers have the function of preventing the growth of drug crystals with liposoluble characteristics in solutions, as is the example of PVA, PEO, and PEG, among others [6,24]. Indomethacin, nifedipine, and tolbutamide are examples of some drugs that can be stabilized through polymers such as PVP and PVP-co-vinyl acetate. Through the incorporation of ritonavir, a low-solubility drug, in PEG-800, it is possible to obtain a significant increase in the concentration of the drug in the gastrointestinal system, contributing to an improvement in its percentage of absorption [6,24,25].

In addition to the applications of polymers in conventional pharmaceutical forms, there are several reports and successful experiences of the use of these raw materials, using their properties in innovative formulations. In the 21st century, research at the interface of polymers and biomedical sciences gave rise to the first drugs based on nano-sized polymers (5-100 nm), the "therapeutic polymers". Therapeutic polymers include rationally designed macromolecular drugs, polymer-drug, and polymer-protein conjugates, in addition to polymer micelles containing covalently linked drugs and polyplexes for DNA delivery [19,26].

Materials Research Forum LLC
https://doi.org/10.21741/9781644903353-10

Controlling the functionality of synthetic polymers allows them to be linked to specific receptors, promoting recognition and triggering the modulation of inflammatory responses and cell adhesion [19,26].

Given the application of polymers for release control, Progestasert® is an intrauterine implant that contains the hormone progesterone and was launched by the American company Alza. In this device, the release of the hormone is made in a controlled manner by non-biodegradable membranes made of poly(ethylene-co-vinyl acetate) (PEVA) [27].

Zoladex® LA is a subcutaneous use drug that contains goserelin acetate as a drug and is dispersed in a PLGA matrix that, after controlling the drug release for 3 months, is biodegraded and eliminated from the patient's body [6,28].

4. Main polymers used in pharmaceutical dosage forms

So far, this chapter has been dedicated to presenting the importance of polymers in pharmaceuticals for health care, including situations involving surgeries, as well as demonstrating their classifications and possible uses. From now on, the main polymers are presented, and their applications are discussed.

It is noteworthy that it is necessary to apply several techniques to characterize the polymers and this knowledge can assist the understanding of the applications of materials in various pharmaceutical products and drug delivery systems [29].

4.1 Carbomers

Carbomers are crosslinked poly(acrylic acid) (PAA) polymers of high molecular weight. Crosslinking is achieved using allyl-sucrose or allyl-pentaerythritol. Most are fluffy white powders that, in the presence of water, swell to form hydrogels. Different levels of crosslinking are available on the market, allowing the achievement of varied characteristics, including the desired controlled drug release profile [30].

Commercially, these polymers are available in powder form or in different grades of granulation, for various applications, under the trade name Carbopol®. Combining powders and granules in one formulation can be used to obtain various release profiles. For oral applications Carbopol® 71G, 971P, 974P, 934P are used. Acrylic acid, the polymer's backbone, provides pH-dependent properties. For example, when carbomers are hydrated in water, they form acidic solutions that become increasingly viscous and gel-like if the pH is adjusted to a more alkaline environment (typically above 4-6) [30].

Some examples of preparations containing PAA are Differin®, a gel for the treatment of acne; Isotrex®, a gel or cream for the treatment of skin photoaging; Cataflam® Emulgel, an anti-inflammatory for topical use.

4.2 Cellulose-derived polymers

Although cellulose itself is insoluble in water, its water-soluble derivatives have found extensive applications in pharmaceutical dosage forms. Cellulose derivatives can be used as binders, agents used to increase the bond between powders, facilitating granulation and compression processes in tablet formulation; disintegrants, agents that facilitate the disintegration of tablets and capsules in biological fluids, in the coating of tablets, in the formation of transdermal films, and as release modulating agents eg in extended-release tablets [21,31,32,33,34].

Methylcellulose, carboxymethylcellulose, and hydroxypropylmethylcellulose are the most common cellulose-based polymers with methyl, carboxymethyl, and hydroxypropyl/methyl substitution, respectively. Hydroxyethylcellulose – HEC, is a synthetic polymer obtained from the reaction of ethylene oxide with alkaline cellulose, and which is compatible with electrolytes. It has a thickening action and maintains its stability in the pH range from 2.0 to 12.0. It is found under the trade names Natrozol® and Cellosize® [35]. In a study by Petrovich and Lopes (1993) [36], the authors used cellulose derivatives in the liquid formulation of ambroxol, obtaining a satisfactory result.

Among numerous bioadhesive formulations, one of the first to be marketed was ORABASE® - prepared from cellulose derivatives and pectin, used to act as an adhesive vehicle for the application of active medication to oral mucosa tissues [37].

Other examples of commercially available bioadhesives are Perioxidin®, a dental gel based on hydroxyethylcellulose that eliminates and prevents the formation of bacterial plaque, helping in the postoperative period on the surgical site in periodontal diseases, and Acidform®, a vaginal gel based on xanthan gum that works as a vaginal microbicide that also has a bioadhesive property [37].

Blends of polymers with different properties allow for an adjustment of formulations for greater control over drug release. Ethylcellulose (EC) (insoluble polymer) was used mixed with acid methacrylate-ethyl acrylate copolymer (Eudragit® L) (enteric polymer) to coat multiparticulate dosage forms to modulate drug release at high and low pH values varying the proportion of polymers in the mixture. A good example of the combination of two different polymers for film coating is the use of a mixture based on ethylcellulose and HPMC [38,39].

4.3 Hydrocolloids

Various hydrocolloids or polysaccharide gums from different sources can be used in the pharmaceutical development field.

Most gums are hydrophilic and contain very long polymer chains as well as different functional groups. These characteristics are very attractive in many pharmaceutical processes such as coating, stabilizing, thickening, binding, solubilizing, and disintegrating [40,41].

Different gums behave differently in aqueous solutions. Almost all exhibit thickening properties, while some show swelling properties. Although thickening is a desirable property for dosage forms such as solutions, suspensions, and emulsions, the swelling property is also used in the encapsulation of drugs for controlled release applications [42].

Gums such as guar gum can provide excellent thickening properties, while gums including alginate and chitosan can offer a gelling property in the presence of ions. Similar to synthetic polymers, gums can be blended to provide superior properties by synergizing their features.

Alginic acid is a linear polysaccharide that consists primarily of the two building blocks of mannuronic acid (M) and guluronic acid (G). Alginic acid and its salts are anionic polymers that can provide gelling properties. Since they contain a carboxyl group, they can easily undergo a complex reaction in the presence of ions [43]. Depending on their source, the contents of mannuronic and guluronic acids may differ. It is known that guluronic acid is responsible for the gelling property, so an alginic acid rich in guluronic acid provides a more viscous gel in the presence of ions, in particular, calcium. In other words, the mechanical properties of this polymer are dependent on its content. For this reason, suppliers offer alginates with different guluronic acid/mannuronic acid ratios [44,45].

An example of practical application is the antibiotic griseofulvin, available as an oral suspension. This medicine contains sodium alginate stabilized with methylparaben [44,45]. Alginic acid, and ammonium calcium alginate, can also be found in metaxalone tablets [44,45].

Chitosan obtained from chitin is the second most abundant natural polymer after cellulose, and it can be found in shrimps, crabs, and lobster shells. Chitosan is a cationic polymer and it has been investigated as an excipient in controlled-release formulations and also in mucoadhesive formulations, due to its gelling properties, adhesiveness, antibacterial activity, and non-toxicity [46].

There are reports that the bitter taste of natural compounds, such as caffeine, can be masked using chitosan [47]. Chitosan can also be used in tablet excipients for disintegrating or coating effects [44,45].

From the point of view of toxicity and safety, low molecular weight chitosan (as an oligosaccharide) has been shown to be safer, even with negligible cytotoxicity in Caco-2 cells [48].

Due to its cationic nature, chitosan can form complexes with negatively charged polymers such as hyaluronic acid (HA) to form a highly viscoelastic polyelectrolyte complex [49].

Chitosan and ovalbumin-based gels have been suggested for pharmaceutical and cosmetic use [50].

In the veterinary field, chitosan can be used to control the release of different drugs such as antibiotics, antiparasitics, anesthetics, analgesics, and growth factors [51].

Other forms of application include, as an absorption enhancer, using protonated chitosan, capable of increasing the paracellular permeability of peptide drugs through mucosal epithelia [52]. Chitosan can also be mixed with nonionic surfactants such as sorbitan esters to make emulsions as solutions or creams [53].

Chitosan has been widely used as a topical dressing in the treatment of wounds, as it has hemostatic, healing stimulation, antimicrobial, biocompatible, and biodegradable properties [54].

Chitosan can be used to prevent or treat infections in wounds and burns, not only because of its intrinsic antimicrobial properties but also because of its ability to provide extrinsic antimicrobial agents to wounds and burns. It can also be used as a slow-release drug delivery vehicle for growth factors to enhance wound healing [54].

4.4 Synthetic polymers

Synthetic polymers are part of our daily lives and represent one of the most versatile classes of materials that exist, with numerous applications, including in the pharmaceutical field [19]. The acceptance of the existence of macromolecules allowed the discovery of many macromolecular substances whose most visible result are the plastics, rubbers, paints, and varnishes that are part of our daily basis. In addition to intentional discoveries such as nylon, polyesters, and isotactic polypropylene, there were also accidental discoveries such as polyethylene and

Materials Research Foundations 172 (2025) 246-262
https://doi.org/10.21741/9781644903353-10

polytetrafluoroethylene. Today, new and interesting macromolecules continue to be created to obtain new mechanical, optical, and electrical properties [55].

Synthetic polymers have been used in therapeutic applications, such as drug administration and tissue regeneration, precisely because of their biocompatibility and adequate mechanical properties. Furthermore, some polymer-drug conjugates have been used as bioactive drugs due to their increased efficacy, solubility, and target specificity compared to small-molecule drugs. Increasing synthetic control of polymer properties has allowed the production of sets of polymers for targeted and controlled drug delivery [19].

Examples of the main synthetic polymers for pharmaceutical use and their applications are:

- Acrylics, such as Carbopol® mainly used as a gelling agent in the preparation of gels; and Eudragit® (Poly(methacrylates), used with varied functions such as organoleptic, stabilizing, protective, release modulator (enteric and sustained);

- Vinyls, for example, PVP (Homopolymer of 1-ethenyl-2-pyrrolidinone) used as a binder, disintegrant, and dissolution enhancer; and Elvanol® (Poly(vinyl alcohol), an emulsion stabilizer and coating agent, especially used in transdermal and ophthalmic forms;

- Esters, such as Purasorb® (Aliphatic Polyester), a biodegradable material, used in the preparation of implantable and injectable drug delivery systems;

- Oxyethylenes, an example is Lutrol® (non-ionic block copolymer of ethylene oxide and propylene oxide) which are Poloxamers® used as emulsifiers, solubilizers, and stabilizers;

- Cellulosics, as examples are Methocell® (polymer of hydroxypropylmethyl cellulose ether) used as viscosity enhancers in ophthalmic preparations, with the main polymer used in matrices to control the release of drugs, Aquacoat® (ethylcellulose) which acts as a viscosity enhancer, as a binder and as a modulator of drug release; sodium or potassium CMC (carboxymethylcellulose) used as a viscosity-increasing agent, as a binder and as a disintegrant in solids; and cellulose acetate (partially acetylated cellulose polymer in different grades) a coating agent for conventional and modified drug release, and use in impermeable membranes in osmotic systems (OROS®) [21].

5. The main dosage forms using polymers

As presented, there are several possibilities of polymers available on the market for formulators of pharmaceutical products, including for the most varied routes of administration. Next, possibilities of polymers to be used in specific pharmaceutical forms are presented.

5.1 Solid pharmaceutical forms

Polymers are present in solid drugs such as tablets, powders, pastilles, and capsules. They act as excipients for the most varied purposes.

In the formulation of tablets, polymers, such as polyvinylpyrrolidone (PVP), carboxymethylcellulose (CMC), methacrylates, vinyl acetate copolymers, acrylic acid polymers, polysaccharides, and cellulose derivatives, are used as binding agents to favor the cohesion of the powders, allowing to ensure the shape and hardness of the tablet.

Materials Research Foundations 172 (2025) 246-262
https://doi.org/10.21741/9781644903353-10

Still in tablets, but also other solid pharmaceutical forms such as capsules, polymers are used as disintegrating agents, aiming to facilitate the disintegration process of the tablets, an essential step to allow distribution and solubilization in biological fluids for subsequent absorption.

Currently, they have also been used to mask an unpleasant taste; promote drug protection; increase bioavailability and ensure controlled release of the drug [23].

Another application of polymers in solid oral medicines is in the form of a complexing agent. An example is the drug Kayexalate®, launched in 1975, which corresponds to a cation exchange resin prepared from polystyrene, for oral or rectal administration in cases of hyperkalemia. When crossing the gastrointestinal tract, the sodium present in the resin is exchanged for potassium, reducing potassium levels [27]. Another example of the use of polymer as a complexing agent is in RenaGel®, a product, administered orally, which is obtained from a poly (allylamine hydrochloride) hydrogel that removes phosphate ions from the body, reducing phosphorus levels [6,56].

Polymers can also be used in solid pharmaceutical form for coating purposes, promoting an improvement in the appearance of the drug, as well as bringing together other properties, such as increased stability and mechanical strength.

5.2 Liquid pharmaceutical forms

In liquid dosage forms, such as syrups, elixirs, and suspensions, polymers are often used as rheological modifiers to regulate the flow, but also as stabilizing and solubilizing agents.

They are used to control the viscosity of aqueous solutions or to stabilize suspensions. Drugs such as indometacin, nifedipine, and tolbutamide can be stabilized through polymeric excipients such as PVP and vinyl coacetate. In the preparation of ritonavir, a drug with limited solubility, PEG-800 can increase the drug concentration in the intestine, improving its bioavailability [24,25].

In injectable drugs, for parenteral use, polymers such as PEG, PEG monooleate, PEG 660, and poly (PPG) are present as solvents and stabilizers for the drugs [57].

5.3 Semi-solid pharmaceutical forms

Among the semi-solid pharmaceutical forms are emulsions, including creams and lotions, ointments, gels, and serums, among others.

Pharmaceutical emulsions can be characterized as two immiscible liquid phases as water in oil (W/O) stabilized by an emulsifying agent, which can be a polymer. Acrylic polymers and copolymers, combined or not with ionic or non-ionic surfactants, form stable oil-in-water (O/W) emulsions. The use of polymeric emulsifiers in place of monomeric surfactants results in the formation of stable oil-in-water emulsions, which can be formulated to be surfactant-free polymeric emulsifiers [58].

The gel pharmaceutical form, on the other hand, is an aqueous-based formulation commonly used nowadays that can be defined as preparations obtained by incorporating gelling agents, such as chitosan, PEO, tragacanth, starch, cellulose derivatives, carboxyvinyl polymers, and double magnesium and aluminum salicylates in water, glycerol or propylene glycol [6,59].

Materials Research Foundations 172 (2025) 246-262 | https://doi.org/10.21741/9781644903353-10

Polymers are used for the preparation of gels due to their properties of imparting viscosity, thickening, and stability in aqueous solutions. Among the main gelling agents used in medicines and cosmetics are carbomers, cellulose derivatives, and other synthetic polymers.

Some examples of medications containing carbomers are topical anti-inflammatory drugs such as Cataflan® Emulgel and Feldene® Gel.

Serums can also be mentioned as semi-solid pharmaceutical forms. Although they do not represent a different dosage form, they are gels or emulsions, with lower viscosity, currently widely used in the cosmetic sector. In other words, a serum is a fluid that has a high concentration of active substances and that, due to its light texture, has fast absorption and action on the skin. Through the association of polymers with other actives, features such as antioxidant, anti-aging, hydration, and toning stand out. Among the main polymers used in serums are xanthan gum, chitin, chitosan, guar gum, and hyaluronic acid [60,61].

5.4 Transdermal pharmaceutical forms – patches

Transdermal therapeutic systems are dosage forms prepared for administration through cutaneous absorption, mainly presented as patches. They aim for the systemic action of the drug, preventing drug loss, maintaining the drug's plasma level for a longer period, and providing controlled and prolonged release of active substances [62,63].

These systems can be classified as monolithic systems or membrane-controlled systems. In the monolithic system, there is the presence of a matrix membrane, where the dispersed drug is found, and which is composed of polymers that will control the drug release during its percutaneous absorption [64]. On the other hand, membrane-controlled systems act as reservoirs containing the drug, with the membrane controlling the release rate [64].

6. Pre- and post-operative topical and oral formulations using polymers

The use of antimicrobials has become essential in the prophylaxis of pre and postoperative infections. During surgical processes there is a need to intervene preventively in post-surgical infections, thus, antibiotics should be administered intraoperatively, that is, starting before and ending soon after surgery [65]. According to the most modern principles of prophylaxis, the first dose of antibiotic should be administered before the start of surgery and in a dose greater than the standard therapeutic dose so that it is in the blood and tissues at the time of contamination [66].

Among the polymers used in prophylactic antibiotic therapy, the incorporation of antibiotics in temporary or permanent implants is cited as an example to reduce infections and improve the organic acceptance of the implanted material, minimizing the chances of rejection [67]. The Septopal® (Merck S.A. laboratory), for example, is a commercially available device, composed of a chain of methyl methacrylate and methyl acrylate (PMMA) copolymer beads, joined by a polyphilic surgical thread. The chains contain gentamicin sulfate, an antimicrobial agent that can be used in the treatment and prophylaxis of infections caused by gentamicin-sensitive microorganisms. This drug is surgically introduced into bones and soft tissues and can be used in the short or long term, being removed manually or through surgical interventions [27].

Among the various medications used in the postoperative period, in addition to antimicrobials, are emulsions that promote better healing, aiding in hydration, accelerating skin regeneration, and

Materials Research Foundations 172 (2025) 246-262
https://doi.org/10.21741/9781644903353-10

preventing the formation of keloids. An example of an emulsion for use in the postoperative period is the one containing calendula extract (*Calendula officinalis* L.) due to its healing activity.

Ointments containing fibrinolysin and chloramphenicol, an antibiotic, that acts by preventing the multiplication of bacteria in the wound and preventing the infection of surgical scars, employ the use of polyethylene as an excipient in their formulation.

Natural polymer hydrogels have bioactive properties that can accelerate the healing process in acute wounds but have a minimal effect on complex acute injuries such as third-degree burns or chronic wounds. The results can be improved with the use of therapeutic bioactive molecules such as cytokines and growth factors capable of promoting re-epithelialization, debridement, granulation, and other elements of paramount importance in wound healing [68].

Peptide-based hydrogels have advantages in wound healing due to their excellent biodegradability, biocompatibility, and similarity to the native extracellular matrix. Furthermore, their modular structures allow them to be modified to control the delivery rates of therapeutic agents, such as cytokines and growth factors, which can promote wound healing and reduce scar formation. Due to their diverse structural designs, peptide-based hydrogels have been used to improve and aggregate in the treatment of diabetic burns and wounds, as well as to facilitate neural regeneration [69].

Studies have shown the efficient development of nitric oxide releasing ointments and their potential in the healing process. When compared to control groups, the nitric oxide ointment showed the most accelerated wound closure with greater re-epithelialization, collagen deposition, and blood vessel formation *in vivo*. Therefore, this nitric oxide-based ointment has promising potential as an efficient skin wound healing strategy [70].

Conclusion

This chapter demonstrates the wide range of applications of polymers, including those of synthetic or natural origin, used in obtaining pharmaceutical products, more specifically producing conventional or technological pharmaceutical forms. The wide variety of existing polymers allows the formulator to choose the most suitable raw material for his needs, controlling rheological aspects such as viscosity, swelling capacity, stability, and compatibility with other components of the formulation. The controlled release of drugs is another property that can be achieved with the use of polymers, which are of great value in the surgical wound repair process, for the administration of antimicrobial agents and for promoting healing. As in any formulation development situation, bench testing and formulation characterization are essential steps, but the data presented in this paper provide initial guidance.

References

[1] B.E.B. Christóforo, D.S. Carvalho, Cuidados de Enfermagem realizados ao paciente cirúrgico no período pré-operatório, Rev. Esc. Enferm. USP. 43 (2009) 14-22. https://doi.org/10.1590/S0080-62342009000100002

[2] M.R. Daian, A. Petroianu, L.R. Alberti, E.E. Jeunon, Estresse em procedimentos cirúrgicos, ABCD Arq. Bras. Cir. Dig. 25 (2012) 118-124. https://doi.org/10.1590/S0102-67202012000200012

[3] J.B.S Garcia, P. Bonilla, D.C. Kraychete, F.C. Flores, E.D.P. de Valtolina, C. Guerrero, Optimizing post-operative pain management in Latin America, Rev. Bras. Anestesiol. 67 (2017) 395-403. https://doi.org/10.1016/j.bjane.2016.04.003

[4] A.D.J. Bombin, N.J. Dunne, H.O Mccarthya, Electrospinning of natural polymers for the production of nanofibres for wound healing applications, Mater. Sci. Eng. C. 114 (2020) 110994. https://doi.org/10.1016/j.msec.2020.110994

[5] H.M Mansour, M. Sohn, A. Al-Ghananeem, P.P. Deluca, Materials for pharmaceutical dosage forms: molecular pharmaceutics and controlled release drug delivery aspects, Int. J. Mol. Sci. 11 (2010) 3298–3322. https://doi.org/10.3390/ijms11093298

[6] J.C.O. Villanova; R.L. Oréfice, S.A. Cunha, Aplicações Farmacêuticas de Polímeros, Polímeros: Ciênc. Tecnol. 20 (2010) 51-64. https://doi.org/10.1590/S0104-14282010005000009

[7] P.J. Sinko, Y. Singh, G.E. Amidon, C.R. Middaugh, H. Omidian, K. Park, T.J. Siahaan, H. Zheng, Martin's physical pharmacy and pharmaceutical sciences: physical chemical and biopharmaceutical principles in the pharmaceutical sciences, sixth ed., New Jersey, Department of Pharmaceutics Ernest Ernest Mario School of Pharmacy Rutgers, The State University of New Jersey, 2006.

[8] Jr. Canevarolo, V. Sebastião, Ciência dos polímeros, second ed., Artliber Editora, São Paulo, 2002.

[9] P.K Dutta, J. Dutta, V.S Tripathi, Chitin and Chitosan: Chemistry, properties and application. J. Sci. Ind. Res. (2004) 20-31.

[10] H.A Pawar, S.R. Kamat, P.D. Choudhary, An Overview of Natural Polysaccharides as Biological Macromolecules: Their Chemical Modifications and Pharmaceutical Applications. Biol. Med. 7 (2015) 1-9.

[11] S.J. Gautier, J. Lecourtier, Structural characterization by ^{13}C nuclear magnetic resonance of hydrolyzed carboxymethylcellulose, Polym. Bull. 26 (1991) 457- 464.

[12] S.H. Barbanti, C.A.C. Zavaglia, E.A.R. Duek, Polímeros Biorreabsorvíveis como Substrato para Cultura de Células e Engenharia Tecidual. Polímeros: Ciênc. Tecnol. 15 (2005) 13. https://doi.org/10.1590/S0104-14282007000400010

[13] F.C. Carvalho, M.L. Bruschi, R.C. Evangelista, M.P.D. Gremião, Mucoadhesive drug delivery systems. Braz. J. Pharm. Sci. 46 (2010) 2. https://doi.org/10.1590/S1984-82502010000100002

[14] G.P Andrews, T.P Laverty, D.S Jones, Mucoadhesive polymeric platforms for controlled drug delivery. Eur. J. Pharm. Biopharm. 71 (2009) 505. https://doi.org/10.1016/j.ejpb.2008.09.028

[15] N.A. Peppas, P. Bures, W. Leobandung, H. Ichikawa, Hydrogeles in pharmaceutical formulations, Eur. J. Pharm. Biopharm. 50 (2000) 27-46. https://doi.org/10.1016/S0939-6411(00)00090-4

[16] C. Lin, A.T Metters, Hydrogels in controlled release formulations: network design and mathematical modeling, Adv. Drug Delivery Rev. 58 (2006) 1379. https://doi.org/10.1016/j.addr.2006.09.004

[17] O. Pillai, R. Panchagnula, Polymers in drug delivery, Curr. Opin. Chem. Biol. 5 (2001) 447. https://doi.org/10.1016/S1367-5931(00)00227-1

[18] H.G. Börner, H. Schlaad, Bioinspired functional block copolymers. Soft Matter. 3 (2007) 394. https://doi.org/10.1039/B615985K

[19] S. Liu; R. Maheshwari, K.L. Kiick, Polymer-Based Therapeutics, Macromolecules. 42 (2009) 3-13. https://doi.org/10.1021/ma801782q

[20] J.C.O. Villanova, V.R. Sá, Excipientes: guia prático para padronização, LMC - Pharmabooks, São Paulo, 2009.

[21] R.C. Rowe, P. Sheskey, S.C. Owen, Handbook Pharmaceutical Excipients, fifth ed., American Pharmacists Association, Washington, 2006.

[22] M.E. Aulton, Delineamento de formas farmacêuticas, second ed., Artmed, Porto Alegre, 2005.

[23] J. Swarbrick, Encyclopedia of Pharmaceutical Technology, third ed., Informa Healthcare, England, 2007.

[24] S. Mallick; S. Pattnaik; K. Swain, Current perspectives of solubilization: potential for improved bioavailability, Drug Dev. Ind. Pharm. 33 (2007) 865. https://doi.org/10.1080/03639040701429333

[25] T. Tadros, T. Polymeric surfactants in disperse systems, Adv. Colloid Interface Sci. 147 (2009) 281-299. https://doi.org/10.1016/j.cis.2008.10.005

[26] K.L. Kiick, Polymer Therapeutics, Science. 317 (2007) 1182-1183.

[27] A.S. Hoffman, The origins and evolution of "controlled" drug delivery systems, J. Control. Release, 132 (2008) 153. https://doi.org/10.1016/j.jconrel.2008.08.012

[28] W.F. Paterson, E. Mcneil, S. Reis, A.S. Hollmann, M.D.C. Donaldson, Efficacy of Zoladex LA (goserelin) in the treatment of girls with central precocious or early puberty, Arch. Dis. Child. 79 (1988) 323. https://doi.org/10.1136/adc.79.4.323

[29] L.S. Nair & C.T. Laurencin, Biodegradable polymers as biomaterials, Progr. Polym. Sci. 32 (2007) 762. https://doi.org/10.1016/j.progpolymsci.2007.05.017

[30] J. Brady, T. Dürig, P.I. Lee, J.-X. Li, Polymer Properties and Characterization, in: Y. Qiu, Y. Chen, G.G.Z. Zhang, L. Yu, R.V Mantri (Eds.), Developing Solid Oral Dosage Forms: pharmaceutical theory and practice, Academic Press, Massachusetts, 2017, pp. 181-223.

[31] A.T. Pham, P.I. Lee, Probing the mechanism of drug release from hydroxypropyl methylcellulose matrices, Pharm. Res. 11 (1994) 1379-1384. https://doi.org/10.1023/A:1018975318805

[32] P. Gao, J.W. Skoug, P.R. Nixon, T.R. Ju, N.L. Stemm, K.C. Sung, Swelling of hydroxypropyl methylcellulose matrix tablets. 2. Mechanistic study of the influence of

formulation variables on matrix performance and drug release, J. Pharm. Sci. 85 (1996) 732-740. https://doi.org/10.1021/js9504595

[33] R. Eyjolfsson, Hydroxypropyl methylcellulose mixtures: effects and kinetics of release of an insoluble drug, Drug Dev. Ind. Pharm. 25 (1999) 667-669. https://doi.org/10.1081/DDC-100102223

[34] M.N.V. Ravi Kumar, N. Kumar, Polymeric controlled drug-delivery systems: perspective issues and opportunities, Drug Dev. Ind. Pharm. 27 (2001) 1-30. https://doi.org/10.1081/DDC-100000124

[35] M.E. Aulton, Pharmaceutics: The science of dosage form design, second ed., Churchill Livingstone, Edinburgh, 2002.

[36] P.R. Petrovick, E.C. Lopes, Emprego da ultrafiltração na análise do ambroxol. UFRGS, Porto Alegre, 1993.

[37] L.V. Allen, N.G. Popovich, H.C. Ansel, Ansel's pharmaceutical dosage forms, ninth ed., Lippincott Williams & Wilkins, Philadelphia, 2005.

[38] W. Gunder, B.H. Lippold, B.C. Lippold, Release of drugs from ethyl cellulose microcapsules (diffusion pellets) with pore formers and pore fusion, Eur. J. Pharm. Sci. 3 (1995) 203–214. https://doi.org/10.1016/0928-0987(95)00009-3

[39] R.B. Oliveira, E.M. Lima, Polímeros na obtenção de sistemas de liberação de fármacos, Rev. Eletrôn. Farm. 3 (2006) 29-35.

[40] P. Burey, B.R. Bhandari, T. Howes, M.J. Gidley, Hydrocolloid gel particles: formation, characterization, and application, Crit. Rev. Food Sci. Nutr. 48 (2008) 361-377. https://doi.org/10.1080/10408390701347801

[41] J. Milani, G. Maleki, Hydrocolloids in Food Industry, in: B. Valdez (Ed.), Food Industrial Processes - Methods and Equipment, BoD-Books on Demand, Norderstedt, 2012, pp. 18-22.

[42] S.P. Vyas, R.K. Khar, Controlled Drug Delivery: Concepts and Advances, first. ed., Vallabh Prakashan, Dehli, 2002.

[43] T.K. Giri, Alginate Containing Nanoarchitectonics for Improved Cancer Therapy, in: A.M. Holban, A.M. Grumezescu (Eds.), Nanoarchitectonics for Smart Delivery and Drug Targeting, William Andrew Publishing, Norwich, 2016, pp. 565-588. https://doi.org/10.1016/B978-0-323-47347-7.00020-3

[44] O. Felt, P. Buri, R. Gurny, Chitosan: a unique polysaccharide for drug delivery, Drug Dev. Ind. Pharm. 24 (1998) 979–993. https://doi.org/10.3109/03639049809089942

[45] A. K. Singla, M. Chawla, Chitosan: some pharmaceutical and biological aspects, J. Pharm. Pharmacol. 53 (2001) 1047–1067. https://doi.org/10.1211/0022357011776441

[46] B. Singh, S. Maharjan, K-H. Cho, L. Cui, I-K Park, Y-J. Choi, C-S. Cho, Chitosan-based particulate systems for the delivery of mucosal vaccines against infectious diseases, Int. J. Biol. Macromol. 110 (2018) 54-64. https://doi.org/10.1016/j.ijbiomac.2017.10.101

[47] A. Binello, G. Crabotto, G.M. Nano, P. Spagliardi, Synthesis of chitosan–cyclodextrin adducts and evaluation of their bitter-masking properties, Flavour Fragr. J. 19 (2004) 394-400. https://doi.org/10.1002/ffj.1434

[48] L. Kudsiova, M.J. Lawrence, A comparison of the effect of chitosan and chitosan-coated vesicles on monolayer integrity and permeability across Caco-2 and 16HBE14o-cells, J. Pharm. Sci. 97 (2008) 3998-4010. https://doi.org/10.1002/jps.21262

[49] J. Potaś, E. Szymańska, K. Winnicka, Challenges in developing of chitosan – Based polyelectrolyte complexes as a platform for mucosal and skin drug delivery, Eur. Polym. J. 140 (2020) 1-11. https://doi.org/10.1016/j.eurpolymj.2020.110020

[50] S. Y. Yu, J. Hu, X. Pan, P. Yao, M. Jiang, Stable and pH-sensitive nanogels prepared by self-assembly of chitosan and ovalbumin, Langmuir. 22(2006) 2754–2759. https://doi.org/10.1021/la053158b

[51] S. Senel, S. J. McClure, Potential applications of chitosan in veterinary medicine, Adv. Drug Deliv. Rev. 56 (2004) 1467–1480. https://doi.org/10.1016/j.addr.2004.02.007

[52] M. Thanou, J. C. Verhoef and H. E. Junginger, Oral drug absorption enhancement by chitosan and its derivatives, Adv. Drug Deliv. Rev. 52 (2001) 117–126. https://doi.org/10.1016/S0169-409X(01)00231-9

[53] J. Grant, J. Cho and C. Allen, Self-assembly and physicochemical and rheological properties of a polysaccharide−surfactant system formed from the cationic biopolymer chitosan and nonionic sorbitan esters, Langmuir. 22 (2006) 4327–4335. https://doi.org/10.1021/la7004166

[54] T. Dai, M. Tanaka, Y. Huang, M.R. Hamblin, Chitosan preparations for wounds & burns: antimicrobial & wound-healing effects, Expert. Rev. Anti Infect. Ther. 9 (2011) 857–879. https://doi.org/10.1586/eri.11.59

[55] E. Wan, E. Galembeck, F. Galembeck, Polímeros sintéticos, in: Química Nova na Escola, PubliSBQ, São Paulo, 2001, pp. 5-8.

[56] E.A. Slatopolsky, S.K. Burke, M.A. Dillon, RenaGel, a nonabsorbed calcium- and aluminum-free phosphate binder, lowers serum phosphorus and parathyroid hormone, Kidney Int. 55 (1999) 299-306. https://doi.org/10.1046/j.1523-1755.1999.00240.x

[57] R.G. Strickley, Solubilizing excipients in oral and injectable formulations, Pharm. Res. 21 (2004) 201. https://doi.org/10.1023/B:PHAM.0000016235.32639.23

[58] A. Franzol, M.C. Rezende, Estabilidade de emulsões: um estudo de caso envolvendo emulsionantes aniônico, catiônico e não-iônico. Polímeros. 25 (2015) 1-9. https://doi.org/10.1590/0104-1428.1669.

[59] Brasil. Agência Nacional de Vigilância Sanitária. Brasília: ANVISA. Farmacopeia Brasileira, volume 2, 2010.

[60] CosmeticsOnline, Fundamentos da Cosmetologia – Cremes e Loções Cosmetics, https://www.cosmeticsonline.com.br/artigo/81 (accessed 20 May 2024).

[61] CosmeticsOnline, Fundamentos da Cosmetologia – Géis Cosméticos, https://www.cosmeticsonline.com.br/artigo/83 (accessed 20 May 2024).

[62] A.M.S Sawamura, S.L. Franco, Sistemas Terapêuticos Transdérmicos, Arquivos de Mudi. 8 (2012) 1-8.

[63] A.R. Gennaro, Remington: the science and practice of pharmacy, Lippincott Williams e Wilkins, Philadelphia, 2006.

[64] M.R. Prausnitz, R. Langer. Transdermal drug delivery. Nat. Biotechnol. 11 (2008) 1261-1268. https://doi.org/10.1038/nbt.1504.

[65] I.D Miziara, O uso da antibioticoterapia no tratamento das doenças bucais, JBC. 2 (1998) 57-67.

[66] F.R.G Seabra, B.G.M Seabra, E.G. Seabra, Antibioticoterapia profilática em cirurgias periodontais, Odont. Clín. Cient. 3 (2004) 11-16.

[67] T.C. Woods, A.R. Marks, Drug-eluting stents, Annu. Rev. Med. 55 (2004) 169-178. https://doi.org/10.1146/annurev.med.55.091902.105243

[68] R. Ghadi, A. Jain, W. Khan, A.J. Domb, Microparticulate polymers and hydrogels for wound healing, in: M.S. Ågren (Ed.), Wound Healing Biomaterials, Woodhead Publishing, 2016, pp. 203-25.

[69] D. Stern, H. Cui, Crafting polymeric and peptidic hydrogels for improved wound healing, Adv. Healthc. Mater. 8 (2019). https://doi.org/10.1002/adhm.201900104

[70] Y. Kang, J. Kim, Y. Mi Lee, S.S. Kim; H. Park, W.J. Kim, Nitric oxide-releasing polymer incorporated ointment for cutaneous wound healing, J. Control. Release. 220 (2015) 624-630. https://doi.org/10.1016/j.jconrel.2015.08.057

Materials Research Foundations 172 (2025) 263-281
https://doi.org/10.21741/9781644903353-11

Chapter 11

Polymers in Drug Delivery

Sinazo Cobongela [1,2,*], Tintswalo Mgwenya[3]

[1]Health Platform, Advanced Materials Division, Mintek, Randburg 2194, South Africa

[2]Department of Science and Innovation (DSI)/Mintek Nanotechnology Innovation Centre (NIC), Advanced Materials Division, Mintek, Randburg 2194, South Africa

[3]Diagnostic Unit, Health Platform, Advanced Materials Division, Mintek, Randburg 2194, South Africa

snazozezethu@gmail.com, sinazoc@mintek.co.za

Abstract

Polymers are crucial in drug delivery systems (DDS), enhancing the efficacy and safety of therapeutics through controlled release, targeted delivery, and improved bioavailability. Natural polymers like alginate and chitosan offer biocompatibility and biodegradability, while synthetic polymers such as polyethylene glycol (PEG) and polylactic-co-glycolic acid (PLGA) provide solubility and controlled release. The incorporation of nanotechnology with polymers has led to advanced nanocarriers, enhancing drug stability and therapeutic effectiveness. Despite challenges like potential toxicity and complex manufacturing, careful selection and modification of polymers enable their application in advanced therapeutic areas like gene therapy and personalized medicine, demonstrating their versatility and importance in modern medicine.

Keywords

Drug Delivery Systems, Polymers, Controlled-Release, Stimuli-Responsive, Biodegradable

Contents

Materials Research Foundations 172 (2025) 263-281
https://doi.org/10.21741/9781644903353-11

1. Introduction

Polymers play a crucial role in drug delivery systems (DDS), offering a versatile platform for enhancing the efficacy and safety of therapeutic agents. Their distinctive characteristics allow for controlled release, targeted delivery, and improved bioavailability of drugs [1]. Various types of polymers, including natural, synthetic, and smart polymers, have been extensively studied for their applications in drug delivery. Natural polymers, such as alginate and chitosan, are increasingly utilized due to their biocompatibility and biodegradability [2]. They provide a favourable environment for drug encapsulation and release, improving patient adherence and treatment results [1,3]. The inherent diversity of natural polymers necessitates a comprehensive comprehension of their surface and volumetric attributes to optimize their functionalities in DDS [2,3]. For instance, alginate-based systems can modulate drug release rates through interactions with drugs and by chemical immobilization [4,5].

Synthetic polymers, particularly polyethylene glycol (PEG), have acquired prominence in drug delivery owing to their exceptional solubility and low immunogenicity [6]. PEG and Poly(vinyl alcohol) (PVA) though they have different molecular structures, they are often used in similar applications due to their similar properties. PEG-based systems are known for their ability to form micelles that can encapsulate hydrophobic drugs, thereby improving their solubility and bioavailability [7,8]. Furthermore, the integration of stimuli-responsive features into synthetic polymers allows for the development of smart DDS that can release drugs in response to specific environmental triggers, such as pH or temperature [9,10]. For example, poly(N-isopropylacrylamide) (PNIPAM) demonstrates a reduced critical solution temperature of approximately 32°C, which may be utilized for temperature-sensitive drug distribute applications [11].

The amalgamation of nanotechnology and polymer science has resulted in the creation of advanced nanocarriers for the administration of drugs. Polymer-coated nanoparticles, including mesoporous silica nanoparticles, have shown promise in achieving controlled drug release while minimizing systemic toxicity [12]. These nanocarriers can be engineered to respond to specific stimuli, facilitating exact regulation of drug release patterns [12,13]. Furthermore, the combination of various polymer architectures, such as dendritic and branched structures, has been examined to enhance the therapeutic efficacy of DDS.

2. Types of polymers used in drug delivery

Polymers play a crucial role in DDS, providing various mechanisms for controlled release and targeted therapy. Amidst the numerous types of polymers utilized, several categories demonstrate distinction due to their unique properties and functionalities.

2.1 Natural polymers

Natural polymers are essential for life, serving functions such as structural support, energy storage, facilitation of biochemical reactions, and encoding of genetic information. Their biodegradability

Materials Research Foundations 172 (2025) 263-281
https://doi.org/10.21741/9781644903353-11

and sustainability render them appealing for diverse industrial applications, as there is a growing global demand for environmentally friendly materials [2]. Chitosan has attracted interest as a natural polymer in pharmaceutical delivery methods. Its distinctive attributes, including as biocompatibility, biodegradability, and non-toxicity, render it an attractive option for pharmaceutical applications [14]. The main reasons for its application in drug delivery are its capacity to augment solubility, promote drug stability, and enable controlled release. One significant advantage of chitosan is mucoadhesive in nature, which helps in drug delivery. This trait enables adherence to mucosal surfaces, hence improving drug absorption across biological membranes [15]. Thus, chitosan has the potential to increase the bioavailability of oral drugs, a frequent obstacle in the development of efficient oral DDS. Additionally, it can undergo chemical modifications to optimize its characteristics, like solubility and release rate, to align with certain drug delivery specifications [14,15].

Besides its physical characteristics, chitosan's capacity to disrupt tight connections between epithelial cells improves drug absorption. This characteristic is particularly advantageous for the conveyance of substantial molecules, including proteins and peptides, which generally have difficulties in penetrating cellular walls [16]. Notwithstanding its myriad benefits, the application of chitosan in drug delivery presents certain obstacles. Variability in its origin and extent of deacetylation may result in performance discrepancies. Furthermore, its solubility is contingent upon pH, constraining its efficacy in specific settings [14].

Like chitosan, alginate, a biopolymer that occurs naturally produced by brown algae, has attracted considerable interest in drug administration owing to its flexibility, biocompatibility, and mucoadhesive characteristics [4]. Alginate comprises α-L-guluronic acid (G) and β-D-mannuronic acid (M) units, endowing it with distinctive physical and chemical features that provide it an excellent material for the advancement of controlled release mechanisms [17]. A key benefit of alginate in drug delivery is its capacity to create hydrogels in the presence of divalent cations including calcium. These hydrogels can encapsulate a diverse array of therapeutic molecules, encompassing pharmaceuticals that are composed of minuscule molecules, large proteins, and nucleic acids, providing protection from degradation and facilitating continuous release [17]. The gelation procedure is generally gentle, frequently conducted at physiological conditions, which is especially advantageous for sensitive medicinal molecules that may deteriorate or denature under severe conditions [18].

Furthermore, the biocompatibility of alginate guarantees a minimal inflammatory response and decreased cytotoxicity, which are essential considerations in the development of medicinal formulations [19]. The mucoadhesiveness, derived from its capacity to stick to mucosal surfaces, improves the bioavailability of medications delivered by mucosal routes such as oral, or nasal administration [4,19]. This sticky characteristic guarantees that the medicine stays near the absorption site, enhancing therapeutic efficiency. In addition to its use in traditional drug delivery, alginate is essential in innovative therapeutic approaches including gene therapy and tissue engineering [20]. In gene therapy, alginate can encapsulate DNA or RNA molecules, protecting them from enzymatic breakdown and enhancing their absorption by target cells. Alginate scaffolds in tissue engineering offer a beneficial milieu for cellular proliferation and tissue regeneration, sometimes combined with continuous release of growth factors to enhance tissue repair [21].

Another natural polymer is gelatin, a biopolymer obtained from collagen. Gelatin-based DDS can be developed for multiple routes of administration, such as oral, injectable, transdermal, and ocular

 Materials Research Forum LLC
https://doi.org/10.21741/9781644903353-11

delivery [22]. Gelatin's versatility is owing to its capacity to be converted into multiple forms, such as films, capsules, and scaffolds [23]. Gelatin's solubility in water and its ability to gel at physiological temperatures make it a suitable candidate for the development of in situ gels that can effectively encapsulate and release drugs at the targeted site of action. Gelatin is susceptible to diverse chemical modifications, enabling the tailoring of its characteristics for certain drug delivery applications. Gelatin can be conjugated with targeted ligands to enhance the selectivity of medication delivery to particular cells or tissues [24]. This can enhance the therapeutic efficacy of drugs while reducing undesirable effects. Moreover, gelatin can be integrated with other polymers or nanoparticles to generate composite systems exhibiting synergistic features [25]. Gelatin nanoparticles and microspheres have been engineered as carriers for various medicinal agents, encompassing small molecule medicines, proteins, and nucleic acids. These particle systems provide benefits including increased stability, better bioavailability, and tailored distribution [26]. Gelatin nanoparticles can be designed to evade immune system detection, therefore extending circulation duration and enhancing drug concentration at the target location [24].

2.2 Synthetic polymers

Synthetic polymers have transformed pharmaceutical administration, provided unparalleled regulation of the issuance and targeting of medications. The adaptability, reliability, and efficacy of synthetic polymers render them optimal choices for the advancement of DDS. In drug delivery, especially for ingested drugs, it is pivotal to use biocompatible polymers, though others like non-biodegradable Poly(methyl methacrylate) (PMMA) are better suited for applications including medical implants, dental prosthetics, and hard tissue replacements due to their strength and durability [27]. A significant category of synthetic polymers utilized in drug delivery encompass biodegradable polymers, including polylactic acid (PLA) and polyglycolic acid (PGA), together with its copolymers such as poly(lactic-co-glycolic acid) (PLGA) [28]. These polymers decompose into biocompatible metabolites that are readily excreted from the body. The degradation rate may be regulated by modifying the polymer composition, molecular weight, and crystallinity [28,29]. This renders them optimal for the formulation of controlled-release DDS, wherein the drug is released progressively over a designated timeframe, hence improving therapeutic efficacy and patient adherence. PLGA has been widely utilized for the transport of small compounds, proteins, and nucleic acids [30].

One prominent class of polymers utilized in pharmaceutical delivery are PEG and its derivatives. PEG is extensively utilized as a benchmark in bioconjugation and nanomedicine to extend blood circulation duration and enhance medication potency [31]. PEG is preferred due to its superior biocompatibility, solubility in physiological environments, and less immunogenicity, rendering it an optimal choice for in vivo applications [32]. The capacity to form aggregates at LCST, facilitates temperature-responsive drug release, offering significant benefits in targeted therapies [33,34]. Additionally, PEG-based polymers can be engineered to enhance the solubility and durability of therapeutic agents, thereby improving their bioavailability [35].

Polycaprolactone (PCL) is a biodegradable polyester that has garnered substantial interest in drug delivery owing to its distinctive characteristics and adaptability. A primary advantage of PCL in DDS is its biocompatibility [36]. It gradually decomposes into non-toxic metabolites, chiefly caproic acid, by hydrolytic degradation [36,37]. This progressive degradation is beneficial for controlled drug release applications, facilitating a prolonged release of the medication over an extended duration [40]. The degradation rate can be adjusted based on molecular weight and

Materials Research Foundations 172 (2025) 263-281
https://doi.org/10.21741/9781644903353-11

crystallinity, permitting customization of DDS to encounter diverse therapeutic specifications. Within medical applications, it does not elicit significant inflammatory responses, making it safe for prolonged usage in the human body [36]. This characteristic is crucial for DDS, since it ensures that the polymer will not cause harm or adverse responses during the administration of pharmaceuticals. PCL is recognized as an exceptionally potential material for drug delivery uses. Its biocompatibility, biodegradability, and property-modifying capabilities render it a flexible and efficacious polymer in the medical domain. As research progresses, the possibility for novel PCL-based DDS increases, providing promises for more effective and tailored therapeutic interventions.

3. Mechanisms of drug delivery using polymers

It is pivotal to maintain a consistent concentration of medication in the blood and tissues for a protracted duration for better therapeutic effectiveness, reducing adverse effects, and enhancing patient adherence by reducing the frequency of dosing. Conventional delivery systems present fluctuating levels of the minimum effective concentration of the drug, leading to peaks and troughs that can result in suboptimal therapeutic effects or increased toxicity [38]. In contrast, polymer-based release systems offer a more controlled, sustained release, preserving medication concentrations internal to the therapeutic aperture for a longer duration, enhancing efficacy and minimizing adverse effects [39]. Controlled DDS provide diminished drug toxicity and reduced drug exposure to the biological milieu, augmented drug efficacy, lower minimal adverse effects, minimal blood plasma concentration fluctuation, reduction in dosing frequency, and improved patience complience [40]. In order to provide all these benefits, controlled DDS have to be carefully designed with consideration of various aspects, including drug solubility, polymer compatibility, and the desired release profile. The design must account for the drug's pharmacokinetics, the degradation rate of the polymer, and the interaction between the drug and polymer to ensure a predictable and sustained release [41]. Moreover, factors like environmental conditions (e.g., pH, temperature), biocompatibility, and potential immune responses must also be considered to enhance treatment efficacy while reducing adverse effects [1]. There are multiple polymer release mechanisms, including controlled release, targeted delivery, and stimuli-responsive delivery (Figure 1) [29,33,42].

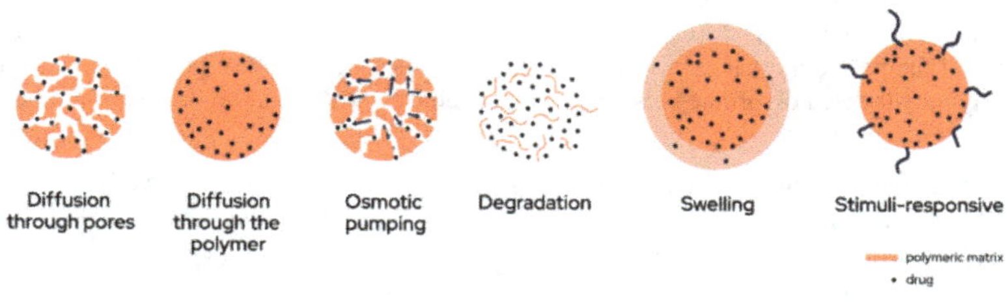

Figure 1. Drug-release mechanisms. Representation obtained from Rheolution Article (May 2021) by Dr. Dimitria Bonizol Camasao with a tittle „What are DDS".

3.1 Controlled polymer drug release

There are different methods in controlled polymer drug release, including degradation (erosion)-controlled release, diffusion-regulated release and swelling-modulated release, osmotic pressure-regulated release [43]. The drug release rate is influenced by factors such as polymer porosity, drug solubility, and the polymer's physicochemical properties, which can be adjusted to attain prolonged or regulated release over a specific duration. Polymers are not limited to a single drug release mechanism; instead, they are versatile, employing a variety of controlled release mechanisms. As the polymer gradually degrades in a degradation-controlled release, new routes and pores open up, allowing the drugs to diffuse, implying that both mechanisms frequently operate together to ensure sustained and controlled release over time. Polymer degradation-regulated pharmacological dissemination is the slow breakdown of the polymer matrix by hydrolytic or enzymatic mechanisms, culminating in the culminating in the scheduled liberation of the encapsulated pharmaceuticals release of the encapsulated pharmaceuticals. Polymer degradation-controlled drug release occurs via surface erosion and bulk degradation [29]. Surface erosion confines degradation to the polymer's surface, maintaining its shape while thinning, whereas bulk degradation involves uniform breakdown throughout the matrix due to water penetration, leading to consistent drug discharge.

In diffusion-controlled release, a drug disperses via the polymer guided through the diffusion coefficient of the pharmaceutical agent within the polymeric matrix. In diffusion-controlled release, a reservoir system contains the drug in a central core enclosed by a polymer membrane that facilitates the diffusion of the medication, while in a matrix system, the drug is dispersed throughout the polymer, and release occurs as the drug diffuses out of the polymer matrix over time. PLGA is one of the polymers that is widely used in diffusion-controlled and degradation-controlled drug release systems[28,44]. It provides sustained release of drugs due to its ability to degrade into lactic acid and glycolic acid through hydrolysis of its ester bonds [28]. This enables regulated the gradual emission of the drug over time as the polymer matrix degrades, allowing encapsulated drugs to diffuse. The rate of PLGA degradation and the ratio of lactic acid to glycolic acid, as well as the molecular weight, and particle size can be adjusted to regulate drug release [28]. Higher glycolic acid content typically accelerates the hydrolysis process due to its greater hydrophilicity, while factors like water presence and pH also influence the degradation rate. PLGA is widely used in DDS for vaccines [45], cancer therapy and diagnostic applications [46,47], and antibiotics [48]. PEG, polyanhydrides, and poly(ethylene-co-vinyl acetate) (PEVA) are some of the polymers that exhibit a diffusion-controlled drug release [49]. Polymers such as PCL, Polyanhydrides, Polyorthoesters, and Polyphosphazenes also exhibit degradation-controlled drug release mechanisms.

The polymer composition, crosslink density, and environmental conditions critically influence the drug release mechanism in swelling-controlled systems. Hydrogels, for instance, absorb water or biological fluids, causing the polymer to swell, which creates diffusion pathways for the drug to be administered in a regulated manner [1]. The ingress of solvent into a drug carrier can affect drug release behaviour by mechanisms such as osmosis- mediated release and swelling- regulated release, which regulate the drug release based on solvent absorption and polymer expansion [50]. This phenomenon is therefore refered to as solvent-controlled release, where the interaction of the polymer with the surrounding solvent influences drug diffusion.

Materials Research Foundations 172 (2025) 263-281 https://doi.org/10.21741/9781644903353-11

3.2 Polymer stimuli-responsive delivery

In some drug-delivery polymer systems, polymers are sensitive to environmental stimuli such as alterations in temperature, pH, redox, electromagnetic, light, or the existence of particular enzymes [51]. These stimuli-responsive polymers undergo structural modifications, such as swelling or degradation, in response to these triggers, which control the timing and location of drug release for enhanced therapeutic effect. Thus stimili-responsive polymers offer spatiotemporal in drug delivery. Smart polymers, which can respond to external stimuli, represent a significant advancement in drug delivery technology. These polymers may experience reversible transformations in their properties, enabling controlled drug liberation in reaction to physiological changes or external stimuli [10,11]. For example, pH-sensitive polymers can enhance drug delivery to tumor sites by releasing their payload in response to the acidic microenvironment typical of tumors [10,11]. Additionally, the use of magnetic nanoparticles in conjunction with polymer matrices has been explored to create localized drug release hotspots, further enhancing the targeting capabilities of DDS [52,53].

Temperature-sensitive polymers, for example, can change their structure when exposed to a specific temperature, allowing drugs to be released only when the body temperature rises, such as during a fever or inflammation. Thermosensitive polymers such as PNIPAM consist of monomers with hydrophilic amide and hydrophobic isopropyl functional groups, which impart temperature-responsive behavior [54,55]. The polymer's structure undergoes a reversible phase shift around its LCST, caused by hydrophobic interactions of the isopropyl groups and hydrogen bonding in the amide moiety. This phase transition allows controlled medication release in response to temperature fluctuations. PNIPAM-based smart hydrogels have become a prominent trend for developing advanced functional biomaterials [56].

4. Advantages of Polymers in Drug Delivery

Polymers have transformed drug delivery, providing many benefits that improve therapeutic efficacy, patient adherence, and overall therapy results. A primary advantage is their capacity for regulated and sustained medication release. Through meticulous design of the polymer architecture, one can achieve the controlled release of the medicine at a specified rate, thereby maintaining stable therapeutic concentrations in the bloodstream for extended durations [97]. This reduces the necessity for frequent dosing, hence improving patient adherence and convenience. Polymers can be engineered for biocompatibility and biodegradability, hence decreasing likelihood of detrimental responses and the necessity for surgical extraction post-therapeutic application [40]. This characteristic is especially advantageous in applications like implants and injectable systems, where the polymer can safely decompose within the body following medication delivery.

Polymers provide diversity in their chemical makeup, enabling the tailoring of DDS to aim at particular tissues or cells. This precise distribution reduces systemic negative effects and enhances the drug's therapeutic index [1,98]. Polymers can be engineered to react to certain biological stimuli, such as alterations in pH or temperature, facilitating the targeted release of the medicine at the location of the incident. Moreover, the application of polymers in drug delivery enables the encapsulation of many therapeutic agents, including tiny molecules, proteins, and nucleic acids [1,31]. This adaptability is essential for the advancement of sophisticated therapeutic approaches, including gene therapy and customized medicine, which necessitate the delivery of intricate biological components. Furthermore, Polymers can improve the stability of sensitive pharmaceuticals, safeguarding them from degradation due to environmental elements for example, light, heat, and enzymes. The stabilization is essential for prolonging the shelf life of pharmaceutical items and maintaining their potency till administration [99]. Polymers are not limited to their natural functions and functional groups. They can also be functionalized to be suited to the drugs loaded. For example, PLGA possesses an ester which can be functionalized with amines for conjugation with the intended active pharmaceutical ingredients (APIs) [100,101]. Polymers can be functionalized with other end-groups such as esters to improve half-life compared to polymers wuth carboxylic-end [102,103].

Polymers can markedly enhance the pharmacokinetic and pharmacodynamic characteristics of pharmaceuticals. They can improve the plasma half-life of biopharmaceuticals, diminish immunogenicity, and enhance the solubility of low molecular weight medications, which is essential for compounds with inadequate water solubility. [35,99]. This improvement is especially advantageous in targeted DDS, where polymers may be designed to release medications at precise locations in the body, therefore reducing systemic adverse effects and optimizing results of therapy [104]. For instance, polymeric nanoparticles can be designed to encapsulate anticancer drugs, allowing for localized delivery and reduced toxicity compared to conventional chemotherapy [105]. The benefits of polymers in DDS are numerous, including controlled release, improved drug characteristics, formulation flexibility, and biocompatibility. These attributes not only augment the effectiveness of therapeutic drugs but also reinforce patient adherence and safety. Therefore, this makes polymers an essential element of modern drug and delivery methods. Table 1 list the most common polymers utilized in drug administration, highlighting their properties, drug release mechanism, and drugs that have been delivered by those polymers. Table 1 provides an overview of commonly used polymers in DDS, detailing their key properties, functional groups, associated drug release mechanisms, and examples of drugs successfully delivered through these polymers.

Table 1. List of natural and synthetic polymers with their properties,

Polymers	Functional groups	Drug release mechanism	Properties	Drugs	References
Chitosan	Amine ($-NH_2$) and hydroxyl (-OH) groups	Ionic interaction-mediated and pH-responsive release	Mucoadhesive Biocompatible Biodegradable	Doxorubicin 5-fluorouracil Paclitaxel Insulin	[57–61]
Alginate	Hydroxyl (-OH) and	Ionic crosslinking-controlled and swelling-controlled release	Hydrophilic Biocompatible Biodegradable	Ibuprofen Insulin Vaccine	[62–65]
PLGA	Ester (-COO-) linkages	Degradation-controlled and diffusion-controlled release	Biocompatibility Biodegradable Tuneable degradation	Ciprofloxacin Dexamethasone Docetaxel Risperidone	[66–69]

PEG	Hydroxyl (-OH) and methoxy (-CH_3) groups	Diffusion-controlled and conjugation-based release	Hydrophobic Biodegradable Non-immunogenic Stealth Effect	Anticancer agents Doxorubicin Paclitaxel Insulin	[70–73]
PCL	Ester (-COO-) backbone	Degradation-controlled and diffusion-controlled release	Hydrophobic Biodegradable Elasticity	Anticancer agents 5-Fluorouracil Paclitaxel Tetracycline	[36,74–76]
PLA	Ester (-COO-) linkages	Degradation-controlled and diffusion-controlled release	Hydrophobic Biodegradable Tunable degradation	Bupivacaine Risperidone Insulin	[77–80]
PNIPAM	Amide (-CONH-) and isopropyl (-$CH(CH_3)_2$) groups	Thermally-induced diffusion-controlled release	Thermosensitive Reversible	Doxorubicin Anticacer agents	[54,55,81,82]
PVA	Hydroxyl (-OH) end groups	Swelling-controlled and diffusion-controlled release	Hydrophilic Biocompatible	Analgesia (Tramadol–Dexketoprofen) Paracetamol Ibuprofen Theophylline	[83–87]
PMMA	Ester (-COO-) groups	Diffusion-controlled release and surface erosion-controlled release	Non-degradable Biocompatible Hard and Durable	Gentamicin Vancomycin Anticancer agents [88–90]	
Hydrogels	Hydroxyl (-OH), carboxyl (-COOH), and amide (-CONH-) groups	Swelling-controlled release, stimuli-responsive release, and diffusion-controlled release		Doxorubicin Insulin Antibacterial agents Anticancer agents Growth factors	[91–96]

5. Limitations of polymer-based drug delivery systems

Although polymers provide significant benefits in drug delivery systems, they also present inherent limitations and challenges that can compromise their overall efficacy. They present challenges such as potential toxicity, slow degradation, and manufacturing complexity. Polymers like PLGA, though widely used, may cause inflammatory responses in some individuals due to their lactic/glycolic acidic degradation by-products [106]. Depending on the formulation, chitosan may also trigger undesired immunological responses or cause cutaneous discomfort [107]. Both natural polymers, chitosan and alginate, present with low mechanical strength, making them less appropriate for applications that require structural integrity. Among the major positive aspects of polymers is controlled release, however, some of the polymers such as PCL may exhibit extremely degradation rates [36]. This makes them unsuitable for rapid drug release and incomplete drug release. Incomplete drug release can be seen in other polymers such as hydrogels and alginate caused by poor diffusion through the polymer matrices. Simultaneously, polymers with high solubility of degradation rate such as PEG may lead to accelerated blood clearance, thereby diminishing the long-term effectiveness of the DDS [108]. These limitations can be managed through material selection and engineering maximizing the advantages of polymer-based DDS while mitigating the dangers. For example, a recent study by Upadhyay *et al.* 2024 [109] reported an increase in mechanical strength when using a chitosan-alginate biopolymer. This development mitigates the structural integrity limitations of chitosan and alginate, rendering it suitable for applications that necessitate strong structural stability

Conclusion

Polymers have established themselves as essential tools in the advancement of DDS due to their versatility, biocompatibility, and capacity to enable controlled and targeted drug release. These materials can improve the therapeutic efficacy of drugs by ensuring stable release over time, reducing the necessity for frequent dosing, and improving compliance of patients. Their biocompatibility and biodegradability also make them ideal for applications like implants and injectables. Polymers offer remarkable control over drug release rates, enhancing therapeutic efficacy through mechanisms including diffusion, degradation, and swelling-controlled release. The integration of smart polymers, responsive to triggers including pH and temperature, further expands the potential for precise, site-specific drug delivery. However, the effectiveness of these systems is often shaped by elements such as polymer degradation rates, mechanical properties, and potential immune responses, which can limit their performance in specific applications. While significant advancements have been made, ongoing research continues to focus on optimizing these materials to overcome limitations and improve drug delivery outcomes across diverse therapeutic areas.

References

[1]N.H. Thang, T.B. Chien, D.X. Cuong, Polymer-Based Hydrogels Applied in Drug Delivery: An Overview, Gels 9 (2023) 523. https://doi.org/10.3390/gels9070523.

[2]G. Satchanska, S. Davidova, P.D. Petrov, Natural and Synthetic Polymers for Biomedical and Environmental Applications, Polymers 16 (2024) 1159. https://doi.org/10.3390/polym16081159.

[3]M.C. Biswas, B. Bush, E. Ford, Glucaric acid additives for the antiplasticization of fibers wet spun from cellulose acetate/acetic acid/water, Carbohydr Polym 245 (2020) 116510. https://doi.org/10.1016/j.carbpol.2020.116510.

Materials Research Forum LLC
https://doi.org/10.21741/9781644903353-11

[4]M.A.S. Abourehab, R.R. Rajendran, A. Singh, S. Pramanik, P. Shrivastav, M.J. Ansari, R. Manne, L.S. Amaral, A. Deepak, Alginate as a Promising Biopolymer in Drug Delivery and Wound Healing: A Review of the State-of-the-Art, Int J Mol Sci 23 (2022) 9035. https://doi.org/10.3390/ijms23169035.

[5]K. Kruk, K. Winnicka, Alginates Combined with Natural Polymers as Valuable Drug Delivery Platforms, Mar Drugs 21 (2022) 11. https://doi.org/10.3390/md21010011.

[6]D. Braatz, M. Cherri, M. Tully, M. Dimde, G. Ma, E. Mohammadifar, F. Reisbeck, V. Ahmadi, M. Schirner, R. Haag, Chemical Approaches to Synthetic Drug Delivery Systems for Systemic Applications, Angew Chem Int Ed Engl 61 (2022) e202203942. https://doi.org/10.1002/anie.202203942.

[7]Y. Li, T. Zhang, Q. Liu, J. He, PEG-Derivatized Dual-Functional Nanomicelles for Improved Cancer Therapy, Front. Pharmacol. 10 (2019). https://doi.org/10.3389/fphar.2019.00808.

[8]M. Ghezzi, S. Pescina, C. Padula, P. Santi, E. Del Favero, L. Cantù, S. Nicoli, Polymeric micelles in drug delivery: An insight of the techniques for their characterization and assessment in biorelevant conditions, J Control Release 332 (2021) 312–336. https://doi.org/10.1016/j.jconrel.2021.02.031.

[9]C.M. Wells, M. Harris, L. Choi, V.P. Murali, F.D. Guerra, J.A. Jennings, Stimuli-Responsive Drug Release from Smart Polymers, J Funct Biomater 10 (2019) 34. https://doi.org/10.3390/jfb10030034.

[10] A. Purohit, S. Jain, P. Nema, H. Vishwakarma, P.K. Jain, Intelligent or Smart Polymers: Advance in Novel Drug Delivery, Journal of Drug Delivery and Therapeutics 12 (2022) 208–216. https://doi.org/10.22270/jddt.v12i5.5578.

[11] X. Xu, Y. Liu, W. Fu, M. Yao, Z. Ding, J. Xuan, D. Li, S. Wang, Y. Xia, M. Cao, Poly(N-isopropylacrylamide)-Based Thermoresponsive Composite Hydrogels for Biomedical Applications, Polymers 12 (2020) 580. https://doi.org/10.3390/polym12030580.

[12] W. Zhang, H. Liu, X. Qiu, F. Zuo, B. Wang, Mesoporous silica nanoparticles as a drug delivery mechanism, Open Life Sci 19 (2024) 20220867. https://doi.org/10.1515/biol-2022-0867.

[13] N.U. Khaliq, J. Lee, J. Kim, Y. Kim, S. Yu, J. Kim, S. Kim, D. Sung, H. Kim, Mesoporous Silica Nanoparticles as a Gene Delivery Platform for Cancer Therapy, Pharmaceutics 15 (2023) 1432. https://doi.org/10.3390/pharmaceutics15051432.

[14] N. Desai, D. Rana, S. Salave, R. Gupta, P. Patel, B. Karunakaran, A. Sharma, J. Giri, D. Benival, N. Kommineni, Chitosan: A Potential Biopolymer in Drug Delivery and Biomedical Applications, Pharmaceutics 15 (2023) 1313. https://doi.org/10.3390/pharmaceutics15041313.

[15] P. Mura, F. Maestrelli, M. Cirri, N. Mennini, Multiple Roles of Chitosan in Mucosal Drug Delivery: An Updated Review, Mar Drugs 20 (2022) 335. https://doi.org/10.3390/md20050335.

[16] V. Mikušová, P. Mikuš, Advances in Chitosan-Based Nanoparticles for Drug Delivery, Int J Mol Sci 22 (2021) 9652. https://doi.org/10.3390/ijms22179652.

[17] R. Abka-khajouei, L. Tounsi, N. Shahabi, A.K. Patel, S. Abdelkafi, P. Michaud, Structures, Properties and Applications of Alginates, Mar Drugs 20 (2022) 364. https://doi.org/10.3390/md20060364.

[18] P. Masson, S. Lushchekina, Conformational Stability and Denaturation Processes of Proteins Investigated by Electrophoresis under Extreme Conditions, Molecules 27 (2022) 6861. https://doi.org/10.3390/molecules27206861.

[19] F. Abasalizadeh, S.V. Moghaddam, E. Alizadeh, E. Akbari, E. Kashani, S.M.B. Fazljou, M. Torbati, A. Akbarzadeh, Alginate-based hydrogels as drug delivery vehicles in cancer treatment and their applications in wound dressing and 3D bioprinting, J Biol Eng 14 (2020) 8. https://doi.org/10.1186/s13036-020-0227-7.

[20] R. Ahmad Raus, W.M.F. Wan Nawawi, R.R. Nasaruddin, Alginate and alginate composites for biomedical applications, Asian J Pharm Sci 16 (2021) 280–306. https://doi.org/10.1016/j.ajps.2020.10.001.

[21] N. Farshidfar, S. Iravani, R.S. Varma, Alginate-Based Biomaterials in Tissue Engineering and Regenerative Medicine, Mar Drugs 21 (2023) 189. https://doi.org/10.3390/md21030189.

[22] F. Milano, A. Masi, M. Madaghiele, A. Sannino, L. Salvatore, N. Gallo, Current Trends in Gelatin-Based Drug Delivery Systems, Pharmaceutics 15 (2023) 1499. https://doi.org/10.3390/pharmaceutics15051499.

[23] I. Zulkiflee, M.B. Fauzi, Gelatin-Polyvinyl Alcohol Film for Tissue Engineering: A Concise Review, Biomedicines 9 (2021) 979. https://doi.org/10.3390/biomedicines9080979.

[24] O.A. Madkhali, Drug Delivery of Gelatin Nanoparticles as a Biodegradable Polymer for the Treatment of Infectious Diseases: Perspectives and Challenges, Polymers 15 (2023) 4327. https://doi.org/10.3390/polym15214327.

[25] S. Afewerki, A. Sheikhi, S. Kannan, S. Ahadian, A. Khademhosseini, Gelatin-polysaccharide composite scaffolds for 3D cell culture and tissue engineering: Towards natural therapeutics, Bioeng Transl Med 4 (2019) 96–115. https://doi.org/10.1002/btm2.10124.

[26] M. Santoro, A.M. Tatara, A.G. Mikos, Gelatin carriers for drug and cell delivery in tissue engineering, J Control Release 190 (2014) 210–218. https://doi.org/10.1016/j.jconrel.2014.04.014.

[27] M.S. Zafar, Prosthodontic Applications of Polymethyl Methacrylate (PMMA): An Update, Polymers (Basel) 12 (2020) 2299. https://doi.org/10.3390/polym12102299.

[28] H.K. Makadia, S.J. Siegel, Poly Lactic-co-Glycolic Acid (PLGA) as Biodegradable Controlled Drug Delivery Carrier, Polymers (Basel) 3 (2011) 1377–1397. https://doi.org/10.3390/polym3031377.

[29] N. Kamaly, B. Yameen, J. Wu, O.C. Farokhzad, Degradable Controlled-Release Polymers and Polymeric Nanoparticles: Mechanisms of Controlling Drug Release, Chem Rev 116 (2016) 2602–2663. https://doi.org/10.1021/acs.chemrev.5b00346.

[30] X. Guo, X. Zuo, Z. Zhou, Y. Gu, H. Zheng, X. Wang, G. Wang, C. Xu, F. Wang, PLGA-Based Micro/Nanoparticles: An Overview of Their Applications in Respiratory Diseases, Int J Mol Sci 24 (2023) 4333. https://doi.org/10.3390/ijms24054333.

[31] T.T. Hoang Thi, E.H. Pilkington, D.H. Nguyen, J.S. Lee, K.D. Park, N.P. Truong, The Importance of Poly(ethylene glycol) Alternatives for Overcoming PEG Immunogenicity in Drug Delivery and Bioconjugation, Polymers (Basel) 12 (2020) 298. https://doi.org/10.3390/polym12020298.

[32] S. Sun, Y. Cui, B. Yuan, M. Dou, G. Wang, H. Xu, J. Wang, W. Yin, D. Wu, C. Peng, Drug delivery systems based on polyethylene glycol hydrogels for enhanced bone regeneration, Front Bioeng Biotechnol 11 (2023) 1117647. https://doi.org/10.3389/fbioe.2023.1117647.

[33] M.A. Rahim, N. Jan, S. Khan, H. Shah, A. Madni, A. Khan, A. Jabar, S. Khan, A. Elhissi, Z. Hussain, H.C. Aziz, M. Sohail, M. Khan, H.E. Thu, Recent Advancements in Stimuli Responsive Drug Delivery Platforms for Active and Passive Cancer Targeting, Cancers (Basel) 13 (2021) 670. https://doi.org/10.3390/cancers13040670.

[34] S.H. Pham, Y. Choi, J. Choi, Stimuli-Responsive Nanomaterials for Application in Antitumor Therapy and Drug Delivery, Pharmaceutics 12 (2020) 630. https://doi.org/10.3390/pharmaceutics12070630.

[35] A.I. Fernandes, Polymers Enhancing Bioavailability in Drug Delivery, 2nd Edition, Pharmaceutics 15 (2023) 2604. https://doi.org/10.3390/pharmaceutics15112604.

[36] A. Bhadran, T. Shah, G.K. Babanyinah, H. Polara, S. Taslimy, M.C. Biewer, M.C. Stefan, Recent Advances in Polycaprolactones for Anticancer Drug Delivery, Pharmaceutics 15 (2023) 1977. https://doi.org/10.3390/pharmaceutics15071977.

[37] N. Geoghegan, M. O'Loughlin, C. Delaney, K.D. Rochfort, M. Kennedy, S. Kolagatla, L. Podhorska, B.J. Rodriguez, L. Florea, S.M. Kelleher, Controlled degradation of polycaprolactone-based micropillar arrays, Biomater. Sci. 11 (2023) 3077–3091. https://doi.org/10.1039/D3BM00165B.

[38] T.A. Ahmed, Pharmacokinetics of Drugs Following IV Bolus, IV Infusion, and Oral Administration, in: Basic Pharmacokinetic Concepts and Some Clinical Applications, IntechOpen, 2015. https://doi.org/10.5772/61573.

[39] A.S. Hoffman, The origins and evolution of "controlled" drug delivery systems, Journal of Controlled Release 132 (2008) 153–163. https://doi.org/10.1016/j.jconrel.2008.08.012.

[40] S. Adepu, S. Ramakrishna, Controlled Drug Delivery Systems: Current Status and Future Directions, Molecules 26 (2021) 5905. https://doi.org/10.3390/molecules26195905.

[41] S. Borandeh, B. van Bochove, A. Teotia, J. Seppälä, Polymeric drug delivery systems by additive manufacturing, Advanced Drug Delivery Reviews 173 (2021) 349–373. https://doi.org/10.1016/j.addr.2021.03.022.

[42] A. Zhang, K. Jung, A. Li, J. Liu, C. Boyer, Recent advances in stimuli-responsive polymer systems for remotely controlled drug release, Progress in Polymer Science 99 (2019) 101164. https://doi.org/10.1016/j.progpolymsci.2019.101164.

Materials Research Forum LLC
https://doi.org/10.21741/9781644903353-11

[43] G.-H. Son, B.-J. Lee, C.-W. Cho, Mechanisms of drug release from advanced drug formulations such as polymeric-based drug-delivery systems and lipid nanoparticles, Journal of Pharmaceutical Investigation 47 (2017) 287–296. https://doi.org/10.1007/s40005-017-0320-1.

[44] R.A. Jain, The manufacturing techniques of various drug loaded biodegradable poly(lactide-*co*-glycolide) (PLGA) devices, Biomaterials 21 (2000) 2475–2490. https://doi.org/10.1016/S0142-9612(00)00115-0.

[45] A.L. Silva, P.C. Soema, B. Slütter, F. Ossendorp, W. Jiskoot, PLGA particulate delivery systems for subunit vaccines: Linking particle properties to immunogenicity, Hum Vaccin Immunother 12 (2016) 1056–1069. https://doi.org/10.1080/21645515.2015.1117714.

[46] X. Shen, T. Li, X. Xie, Y. Feng, Z. Chen, H. Yang, C. Wu, S. Deng, Y. Liu, PLGA-Based Drug Delivery Systems for Remotely Triggered Cancer Therapeutic and Diagnostic Applications, Front Bioeng Biotechnol 8 (2020) 381. https://doi.org/10.3389/fbioe.2020.00381.

[47] D. Zhang, L. Liu, J. Wang, H. Zhang, Z. Zhang, G. Xing, X. Wang, M. Liu, Drug-loaded PEG-PLGA nanoparticles for cancer treatment, Front Pharmacol 13 (2022) 990505. https://doi.org/10.3389/fphar.2022.990505.

[48] A. Shariati, Z. Chegini, E. Ghaznavi-Rad, E.N. Zare, S.M. Hosseini, PLGA-Based Nanoplatforms in Drug Delivery for Inhibition and Destruction of Microbial Biofilm, Front Cell Infect Microbiol 12 (2022) 926363. https://doi.org/10.3389/fcimb.2022.926363.

[49] Y. Fu, W.J. Kao, Drug Release Kinetics and Transport Mechanisms of Non-degradable and Degradable Polymeric Delivery Systems, Expert Opin Drug Deliv 7 (2010) 429–444. https://doi.org/10.1517/17425241003602259.

[50] J.H. Lee, Y. Yeo, Controlled Drug Release from Pharmaceutical Nanocarriers, Chem Eng Sci 125 (2015) 75–84. https://doi.org/10.1016/j.ces.2014.08.046.

[51] Y. Zhao, Y. Guo, L. Tang, Engineering cancer vaccines using stimuli-responsive biomaterials, Nano Res. 11 (2018) 5355–5371. https://doi.org/10.1007/s12274-018-2162-1.

[52] E. Cazares-Cortes, M. Nerantzaki, J. Fresnais, C. Wilhelm, N. Griffete, C. Ménager, Magnetic Nanoparticles Create Hot Spots in Polymer Matrix for Controlled Drug Release, Nanomaterials 8 (2018) 850. https://doi.org/10.3390/nano8100850.

[53] A. Pusta, M. Tertis, I. Crăciunescu, R. Turcu, S. Mirel, C. Cristea, Recent Advances in the Development of Drug Delivery Applications of Magnetic Nanomaterials, Pharmaceutics 15 (2023) 1872. https://doi.org/10.3390/pharmaceutics15071872.

[54] L. Tang, L. Wang, X. Yang, Y. Feng, Y. Li, W. Feng, Poly(*N*-isopropylacrylamide)-based smart hydrogels: Design, properties and applications, Progress in Materials Science 115 (2021) 100702. https://doi.org/10.1016/j.pmatsci.2020.100702.

[55] B. Kalkan, N. Orakdogen, Negatively charged poly(N-isopropyl acrylamide-*co*-methacrylic acid)/polyacrylamide semi-IPN hydrogels: Correlation between swelling and compressive elasticity, Reactive and Functional Polymers 174 (2022) 105245. https://doi.org/10.1016/j.reactfunctpolym.2022.105245.

Materials Research Forum LLC
https://doi.org/10.21741/9781644903353-11

[56] N.A. Shaibie, N.A. Ramli, N.D.F. Mohammad Faizal, T. Srichana, M.C.I. Mohd Amin, Poly(N-isopropylacrylamide)-Based Polymers: Recent Overview for the Development of Temperature-Responsive Drug Delivery and Biomedical Applications, Macromolecular Chemistry and Physics 224 (2023) 2300157. https://doi.org/10.1002/macp.202300157.

[57] M. Yadav, B. Kaushik, G.K. Rao, C.M. Srivastava, D. Vaya, Advances and challenges in the use of chitosan and its derivatives in biomedical fields: A review, Carbohydrate Polymer Technologies and Applications 5 (2023) 100323. https://doi.org/10.1016/j.carpta.2023.100323.

[58] M. Ashrafizadeh, K. Hushmandi, S. Mirzaei, S. Bokaie, A. Bigham, P. Makvandi, N. Rabiee, V.K. Thakur, A.P. Kumar, E. Sharifi, R.S. Varma, A.R. Aref, M. Wojnilowicz, A. Zarrabi, H. Karimi-Maleh, N.H. Voelcker, E. Mostafavi, G. Orive, Chitosan-based nanoscale systems for doxorubicin delivery: Exploring biomedical application in cancer therapy, Bioeng Transl Med 8 (2022) e10325. https://doi.org/10.1002/btm2.10325.

[59] K.R. Karnati, Y. Wang, Understanding the co-loading and releasing of doxorubicin and paclitaxel using chitosan functionalized single-walled carbon nanotubes by molecular dynamics simulations, Phys Chem Chem Phys 20 (2018) 9389–9400. https://doi.org/10.1039/C8CP00124C.

[60] M.U. Akbar, S. Khattak, M.I. Khan, U.A.K. Saddozai, N. Ali, A.F. AlAsmari, M. Zaheer, M. Badar, A pH-responsive bi-MIL-88B MOF coated with folic acid-conjugated chitosan as a promising nanocarrier for targeted drug delivery of 5-Fluorouracil, Front Pharmacol 14 (2023) 1265440. https://doi.org/10.3389/fphar.2023.1265440.

[61] S. Sarkar, D. Das, P. Dutta, J. Kalita, S.B. Wann, P. Manna, Chitosan: A promising therapeutic agent and effective drug delivery system in managing diabetes mellitus, Carbohydr Polym 247 (2020) 116594. https://doi.org/10.1016/j.carbpol.2020.116594.

[62] I.P.S. Fernando, W. Lee, E.J. Han, G. Ahn, Alginate-based nanomaterials: Fabrication techniques, properties, and applications, Chemical Engineering Journal 391 (2020) 123823. https://doi.org/10.1016/j.cej.2019.123823.

[63] O.J. Olayemi, Y.E. Apeji, C.Y. Isimi, Formulation and Evaluation of Cyperus esculentus (Tiger Nut) Starch-Alginate Microbeads in the Oral Delivery of Ibuprofen, J Pharm Innov 17 (2022) 366–375. https://doi.org/10.1007/s12247-020-09509-2.

[64] W. Yu, G. Jiang, Y. Zhang, D. Liu, B. Xu, J. Zhou, Polymer microneedles fabricated from alginate and hyaluronate for transdermal delivery of insulin, Mater Sci Eng C Mater Biol Appl 80 (2017) 187–196. https://doi.org/10.1016/j.msec.2017.05.143.

[65] F. Sarei, N.M. Dounighi, H. Zolfagharian, P. Khaki, S.M. Bidhendi, Alginate nanoparticles as a promising adjuvant and vaccine delivery system, Indian J Pharm Sci 75 (2013) 442–449. https://doi.org/10.4103/0250-474X.119829.

[66] M. Ramchandani, D. Robinson, In vitro and in vivo release of ciprofloxacin from PLGA 50:50 implants, J Control Release 54 (1998) 167–175. https://doi.org/10.1016/s0168-3659(97)00113-2.

[67] H. Gasmi, F. Siepmann, M.C. Hamoudi, F. Danede, J. Verin, J.-F. Willart, J. Siepmann, Towards a better understanding of the different release phases from PLGA microparticles:

Dexamethasone-loaded systems, Int J Pharm 514 (2016) 189–199. https://doi.org/10.1016/j.ijpharm.2016.08.032.

[68] Y. Zheng, F. Sheng, Z. Wang, G. Yang, C. Li, H. Wang, Z. Song, Shear Speed-Regulated Properties of Long-Acting Docetaxel Control Release Poly (Lactic-Co-Glycolic Acid) Microspheres, Front Pharmacol 11 (2020) 1286. https://doi.org/10.3389/fphar.2020.01286.

[69] J. Shen, S. Choi, W. Qu, Y. Wang, D.J. Burgess, In vitro-in vivo correlation of parenteral risperidone polymeric microspheres, J Control Release 218 (2015) 2–12. https://doi.org/10.1016/j.jconrel.2015.09.051.

[70] S.S. Banerjee, N. Aher, R. Patil, J. Khandare, Poly(ethylene glycol)-Prodrug Conjugates: Concept, Design, and Applications, Journal of Drug Delivery 2012 (2012). https://doi.org/10.1155/2012/103973.

[71] P.G. Rose, Pegylated liposomal doxorubicin: optimizing the dosing schedule in ovarian cancer, Oncologist 10 (2005) 205–214. https://doi.org/10.1634/theoncologist.10-3-205.

[72] H. Hashemzadeh, H. Raissi, The functionalization of carbon nanotubes to enhance the efficacy of the anticancer drug paclitaxel: a molecular dynamics simulation study, J Mol Model 23 (2017) 222. https://doi.org/10.1007/s00894-017-3391-z.

[73] C. Yang, D. Lu, Z. Liu, How PEGylation enhances the stability and potency of insulin: a molecular dynamics simulation, Biochemistry 50 (2011) 2585–2593. https://doi.org/10.1021/bi101926u.

[74] G.Ö. Kayan, A. Kayan, Polycaprolactone Composites/Blends and Their Applications Especially in Water Treatment, ChemEngineering 7 (2023) 104. https://doi.org/10.3390/chemengineering7060104.

[75] S. Iqbal, M.H. Rashid, A.S. Arbab, M. Khan, Encapsulation of Anticancer Drugs (5-Fluorouracil and Paclitaxel) into Polycaprolactone (PCL) Nanofibers and In Vitro Testing for Sustained and Targeted Therapy, J Biomed Nanotechnol 13 (2017) 355–366. https://doi.org/10.1166/jbn.2017.2353.

[76] E.R. Essien, V.N. Atasie, S.T. Scott, O.A. Ajayi, A tetracycline hydrochloride-loaded SiO2/polycaprolactone composite from bamboo stem for controlled drug release study, South African Journal of Science 118 (2022) 1–6. https://doi.org/10.17159/sajs.2022/12560.

[77] S.K. Saha, H. Tsuji, Effects of molecular weight and small amounts of d-lactide units on hydrolytic degradation of poly(l-lactic acid)s, Polymer Degradation and Stability 91 (2006) 1665–1673. https://doi.org/10.1016/j.polymdegradstab.2005.12.009.

[78] J. Xu, Y. Bai, X. Li, Z. Wei, L. Sun, H. Yu, H. Xu, Porous Core/Dense Shell PLA Microspheres Embedded with High Drug Loading of Bupivacaine Crystals for Injectable Prolonged Release, AAPS PharmSciTech 22 (2021) 27. https://doi.org/10.1208/s12249-020-01878-8.

[79] S. Nanaki, P. Barmpalexis, A. Iatrou, E. Christodoulou, M. Kostoglou, D.N. Bikiaris, Risperidone Controlled Release Microspheres Based on Poly(Lactic Acid)-Poly(Propylene Adipate) Novel Polymer Blends Appropriate for Long Acting Injectable Formulations, Pharmaceutics 10 (2018) 130. https://doi.org/10.3390/pharmaceutics10030130.

[80] R. Hashide, K. Yoshida, Y. Hasebe, M. Seno, S. Takahashi, K. Sato, J.-I. Anzai, Poly(lactic acid) microparticles coated with insulin-containing layer-by-layer films and their pH-dependent insulin release, J Nanosci Nanotechnol 14 (2014) 3100–3105. https://doi.org/10.1166/jnn.2014.8562.

[81] Y. Yar, R. Khodadust, Y. Akkoc, M. Utkur, E.U. Saritas, D. Gozuacik, H. Yagci Acar, Development of tailored SPION-PNIPAM nanoparticles by ATRP for dually responsive doxorubicin delivery and MR imaging, J Mater Chem B 6 (2018) 289–300. https://doi.org/10.1039/c7tb00646b.

[82] P. Baipaywad, N. Ryu, S.-S. Im, U. Lee, H.B. Son, W.J. Kim, H. Park, Facile preparation of poly(N-isopropylacrylamide)/graphene oxide nanocomposites for chemo-photothermal therapy, Des Monomers Polym 25 (n.d.) 245–253. https://doi.org/10.1080/15685551.2022.2111854.

[83] J.C. Flores-Arriaga, D. Chavarría-Bolaños, A. de J. Pozos-Guillén, V.A. Escobar-Barrios, B.I. Cerda-Cristerna, Synthesis of a PVA drug delivery system for controlled release of a Tramadol–Dexketoprofen combination, J Mater Sci Mater Med 32 (2021) 56. https://doi.org/10.1007/s10856-021-06529-3.

[84] A. Goyanes, P. Robles Martinez, A. Buanz, A.W. Basit, S. Gaisford, Effect of geometry on drug release from 3D printed tablets, International Journal of Pharmaceutics 494 (2015) 657–663. https://doi.org/10.1016/j.ijpharm.2015.04.069.

[85] S. Das, U. Subuddhi, Controlled delivery of ibuprofen from poly(vinyl alcohol)−poly(ethylene glycol) interpenetrating polymeric network hydrogels, J Pharm Anal 9 (2019) 108–116. https://doi.org/10.1016/j.jpha.2018.11.007.

[86] H. Asci, M. Savran, F. Cengiz Callıoglu, S. Sahin, N. Hasseyid, M. Kaynak, N. Izat, H. Kesici Guler, Supralingual administration of paracetamol embedded in polyvinyl alcohol nanofibers: A pharmacokinetic study, Journal of Drug Delivery Science and Technology 67 (2022) 102948. https://doi.org/10.1016/j.jddst.2021.102948.

[87] R.C. DiLuccio, M.A. Hussain, D. Coffin-Beach, G. Torosian, E. Shefter, A.R. Hurwitz, Sustained-release oral delivery of theophylline by use of polyvinyl alcohol and polyvinyl alcohol-methyl acrylate polymers, J Pharm Sci 83 (1994) 104–106. https://doi.org/10.1002/jps.2600830124.

[88] A. Kowalewska, A.S. Herc, J. Bojda, M. Nowacka, M. Svyntkivska, E. Piorkowska, W. Kaczorowski, W. Szymański, Phase Structure and Properties of Ternary Polylactide/Poly(methyl methacrylate)/Polysilsesquioxane Blends, Polymers (Basel) 13 (2021) 1033. https://doi.org/10.3390/polym13071033.

[89] E. Bertazzoni Minelli, C. Caveiari, A. Benini, Release of antibiotics from polymethylmethacrylate cement, J Chemother 14 (2002) 492–500. https://doi.org/10.1179/joc.2002.14.5.492.

[90] F.A. Khan, S. Akhtar, D. Almohazey, M. Alomari, S.A. Almofty, I. Badr, A. Elaissari, Targeted delivery of poly (methyl methacrylate) particles in colon cancer cells selectively attenuates cancer cell proliferation, Artif Cells Nanomed Biotechnol 47 (2019) 1533–1542. https://doi.org/10.1080/21691401.2019.1577886.

[91] S. Bashir, M. Hina, J. Iqbal, A.H. Rajpar, M.A. Mujtaba, N.A. Alghamdi, S. Wageh, K. Ramesh, S. Ramesh, Fundamental Concepts of Hydrogels: Synthesis, Properties, and Their Applications, Polymers (Basel) 12 (2020) 2702. https://doi.org/10.3390/polym12112702.

[92] H. Wu, S. Liu, L. Xiao, X. Dong, Q. Lu, D.L. Kaplan, Injectable and pH-Responsive Silk Nanofiber Hydrogels for Sustained Anticancer Drug Delivery, ACS Appl Mater Interfaces 8 (2016) 17118–17126. https://doi.org/10.1021/acsami.6b04424.

[93] N.S.V. Capanema, I.C. Carvalho, A.A.P. Mansur, S.M. Carvalho, A.P. Lage, H.S. Mansur, Hybrid Hydrogel Composed of Carboxymethylcellulose–Silver Nanoparticles–Doxorubicin for Anticancer and Antibacterial Therapies against Melanoma Skin Cancer Cells, ACS Appl. Nano Mater. 2 (2019) 7393–7408. https://doi.org/10.1021/acsanm.9b01924.

[94] T. Iimaa, T. Hirayama, N. Shirakigawa, D. Imai, T. Yamao, Y.-I. Yamashita, H. Baba, H. Ijima, Antibacterial-Agent-Immobilized Gelatin Hydrogel as a 3D Scaffold for Natural and Bioengineered Tissues, Gels 5 (2019) 32. https://doi.org/10.3390/gels5020032.

[95] S. Mansoor, P.P.D. Kondiah, Y.E. Choonara, Advanced Hydrogels for the Controlled Delivery of Insulin, Pharmaceutics 13 (2021) 2113. https://doi.org/10.3390/pharmaceutics13122113.

[96] B.-H. Shan, F.-G. Wu, Hydrogel-Based Growth Factor Delivery Platforms: Strategies and Recent Advances, Adv Mater 36 (2024) e2210707. https://doi.org/10.1002/adma.202210707.

[97] I. Negut, B. Bita, Polymersomes as Innovative, Stimuli-Responsive Platforms for Cancer Therapy, Pharmaceutics 16 (2024) 463. https://doi.org/10.3390/pharmaceutics16040463.

[98] W.B. Liechty, D.R. Kryscio, B.V. Slaughter, N.A. Peppas, Polymers for drug delivery systems, Annu Rev Chem Biomol Eng 1 (2010) 149–173. https://doi.org/10.1146/annurev-chembioeng-073009-100847.

[99] M. Akbarian, S.-H. Chen, Instability Challenges and Stabilization Strategies of Pharmaceutical Proteins, Pharmaceutics 14 (2022) 2533. https://doi.org/10.3390/pharmaceutics14112533.

[100] P. Gentile, V. Chiono, I. Carmagnola, P.V. Hatton, An Overview of Poly(lactic-co-glycolic) Acid (PLGA)-Based Biomaterials for Bone Tissue Engineering, Int J Mol Sci 15 (2014) 3640–3659. https://doi.org/10.3390/ijms15033640.

[101] J.J. Yoon, S.H. Song, D.S. Lee, T.G. Park, Immobilization of cell adhesive RGD peptide onto the surface of highly porous biodegradable polymer scaffolds fabricated by a gas foaming/salt leaching method, Biomaterials 25 (2004) 5613–5620. https://doi.org/10.1016/j.biomaterials.2004.01.014.

[102] X.S. Wu, N. Wang, Synthesis, characterization, biodegradation, and drug delivery application of biodegradable lactic/glycolic acid polymers. Part II: biodegradation, J Biomater Sci Polym Ed 12 (2001) 21–34. https://doi.org/10.1163/156856201744425.

[103] L. Lu, S.J. Peter, M.D. Lyman, H.L. Lai, S.M. Leite, J.A. Tamada, J.P. Vacanti, R. Langer, A.G. Mikos, In vitro degradation of porous poly(L-lactic acid) foams, Biomaterials 21 (2000) 1595–1605. https://doi.org/10.1016/s0142-9612(00)00048-x.

[104] T.C. Ezike, U.S. Okpala, U.L. Onoja, C.P. Nwike, E.C. Ezeako, O.J. Okpara, C.C. Okoroafor, S.C. Eze, O.L. Kalu, E.C. Odoh, U.G. Nwadike, J.O. Ogbodo, B.U. Umeh, E.C. Ossai, B.C. Nwanguma, Advances in drug delivery systems, challenges and future directions, Heliyon 9 (2023) e17488. https://doi.org/10.1016/j.heliyon.2023.e17488.

[105] X. Xiao, F. Teng, C. Shi, J. Chen, S. Wu, B. Wang, X. Meng, A. Essiet Imeh, W. Li, Polymeric nanoparticles—Promising carriers for cancer therapy, Front. Bioeng. Biotechnol. 10 (2022). https://doi.org/10.3389/fbioe.2022.1024143.

[106] E.J. Go, E.Y. Kang, S.K. Lee, S. Park, J.H. Kim, W. Park, I.H. Kim, B. Choi, D.K. Han, An osteoconductive PLGA scaffold with bioactive β-TCP and anti-inflammatory Mg(OH)2 to improve in vivo bone regeneration, Biomater. Sci. 8 (2020) 937–948. https://doi.org/10.1039/C9BM01864F.

[107] S.L. Reay, E.L. Jackson, D. Salthouse, A.M. Ferreira, C.M.U. Hilkens, K. Novakovic, Effective Endotoxin Removal from Chitosan That Preserves Chemical Structure and Improves Compatibility with Immune Cells, Polymers (Basel) 15 (2023) 1592. https://doi.org/10.3390/polym15071592.

[108] G. Miao, Y. He, K. Lai, Y. Zhao, P. He, G. Tan, X. Wang, Accelerated blood clearance of PEGylated nanoparticles induced by PEG-based pharmaceutical excipients, Journal of Controlled Release 363 (2023) 12–26. https://doi.org/10.1016/j.jconrel.2023.09.003.

[109] U. Upadhyay, S. Kolla, S. Maredupaka, S. Priya, K. Srinivasulu, L.K. Chelluri, Development of an alginate–chitosan biopolymer composite with dECM bioink additive for organ-on-a-chip articular cartilage, Sci Rep 14 (2024) 11765. https://doi.org/10.1038/s41598-024-62656-1.

Materials Research Forum LLC
https://doi.org/10.21741/9781644903353-12

Chapter 12

Polymers in Wound Care and Dressings

Ingrid Larissa da Silva Soares, Maria Júlia Araújo Dias, Alessandra Teixeira, Mariana Rillo Sato, João Augusto Oshiro-Júnior*

Pharmaceutical Sciences Postgraduate Program, Center for Biological and Health Sciences, State University of Paraíba - UEPB, Campina Grande, Paraíba, Brazil

*joaooshiro@yahoo.com.br

Abstract

Polymers, both of natural and synthetic origin, have been essential in the evolution of chronic and acute wound healing and tissue regeneration, thanks to their unique and adaptable properties. Their applicability covers a wide range of areas, including industry. Biopolymers such as alginate, chitosan and cellulose are valued for their biocompatibility and ability to form gels, promoting favorable conditions for wound healing. However, their rapid degradation and limited mechanical resistance are disadvantages that can be overcome by combining them with synthetic polymers. Thus, synthetic polymers such as PVA, PLA and polyurethane offer greater structural stability, as well as allowing controlled drug release. In addition, hybrid polymers are emerging as innovative materials with improved absorption properties, mechanical resistance and antimicrobial functionalities, essential for the development of dressings, membranes, films and ova. This chapter covers the characteristics and applications of the main polymers, highlighting the most recent scientific and market contributions that act in the treatment of complex wounds, bone regeneration, as well as in the control of inflammatory and infectious processes.

Keywords

Smart Materials, Biocompatible Membrane, Wound Healing, Occlusive Dressings, Hydrocolloid

Contents

 Materials Research Forum LLC
https://doi.org/10.21741/9781644903353-12

1. Introduction

Wounds can be defined as lesions that structurally and functionally compromise tissues, especially the skin. Their appearance is linked to mechanical or thermal damage or to some physiological condition. These injuries can be classified according to their healing time and can be called chronic or acute. When each phase of healing (homeostasis, inflammation, proliferation and remodeling), as shown in Figure 1, occurs in an efficient and orderly manner, wounds heal quickly [1-3].

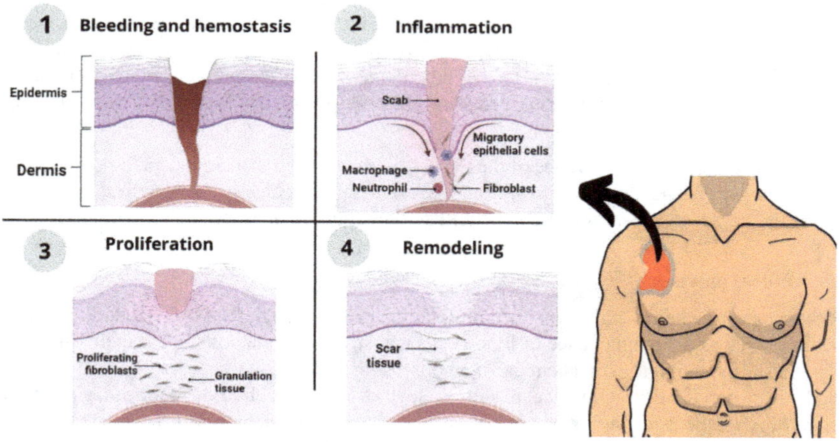

Figure 1 - (1) Hemostasis; (2) Inflammatory phase: neutrophils release pro-inflammatory cytokines that trigger the appearance of inflammatory cells in the injured tissue. Local proliferation of fibroblasts and endothelial cells is encouraged; (3) Proliferation phase: excessive migration and proliferation of cells and synthesis of granulation tissue (4) Remodeling phase: Final remodeling phase of wound healing, granulation tissue is replaced by a permanent scar. Image generated using Biorender.

Otherwise, they become chronic, taking 8-12 weeks to complete the healing process. Due to the long period and difficulty in the healing process, these types of wounds represent a growing challenge in global health. Causing enormous physical and financial inconvenience to health systems, this type of injury affects an average of 6 million people in the United States and 1.5 million on the European continent [1,3,4].

Materials Research Foundations 172 (2025) 282-298
https://doi.org/10.21741/9781644903353-12

During treatment, in most cases, conventional dressings are used, such as ordinary gauze, usually cotton-based, bandages based on synthetic polymers, such as polyesters, hydrogels and foams, which serve as a physical barrier to isolate the affected area during its healing process, as well as absorbing the exudate released at the site and speeding up wound closure. In addition to these, in the search for better results, technological advances have led to the creation of various types of dressings that treat numerous types of wounds, focusing on the four phases of healing, including so-called smart dressings that can interact with wounds [5,6]. However, for dressings to be considered ideal, they must have characteristics that allow the affected tissue to regenerate, and their performance depends directly on their action on the tissue (adherent, occlusive or absorbent), the type of polymer used in their production and their physical form, which can be presented in the form of film, gel, foams or nanofibers. As essential characteristics, they must have biocompatibility, biodegradability, non-immunogenicity, conformability to the wound, hydration capacity, stimulation of cell migration and proliferation, maintenance of adequate tissue temperature, prevention of infections by pathogens and cost-effectiveness [1,7].

In this context, wound dressings composed of polymeric matrices of natural or synthetic origin have aroused great interest in the treatment of chronic wounds due to their excellent biomedical properties, for example, high porosity, biocompatibility, low immunogenicity, atoxity, antibacterial and healing action, as well as the ability to incorporate bioactive agents. In addition, combinations of synthetic polymers with natural polymers have been widely used to overcome the shortcomings common to both types, resulting in hybrid bases, for example, ureasil-polyether. [8,9].

Therefore, considering the relevance of the topic, this chapter will discuss the main polymers used to produce dressings applied to wound treatment, addressing their physical-chemical characteristics, advantages, disadvantages, among other important aspects.

2. Natural polymers

The category of natural polymers includes components that are part of essential biological functions, such as the maintenance of living cells and structural functions like proteins. They can be produced by biological or chemical methods using biological resources. Natural polymers include those derived from biological resources, such as cellulose, chitosan, collagen, alginate and starch, which depend on a variety of natural precursors and metabolic mechanisms for their synthesis [10], as shown in figure 2. Consequently, because they have numerous applications, easy availability and low cost, they have become highly speculated products for the formulation of hydrogels, films, membranes, medical treatments and functions in other industries, such as food and textiles, for example [11,12].

Materials Research Forum LLC
https://doi.org/10.21741/9781644903353-12

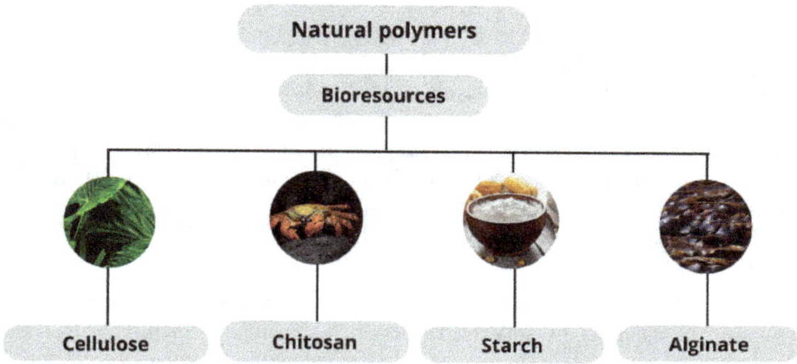

Figure 2 - Natural polymers.

2.1 Cellulose

Cellulose is a biopolymer synthesized in plants, fungi and some bacteria. Within the classification of natural polymers, it is defined as a polysaccharide macromolecule with linear chains of thousands of D-glucose linked to β (1→4). Cellulose produced by bacteria has remarkable mechanical characteristics, high affinity for water, crystallinity and polymerization, as well as being simpler to purify. However, despite presenting favorable conditions for use mainly in tissue engineering applications, its mechanical properties are still not ideal, such as tensile modulus, Young's modulus and compressive stress [13].

The great challenge in relation to the properties of cellulose is its insolubility in water and low solubility in a variety of organic solvents, so to be able to take advantage of its advantages, it becomes necessary to facilitate the processability of this component without resulting in the degradation of relevant properties. In this sense, studies carried out using chemical reactions of carboxymethylation and hydroxypropylation at the main sites have shown that the physical-chemical alteration of cellulose has resulted in an improvement in its solubility in common solvents and in water [14,15].

This type of polymer is already found in traditional dressings, for example Membracel® (Vuelo Pharma, Brazil), due to its ability to absorb wound exudate. Together with gauze fabrics, made from cotton fibers or combinations with other synthetic fibers such as nylon and polyester, they are used to keep the wound dry and protected. Although they offer some protection against bacteria, their effectiveness decreases when moistened. In addition, they can stick to the wound, making removal painful. To improve this characteristic, some gauze is impregnated with paraffin [1].

In search of new applications, Kiiskinen *et al* [16] analyzed the potential of natural polymer matrix dressings, using cellulose of plant origin, as a support for mesenchymal stem/stromal cells derived from human adipose tissue (hASCs). The main objective was to develop an innovative method of cell transplantation without animal components for the treatment of wounds. As a result, over the course of 2 weeks, the researchers observed high cell viability with cell densities of 300 K and 500

Materials Research Forum LLC
https://doi.org/10.21741/9781644903353-12

K in the dressings produced. On days 4 and 7, at a density of 300 K, they showed no significant cytotoxicity, with 12.47 ± 1.61% and 4.35 ± 0.88%, respectively. In addition, more than 97% of the cells that were seeded adhered within the first 24 hours of culture, with 98.04 ± 1.50% for the cells cultured on the cellulose-based dressing and ± 2.49% for the control groups. Compared to the control cells, cell adhesion was statistically significant (*p<0.05) only within 1 hour. In addition, the authors point out that there was no statistically significant increase in the expression levels of tumor necrosis factor (TNF-α), interleukin (IL-6) and fibroblast growth factor (FGF)-2 at high cell densities.

2.2 Chitosan

Chitosan, the second most abundant biopolymer in nature, results from the deacetylation of chitin and belongs to the category of polysaccharides. It has a linear structure, with bonds (β1→4) between residues of N-acetyl-2-amino-2-deoxy-D-glucose (glucosamine, GlcN) and 2-amino-2-deoxy-D-glucose (N-acetyl-glucosamine, GlcNAc). This biopolymer is soluble in acidic aqueous solutions due to the protonation of its primary amine groups, unlike chitin, which contains acetylated groups in its composition. [17,18].

After chemical alterations to their functional groups, this type of polymer presents products with better solubility and biodegradability, using different processes to reach the ideal point for the intended activities. For example, phosphorylated chitosan has shown favorable results in corrosion inhibition and other derivatives have demonstrated biological activities that offer therapeutic possibilities, including antioxidant activity, anti-inflammatory, antimicrobial and healing properties [18].

In search of these applications, hydrogels were developed using chitosan with a high dissolving capacity, high mechanical strength and on demand for the production of dressings used in wound treatment. The gels were obtained from different concentrations of chitosan (2%, 3%, 4% and 5%). As a result, among the concentrations tested by the authors, the gel containing 3% chitosan proved to have excellent mechanical properties and a short time to complete dissolution; consequently, this concentration was applied in the following tests of the study. In relation to the wound closure test, a faster closure of the lesion was observed in rats treated with the gels containing 3% chitosan and, after 14 days, the lesion was completely closed, with a percentage of 90% closure of the affected area, while the group that did not receive treatment with the chitosan hydrogel showed only 73% [19].

However, chitosan has some pharmacotechnical limitations that make it difficult to apply on a large scale, including its rapid degradation. To overcome these limitations, it is commonly found in combination with other polymers, such as synthetic polymers like polyvinyl alcohol (PVA), which act synergistically to improve its mechanical strength. This synergism has already been described in the literature, where it has been shown that this association stimulates the wound healing process in diabetic patients by acting on the elimination of reactive oxygen species (ROS) and antibacterial action. In cell viability tests using L929 cells, it was shown that this association was able to increase the viability of these cells by 90% after the third and fifth day of the experiment, increasing the number of cells in the samples approximately five times compared to the beginning of the test. In addition, this material proved to be non-toxic to L929 cells and showed the potential to reduce ROS by 90% in the tissues analyzed [20].

Materials Research Forum LLC
https://doi.org/10.21741/9781644903353-12

2.3 Alginate

The alginate polysaccharide is made up of β-D-mannuronic acid and α-L-guluronic acid joined by β-(1→4) glycosidic bonds. It is produced from the acidification of marine plants that produce sodium alginate, which is soluble in water and therefore more flexible to changes that allow it to be used in industry. However, when subjected to *in vivo* tests, unfavorable responses were observed, probably due to the impurity of the sample. Therefore, for its correct application, a purification process must be carried out without altering the properties of the component [15].

One of the main limitations of alginate is the need for a rigorous purification process to remove impurities, which can provoke adverse immune responses in in vivo applications. Studies have shown that unpurified alginate can trigger inflammatory reactions, which limits its use in biomedical applications. However, when properly purified, alginate has been shown to be biocompatible and biodegradable, characteristics that are highly desirable in materials for dressings, tissue engineering and other medical devices [21].

In the context of dressings, alginate is particularly valued for its ability to form gels in contact with wound exudate, promoting a moist environment that favors healing and non-traumatic removal. In addition, alginate-based dressings have hemostatic properties and are effective in containing moderate to severe bleeding, such as Biatain® (Coloplast, Denmark). However, to guarantee their efficacy and safety, as well as ensuring the purity of the alginate used, it is necessary to investigate its behavior in different physiological conditions and in numerous types of wounds [22].

Compared to conventional saline dressings, alginate-containing dressings have already been described in the literature as excellent in the treatment of diabetic wounds, including diabetic foot ulcers, since they showed positive results in reducing the superficial area of the lesion from 9.07 to 4.68, while the groups containing conventional dressings only reduced it from 9.25 to 6.50. In addition, it has already been found that dressings containing this type of polymer help to significantly reduce exudate ($p<0.05$) when compared to conventional treatments. In the evaluation of wound cultures in patients with this type of injury, approximately 13.6% of patients had a positive culture in the alginate group, while the group that received only a solution containing 0.9% sodium chloride had 68.2%. Thus, the studies confirm that alginate dressings stand out when compared, for example, with solutions containing 0.9% sodium chloride, which are commonly used to treat this type of wound. These characteristics can be explained by alginate's ability to absorb substances and its easy facial modeling [23].

In this context, associations of biomedical natural latex with alginate showed high biocompatibility. This shows that as well as being excellent at controlling exudate and healing wounds, this type of polymer can be used in conjunction with other materials and serve as a matrix for the production of numerous biomedical materials, including wound treatment dressings. Studies in the literature have shown that natural latex in association with 15% alginate was able to increase the metabolic activity of fibroblasts by 218% and keratinocytes by 464%, compared to groups that received conventional treatment. In addition, an increase in tensile strength was observed when concentrations of alginate greater than 15% were added to natural latex. At this same concentration, higher proliferation rates (526%) were observed when compared to the material without alginate, leading to the understanding that this natural polymer aids in the release of biomedical natural latex [24].

Materials Research Foundations 172 (2025) 282-298
https://doi.org/10.21741/9781644903353-12

3. Synthetic polymers

The category of synthetic polymers is one that is attracting increasing interest from the industrial sector for biomedical applications. This phenomenon is explained by the unique characteristics of this type of polymer, such as easy reproduction, good compatibility with human tissues, reasonable costs, applicable in numerous production techniques, high availability on the market and excellent in the delivery of active pharmaceutical ingredients (APIs). For example, they serve as matrices for the production of nanofibers, hydrogels, 3d materials, nanocomposites and films applied to wound treatment [25,26]. Among the most commonly used are polyvinyl alcohol (PVA), polyethylene oxide (PEO), poly (ε-caprolactone) (PCL), polyurethane (PU), polylactic acid (PLA), polyvinylpyrrolidone (PVP) and polyglycolic acid [25], represented structurally in figure 3.

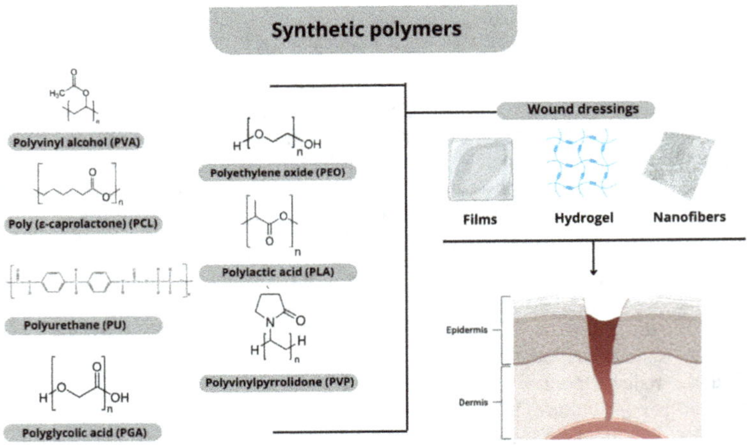

Figure 3 - Synthetic polymers. Image generated using Biorender

3.1 Aliphatic polyesters

Aliphatic polyesters, such as polylactic acid (PLA), polyglycolide (PGA) and polyethylene glycol (PEG), are produced from some cyclic lactones found in nature (such as lactide, glycolide and ε-caprolactone) through a ring-opening polymerization (ROP) process to form polylactic acid (PLA), polyglycolic acid (PGA) and ε-polycaprolactone (PCL), respectively. As well as promoting polymerization, this process also facilitates the biodegradation of these polymers formed by the catalytic reaction [27,28]. Consequently, as a result of this process, they are considered to be the main biodegradable synthetic polymers used in biomedical areas, with approval from the US Food and Drug Administration (FDA) [28].

In this context, different concentrations of PLGA/gelatin were analyzed in order to understand how this type of polymer can influence the physical, biological and structural properties of dressings that are used to treat wounds. As a result, it was described that dressings containing 7:3 and 5:5 (v/v) PLGA/gelatin have higher affinities in aqueous media, high swelling rates and higher degradation compared to PLGA alone. At these same concentrations, complete degradation and absence of inflammatory reactions were observed, demonstrating good biocompatibility for tissue applications, especially in epithelial tissue. It can therefore be seen that this type of polymer can work effectively with other types of material without losing its excellent characteristics for application in the production of dressings [29].

In addition to being biodegradable, these polymers also have excellent biocompatibility, low immunogenicity and a variety of mechanical and physical properties, making it possible to produce 3D materials, films, nanofibers, dressings, nanoparticles and nanocomposites. Due to these characteristics, they are often used in various types of dressings available on the market for the treatment of wounds, whether chronic or acute. Some examples include Vliwaktiv AG® (Lohmann & Rauscher, UK), Acticoat® (Smith & Nephew, UK), Melolin® (Smith & Nephew, UK), UrgoStart Contact® (Urgo, France) Hydrosorb® (Hartmann, Germany), Intrasite® (Smith & Nephew, UK) and Telfa® (Covidien & Medtronic, Ireland) [27,30].

Thus, in search of new application options for this type of polymer in wound treatment, a study carried out by Ciftci *et al* [31] sought to understand the synergistic effects of clarithromycin in association with poly(lactic acid) (PLA), selenium (Se) and graphene oxide (GO) in the wound healing process. To this end, solutions were prepared containing PLA (8%) + 1% by weight of GO (PLA-G), PLA (8%) + 1% by weight of GO + 1% by weight of Se (PLA-G1), PLA (8%) + 1% by weight of GO + 1% by weight of clarithromycin (PLA-G2) and PLA (8%) + 1% by weight of GO + 1% by weight of Se + 1% by weight of clarithromycin (PLA-G3). As a result, it was observed that the PLA-G1 group had the smallest diameter (354 ± 154.12 nm), while the PLA-G3 group had diameters of 642 ± 173.79 nm, thus demonstrating that the size of the diameter is directly proportional to the concentration. In addition, the group with the largest diameter showed higher viscosity and surface tension values than the other concentrations. In relation to the antimicrobial analysis, the authors point out that the groups with the largest diameters were more effective in inhibiting *Escherichia coli*, *Staphylococcus aureus* and *Bacillus cereus*, i.e. showing that the diameter of the dressings produced for wound treatment applications can have a direct influence on the treatment. From the analysis of wound healing *in vitro* using L929 cells, it was observed that all concentrations contributed to cell adhesion and, after exposure, the samples showed faster cell migration compared to the control group.

3.2 Polyvinyl alcohol (PVA)

Polyvinyl alcohol (PVA) is a synthetic polymer derived from vinyl acetate and is widely used in the production of films to aid wound healing. Its advantages include a simple chemical structure, high degree of expansion, ability to dissolve in water, low toxicity, adhesive properties, biocompatibility, elasticity and chemically inert due to its crystallization. Normally, the polymer is dissolved in an aqueous medium with heating (80-95°C), resulting in odorless, transparent and uniform films after the water has evaporated. PVA has low solubility in ethanol and is insoluble in organic solvents, and various factors such as pH, viscosity, melting point and refractive index are associated with its molecular mass and degree of hydrolysis [32,33].

Consequently, they are widely used in the biomedical field as matrices for incorporating different active pharmaceutical ingredients, resulting in multifunctional materials applied to the treatment of chronic or acute wounds [34]. Commercially, this polymer is already found as a matrix to produce dressings, including Exufiber® and Exufiber Ag® (Mölnlycke Health Care, Sweden), which are widely used in various types of exuding wounds, such as leg and foot ulcers, pressure injuries and thick burns [35]. In this context, recent research has focused on the synergy of PVA with additional functional constituents, such as chitosan, gelatin, growth peptides, alginate and oxidized cellulose, as a means of improving wound healing [34].

Considering this fact, Bahadoran, Shamloo and Nokoorani [36] decided to develop a nanocomposite containing PCL microspheres/PVA hydrogel and sodium alginate, in which the microspheres were incorporated with fibroblast growth factors. For this, a solution containing 10% (w/v) and 3% (w/v) sodium alginate was used. Subsequently, different proportions of 10, 30 and 50% v/v of AS/PVA combinations were prepared to analyze the relationship between concentration and effect. As a result, it was observed that formulations containing 10% PVA and 50% sodium alginate showed more porous structures with approximately 52% porosity and diameters of 62 ± 16 μm, facilitating the process of fibroblast intrusion into the tissue.

Furthermore, in the *in vivo* test using male Wistar rats, it was observed that from the 14th day of treatment there was a reduction in the wound area of the animals treated with the bFGF-encapsulated microspheres/hydrogel (H-bFGF) compared to the control group. As a result, the study showed that on the 14th day more than 79% of the wound had healed and, on the 21st day, approximately 95% of the lesion had healed. The authors attribute these results to the bioactivity of bFGF within the H-bFGF structure, which led to better regeneration of the affected tissues [36].

Seeking even more results, Fraga *et al.* [37] developed a study based on the production of dressings from PVA nanostructures containing chloramphenicol. They used 8% (w/v) PVA dissolved in an aqueous solution, 10% (v/v) ethanol, 10% (w/w) chloramphenicol (CLF) and 3% (w/w) citric acid (CA). The scanning electron microscopy (SEM) test showed that the PVA solution without CLF was more homogeneous. In addition, using SEM, it was possible to observe a variation of 681.42 to 604.73 nm in the diameters of the materials obtained. In terms of mass loss, a 60% loss of mass was observed for the non-cross-linked membranes, while for the cross-linked membranes, this loss was 4.02% for the (PVA/GLA) membrane and 0.7% for (PVA/CA). Those containing CLF showed losses of 9.11% and 9.14% for a concentration of 10% CLF. In addition, there was a reduction in the maximum release of the cross-linked membranes, reaching 84.4% and 75.4% for GLA and CA, respectively.

3.3 Polyurethane

Created by Dr. Otto Bayer and his team in 1937, polyurethane stands out for its versatility compared to other polymers of the time. Its simple production and exceptional properties have made it one of the most sought-after materials. From a chemical point of view, it is a condensation polymer group that includes a urethane group in its composition. Synthesis is carried out by means of step-growth polymerization reactions, which occur between isocyanates (such as diisocyanates or polyisocyanates) and polyols (which are low molecular weight oligomeric polyols with hydroxyl groups at the ends). This reaction takes place in the presence of an appropriate catalyst and results in the formation of polymers that can be linear, branched or cross-linked. However, despite their excellent performance, as polyols and isocyanates are derived from petroleum, researchers are looking for new alternatives to avoid dependence on petrochemical resources [28,38].

Polyurethane is among the most significant classes of polymers applied to industrial applications and is growing at a faster rate in the global market. This trend is due to the versatility of adapting basic chemicals, resulting in a variety of polyurethanes. Currently, the global polyurethane market is valued at more than US$65 billion and is projected to grow at a rate of 3.2% by 2027. In the dressings market, some types of dressings are available, especially in the form of foams, using polyurethane. For example, Curatec® (Urgo, France) and PolyMem® (Ferris mfg, USA) [39].

However, with a view to new strategies for the production of dressings used in wound treatment, researchers are developing studies with this type of polymer. These include a study carried out by Li *et al* [40] which presents a new strategy for forming foams containing lignin-based polyurethane (LPUF) incorporated with silver nanoparticles (AgNps). As a result, the authors found that Ag NP-LPUF has an antibacterial rate of over 99% against *Escherichia coli* in 1h and *Staphylococcus aureus* in 4h. With regard to evaluations of wound closure *in vivo*, using ICR mice, the results showed that wounds treated with Tegaderm film (control group), PUF, LPUF and LPUF-7.5Ag showed significant contraction on day 7 and 10, and the LPUF-7.5Ag group showed greater wound healing than the other groups. A significant difference ($p<0.05$) was also observed between the Tegaderm film group and the other groups tested.

In addition, a study conducted by Khodabakhshi *et al* [41] sought to evaluate the *in vitro* and *in vivo* performance of a propolis-coated polyurethane dressing with high porosity and antibacterial efficacy. To do this, they used a concentration of 12% w/v polyurethane (PU) dissolved in tetrahydrofuran (THF) and dimethylformamide (DMF) and different concentrations of propolis (WEP) (0%, 10%, 20% and 30%). As a result, the researchers observed that the foams containing PU and PU/WEP showed no antibacterial activity against *Staphylococcus epidermidise* and *Pseudomonas aeruginosa.* However, against *Escherichia coli* and *Staphylococcal aureus* strains, they showed values of 3.55 ± 0.47 mm and 3.89 ± 0.31 mm, respectively. In relation to the *in vivo* test, using female Wistar rats of approximately 8 weeks of age, it was observed that the group containing 30% WEP showed better healing compared to the control and PU groups on the 5th and 10th. Particularly in these periods, the authors emphasize that these findings help in the cellular hemostasis and inflammation stages of the injury repair process, and on the 15th day, a rate of 5.68% was recorded.

4. Hybrid polymers

In recent years, organic-inorganic hybrid materials have attracted great interest due to their versatility, wide range of functionalities and advanced applications in technological development, especially biomedical. This type of material is formed from the interaction of organic polymer chains with inorganic components such as metals, oxides or salts, giving them interesting mechanical properties, better thermal and chemical stability, as well as selective permeation [15].

Thus, the organic phase can be bonded to the inorganic phase through physical interactions such as *van der Waals* forces, electrostatic, hydrogen bonding, dipole-dipole or dipole-induced, in which an organic modifier with chemical fractions similar to those of the polymer structure is generally used to increase compatibility. The inorganic-organic networks can be joined by covalent chemical bonds, which allows for greater interaction and coupling between the phases; essential for certain applications [42].

4.1 Ureasil-polyether

The use of organic-inorganic hybrid polymers has been explored for numerous applications, including dressings to modulate the microenvironment of complex wounds (burns and ulcers), with the aim of reducing constant dressing changes, facilitating wound visualization, offering prolonged protection and support during the healing process, to promote an effective, painless and less invasive treatment [9].

Based on these characteristics, randomized clinical trials comparing a hybrid-scale synthetic fibre matrix (Restrata®) as a standard for treating diabetic foot ulcers and a living cell skin substitute (Apligraf®) have been shown to epithelialize wounds and revascularize patients' legs (43). Hybrid membranes made of ureasil-polyether, specifically ureasil-PEO (polyethylene oxide) and ureasil-PPO (polypropylene oxide) of different molecular weights (400, 500 and 1900 g/mol), have emerged as a viable strategy applied to tissue regeneration. In addition to promoting the controlled release of hydrophilic and lipophilic drugs and metals, they have crucial characteristics such as improved mechanical resistance, flexibility, biocompatibility and biodegradability [44,45].

Oshiro-Junior *et al.* [46] developed hybrid membranes for bone regeneration from mixtures of ureasil-PPO400, ureasil-PEO500 and ureasil-PEO1900 with osteogenic growth peptide (OGP at 2. 65 x 10-5 mol/L), named ureasil-PPO400/PEO500 (at concentrations of 80:20; 50:50 and 20:80 w/w) and ureasil-PPO400/PEO1900 (80:20 and 50:50 w/w). In the low-angle X-ray scattering test, the incorporation of OGP into the membranes resulted in the structural homogeneity of the material with swelling of 14.87% and 7.36% for the ureasil-PPO400/PEO1900 (80/20) and ureasil-PPO400/PEO500 (50/50) mixtures, respectively after 60 minutes. The ureasil-PPO400/PEO1900 membranes (21.76%) showed a greater release of OGP compared to PPO400/PEO500 (7.79%) after 48 hours, which exhibited a zero-order release mechanism. In the animal test using norvegicus albinus rats (Holtzman), the hybrid membranes containing OGP ureasil-PPO400/PEO500 (50/50) and ureasil-PPO400/PEO1900 (80/20) showed significant bone formation at 60 days ($p < 0.05$) of 73% and 77%, respectively, higher than the negative control (24%) and close to the commercial material (90%); demonstrating the therapeutic potential of osteoinductive biomaterials.

Polymeric membranes of PPO4000 and PEO500 in modified alkoxide (molar ratio 1:2 (polymer:alkoxide) containing incorporated glucose (6%, w/w), synthesized by the sol-gel process were obtained by Silva *et al.* [47]. The results showed that the U-PPO4000 and U-PEO500 membranes were mechanically resistant and modulated when subjected to a high compressive force (15 N) with 42% and 6% deformation, respectively, obtained by dynamic mechanical analysis. In the wettability assessment, U-PEO500 showed a low contact angle, measuring approximately 50°. Both proved to be biocompatible and relatively hemolytic *in vitro* with hemolysis rates of 2 to 5% among the blood types tested. Among them, the one with the highest concentration of U-PEO500 showed a controlled release of glucose of 70.5% in 14 days, and indicated that the kinetic mechanism of release occurred by anomalous transport; certainly contributing to the advancement of regenerative medicine for bone defects.

The effectiveness of organic-inorganic ureasil-polyether hybrid polymers synthesized by the sol-gel method was investigated by Fonseca and colleagues [48]. In an innovative way, the authors showed the influence of laponite (1.5% w/w) dispersed in the U-PEO500 matrix containing D-glucose (3% w/w). The membranes remained mechanically stable, absent of precipitates and without loss of elasticity for 9 months. Contact angle analysis showed a low contact angle and a greater tendency for the droplet to disperse around 45°. The release of D-glucose from the polymer matrix showed a faster profile of around 20% in 720 minutes, with the Korsmeyer-Peppas kinetic model providing the best fit for the experimental data; the positive results therefore support various therapies, including bone regeneration.

Among the polymeric materials reported in the last decade, ureasil-polyether hybrid films have the potential to act as platforms for controlled release systems of synthetic and natural drugs for the treatment of skin wounds. The transdermal route offers advantages compared to the oral, intravenous and intramuscular routes of administration, as it prevents the hepatic first-pass effect (in the case of non-steroidal anti-inflammatory drugs, for example) and is non-invasive and painful [49].

Oshiro-Junior and colleagues [50] synthesized hybrid films via sol-gel from PEO (Mw = 500-1900 gmol/L), PPO (Mw = 400-2000 gmol/L) and modified alkoxide IsOTrEOS (1:2 polymer:alkoxide), with acidic and basic catalysts such as HCl (2M) and NH_4F (0.1M). Morphological analysis revealed that the acid catalyst promoted the formation of more linear and less branched polymer chains than the basic catalyst. The U-PEO films showed higher bioadhesive strengths than the U-PPO films, regardless of the molecular weight or the catalyst used. They showed a cell viability of over 75%, indicating the low toxicity of the films, and no significant changes in viability between the films catalyzed with HCl and NH_4F; confirming that the catalysts were capable of forming highly adhesive, safe and uniform hybrid films for skin application.

Meanwhile, the study by Araújo *et al* [51] explored the *in vitro* and *in vivo* anti-inflammatory action of a plant extract in a new transdermal hybrid matrix of ureasil-polyether PEO (Mw = 500 g/mol and modified alkoxide IsOTrEOS (1:2 polymer:alkoxide), synthesized by the sol-gel method. In the topographic surface of the U-PEO matrix with *Annona muricata* extract showed greater roughness (Ra) and porosity than the pure matrix, whose Ra was 6.89 and 1.90 nm, and diameter ≈ 393.75 and 928.80 nm, respectively. The release profile of the extract incorporated into the U-PEO matrix showed no burst effect, reaching a release of 23.72% after 24 hours, suggestive of a Weibull kinetic model in which the drug mechanism tends to be predominantly controlled by diffusion through the matrix. The test on paw oedema using adult male/female Swiss albino mice showed that the 10 mg/kg dose of *A. muricata* extract inhibited oedema between 1 and 6 hours after application, with efficacy in the treatment of acute inflammation comparable to the commercial anti-inflammatory drug used as a positive control.

With an innovative proposal, Bezerra *et al* [52] investigated the coadministration of curcumin (CUR) and miconazole (MCZ) in U-PEO vaginal ovules to contribute to the treatment of vulvovaginal candidiasis. The hybrid precursor was synthesized by the sol-gel method between PEO (Mw = 500 g/mol) and a functionalized alkoxide (IsOTrEOS, 1:2 polymer:alkoxide) containing CUR and MCZ (2.7% w/w) incorporated into the polymer matrix. The U-PEO ovules were homogeneous, transparent and free of cracks and fissures. The combination of CUR and MCZ in U-PEO showed a prolonged release profile of 42% and 61% in 24 and 48 hours, respectively for CUR, and 57% and 70% in these periods for MCZ. The release kinetics of CUR from the matrix followed a Weibull model, while MCZ was governed by pseudo-Fickian diffusion. The antifungal activity evaluated against Candida albicans (ATCC 10231) revealed a minimum inhibitory concentration (MIC) of 256 μg/mL and 2.5 μg/mL for CUR and MCZ, respectively.

Materials Research Foundations 172 (2025) 282-298
https://doi.org/10.21741/9781644903353-12

After combining the drugs, the MIC value obtained in the Checkerboard test was reduced fourfold (0.625 μg/mL) for MCZ, and eightfold (32 μg/mL) for CUR, confirming the synergistic antifungal activity between the drugs. The agar diffusion analysis (absent PBS) showed that the U-PEO/CUR, U-PEO/MCZ and U-PEO/CUR/MCZ ovules formed inhibition halos of 2.25, 22.5 and 18.5 mm, respectively, proving their antifungal efficacy, although in the first case, CUR remained retained in the hybrid matrix.

Conclusion

The treatment of wounds is a complex process that involves various factors, such as the nature of the injury, the presence of infections, the general condition of the patient and the body's response to the healing process. It is important to note that dressings are remarkably effective at facilitating wound healing, especially those made from polymers.

In this regard, despite the wide range of applications, the large number of studies, the excellent results and the easy accessibility, natural polymers are less common in the dressing market compared to synthetic polymers. This phenomenon may be related to the cost of production, since synthetic polymers are generally cheaper and easier to produce on a large scale than natural polymers.

In addition, specific properties such as resistance, durability, flexibility and versatility have drawn the industry's attention to their use. Meanwhile, hybrid polymers, which combine the characteristics of organic polymers with inorganic polymers, are attracting market attention as a new alternative material for the production of dressings, with improved properties, versatility, functionality and sustainability. These properties are ideal for the production of biomedical materials, for example, dressings as a drug release system, coatings and sensors.

References

[1] Aulton ME, Taylor KMG. Delineation_of_Pharmaceutical. 4th ed. Vol. 4, GEN Guanabara Koogan. Rio de Janeiro: Elsevier, 2; 2016. 872 páginas.

[2] Eriksson E, Liu PY, Schultz GS, Martins-Green MM, Tanaka R, Weir D, et al. Chronic wounds: Treatment consensus. Wound Repair and Regeneration. 2022 Mar 7;30(2):156-71. https://doi.org/10.1111/wrr.12994

[3] Firlar I, Altunbek M, McCarthy C, Ramalingam M, Camci-Unal G. Functional Hydrogels for Treatment of Chronic Wounds. Gels. 2022 Feb 17;8(2):127. https://doi.org/10.3390/gels8020127

[4] Mota FAR, Passos MLC, Santos JLM, Saraiva MLMFS. Comparative analysis of electrochemical and optical sensors for detection of chronic wounds biomarkers: A review. Biosens Bioelectron. 2024 May;251:116095. https://doi.org/10.1016/j.bios.2024.116095

[5] Dong R, Guo B. Smart wound dressings for wound healing. Nano Today. 2021 Dec;41:101290. https://doi.org/10.1016/j.nantod.2021.101290

[6] Gruppuso M, Turco G, Marsich E, Porrelli D. Polymeric wound dressings, an insight into polysaccharide-based electrospun membranes. Appl Mater Today. 2021 Sep;24:101148. https://doi.org/10.1016/j.apmt.2021.101148

[7] Gruppuso M, Turco G, Marsich E, Porrelli D. Polymeric wound dressings, an insight into polysaccharide-based electrospun membranes. Appl Mater Today. 2021 Sep;24:101148. https://doi.org/10.1016/j.apmt.2021.101148

[8] Alven S, Peter S, Mbese Z, Aderibigbe BA. Polymer-Based Wound Dressing Materials Loaded with Bioactive Agents: Potential Materials for the Treatment of Diabetic Wounds. Polymers (Basel). 2022 Feb 14;14(4):724. https://doi.org/10.3390/polym14040724

[9] Aderibigbe BA. Hybrid-Based Wound Dressings: Combination of Synthetic and Biopolymers. Polymers (Basel). 2022 Sep 12;14(18):3806. https://doi.org/10.3390/polym14183806

[10] Ahmed S, Sharma P, Bairagi S, Rumjit NP, Garg S, Ali A, et al. Nature-derived polymers and their composites for energy depository applications in batteries and supercapacitors: Advances, prospects and sustainability. J Energy Storage. 2023 Aug;66:107391. https://doi.org/10.1016/j.est.2023.107391

[11] Silva ACQ, Silvestre AJD, Vilela C, Freire CSR. Natural Polymers-Based Materials: A Contribution to a Greener Future. Molecules. 2021 Dec 24;27(1):94. https://doi.org/10.3390/molecules27010094

[12] Eichhorn SJ, Gandini A. Materials from Renewable Resources. MRS Bull. 2010 Mar 31;35(3):187-93. https://doi.org/10.1557/mrs2010.650

[13] Bouhlouli M, Pourhadi M, Karami F, Talebi Z, Ranjbari J, Khojasteh A. Applications of Bacterial Cellulose as a Natural Polymer in Tissue Engineering. ASAIO Journal. 2021 Jul 11;67(7):709-20. https://doi.org/10.1097/MAT.0000000000001356

[14] Matei E, Predescu AM, Râpă M, Țurcanu AA, Mateș I, Constantin N, et al. Natural Polymers and Their Nanocomposites Used for Environmental Applications. Nanomaterials. 2022 May 17;12(10):1707. https://doi.org/10.3390/nano12101707

[15] Sanfelice, R. C., Pavinatto, A., Correa, D., & Correa, D. S. (2022). Nanotechnology applied to polymers.

[16] Kiiskinen J, Merivaara A, Hakkarainen T, Kääriäinen M, Miettinen S, Yliperttula M, et al. Nanofibrillar cellulose wound dressing supports the growth and characteristics of human mesenchymal stem/stromal cells without cell adhesion coatings. Stem Cell Res Ther. 2019 Dec 23;10(1):292. https://doi.org/10.1186/s13287-019-1394-7

[17] Matei E, Predescu AM, Râpă M, Țurcanu AA, Mateș I, Constantin N, et al. Natural Polymers and Their Nanocomposites Used for Environmental Applications. Nanomaterials. 2022 May 17;12(10):1707. https://doi.org/10.3390/nano12101707

[18] Aranaz I, Alcántara AR, Civera MC, Arias C, Elorza B, Heras Caballero A, et al. Chitosan: An Overview of Its Properties and Applications. Polymers (Basel). 2021 Sep 24;13(19):3256. https://doi.org/10.3390/polym13193256

[19] Luo Y, Cui L, Zou L, Zhao Y, Chen L, Guan Y, et al. Mechanically strong and on-demand dissoluble chitosan hydrogels for wound dressing applications. Carbohydr Polym. 2022 Oct;294:119774. https://doi.org/10.1016/j.carbpol.2022.119774

[20] Liu H, Chen R, Wang P, Fu J, Tang Z, Xie J, et al. Electrospun polyvinyl alcohol-chitosan dressing stimulates infected diabetic wound healing with combined reactive oxygen species scavenging and antibacterial abilities. Carbohydr Polym. 2023 Sep;316:121050. https://doi.org/10.1016/j.carbpol.2023.121050

[21] Pires ALR, Bierhalz ACK, Moraes ÂM. BIOMATERIALS: TYPES, APPLICATIONS, AND MARKET. Quim Nova. 2015; https://doi.org/10.5935/0100-4042.20150094

[22] Garcia T de F, Silva PGA, Barcelos BJ, Miranda M das GR de, Alonso C da S, Abreu MNS, et al. Criteria to evaluate the quality of alginate wound dressings. Rev Bras Enferm. 2021;74(4). https://doi.org/10.1590/0034-7167-2020-1091

[23] Chitrambalam TG, Christopher PJ, Sundaraj J, Paladugu R, Selvamuthukumaran S. Comparison of Efficacy of Alginate Filler Dressings with Conventional Saline Dressings for Cavity Wounds in Diabetic Foot Ulcer- A Prospective Cohort Study. JOURNAL OF CLINICAL AND DIAGNOSTIC RESEARCH. 2020;14(12):1-1. https://doi.org/10.7860/JCDR/2020/46448.14221

[24] Barros NR, Ahadian S, Tebon P, Rudge MVC, Barbosa AMP, Herculano RD. Highly absorptive dressing composed of natural latex loaded with alginate for exudate control and healing of diabetic wounds. Materials Science and Engineering: C. 2021 Feb;119:111589. https://doi.org/10.1016/j.msec.2020.111589

[25] Prete S, Dattilo M, Patitucci F, Pezzi G, Parisi OI, Puoci F. Natural and Synthetic Polymeric Biomaterials for Application in Wound Management. J Funct Biomater. 2023 Sep 3;14(9):455. https://doi.org/10.3390/jfb14090455

[26] Terzopoulou Z, Zamboulis A, Koumentakou I, Michailidou G, Noordam MJ, Bikiaris DN. Biocompatible Synthetic Polymers for Tissue Engineering Purposes. Biomacromolecules. 2022 May 9;23(5):1841-63. https://doi.org/10.1021/acs.biomac.2c00047

[27] Chin AL, Wang X, Tong R. Aliphatic Polyester-Based Materials for Enhanced Cancer Immunotherapy. Macromol Biosci. 2021 Jul 28;21(7). https://doi.org/10.1002/mabi.202100087

[28] Barros RM. Polymers in Guided Bone Regeneration. In 2022. p. 195-215. https://doi.org/10.21741/9781644901892-8

[29] Vázquez N, Sánchez-Arévalo F, Maciel-Cerda A, Garnica-Palafox I, Ontiveros-Tlachi R, Chaires-Rosas C, et al. Influence of the PLGA/gelatin ratio on the physical, chemical and biological properties of electrospun scaffolds for wound dressings. Biomedical Materials. 2019 May 3;14(4):045006. https://doi.org/10.1088/1748-605X/ab1741

[30] Stepanova M, Korzhikova-Vlakh E. Modification of Cellulose Micro- and Nanomaterials to Improve Properties of Aliphatic Polyesters/Cellulose Composites: A Review. Polymers (Basel). 2022 Apr 5;14(7):1477. https://doi.org/10.3390/polym14071477

[31] Ciftci F, Ayan S, Duygulu N, Yilmazer Y, Karavelioglu Z, Vehapi M, et al. Selenium and clarithromycin loaded PLA-GO composite wound dressings by electrospinning method. International Journal of Polymeric Materials and Polymeric Biomaterials. 2022 Aug 13;71(12):898-909. https://doi.org/10.1080/00914037.2021.1925276

[32] Saraiva MM, Campelo M da S, Câmara Neto JF, Lima ABN, Silva G de A, Dias AT de FF, et al. Alginate/polyvinyl alcohol films for wound healing: Advantages and challenges. J Biomed Mater Res B Appl Biomater. 2022 Jan 12;111(1):220-33. https://doi.org/10.1002/jbm.b.35146

[33] Barros Araújo CB, da Silva Soares IL, da Silva Lima DP, Barros RM, de Lima Damasceno BPG, Oshiro-Junior JA. Polyvinyl Alcohol Nanofibers Blends as Drug Delivery System in Tissue Regeneration. Curr Pharm Des. 2023 Apr;29(15):1149-62. https://doi.org/10.2174/1381612829666230508144912

[34] Prete S, Dattilo M, Patitucci F, Pezzi G, Parisi OI, Puoci F. Natural and Synthetic Polymeric Biomaterials for Application in Wound Management. J Funct Biomater. 2023 Sep 3;14(9):455. https://doi.org/10.3390/jfb14090455

[35] Lustig A, Alves P, Call E, Santamaria N, Gefen A. The sorptivity and durability of gelling fibre dressings tested in a simulated sacral pressure ulcer system. Int Wound J. 2020 Apr 25;18(2):194-208. https://doi.org/10.1111/iwj.13515

[36] Bahadoran M, Shamloo A, Nokoorani YD. Development of a polyvinyl alcohol/sodium alginate hydrogel-based scaffold incorporating bFGF-encapsulated microspheres for accelerated wound healing. Sci Rep. 2020 Apr 30;10(1):7342. https://doi.org/10.1038/s41598-020-64480-9

[37] Fraga GN, Dragunski DC, Vilsinski BH, da Silva Azevedo CK, Wiggers HJ, Faria MGI, et al. Dressings produced from PVA nanofibers containing chloramphenicol. Colloids Surf A Physicochem Eng Asp. 2023 Oct;674:131760. https://doi.org/10.1016/j.colsurfa.2023.131760

[38] Das A, Mahanwar P. A brief discussion on advances in polyurethane applications. Advanced Industrial and Engineering Polymer Research. 2020 Jul;3(3):93-101. https://doi.org/10.1016/j.aiepr.2020.07.002

[39] de Souza FM, Kahol PK, Gupta RK. Introduction to Polyurethane Chemistry. In 2021. p. 1-24. https://doi.org/10.1021/bk-2021-1380.ch001

[40] Li S, Zhang Y, Ma X, Qiu S, Chen J, Lu G, et al. Antimicrobial Lignin-Based Polyurethane/Ag Composite Foams for Improving Wound Healing. Biomacromolecules. 2022 Apr 11;23(4):1622-32. https://doi.org/10.1021/acs.biomac.1c01465

[41] Khodabakhshi D, Eskandarinia A, Kefayat A, Rafienia M, Navid S, Karbasi S, et al. In vitro and in vivo performance of a propolis-coated polyurethane wound dressing with high porosity and antibacterial efficacy. Colloids Surf B Biointerfaces. 2019 Jun;178:177-84. https://doi.org/10.1016/j.colsurfb.2019.03.010

[42] Jena KK, Narayan R, Alhassan SM. Highly branched graphene siloxane–polyurethane-urea (PU-urea) hybrid coatings. Prog Org Coat. 2017 Oct;111:343-53. https://doi.org/10.1016/j.porgcoat.2017.06.020

[43] Clinical Trials. Assessment of Wound Closure Comparing Synthetic Hybrid-Scale Fiber Matrix With Standard of Care in Treating Diabetic Foot Ulcers (DFU) and With Living Cellular Skin Substitute in Treating Venous Leg Ulcers (VLU). https://clinicaltrials.gov/study/NCT04927702?tab=results, 2024 (accessed 01 september).

Materials Research Forum LLC
https://doi.org/10.21741/9781644903353-12

[44] Oshiro Junior JA, Mortari GR, de Freitas RM, Marcantonio-Junior E, Lopes L, Spolidorio LC, et al. Assessment of biocompatibility of ureasil-polyether hybrid membranes for future use in implantodontology. International Journal of Polymeric Materials and Polymeric Biomaterials. 2016 Sep 2;65(13):647-52. https://doi.org/10.1080/00914037.2016.1157796

[45] Oshiro JA, Scardueli CR, de Oliveira GJPL, Marcantonio RAC, Chiavacci LA. Development of ureasil-polyether membranes for guided bone regeneration. Biomed Phys Eng Express. 2017 Jan 31;3(1):015019. https://doi.org/10.1088/2057-1976/aa56a6

[46] Oshiro-Junior JA, Barros RM, Silva CG, Souza CC, Scardueli CR, Marcantonio CC, et al. In vivo effectiveness of hybrid membranes with osteogenic growth peptide for bone regeneration. J Tissue Eng Regen Med. 2021 Aug 3;15(8):722-31. https://doi.org/10.1002/term.3226

[47] da Silva CG, Monteiro JR, Oshiro-Júnior JA, Chiavacci LA. Hybrid Membranes of the Ureasil-Polyether Containing Glucose for Future Application in Bone Regeneration. Pharmaceutics. 2023 May 12;15(5):1474. https://doi.org/10.3390/pharmaceutics15051474

[48] da Fonsêca NF, de Oliveira Alves-Júnior J, de Oliveira GD, Costa KMN, de Melo DF, Rolim Neto PJ, et al. Modified release of D-glucose incorporated into laponite/ureasil-poly(ethylene oxide) hybrid nanocomposite. J Solgel Sci Technol. 2024 Jun 10;110(3):705-19. https://doi.org/10.1007/s10971-024-06387-9

[49] Silva JA; AAC; SMSR, DBPGL, MACD. Cutaneous drug delivery: challenges and strategies for the development of transdermal formulations. Journal of Basic and Applied Pharmaceutical Sciences. 2010 Sep;31(3):125-31.

[50] Oshiro Junior JA, Carvalho FC, Soares CP, Chorilli M, Chiavacci LA. Development of Cutaneous Bioadhesive Ureasil-Polyether Hybrid Films. Int J Polym Sci. 2015;2015:1-7. https://doi.org/10.1155/2015/727324

[51] Araújo CBB, Alves Júnior J de O, Sato MR, Costa KMN, Lima JR, Damasceno BPG de L, et al. The Development and Pre-Clinical Anti-Inflammatory Efficacy of a New Transdermal Ureasil-Polyether Hybrid Matrix Loaded with Flavonoid-Rich Annona muricata Leaf Extract. Pharmaceutics. 2024 Aug 21;16(8):1097. https://doi.org/10.3390/pharmaceutics16081097

[52] Bezerra BMS, y Araújo SED de M, Alves-Júnior J de O, Damasceno BPG de L, Oshiro-Junior JA. The Efficacy of Hybrid Vaginal Ovules for Co-Delivery of Curcumin and Miconazole against Candida albicans. Pharmaceutics. 2024 Feb 23;16(3):312. https://doi.org/10.3390/pharmaceutics16030312

Applications of Polymers in Surgery II
Materials Research Foundations 172 (2025) 299-328
Materials Research Forum LLC
https://doi.org/10.21741/9781644903353-13

Chapter 13

Polymers in Implant Medical Devices

Rohini Kharwade[1], Purushottam Gangane[1], Nilesh Mahajan[1]*, Sachin More[1]
[1]Department of Pharmaceutics, Dadasaheb Balpande College of Pharmacy, Besa, Nagpur
nmmahajan78@gmail.com

Abstract

This chapter provides a comprehensive overview, emphasizing both the potential and challenges of polymers in medical implants. Polymers have become essential materials in the development of implantable medical devices due to their unique properties, including biocompatibility, flexibility, and adaptability. This chapter explores the roles of both biodegradable and non-biodegradable polymers in medical implants, focusing on their applications in devices ranging from orthopedic and cardiovascular implants to drug-delivery systems. Biodegradable polymers, such as polylactic acid (PLA) and polyglycolic acid (PGA), are highlighted for their ability to degrade safely within the body, making them ideal for temporary implants and controlled-release drug systems. Conversely, non-biodegradable polymers, including silicone and polyethylene, offer long-term stability and are suited for permanent implants.

Keywords

Medical Implants, Polymers in Implants, Applications of Implants, Controlled Release Implants, Polymers Selection Criteria

Contents

1. Introduction

The conventional drug delivery systems including oral, sublingual, parenteral, transdermal, nasal, rectal, and vaginal offer several benefits. However, the conventional dosage forms create some challenges and propose a limited therapeutic window in various disease conditions and for many drugs. For example, the oral route is unsuitable for drugs that can degrade acidic or alkaline pH in the stomach. In addition, first-pass metabolism is a major issue in achieving bioavailability [1]. These problems may be overcome by the parenteral route. However, sterility and stability make limitations in the storage of parenteral. Subsequently, professional health care is required for the administration of drugs by parenteral route. For the transdermal route of administration, stratum corneum is a major obstacle that prevents ideal drug delivery. Drugs with lower molecular weight, high partition coefficient, and log P value 1-3 are only delivered through the transdermal route. These unwanted outcomes of conventional dosage form hinder the medications and low adherence to therapeutic concentration. In addition, most of the new drug moiety is unable to pass from the first phase of the clinical trial [2].

Due to this, it is critically important to surpass both the therapeutic and psychological barriers of drugs concerning treatment. Therefore, it is a need to optimize the alternative forms of drug delivery that achieve controlled, and target-specific drug release. In this concern, implants have significant potential for better pharmacologic outcomes in precise doses. It lowers the off-target drug delivery and reduces the possible side effects. Therefore, it is possible to enhance the bioavailability and therapeutic window of the drug which is ordinarily inconvenient through conventional dosage form. The implants are devices used for various medical applications and are introduced into patient tissue. It can control the release rate, time, and place of drugs which supports enhancing the therapeutic efficacy of the drug. These are synthesized by synthetic or natural polymers which are either under the category of biodegradable or non-biodegradable. These polymers represent the smart barrier between the targeted tissue and drugs [3].

Implants are novel drug delivery systems that are comprised of a combination of drug or active components, biological products, and metallic or polymeric frameworks. Together it works as a single medical device for different medical applications. The advantages include controlled drug release, targeted drug distribution, enhanced bioavailability, minimized first-pass metabolism and side effects, reduced off-target drug release, and improved patient compliance. Smart implants have recently been furnished with actuators and bio-sensors to control the drug release kinetics [4]. Therefore, implants showed widespread medical applications in the diagnosis, treatment, and prevention of disease. The smart knee implants approved by the FDA include Orthosensor,

 Materials Research Forum LLC
https://doi.org/10.21741/9781644903353-13

Stryker, and Persona. Other smart implants include Dentsply, Osstem for dental, Vagus nerve stimulator for preventing epileptic seizures, Beta Bionics iLet to manage blood glucose levels, scaffolds for tissue repair, drug-eluting stents, ocular inserts, and heart valves [5].

Reasonable intra-canalicular steroid ocular inserts Dextenza® approved by FDA to relieve post-operative pain. It contains a dexamethasone punctum plug which releases the drug in a controlled manner for up to 30 days for managing ocular pain after cataract surgery. From the case study, the patients who were treated with Dextenza were pain-free within 8 days by uploading a preservative-free single insert of steroid [6]. Similarly, bioinspired 3D-printed non-biodegradable silicon heart valves were developed for excellent hemodynamic performance under physiologic heart conditions [7].

Clustered Regularly Interspaced Short Palindromic Repeats is a bacterial defense system that has been adapted by a genome editing technology known as CRISPR-Cas. The CRISPR-based implant strategies provide support in the efficient diagnosis of infectious disease conditions. Recently, nanomedicines-based CRISPR-Cas techniques have been implemented for accurate diagnosis and promoting more accurate treatment therapy. It released drugs in a controlled and targeted manner up to 3-5 years concerning patient medical history and genomics [8].

However, major drawbacks of implants include the requirement of surgical procedures for inserting and host immune rejection. When the implants are inserted into the body tissue, the immune system distinguishes it as a foreign body and activates innate immunity. Subsequently, activates inflammatory interleukins and T-cells observed various hypersensitive immune reactions that lead to fibrotic encapsulation [9]. This issue can be solved by surface modification or by using stimuli-responsive diffusive systems.

In advanced technology, noninvasive *in situ* gel implants that are composed of polylactic-co-glycolic acid are administered using injectable solid implants [10, 11]. Bayan *et al.* developed the smart pellets of 5-fluorouracil by using methacrylic acid and hydroxyethyl methacrylate for the potential treatment of colorectal cancer. It was permitted for the drug release in a controlled manner for up to 14 days. The highest release was shown after 24 hrs and 100 % of the drug release was observed after 14 days [12].

The host hypersensitivity rejection implant issue is also solved through the development of stimuli-responsive expandable implants. These implants are formed by shape memory polymers (SMP) such as polyurethane, ethylene-vinyl acetate, and thermoset-shape memory polymer. These are the class of smart polymers that provisionally modify their shape and surface area. The release of drugs through these implants inspired pH, temperature, magnetic field, infrared, or ultrasound. Polyurethane is widely used non-biodegradable SMP for the manufacturing of scaffolds, stents, sutures, pumps, and bandages [13]. Actuator and biosensor-based polyurethane-based implant used in the treatment of heart disease that is represented in **Fig. 1**.

Materials Research Forum LLC
https://doi.org/10.21741/9781644903353-13

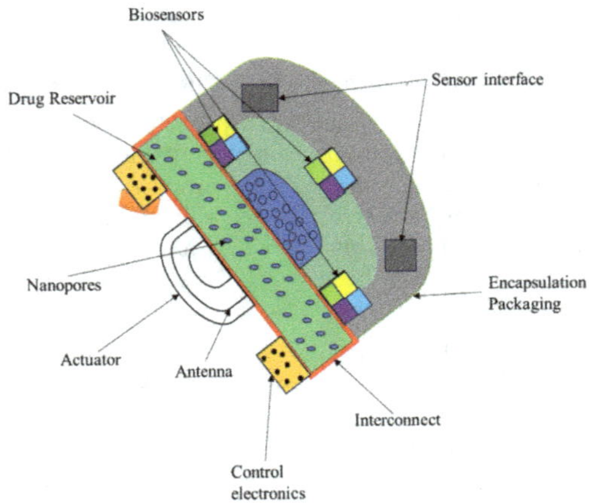

Fig. 1: Diagrammatic representation of actuator and biosensor-based implant.

2. Types of implants in medical devices

The classification of a polymeric implantable drug delivery system is complex and considered as a hybrid category. Based on the release mechanism of drugs, these are broadly categories into passive implants and active implants. The passive implants use biodegradable and non-biodegradable polymers to control the drug release at the targeted site. Therefore, passive implants are categorized as biodegradable and nonbiodegradable implants. Biodegradable implants are the most convenient form because the degradation products of biodegradable polymers include biocompatible traces. Therefore, there is no need to remove them after the completion of treatment. However, non-biodegradable implants require surgery to remove it after the treatment. On the other hand, an energy-dependent driving force such as an osmotic pressure gradient is needed to release the drug from active implants. It mainly includes metallic implants [14].

2.1 Passive implants

Passive implants are simple and uniform that are specifically composed of biodegradable or nonbiodegradable polymers. They are comprised of a drug and polymeric matrix, from which the drug release follows passive diffusion as per Fick's diffusion law. The drug release profile depends on various factors including the physicochemical properties of the drug, the blending of polymers, the design and shape of the implants, and the surface properties of implants. Passive implants are subcategorized into biodegradable and non-biodegradable systems [15].

2.1.1 Biodegradable implant system

Biodegradable implants are more acceptable than non-biodegradable systems. The most acceptable bio-degradable inert polymers include polylactic acid (PLA), polyglycolic acid (PGA), polylactic-co-glycolic acis (PLGA) and polycaprolactone (PCL). These polymers split into small traces and are further absorbed into the body or eliminated from the body. Therefore, after the completion of treatment, there is no need to remove it from the body through surgery which increases patient compliance and acceptance [16].

The disintegration or erosion of polymer makes it possible to release the medicament from implants that depend on various factors. The changes in pH or body temperature abruptly change the disintegration rate of the polymer and ultimately change the drug release. Additionally, the erosion rate of the polymer depends on the surface area of the polymeric implant. If the surface area increases, the erosion rate is decreased. The eroded polymeric implant changes its shape and surface area. The uniform and constant release of medicament is attained by a geometrical shape whose surface area does not change with time. Thus, it's important to focus on the shape of the formulation design which does not change its *in vivo* shape after erosion. To minimize this problem, recently the implant systems have been designed with bio-erodible inert cores and coated with active drug matrix. In this regard, a flat slab design of the implant offers uniform and constant zero-order drug release. Another problem also found in biodegradable implant systems is the slower diffusion rate of the drug than the erosion rate of the polymer. This problem can be solved by the formulation of bio-erodible implants for the extended release of drugs [17].

Biodegradable implants are categorized into reservoir implants and monolithic implants. Both systems consist of exterior polymeric membranes that degrade slowly compared to drug diffusion through the membrane. In reservoir systems, the drug is compressed as a core into the polymeric membrane which is adhered up to the complete release of the drug. Finally, the polymeric matrix is degraded *in vivo* and excreted from the body. In the monolithic type, drugs are dispersed in a polymeric substance and gradually eroded in a controlled release manner [18].

Gao *et al.* assess the intrahepatic penetration of developed PLGA-based doxorubicin-loaded implants. The implant was prepared by melt molding method and characterized for drug content, morphology, drug release profiles, and drug-polymer interaction by differential scanning calorimetry, and FTIR. Further, local intrahepatic tissue penetration of the doxorubicin PLGA implant was checked on minipigs via abdominal surgery. UPLC-MS method determines the concentration of doxorubicin in liver tissue. The result showed that doxorubicin-loaded PLGA implants delivered a high concentration of doxorubicin in a sustained release manner at the local intrahepatic tissue [19].

2.1.2 Non-biodegradable implant system

The most common modified non-biodegradable implant systems are monolithic and reservoir-controlled. In a monolithic implant system drug is uniformly distributed into the non-biodegradable polymeric matrix. However, in a reservoir-controlled system, the compressed drug core is covered by the permeable non-biodegradable polymeric membrane. The drug release through this membrane specifically depends on membrane thickness and permeability coefficient [20] as shown in **Fig. 2**. The preclinically and clinically documented non-biodegradable polymers are applied in these implant systems such as silicones, poly(urethanes), poly(acrylates), or copolymers such as poly (ethylene vinyl acetate) [21].

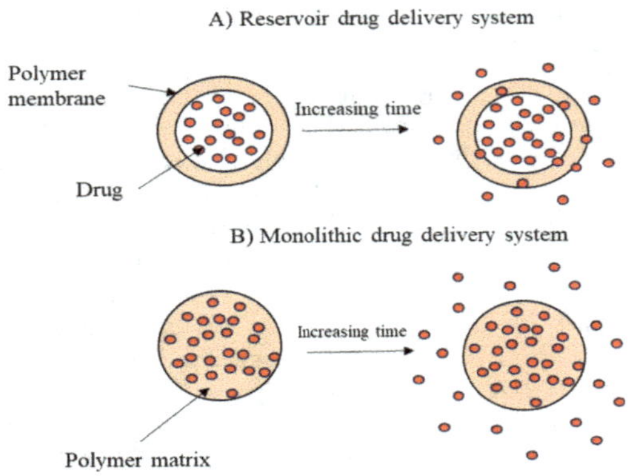

Fig. 2: Schematic representation of drug release from reservoir and monolithic type implant.

Initially, non-biodegradable implants are extensively used to administer the anticancer drug doxorubicin. Mostly poly (urethane) developed the stable implant formulation for different anticancer drugs such as cyclophosphamide, doxorubicin, 5-fluorouracil, methotrexate, and paclitaxel. Poly (urethane) itself has local cytotoxic action and support to improve the pharmacokinetics of anticancer drugs by passively targeting the tumor's local area. The microcapsules of poly (urethane) compressed doxorubicin compared with PLGA biodegradable implant of doxorubicin. From the result, it was concluded that poly (urethane) showed a more prominent effect on controlled drug release as compared to PLGA. In addition, polyurethane did not observe any toxic reaction within the body. However, in recent years biodegradable polymeric implants have been preferable due to their more biocompatibility and single-surgery option [22].

Nowadays, contraceptive drug delivery systems are broadly developed into non-biodegradable implant systems. Because they are structurally dynamic and robust until their shelf life. These implants have shown long-term biocompatibility however, insertion surgery causes tissue damage and infection. Therefore, it is necessary to remove it after releasing the complete drug to prevent any side effects or adverse effects. Norplant contraceptive is the most widely popular non-biodegradable implant which has six thin, elastic silicone capsules each containing 36 mg levonorgestrel. It is typically administered subcutaneously to the upper arm of female user and released drug up to 5 years. Its effectiveness was successfully assessed by more than 50 million women and therefore approved in almost 60 countries. In 2006, the FDA approved one more implantable contraceptive implant containing 68 mg etonogestrel encapsulated in polyethylene vinyl acetate (PEVA). It releases the etonogestrel for up to 3 years in a controlled manner and protects from pregnancy [23].

Magnetically controlled release implant is also an alternative nonbiodegradable system. The drug release from this system is manipulated by a uniformly distributed small magnetic bead as shown in **Fig 3.** When the system is exposed to a physiological system, the drug is normally diffused by using the principle of concentration gradient. However, the external magnetic field immediately releases the drug through implants and manipulates the drug release kinetic according to the need [24].

Materials Research Forum LLC
https://doi.org/10.21741/9781644903353-13

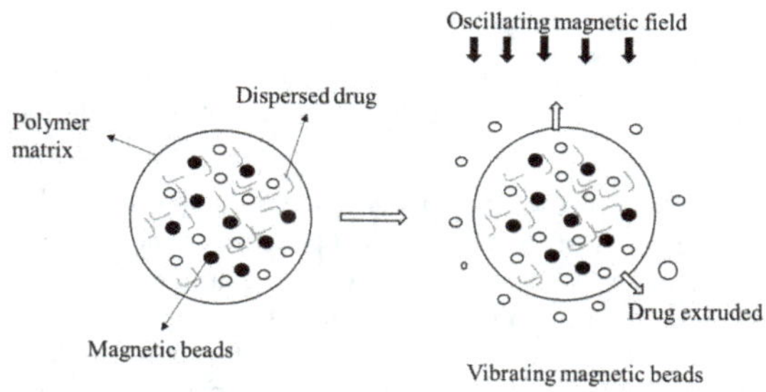

Fig. 3: Schematic representation of magnetically controlled drug delivery implant influenced by magnetic field.

2.2 Active implants

These implants require an outer energy source or positive driving energy for releasing the drug through implants. Therefore, it followed zero order release kinetic but its complex structure makes it exorbitant. The majority of active implant devices are electronic and made up of metallic materials such as pumping devices. The common example of an active implant is the osmotic pump in which an osmotic agent is filled into the semipermeable membrane of a non-biodegradable polymer. The drug reservoir is normally placed in the flexible impermeable membrane which has an orifice and collapses concerning an increase in hydrostatic pressure [25] as shown in **Fig.4.**

The semipermeable membrane permits an inward movement of aqueous fluids by an osmosis mechanism. The osmotic gradient allows a steady flow of fluid within the implant and increases osmotic pressure within the system. This hydrostatic pressure will invariable the ejection of the drug through the orifice. It also controlled the discharge of any reservoir fraction that was altered by modifying the structure of the semipermeable membrane. This design provides zero order drug release still reservoir load has been ended but limited drug loading [26].

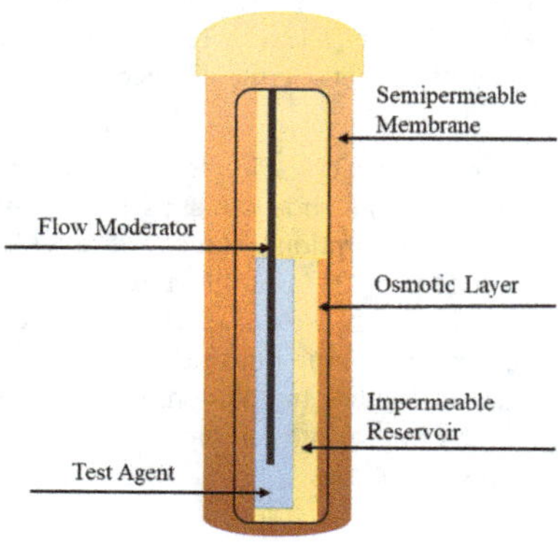

Fig. 4: Diagrammatic representation of osmotic pump implant

Materials Research Foundations 172 (2025) 299-328
https://doi.org/10.21741/9781644903353-13

3. Drug release mechanism from polymeric implants

Drug release mechanisms from implants specifically include polymeric matrix degradation, controlled release through polymer swelling, osmotic pressure pumping (active diffusion), and passive diffusion [27].

In the polymer swelling mechanism, the solvent penetrates into the polymeric matrix which maintains the drug release rate. However, the solvent penetration rate into the polymer is sluggish than the drug diffusion rate. Therefore, it follows prolonged drug release until the shelf life of the system. The drug release from the implant is dependent on the swelling rate of the polymeric matrix, while its degradation rate contributes to the effectiveness of the system. The osmotic pressure pumping and passive diffusion follow the Higuchi pattern of drug release kinetics in a controlled manner. It means initially drug release follows a burst release mechanism further it follows the sustained release and is directly proportional to the square root of time [28,29]. In osmosis, solvent moves from dilute solution toward more concentrated solutions through a semipermeable membrane and develops hydrostatic pressure. It helps to control the drug release rate in distinct conditions. In passive diffusion, drug molecules transfer from high concentration to low until the development of equilibrium through the concentration gradient. These mechanisms also depend on the permeability coefficient of the polymer, diffusion coefficient of the drug through the polymer, % drug loading, encapsulation efficiency of implant, and rate of bio-degradation of the polymer [30].

3.1 Drug release mechanism through non-biodegradable polymeric implants

The non-biodegradable polymeric implants have been broadly used for transdermal drug delivery as well as the administration of contraceptives. It includes membrane-enclosed reservoirs and monolithic/matrix implants. As discussed above, in the reservoir system, the drug core is compressed into the polymeric membrane and forms an intact barrier for drug diffusion. Therefore, reservoir-type implants are not affected by the concentration gradient and maintain a constant release rate. However, the width and porosity of the polymeric membrane control the release rate in zero-order kinetics. Whereas, in matrix system drugs are homogeneously dispersed in a polymeric membrane. In this system, drug release follows desorption from the polymeric matrix which is mediated through Fickian diffusion. The drug release mechanism from the matrix system is strongly affected by the concentration gradient, the diffusion rate of the drug, and the swelling rate of the polymer [31,32]

Sureskin II silver is the FDA-approved transdermal implant of polyurethane foam for the treatment of post-operative wounds, dermal ulcers and abrasion. Drug loading into polyurethane implant is done by polymerization or equilibrium absorption and the release profile confirms the initial burst release and diffusion via driving force [33]. Herold *et al.* evaluated the release kinetics of metronidazole from cross-linked polydimethylsiloxane. They observed that drug release from the matrix followed zero order and depended on water uptake. This happened due to most of the drugs in the matrix existed in the solid-state matrix. In addition, poly (ethylene vinyl) acetate is also a diffusion-controlled polymer and developed into microporous film to achieve zero-order release for sustained release over a definite time [34].

3.2 Drug release mechanism through biodegradable polymeric implants

As discussed earlier, biodegradable implants are either drug reservoir type or drug-polymer mixture. It permits the release of drug in a controlled manner through polymeric biodegradation. The diffusion coefficient of the drug, swelling index, and erosion rate of the polymer control the drug release via degradable polymer. Rate of drug dissolution and diffusion through the polymeric wall are also involved in the drug release mechanism. Various factors also affect the release of drugs including solubility and permeability of drugs through the polymer, % drug loading, and degradation rate of the polymer [35]. The drug release through biodegradable implants involves a complex mechanism including hydrolysis of polymeric backbone ester bonds, and hydrolytically susceptible amide bonds. Therefore, water interference in implant devices is significantly important in studying polymer degradation and drug release kinetics. It occurs in two stages first, water is infused into an amorphous region of polymer, and random hydrolytic scission of weak ester bonds. In the next stage, when most of the ester bonds break, the polymeric chain is completely scissored and changes molecular weight. The degradation process is characterized through gel permeation chromatography. The typical drug release profile through PLGA followed initial burst release followed by zero order [36].

Kranz and Bodmeier evaluate the *in vitro* drug release of diltiazem hydrochloride and buserelin acetate from PLGA implants. It showed a triphasic release pattern, surface-adsorbed drug molecule showed the initial burst release. In the second phase, drug release is due to the degradation of the polymer, and the final phase occurs due to polymer erosion. The release profile of the drug also showed that its release kinetics were not simply driven by degradation but by concentration gradient and the shape of the implant was also profound equally. In addition to this, release behavior of the drug is affected by the structure and physicochemical properties of the drug [37].

4. Types of polymers

Different varieties of natural and synthetic polymers can be applied and investigated in the formulation and packing of implant devices. Synthetic polymers are generally nontoxic with precise physicochemical properties. Natural polysaccharides can also make implants as part of the polymeric blend. The biodegradability and biocompatibility are also important in the development of implants. Any polymeric material used in implant formulation must be biocompatible and its degradation rate is completely investigated. Therefore, polymers are also divided mainly into biodegradable and non-biodegradable.

As we already discussed implants made up of nonbiodegradable polymers need to be removed after a specific period by minor surgery otherwise it may lead to accumulation in the body which is more painful compared to insertion [35]. Jacques *et al.* investigated the traumatic patient condition at the time of surgical removal of the non-biodegradable contraceptive implant 'Nexplanon'. They observed multidisciplinary teams with a peripheral surgeon plays a beneficial role in the effective removal of implants. Therefore, non-biodegradable implants are complicated to use [38]. On the other hand, biodegradable polymers do not require surgical removal after use because they are degraded naturally and excreted from the body. These are fully biocompatible, highly reproducible, easily metabolized, form non-toxic bi-products, free from inflammation, and excreted physiologically. Polymers are also applied as a protective coating and adhesive to seal the interface between two drug substrates as well as the orifice. For example, to monitor the pH level in the gastrointestinal system, implantable sensors are used and prevent the system from corrosion against GIT acidic fluid. Thus, it's a need to apply corrosion-resistant polymer for packaging on the sensor [39].

4.1 Biodegradable polymers

Biodegradable polymers undergo degradation by two mechanisms, either hydrolytic cleavage or enzymatic degradation of biomaterials. On the basis of the degradation mode of polymers, formerly known as hydrolytically degradable polymers, and later called as enzymatically degradable polymers. The preferred mechanism of degradation of biodegradable polymers is enzymatic degradation [40].

4.1.1 Natural biodegradable polymers

Natural biodegradable polymers are mostly synthesized by living organisms such as animals and plants, consisting of repeated units of monomers connected with covalent bonds. These natural polymers are found in the main three types: polysaccharides such as cellulose and chitin, proteins like collagen and gelatin and lipids [41]. Natural biodegradable polymers are mostly suitable for short term implants. These polymers are used for a long time in biomedical applications on account of their suitable chemical, physical and biological properties.

Natural biodegradable polymers are having advantage of similarity to biological systems, no toxicity, compatibility to host tissue, shows no or less noninflammatory action and most important is they are enzymatically degradable. While natural polymers are degraded before reaching the melting point because of its temperature sensitivity. Natural polymers are obtained from plant and animal resources, so there is the possibility to transmit the disease to the humans [42].

a. Collagens

Collagen is a complex macroprotein protein that is most abundantly found in humans and animals. In the body it is mostly found in skin, ligaments, cartilage tendons, blood vessels and bones. Collagen is an important component of extracellular matrix (ECM) and consists of one third of all proteins present in the human body, out of this 90 % collagen present in the skin. Collagen involved in cell growth and cell proliferation and migration; it also supports the structural processes. As a component of the human body, it is a biocompatible, biodegradable and nontoxic for being used as material for development of implants [43].

Collagens are mostly used as biomaterials for manufacturing scaffolds, hemostatic agents, for formulation of sutures, drug delivery systems, and tissue engineering.

Various waste materials such as poultry waste like feathers, chicken skin, bones and waste from the leather industry are used as source for isolation of collagen. Apart from this skin of various species of fish like Nile perch, threadfin bream, Baltic cod and bigeye snapper are also used as source of collagen. Collagen is also extracted and isolated from bones, skin, horn and cartilage of slaughtered cattle [35]. Type I collagen occupies 80 % of total collagen and is mainly obtained in normal human dermis, while 20 % of type III collagen found in human body. Integrity and strength of dermis is due to type I collagen and type III collagen is responsible for its flexibility [44].

The collagen molecule consisted of three polypeptide alpha chains, twined around one another in a triple helical structure. Each polypeptide chain in a helical structure are consisted of 1000 basic amino acids in sequence, glycine, proline and hydroxy proline. These amino substances are essential for tissue connection and providing elasticity to the skin and creating tension. Increased

Materials Research Foundations 172 (2025) 299-328
https://doi.org/10.21741/9781644903353-13

in glycine content provides more flexibility to the collagen [40]. This triple helical structure is connected by hydrogen bond between adjacent CO and NH groups and covalent bonds. Collagen is a rod shape polymer with an approximate length of 300 nm approximate MW of 300 kDa. Different types of collagens are found, and their type is based on polypeptide chains, lengths and interruptions of their helix structure, and difference in the helix terminations [41].

On account of its biocompatibility and biodegradable nature, it has been used as a polymer for implants for four decades. Collagenases and metalloproteinases, these two enzymes play the role in the degradation of collagen in a body that leads to its conversion into simple units like amino acids. This enzymatic degradation, and suitable mechanical, biological and favorable physicochemical properties make it choice of polymer in biomedical devices. It can be shaped in various designs like injectable rods, nanofibrous matrices, sponges for wound treatment, foams, sheets and in tubular forms [40]. Japanese researchers intensively studied collagen minipellets, prepared by molding of collagen solution followed by drying and cutting in a rod-shaped implant, nearly 1 mm in diameter and 15 mm in length. This rod-shaped collagen implants can be administered through injections using a needle with a plunger. Interleukin-2 minipellets of collagen showed good *in-vivo* activity [45].

Safety and efficacy data of collagen implant (Zyderm) was evaluated by Fred F. Castrow II *et al.* Zyderm collagen injectable implant was approved by US FDA in 1981, it is highly purified bovine collagen implant and used for corrections of scars and aging skin lines. As per their study, this injectable implant showed less adverse reactions and the severity of reaction was also not serious [43,46].

Collagen is nonimmunogenic in nature, it acts as a natural substrate for cell adhesion and resistant to proteolysis. Due to these properties, collagen is used as an ideal biomaterial for soft tissue augmentation. To correct small skin defects and facial wrinkles, bovine injectable collagen implants like Zyderm I®, Zyderm II® and Zyplast ®(manufactured by Inamed Corp., Santa Barbara, CA) are used in US, Europe, Japan and in Hong Kong [44]. Bovine collagen is mostly used for the preparation of implants and shows slight immunogenic response in humans as it is foreign material for the human body. Some patients show hypersensitivity reactions like swelling and redness at the site of application. Almost 3 % of patients treated with Zyderm I and Zyderm II bovine collagen implants show the signs of hypersensitivity reactions, while human collagen implants do not induce any allergic reactions [45].

Recently collagen like peptide (CLPs), also called collagen mimetic peptide (CMPs) was found to be better alternative to human collagen, because of its property to self-assemble and generate a triple helical structure of nanofibers like collagen. Mirazul *et al.* investigated the application of self- assembled CLP-PEG polymer as corneal implants [47]. Collagen implants (plugs) are also used to relieve the symptoms of dry eye syndrome. Punctal occlusion is one of the methods which uses a punctal plug which reversibly blocks the nasolacrimal duct. This blockage at the punctum by collagen implant, blocks the tear flow and increases the retention of natural and artificial tears on ocular surface. Presence of continuous tear file in the precorneal region leads to relief from dry eye syndrome. This blockage of punctum is temporary and cane be reversible after removal of collagen plug [48].

Collagen is quite a popular polymer for ocular drug delivery systems. Ocular collagen inserts loaded with drug are designed in rod shape or in the form of a film and drug release was expected for a prolonged period of time, which help in reducing dosing frequency. Collagen shields are also developed with or without drug. Collagen shields cover the corneal surface and act as a corneal bandage, mainly used after corneal transplantation or radial keratotomy to promote the wound healing. Insertion of drug loaded ophthalmic *in situ* gels of a collagen increases residence time in cul-de-sac.

b. Chitosan

Chitosan is a non-toxic, biodegradable polymer with a high molecular weight. It is the N-deacetylated derivative of chitin. Chitosan is safe for oral use, and it is approved by FDA as a food additive. Second to cellulose, chitin is the most abundantly obtained natural polymer with a huge annual production of 10^{10}–10^{12} tons and same chemical structure as that of cellulose. It is found in both plant and animal kingdom and isolated first time from fungi by Braconnot.

It is naturally present as three polymorphs, known as α, β, and γ chitin, depending on the orientations of the microfibrils. Chitin is primarily obtained from crustaceans, hard exterior aquatic invertebrates and arthropods, mostly obtained in the α -form [49]. For commercial production of chitin crab, shrimp and crayfish are mostly preferred, it can be also produced from fungi and insects.

Chitosan is a fiber like compound, basically it is a linear polysaccharide, a homopolymer of β- (1-4)-linked N-acetyl-D-glucosamine. It is obtained from derivatization of chitin; the process involved the removal of few numbers of acetyl groups (CH_3–CO) to solubilize it in diluted acids like acetic acid, ascorbic acid and propionic acid. The main variation between cellulose and chitosan is presence of the hydroxyl (-OH) group at C-2 position in cellulose and amine (-NH_2) group found in chitosan at the same position. Chitosan shows antioxidative and scavenging activity against radicals. Chitosan is obtained by extraction process like chemical, biological or microwave. assisted extraction and extraction is carried out from different sources, but most abundantly from crustacean shells.

Chitosan is a fully acetylated polymer obtained from chitin which is present in exoskeleton of arthropod. In *in vitro* chitosan is degraded by chitosanase and lysozyme enzymes, while *in vivo* degradation mostly occurred due to lysozyme. The degradation pathway includes hydrolysis of acetylated portion. The less the degree of acetylation, higher will be the degradation rate of chitosan. Rate of degradation is also inversely proportional to degree of crystallinity of chitosan.

Chemical modification of chitosan leads to change in its degradation rate and solubility. Predetermined rate of degradation can be achieved by substitution of chitosan like isobutyl substituted chitosan provides the faster degradation in mice. Faster degradation may occur due to destabilization of hydrogen bonding in chitosan molecule.

Park et al. developed graphene-chitosan hybrid dental implant by using different concentration of graphene. These hybrid implants also promote osteoblast proliferation and decreasing the rate of biofilm formation [50].

Further surface modification of these implants with chitosan overcome the issue of an incomplete osseointegration between implant surface and the site of implantation. Implants application may cause implant -associated infections due to deposition of microorganism and biofilm formation on the implant surface. Commonly *E. coli, Staphylococci and Pseudomonas* species are involved in formation of film on the implant surface. Chitosan and its derivatized polymers due to its antimicrobial properties prevent the formation of biofilm. Chitosan also promotes osteoblast

Materials Research Foundations 172 (2025) 299-328
https://doi.org/10.21741/9781644903353-13

formation and so promotes bone formation [51]. Chitosan implants also promote mineralization and osteogenesis of bones [52].

c. Gelatin

Gelatin is a natural biodegradable polymer obtained by alkaline, acid or enzymatic hydrolysis of collagen. It is biocompatible and can be produced on a large scale and having less cost. Major sources of gelatin are porcine and bovine skin. Gelatin is water soluble polymer and used as stabilizers in rubella and measles vaccines. Gelatin is also used in medical devices like matrix for implants. In pharmaceuticals it is used as a carrier for delivery of drug in gels, nanoparticles and also used as a coating material in microencapsulation. Thermal and mechanical properties of gelatin are depended on source of collagen, extraction method used, degree of crosslinking and extent of thermal denaturation. Mechanical properties of gelatin can be modified via crosslinking. Large molecular weight substances like proteins can be delivered in the form of gelatin microparticles and gelatin nanodrug delivery is used for intravenous administration and for brain delivery of drug. In biomedical field gelatin is used in tissue engineering as a bone healing material and for wound dressings [53].

Different geometric shapes of 3D printed implants are designed from gelatin with modified glycerol gelatin soft hydrogel material, that is better option to regularly used rigid polymers. Like gelatin, glycerol gelatin hydrogel also needs to be crosslinked to improve its mechanical properties, to avoid its softening when administered as an implant [54].

Gelatin is used for coating dental implants because of its high cellular affinity. Polymerization of gelatin with methacrylate and further fabricating it in a micro/nanopillars of 1 μm or 250 nm in diameter for dental treatment was reported by Rizwan et al. [55] These micro/nanopillars showed better proliferation of endothelial cells as compared to plane gelatin surface. Gelatin is also used to prepare ophthalmic sheets/films with different biological absorption times for surgical purposes. Gelatin is found to be less immunogenic and produces less inflammatory response. These gelatin implants are used in the treatment of retinal detachment [56].

Makita *et al.* have reported that gelatin and methacrylate combined micro/nanopillars could be fabricated through a combination of molding and polymerization. As compared to planer gelatin surface, these micro/nanopillars, with diameters of 1 μm or 250 nm, were increases the proliferation of endothelial cells at better rate [57]. In dental implants of titanium, gelatin-calcium phosphate (Gel-CaP) composite coating is provided on titanium surfaces to improve the process of osseointegration. It is a biocompatible coating which is able to accelerate osseointegration due to chemical interaction with the substrate. Apart from this amorphous calcium phosphate act as a precursor for the mineralization in human being [58].

For synergistic effect, along with other natural, synthetic and inorganic material, gelatin is used in combination in bone tissue engineering to support the physiological process of bone repair. 3D gelatin, apart from providing a mechanical support and microenvironment for osteoblasts, also promotes the new bone generation by providing a cargo for administration of progenitor cells cytokines and material that help in bone regeneration. These 3D scaffolds release the cytokine and other biological molecules for the prolonged time [59].

4.1.2 Synthetic biodegradable polymers

Synthetic biodegradable polymers have several advantages over natural biodegradable polymers, they are nonimmunogenic in nature, reproducible, and large-scale production is possible. Synthetic biomaterials show batch-to-batch uniformity, their properties can be modified for special applications [60]. Synthetic polymers degraded by hydrolysis mechanism are mostly preferred to enzymatically degradable synthetic polymers. The first synthetic biodegradable polymer poly (glycolic acid) was discovered in the 1960s has started the development of various synthetic polymeric implants. Synthetic biodegradable polymers and co-polymers with different release kinetics are developed nowadays. Synthetic biodegradable implants are preferred over metallic implants to avoid again surgical operation to remove the implant [40].

a. Poly (lactic acid)

Poly (lactic acid) (PLA) is synthesized by polymerization of natural starch/lactic acid including corn, rice, and potato starch, and treated by a variety of thermal processes. PLA is synthesized from pure lactic acid by polycondensation or ring-opening polymerization method. PLA is a white powder at room temperature. The melting point of PLA is 175°C and glass transition temperature is 55 °C [61].

PLA can be synthesized by using D and L monomers of lactic acid. The PLA which is synthesized by the D enantiomer of lactic acid is crystalline with an even chain assembly unit. While the PLA which is produced by the L enantiomer of lactic acid showed a texture like igneous rock [62]. However, when it is synthesized by the blend of both enantiomers of lactic acid is amorphous. The synthesis of PLA requires less non-renewable resource energy such as oil, coal, and natural gas compared to other polymers. Due to its hydrophobic nature, it is solubilized in various organic solvents while insoluble in various hydrophilic solvents. Its water and oxygen-carrying capacity nake PLA more biocompatible and biodegradable at physiological temperature and pH. The orous structure of PLA improves its bio-degradation by hydrolysis of the ester group. The PLA plants when inserted into the body, immune cells are activated and absorbed from the site of ninistration. Further, it supports secrete the enzymes of lactate dehydrogenase and acid phatase that convert PLA into lactic acid and lactic acid oligomers with carboxylic end groups. degraded products are solubilized in water and excreted through urine [63].

chanical strength of PLA makes it a potential member in the development of medical and the second most traded polymer. The US FDA approved the PLA as a generally d as safe (GRAS) material and for direct contact with biological fluid. Heba *et al.* d the *in-situ* implants of doxycycline hydrochloride and secnidazole for the treatment of s by direct periodontal administration. For the synthesis biodegradable polymer poly was used in 25% w/w and *in vitro* drug release and antimicrobial activity was e results showed it exhibited slower drug release up to 80 hrs with significant ctivity [64].

he form of screw, pins, rods, mesh and orthopedic medical implant. Viktoriya N. developed the PLA-PEG- based ocular implants for prolonged delivery of everolimus for wound healing modification after glaucoma surgery without plications [65]. In tissue engineering science PLA is used as a biocompatible material. Linn et al. formulated a chitosan coated PLA Nanofibers by ique and further surface mineralized with phosphate and calcium for regeneration of bone. Lin *et al.* also studied the application of polylactic as a enzymes, hormones, vaccines and other biological products containing semipermeable microcapsule [66].

b. Poly (glycolic acid) (PGA)

PGA was first used for the manufacturing of surgical sutures in 1960s marketed by the brand name DEXON [67]. PGA is a highly crystalline synthetic aliphatic polyester having low solubility in organic solvents and a high rate of degradation. After implantation, it showed poor mechanical strength due to its high degradation rate. It is synthesized from glycolic acid as a starting material by ring-opening polymerization method. Ring-opening polymerization method leads to the production of a high molecular weight product. The glass transition temperature of PGA is between 35 to 40°C and it melts above 200°C. Apart from these other methods used for the synthesis of PGA are polycondensation, solid-state polycondensation of halogen acetates, and Bronsted acid catalyzed polymerization of carbon monoxide and formaldehyde. Polycondensation produces low molecular weight product [67]. In the body degradation of PGA occurs by ester hydrolysis of the polymer chain followed by enzymatic degradation. Finally, it is converted into glycine and either secreted in urine or enters the citric acid cycle [68].

Normal body metabolism also produces glycolic acid which is known as hydroxyacetic acid. Subsequently, the polymerization of glycolic acid units formed the polyester poly (glycolic acid) (PGA) or polyglycolide. PGA is the only polyester that exists in the highly crystalline state with a melting point of 200°C and mechanical strength greater than poly (lactic acid). It was the first biodegradable polymer used in biomedical devices. Biodegradable poly (glycolic acid) sutures were the first synthetic, and absorbable sutures that were introduced in the early 1970s. It was 100% synthesized from a homo-polymer of glycolide and coated with equal parts of polycaprolactone and calcium stearate. PGA is insoluble in various solvents and showed bulk degradation via breakup of its ester backbone. Its osteosynthesis implants have been particularly suitable for the fixation of cancellous bone, trans-epiphyseal, and malleolar fractures in children. It has been reported that very small granulation tissue is developed at the administered site of the PGA implant which is easily tolerated by surrounding tissue [69].

The degradation product forms glycine which is excreted through urine. However, its acidic by-products can produce inflammation in the locally administered area and limit its use in biomedical applications [70]. Therefore, Naser *et al.* highlighted the pivotal role of poly (glycolic acid) along with albumin in tissue regeneration therapy by addressing the associated challenges. The novel filament sutures were developed by the hot-melt extrusion process and used for wound closure with promising results. The prepared sutures are also applicable as a 3D printer filament to print custom scaffolds, nails, and plates observed that a smaller diameter is required for future works [71].

PGA is used for tissue engineering scaffolds; PGA nanofibers are used in various medical applications. PGA is used along with natural polymers like collagen to prepare nanocomposite, as reported by Ghazali *et al.* Nureddin *et al.* formulated a self-reinforced PGA (SR-PGA) screws and implanted in the spongy bones of rabbits. They were investigated incompatibility and the study showed that implants are safe for use. Also, SR-PGA rods, when sub-perichondrially implanted in place of resected costal cartilage of rabbits found that it helps in the growth of cartilage and in bone formation. SR-PGA membranes were also used as an implant in maxillofacial repairs [68].

Materials Research Forum LLC
https://doi.org/10.21741/9781644903353-13

c. Poly (lactic-co-glycolic acid)

Poly (lactic-co-glycolic acid) is a biodegradable copolymer of polylactic acid (PLA) and polyglycolic acid (PGA) and generally known as PLGA. It is prepared by direct polycondensation of lactic acid and glycolic acid. By using this method, it is difficult to achieve high molecular weight (MW) PLGA. Modified melt/solid polycondensation method, can be able to produce the PLGA with high MW compared to the PLGA produced by the ring opening polymerization of the cyclic diesters [72]. Ring opening copolymerization method produced a PLGA of high molecular weight, metal catalysts are used in this synthesis process and temperature is maintained at 130-220 °C. In the contact of an aqueous environment PLGA undergoes hydrolysis of ester linkage via heterogenous or bulk erosion and its end products are PLA and PGA. Hydrolysis involves the breakdown of van der Waals forces and hydrogen bonds followed by cleavage of covalent bonds leads to initial degradation. Further degradation is continued by carboxylic end groups and scission of backbone covalent bonds and loss of integrity. Finally, further breakdown of fragments into molecules and solubilize in the surrounding aqueous fluid [73]. The degradation rate of PLGA is depends on molar ratio of two monomers PLA and PGA, molecular weight, degree of crystallinity and glass transition temperature of polymer [74].

Polylactic acid contains asymmetric alpha-carbon atoms in D and L form or also present in R and S form respectively. PLGA is a mixture of poly D and L-lactic acid that contains D and L-lactic acid in a 50:50 ratio. Its composition directly influences the crystallinity, mechanical properties, and degradation rate of the polymer. The side methyl group in L-lactic acid is responsible for the hydrophobicity of PLGA therefore, the content of L-lactic acid increases the hydrophobicity of PLGA and decreases its rate of degradation [75,76].

It is best defined as an FDA-approved biodegradable and most biocompatible biomaterial. Therefore, it has been mostly used in various drug delivery systems to deliver drugs, phytoconstituent proteins, and other macromolecules. The hydrolytic degradation of PLGA in the body forms lactic acid and glycolic acid. The molecular weight and the ratio of lactic acid to glycolic acid, defines the physical properties of PLGA. Therefore, it has shown significant potential in biodegradable implantable drug delivery systems concerning design and performance. Its glass transition temperature was found to be above 37 °C and exhibits a fairly rigid structure which is important in implants [77].

Yang *et al.* developed the PLGA-based sustained drug release implant using the hot melt extrusion technique. They observed that the morphology and microstructure of the implant depended on the physicochemical properties of PLGA. The higher plasticity of Viatel™ PLGA showed smooth surfaces and strong PLGA-drug interaction in synthesized implants than in Viatel™ Ultrapure PLGA implants. Viatel™ ultrapure PLGA showed burst release of the drug than Viatel™ PLGA. However, Viatel™ PLGA shows a short onset of time and a substantial release phase.

Commercially there are two matrix implants available in the market such as Zoladex and Ozurdex by PLGA copolymer. The AstraZeneca manufacturer formulated a subcutaneous depot implant of Goserelin as a brand name Zoladex for the localized treatment of prostate cancer. It reduces or maintains the level of testosterone and oestradiol production. In formulation, the drug is dispersed in a PLGA matrix and packed in moisture-protected aluminum foil. Allergan marketed an Ozurdex comprised of 0.7 mg dexamethasone. It is used for the treatment of macular edema and administered in the vitreous cavity of the eye. For administration purposes, it required a health professional of an eye [78]. Bassand *et al.* by using hot melt extrusion method, prepared various diameters of ibuprofen-loaded implants. It was observed that the increase in diameter of implants leads to decrease in release of ibuprofen [79]. PLGA is also used for in-situ forming implants (ISFI), upon intramuscular or subcutaneous injections, it converts into solid implants, which can able to release the drug for an extended period of time [80].

However, PLGA polymeric implants still have some issues. The acidic biodegradable products of PLGA may accumulate at the local site of administration lowering the pH at that local site and accelerating the rate of degradation which alters the pattern of drug release [18].

d. Poly(caprolactone)

Poly (caprolactone) (PCL) is a potential member of polymeric biodegradable implant systems and parenteral carriers on account of their biocompatibility, biodegradability, nontoxicity, and economics. Mechanical properties and thermal stability make it a potential member in developing sutures, ligatures, sub-dermal implants, scaffolds, and wound dressings [81].

It has a melting point of 55-60 °C and Tg between -65 °C and -60 °C. Physical, thermal and mechanical properties of PCL mostly depend on its MW and degree of crystallinity.

Its degradation is completely dependent on the hydrolysis of the aliphatic ester bond and degradation products are metabolized by tricarboxylic acid and excreted through renal pathways. Because of the hydrophobic nature of PCL, it takes more time to degrade than PLA, PGA and PLGA. The rate of degradation of PCL or hydrolysis rate of ester linkage depends on the molecular weight of PCL and other factors like temperature, pH, and enzymes. Its MW significantly elaborates the chain length of the present monomer. It requires more time to break down during degradation and form a water-soluble monomer and oligomer [82]. PCL is a very diverse material and compatible with other polymers that create a blend with unique properties. Its hydrophilicity, biocompatibility, and biodegradability are improved when copolymerized with poly (ethylene glycol) (PEG). The PCL-PEG blend lowers the glass transition temperature and crystallinity, which also enhances the degradation [83].

Sarah *et al.* developed PCL-coated 3D biodegradable implants for prolonged drug delivery of hydrophilic moiety. Initially, by fused filament fabrication they prepared the hollow implant from PLA and poly (vinyl alcohol) (PVA). Further, it was coated with a PCL- PEG solution casting mixture [5]. Yang *et al.* developed the 3D-printed PCL-chitosan filament implant by fused deposition molding (FDM). It was formulated for prolonged release of drugs as well as for self-application. Here, the investigator specifically determined the drug release rate, behavior, mechanism, and biocompatibility. They used the model drug Ibuprofen and loaded it into the prepared filaments by hot melting extrusion. In this 3D print implant, PLC was used as a drug carrier and chitosan was used to release the drug in a shape-controlled manner. They observed that the therapeutic efficacy of the drug strictly depends on the processing temperature and method of preparation of the filaments. Notably, the tensile strength of the filament decreased by increasing drug and chitosan content. There was no degradation, and chemical changes were observed during the complete formulation. The results showed that 99% drug was released up to 120 hrs through the mechanism of diffusion-erosion [84].

4.2 Non-biodegradable polymers.

Different types of non-biodegradable polymers are used for the preparation of biomedical devices/implants. These implants are used for different applications in human body. Non-

Materials Research Foundations 172 (2025) 299-328 https://doi.org/10.21741/9781644903353-13

biodegradable implants are particularly used as a pacemaker, artificial blood vessels, mechanical heart valves, dental implants, nasal implants and breast implants. Pacemaker and artificial blood vessels improved the efficiency of body organs and some implants provide physical support to the person, like knee implants. Non-biodegradable implants are intended to be used for working prolonged time or permanently in human body. For efficient working of the implant, it must be properly placed at the site of implantation. The size of implants should be such that it does not exert pressure on surrounding tissue, muscles or bones and minimum invasive procedure is required for its implantation [85]. Various types of non-biodegradable polymers like polyethylene, polypropylene, polyamide, polytetrafluorethylene, silicone and polydimethylsiloxane are used for the fabrication of implants [14].

a. Polyethylene (PE)

Polyethylene is a biocompatible and non- biodegradable polymer. PE is prepared from polymerization of ethylene monomer (-C_2H_4-). American society for testing and materials classified PE into five major categories depending on its density: low density, linear low density, medium density, high density PE (HDPE) and ultra-high molecular weight PE (UHMWPE) [86]. On account of change in molecular weight and branching its density found in the range of 0.88 to 0.93 g/cm^3. UHMWPE were introduced in market in early 1960s [87]. Porous HDPE is elastic at room temperature and becomes malleable when heated in hot water. Porous HDPE has high tensile strength easily available material. Porous HDPE implants are routinely used for primary construction of orbital fractures. This material has great utility in cranial and temporal defects [88].

PE is mostly used in tissue engineering. Surface properties of PE need to be favorable for long term adherence and integration with tissues. PE implant surface wettability is required for attachment and proliferation of cell. PE surface is hydrophobic due to its intrinsic low surface energy. So various techniques are used to increase its surface energy and to make its surface hydrophilic. One of the techniques is plasma technique in which exposure of surface with ionized gases in a high voltage environment. Other techniques used are acid-base etching techniques and chemical etching techniques to modify surface hydrophilicity. Porosity of the PE implant should be optimum as mechanical strength is dependent on porosity. Higher porosity leads to rapid tissue in-growth but at same time decrease in its mechanical strength. PE is popularly used in total joint arthroplasty [89].

UHMWPE implants have a wearing issue that leads to osteolysis and aseptic loosening and decreases the survival rate of implant. Wearing PE implants can be resolved by introducing highly crosslinked PE (HXLPE) in 1990s. Crosslinking is done in three steps: high energy radiation (gamma or electron beam), post irradiation stabilization and end point sterilization. Wear resistance of HXLPE is increased by higher crosslinking occurred due to increasing the dose of irradiation. Irradiated HXLPE contains free radicals that react with oxygen during shelf-storage and even in clinical use leads to decrease its stability. Removal of free radicals is carried out by remelting and annealing [90].

Remelting significantly decreases the free radicals present in irradiated PE but decreases the strength of polymer. While annealing is not effective as that of remelting to reduce the number of free radicals and has low oxygen stability, but better mechanical strength. The issue of mechanical strength and stability due to oxidation is resolved in second generation HXLPE. In this process repetitive irradiation is done at low doses (approx. 30 kGy) followed by annealing. This modification in the process significantly reduces the number of free radicals and increase in stability of the implants. James William *et al.* studied the efficiency of porous high density PE implants in experimental animals to tolerate exposure and support skin grafts while removing defects in auricular cartilage [91]. 3D printing strategy is also used to develop patient specific implants. Selective laser sintering, stereolithography, extrusion-based 3D printing techniques like fused deposition modelling (FDM) is widely used techniques for the manufacturing of patient specific implants.

b. Polypropylene (PP)

Polypropylene is a thermoplastic polymer similar to polyethylene but slightly harder to PE. It has low density and high resistance and molecular formula of $(C_3H_6)_n$. It is synthesized from propene monomers by polymerization. To methods are mostly used for synthesis are Ziegler-Natta polymerization and Metallocene catalysis polymerization. It is categorized into homopolymer and in random and intact copolymer on the basis variable densities. PP is used to support weakened tissues as a surgical mesh. PP is also served as a scaffold for fibro-collagenous tissues to grow on the mesh itself. It is used for the treatment of stress urinary continence and pelvic organ prolapse in urogynecology [86]. On the basis of arrangement of CH_3 group PP is classified into three types: isotactic, syndiotactic and atactic polypropylene. PP is used in pelvic floor repair and also in repair of abdominal wall and inguinal hernia. PP has high melting points, and this makes it suitable polymer for fabrication of biomedical material as it can be sterilized at high temperature by autoclaving. Like PE, it is chemically resistant but less resistant to oxidation, chemical degradation and crosslinking as compared to PE [92]. PP can be sterilized by both autoclaving and gamma radiation. Both processes make changes in their molecular structure. Repetitive heat sterilization also reduces its mechanical strength. Sterilization by gamma radiation increases the susceptibility to oxidation. Free radicals generated during exposure are trapped in the crystalline structure of PP and lead to its molecular disruption. Heat sterilization increases the diffusion of toxic material to the surface of the device [93].

PP based material is used as bone cement. Instead of using human bones, polymeric bone grafts are more suitable as it does not transfer any disease to a person and also it is difficult to adjust the human bone in position of defective bone. Polypropylene combined with different materials like hexagonal boron nitride and n -hyaluronic acid is used in human bone replacement. Polypropylene carbonate and PLA composite nanofibers were prepared by Park *et al.* by sol-gel electrospinning technique. These microfibers are used for the manufacturing of medical devices [92].

Surface modification of PP is carried with polyethylene glycol to decrease surface free energy that leads to improvement of its biocompatibility [94].

PP is used still as a popular material in developments of mesh devices in pelvic floor disorder. Michelle *et al.* reviewed the research work of various researchers and concluded that as compared to other synthetic polymers PP induce less inflammatory or similar host responses. Surface area of these implants is also proportional to host response produced.

c. Polyethylene terephthalate (PET)

PET is a thermoplastic, linear, partially aliphatic and aromatic polymer. PET is synthesized by reaction of terephthalic acid with ethylene glycol. It is a semicrystalline polymer. PET is thermally stable, good mechanical strength at high temperature, impermeable to gases, light weight, glossy, transparent and high hydrolytic stability due to presence of aromatic ring. It is prepared from less

Materials Research Foundations 172 (2025) 299-328
https://doi.org/10.21741/9781644903353-13

expensive raw material. PET is popularly used in the medical field due to its biocompatibility, mechanical strength, high uniformity and resistance to abrasion and chemicals [63].

Due to limited ionic groups and high crystallinity PET is resistant to dye staining and this property makes it suitable polymer in dentistry, because change in colour of dental implant is not acceptable. Geyson *et al.* studied the physical, mechanical and morphological properties of PET compared with different polymers in new application as prosthetic component for overdentures implant retained. In this study a capsule for overdenture retention was prepared from PET and other polymers and studied for various parameters like compressive strength, surface hardness, surface roughness, pullout test, fatigue resistance tests and overall study suggest the PET has potential to be used as material for making a capsule for overdentures [95].

PET is mostly used for manufacturing sewing cuffs and heart valves. It is also used in medical devices like sutures and implantable textiles. Vascular grafts, venous grafts in kidney are manufactured from PET. PET fabricated medical devices when implanted in human body may lead to deposition of microorganism on its surface. Microbial deposition can be avoided by surface modification of the device either by making PET surface resistant to microbial adhesion or making it biocidal. Surface properties like surface energy, surface charge, hydrophilicity and roughness of the surface are mainly responsible for microbial adhesion. The presence of positive or negative charge on the surface of polymer can resist microbial growth. Lower surface free energy and reduced contact angle, hydrophilic surface, and smooth surface of the device leads to lower microbial adhesion to device surface [96].

Different surface modification techniques are used to develop PET antibacterial surface, functionalization, grafting, surface topography modification and coating. Functionalization is mostly carried out by plasma treatment, radiation, oxidation, ozone treatment, enzymatic modification and fluorine treatment. In grafting techniques, enzymes like lysozyme or cationic polymer like chitosan is use for grafting on the surface of PET, both prevents the bacterial adhesion on surface. Lithography is one of the methods used to change surface topography that leads to decrease microbial adhesion. Coating technique includes the formation of thin film of polyelectrolyte around the implant by reacting cationic and anionic polymers via ionic bonds [95]. Prosthetic valves of PET can lead to complications like thromboembolism and endocarditis. Wang *et al.* studied the antibacterial potency of PET surface modified by silver ion treatment. These silver coated PET prosthetic valves have antibacterial activity and may protect against prosthetic endocarditis [97].

d. Polydimethylsiloxane (PDMS)

PDMS is a weakly crosslinked thermostable, biocompatible, synthetic elastomeric polymer. Basically, it is silicone elastomer. PDMS is synthesized from silicone and methyl chloride. It has repeating units of $(CH_3)_2SiO$ and depending on these repeating units in the polymer chain and degree of crosslinking, the product is generated in different physical forms like emulsion, resins, elastomer and rubber [98].

It is a highly stretchable material and widely used in manufacturing of various implants and medical devices due to its low cost and good microstructural ability. PDMS is optically transparent, thermally stable and chemically stable polymer. It has a hydrophobic surface and so possess surface wetting problem. Surface hydrophilicity of PDMS can be improved by oxygen plasma treatment, but this treatment is not durable. PDSM has better permeability than other

Materials Research Foundations 172 (2025) 299-328 https://doi.org/10.21741/9781644903353-13

elastomers and it is helpful for separation of gases from the material during fabrication of artificial skin. High elasticity of PDMS is due to Si-O structure than C-C structure, generally present in other polymers [99]. PDMS is highly resistant to biodegradation, biocompatible, good mechanical strength and fabricated easily by replica moulding. Due to these properties, it is widely used in manufacturing of micropumps, catheter surface, dressing and bandages, optical system, microvalves, various implants [100].

PDMS is the most widely used elastomer in microelectromechanical system (MEMS) particularly in microfluidic and bioMEMS [98]. MEMS based medical devices are widely used for therapeutic and diagnostic purpose [100]. PDMS is also used as a material of choice in manufacturing of implants for joints, nasal bones, artificial blood vessels, teeth, sutures. Tayyaba *et al.* fabricated the PDMS microchannel for biomedical applications. These microchannel used for the implantation of varicose veins to avoid tissue damage. PDMS is also used for improving the beauty or expansion of breast, cheeks or nasal bones. PDMS is applicable in facial implantation, retinal implant in surgery of tendons and ligaments. PDMS is also used in knee treatment and replacement [101].

PDMS is used to fabricate prosthetic implant for bone repair. Bone related implants are mostly prepared for bone graft and artificial joints. PDMS possess good mechanical strength and produce long term biological interaction with surrounding biological tissues. Jia *et al.* fabricated the PDMS incorporated carbon/carbon composite for bone tissue repair. These composites were further coated with the phytoconstituent, Rhein to reduce the inflammation. PDMS present in this implant avoid the formation carbon debris and increase the biocompatibility of carbon/carbon composite [102]. Peng *et al.* fabricated the PDMS base implant modified with hyaluronic acid and polydopamine composite. This PDMS implant indicates an improved hemocompatibility related to unmodified PDMS implant or solely coated with hyaluronic acid or polydopamine [103].

Conclusion

In conclusion, polymers have become indispensable in the field of implantable medical devices due to their versatility, biocompatibility, and tunable properties. Through advances in material science, polymers have been tailored to meet specific mechanical, chemical, and biological requirements for various implants, including orthopedic, cardiovascular, and neural applications. Innovations in polymer chemistry and processing techniques have enabled the development of implants that can mimic natural tissues, promote tissue integration, and offer controlled drug delivery functionalities, significantly improving patient outcomes.

The evolution of polymers for implantable medical devices holds significant promise, driven by advances in smart materials, tissue engineering, and sustainability. These materials can vastly improve device integration, longevity, and functionality. However, challenges remain in material stability, biodegradability control, surface modification durability, and regulatory hurdles. Overcoming these challenges will require interdisciplinary research, collaboration, and innovation, which together can unlock the full potential of polymers in transforming medical implant technology and patient care.

Materials Research Foundations 172 (2025) 299-328
https://doi.org/10.21741/9781644903353-13

Future prospective

The future of polymers in implantable medical devices is poised to be transformative, with numerous opportunities to enhance biocompatibility, functionality, and patient outcomes. However, significant challenges must also be addressed to fully realize their potential. The development of smart polymers that can respond to environmental changes like temperature, pH, or biochemical signal offers exciting possibilities for dynamic, real-time interaction with biological tissues. These polymers could release drugs on demand, adjust their mechanical properties based on body activity, or provide biofeedback. Designing materials with reliable and reversible responses while maintaining biocompatibility remains complex. Additionally, extensive testing is needed to ensure safety and efficacy over extended periods, especially in the fluctuating conditions of the human body. Bioresorbable polymers can be used in temporary implants, which dissolve over time, eliminating the need for secondary surgery. These are especially useful in applications like orthopaedic fixation, cardiovascular stents, and sutures, where the implant's function is only temporarily required. Controlling degradation rates to match tissue healing times is challenging. Additionally, ensuring that degradation by-products are non-toxic and efficiently processed by the body is crucial to prevent inflammation or adverse reactions. Advances in tissue engineering are creating polymeric scaffolds that not only provide structural support but also encourage tissue regeneration and integration. These scaffolds can mimic the extracellular matrix, promoting cell attachment, growth, and differentiation, paving the way for regenerative implants that integrate seamlessly with native tissues. Achieving precise control over scaffold architecture and bioactivity remains difficult. Furthermore, the risk of immune responses or rejection needs to be mitigated, especially in cases where the scaffold is intended to be gradually replaced by the body’s own tissue. Combining polymers with nanoparticles, ceramics, or metallic materials can result in hybrid materials that benefit from the advantages of both classes. These composites can offer enhanced mechanical properties, bioactivity, and conductivity, ideal for applications like bone repair or neural implants. Compatibility between different materials and ensuring stable bonding between components are significant obstacles. Additionally, regulatory challenges are higher for composites due to the complex nature of multi-material interactions within the body. Surface engineering such as coating with bioactive molecules or designing anti-fouling surfaces can reduce immune response and improve cell adhesion, resulting in better tissue integration and longevity of implants. Surface modifications must remain stable over time despite the mechanical and chemical stresses of the body. Additionally, achieving uniform, durable, and reproducible surface coating remains a technical hurdle. As awareness grows regarding the environmental impact of medical waste, there is a demand for sustainable, eco-friendly polymer options. This includes the development of bio-based and recyclable polymers for implants, contributing to a circular economy in healthcare. Bio-based polymers often need additional modifications to meet the stringent demands of medical implants. Additionally, ensuring that sustainable materials can still achieve regulatory approval and maintain cost-effectiveness is a significant challenge.

With the introduction of highly innovative materials, there is an opportunity to refine regulatory frameworks to better evaluate and facilitate the use of advanced polymeric materials in implants. Regulatory approval for novel polymers, especially bioactive or hybrid ones, involves extensive testing, which can be time-consuming and costly. Ethical considerations also arise with permanent or degradable implants in vulnerable populations, underscoring the need for robust clinical trials and long-term safety data.

Conflict of interest

There was no conflict of interest, financial or otherwise.

Acknowledgment

The authors would like to acknowledge the Dadasaheb Balpande College of Pharmacy, Rashtrasant Tukdoji Maharaj Nagpur University, Nagpur, (MH)-440037, India for providing all necessities and technical research support.

References

[1]J. Lou, H. Duan, Q. Qin, Z. Teng, F. Gan, X. Zhou, X. Zhou, Advances in Oral Drug Delivery Systems : Challenges and Opportunities, Pharmaceutics. 484 (2023) 1–22.

[2]A. Sultana, M. Zare, V. Thomas, T.S.S. Kumar, S. Ramakrishna, Nano-based drug delivery systems: Conventional drug delivery routes, recent developments and future prospects, Med. Drug Discov. 15 (2022) 100134–100151. https://doi.org/10.1016/j.medidd.2022.100134

[3]S. Amreen, S.M. Shahidulla, A. Sultana, N. Fatima, Journal of Drug Delivery and Therapeutics Implantable Drug Delivery System : An Innovative Approach, 13 (2023) 98–105.

[4]J.K. Patra, G. Das, L.F. Fraceto, E.V.R. Campos, M.D.P. Rodriguez-Torres, L.S. Acosta-Torres, L.A. Diaz-Torres, R. Grillo, M.K. Swamy, S. Sharma, S. Habtemariam, H.S. Shin, Nano based drug delivery systems: Recent developments and future prospects Nanobiotechnology. 16 (2018) 1–33. https://doi.org/10.1186/s12951-018-0392-8

[5]S.A. Stewart, J. Dom, R.F. Donnelly, E. Larrañeta, Implantable Polymeric Drug Delivery Devices : Clinical Applications, Polymers (Basel). 10 (2018) 1–24. https://doi.org/10.3390/polym10121379

[6]A.S. Alsudais, W.M. Alshehri, A.M. Alrehaili, R.K. Albeladi, M. Khoshhal, A. Albelowi, R.S. Alzahrani, A. Alnabihi, B. Bashrahil, O. Alabbasi, The Efficacy and Safety of Dexamethasone Intracanalicular Insert Use in Patients with Chronic Seasonal / Perennial Allergic Conjunctivitis : A Systematic Review and Meta-Analysis, Clin. Ophthalmol. 18 (2024) 2657–2666.

[7]F.B. Coulter, J.A. Faber, P. Zilla, D. Bezuidenhout, R. Studart, F.B. Coulter, M. Schaffner, J.A. Faber, A. Rafsanjani, R. Smith, H. Appa, P. Zilla, D. Bezuidenhout, Bioinspired Heart Valve Prosthesis Made by Silicone Additive Manufacturing Bioinspired Heart Valve Prosthesis Made by Silicone Additive Manufacturing, Matter. 1 (2019) 266–279. https://doi.org/10.1016/j.matt.2019.05.013

[8]C. Kuscu, C. Kuscu, A. Bajwa, J.D. Eason, D. Maluf, R. Mas, Applications of CRISPR Technologies in Transplantation, Ameriacan J Ournal Transpl. 20 (2021) 3285–3293. https://doi.org/10.1111/ajt.16095.Applications

[9]D. Lambade, P. Lambade, S. Gundawar, Implant Supported Mandibular Overdenture : A Viable Treatment Option for Edentulous Mandible, J. Clin. Diagnostic Res. 8 (2014) 6–8. https://doi.org/10.7860/JCDR/2014/7711.4332

[10] Z. Li, H. Mu, S.W. Larsen, H. Jensen, Ø. Jesper, An in vitro gel-based system for characterizing and predicting the long-term performance of PLGA in situ forming implants, Int. J. Pharm. 609 (2021) 121183–121198. https://doi.org/10.1016/j.ijpharm.2021.121183

[11] R. Kharwade, N. Ali, P. Gangane, K. Pawar, S. More, of Tioconazole-Loaded Transferosomal Hydrogel for the Effective Treatment of Atopic Dermatitis : In Vitro and, Gels. 9 (2023) 1–20.

[12] M.F. Bayan, A. Jaradat, M.H. Alyami, Smart Pellets for Controlled Delivery of 5-Fluorouracil, Molecules. 28 (2023) 1–10.

[13] X. Ju, R. Xie, L. Yang, L. Chu, Biodegradable ‘ intelligent ’ materials in response to chemical stimuli for biomedical applications, Expert Opin. 19 (2009) 683–696.

[14] A.K. Dash, G.C.C. Ii, Therapeutic Applications of Implantable Drug Delivery Systems, J. Pharmacol. Toxicol. Methods. 12 (1999) 1–12.

[15] E. Magill, S. Demartis, E. Gavini, A. Dian, R. Raj, S. Thakur, M. Faris, D. Waite, K. Glover, C.J. Picco, A. Korelidou, U. Detamornrat, L.K. Vora, L. Li, Q. Kurnia, R.F. Donnelly, J. Domínguez-robles, E. Larrañeta, Solid implantable devices for sustained drug delivery, Adv. Drug Deliv. Rev. 199 (2023) 114950–114967. https://doi.org/10.1016/j.addr.2023.114950

[16] G. Haoyan, Zhou., Christopher, Hernandez., Monika, Goss., Anna, Biomedical Imaging in Implantable Drug Delivery Systems, Curr. Drug Target. 16 (2016) 672–682.

[17] A. Gopferich, Bioerodible implants with programmable drug release, J. Control. Release. 44 (1997) 271–281.

[18] C. Dharmayanti, T.A. Gillam, D.B. Williams, A. Blencowe, Drug-Eluting Biodegradable Implants for the Sustained Release of Bis phosphonates, Polymers (Basel). 12 (2020) 1–14.

[19] L. Gao, Q. Li, J. Zhang, Y. Huang, L. Deng, C. Li, G. Tai, B. Ruan, Local penetration of doxorubicin via intrahepatic implantation of PLGA based doxorubicin-loaded implants, Drug Deliv. 26 (2019) 1049–1057. https://doi.org/10.1080/10717544.2019.1676842

[20] M. Aqil, A. Ali, Monolithic matrix type transdermal drug delivery systems of pinacidil monohydrate : in vitro characterisation, Eur. J. Pharm. Biopharm. 54 (2002) 161–164.

[21] V.P. Chavda, G. Jogi, A.C. Paiva-santos, A. Kaushik, Biodegradable and removable implants for controlled drug delivery and release application, Expert Opin. Drug Deliv. 19 (2022) 1177–1182. https://doi.org/10.1080/17425247.2022.2110065

[22] V.S. Chudinov, I.N. Shardakov, V. V Litvinov, S.Y. Solodnikov, E.Y. Chudinova, I. V Kondyurina, A. V Kondyurin, Foreign Body Reaction to Ion-Beam-Treated Polyurethane Implant, Materials (Basel). 17 (2024) 1–23.

[23] M.M. Isley, A. Edelman, Contraceptive Implants : An Overview and Update, 34 (2007) 73–90. https://doi.org/10.1016/j.ogc.2007.01.002

[24] A.S. Komlev, R.R. Gimaev, V.I. Zverev, Smart magnetocaloric coatings for implants : Controlled drug release for targeted delivery, Phys. Open. 7 (2021) 100063. https://doi.org/10.1016/j.physo.2021.100063

[25] Y. Almoshari, Osmotic Pump Drug Delivery Systems — A Comprehensive Review, Pharmaceuticals. 15 (2022) 1–16.

[26] R.A. Keraliya, C. Patel, P. Patel, V. Keraliya, T.G. Soni, R.C. Patel, M.M. Patel, Osmotic Drug Delivery System as a Part of Modified Release Dosage Form, ISRN Pharm. 2012 (2012) 1–12. https://doi.org/10.5402/2012/528079

[27] F.P. Pons-faudoa, A. Ballerini, J. Sakamoto, H. Sciences, T. De Monterrey, A.E. Garza, H.M. Hospital, H.M. Hospital, Advanced implantable drug delivery technologies: transforming the clinical landscape of therapeutics for chronic diseases, Biomed Micro. 21 (2020) 47–89. https://doi.org/10.1007/s10544-019-0389-6.Advanced

[28] D.R. Paul, Elaborations on the Higuchi model for drug delivery, Int. J. Pharm. 418 (2011) 13–17. https://doi.org/10.1016/j.ijpharm.2010.10.037

[29] R. Kharwade, N. Mahajan, S. More, A. Warokar, A. Dhobley, D. Palve, Effect of PEGylation on drug uptake , biodistribution , and tissue toxicity of efavirenz – ritonavir loaded PAMAM G4 dendrimers, Pharm. Dev. Technol. 24 (2023) 1–19. https://doi.org/10.1080/10837450.2023.2173230

[30] S. Herrlich, S. Spieth, S. Messner, R. Zengerle, Osmotic micropumps for drug delivery, Adv. Drug Deliv. Rev. 64 (2012) 1617–1627. https://doi.org/10.1016/j.addr.2012.02.003

[31] M.U. Ghori, B.R. Conway, Hydrophilic Matrices for Oral Control Drug Delivery, Am. J. Pharmacol. Sci. 3 (2015) 103–109. https://doi.org/10.12691/ajps-3-5-1

[32] Y. Fu, W.J. Kao, Drug release kinetics and transport mechanisms of non-degradable and degradable polymeric delivery systems, Expert Opin. Drug Deliv. 7 (2010) 10272-10286. https://doi.org/10.1517/17425241003602259

[33] M. Shevtsov, D. Gavrilov, N. Yudintceva, E. Zemtsova, A. Arbenin, V. Smirnov, I. Voronkina, P. Adamova, M. Blinova, N. Mikhailova, O. Galibin, M. Akkaoui, M. Pitkin, Protecting the skin-implant interface with transcutaneous silver-coated skin-and-bone-integrated pylon in pig and rabbit dorsum models, J. Biomed Matter Res. 109 (2022) 1–21. https://doi.org/10.1002/jbm.b.34725.Protecting

[34] S.E. Herold, A.J. Kyser, M.G. Orr, M.Y. Mahmoud, W.G. Lewis, A.L. Lewis, J.M. Steinbach-rankins, H.B. Frieboes, Biomedical Engineering Advances Release kinetics of metronidazole from 3D printed silicone scaffolds for sustained application to the female reproductive tract, Biomed. Eng. Adv. 5 (2023) 100078. https://doi.org/10.1016/j.bea.2023.100078

[35] J.S. Boateng, K.H. Matthews, H.N.E. Stevens, G.M. Eccleston, Wound healing dressings and drug delivery systems: A review, J. Pharm. Sci. 97 (2008) 2892–2923. https://doi.org/10.1002/jps.21210

[36] R. Rajan, D. Vasudevan, V. Biju Mukund, S. Jose, Transferosomes - A vesicular transdermal delivery system for enhanced drug permeation, J. Adv. Pharm. Technol. Res. 2 (2011) 138. https://doi.org/10.4103/2231-4040.85524

[37] H. Kranz, R. Bodmeier, A novel in situ forming drug delivery system for controlled parenteral drug delivery, 332 (2007) 107–114. https://doi.org/10.1016/j.ijpharm.2006.09.033

[38] T. Jacques, C. Brienne, S. Henry, H. Baffet, G. Giraudet, X. Demondion, A. Cotten, Minimally invasive removal of deep contraceptive implants under continuous ultrasound guidance is effective , quick , and safe, Eur. Radiol. (2022) 1718–1725. https://doi.org/10.1007/s00330-021-08263-4

[39] T. Goldberg, B.M. Maoz, Current state of the art and future directions for implantable sensors in medical technology : Clinical needs and engineering challenges, 031506 (2023) 1–29. https://doi.org/10.1063/5.0152290

[40] L.S. Nair, C.T. Laurencin, Biodegradable polymers as biomaterials, Prog. Polym. Sci. 32 (2007) 762–798. https://doi.org/10.1016/j.progpolymsci.2007.05.017

[41] M. Abdelgawad, M.A. Elkodous, W.M.A. El Rouby, Biodegradable Polymers in Biomedical Applications: A Focus on Skin and Bone Regeneration, Handbook of Biodegradable Material,(2023)1015-1043. https://doi.org/10.1007/978-3-031-09710-2_45

[42] F. Asghari, M. Samiei, K. Adibkia, A. Akbarzadeh, S. Davaran, Biodegradable and biocompatible polymers for tissue engineering application: a review, Artif. Cells, Nanomedicine Biotechnol. 45 (2017) 185–192. https://doi.org/10.3109/21691401.2016.1146731

[43] F.F. Castrow, E.A. Krull, Injectable collagen implant—update, J. Am. Acad. Dermatol. 9 (1983) 889–893. https://doi.org/10.1016/S0190-9622(83)70204-5

[44] B. Liu, Z. Xu, R. Yu, J. Wang, Z. Wang, C.R. Harrell, The Use of Type I and Type III Injectable Human Collagen for Dermal Fill: 10 Years of Clinical Experience in China, Semin. Plast. Surg. 19 (2005) 241–250. https://doi.org/10.1055/s-2005-919719

[45] W. Friess, Collagen - Biomaterial for drug delivery, Eur. J. Pharm. Biopharm. 45 (1998) 113–136. https://doi.org/10.1016/S0939-6411(98)00017-4

[46] M. Osorio, E. Martinez, T. Naranjo, C. Castro, Recent advances in polymer nanomaterials for drug delivery of adjuvants in colorectal cancer treatment: A scientific-technological analysis and review, Molecules. 25 (2020) 256-269. https://doi.org/10.3390/molecules25102270

[47] M.M. Islam, R. Ravichandran, D. Olsen, M.K. Ljunggren, P. Fagerholm, C.J. Lee, M. Griffith, J. Phopase, Self-assembled collagen-like-peptide implants as alternatives to human donor corneal transplantation, RSC Adv. 6 (2016) 55745–55749. https://doi.org/10.1039/c6ra08895c

[48] D.S. Rootman, J.Y.F. Ku, S.N. Yeung, Ocular Surface Disease: Surgical Management, Elsevier Inc., 2013. https://doi.org/10.1016/B978-1-4557-2876-3.00036-5

[49] N.A.Z. Abidin, F. Kormin, N.A.Z. Abidin, N.A.F.M. Anuar, M.F.A. Bakar, The potential of insects as alternative sources of chitin: An overview on the chemical method of extraction from various sources, Int. J. Mol. Sci. 21 (2020) 1–25. https://doi.org/10.3390/ijms21144978

[50] S. Park, H. Kim, K.S. Choi, M.K. Ji, S. Kim, Y. Gwon, C. Park, J. Kim, H.P. Lim, Graphene-chitosan hybrid dental implants with enhanced antibacterial and cell-proliferation properties, Appl. Sci. 10 (2020) 1–10. https://doi.org/10.3390/app10144888

 Materials Research Forum LLC
https://doi.org/10.21741/9781644903353-13

[51] L. Polo-Corrales, M. Latorre-Esteves, J.E. Ramirez-Vick, Scaffold design for bone regeneration, J. Nanosci. Nanotechnol. 14 (2014) 15–56. https://doi.org/10.1166/jnn.2014.9127

[52] F. Lopez-Moya, M. Suarez-Fernandez, L.V. Lopez-Llorca, Molecular mechanisms of chitosan interactions with fungi and plants, Int. J. Mol. Sci. 20 (2019) 1–23. https://doi.org/10.3390/ijms20020332

[53] R. Song, M. Murphy, C. Li, K. Ting, C. Soo, Z. Zheng, Current development of biodegradable polymeric materials for biomedical applications, Drug Des. Devel. Ther. 12 (2018) 3117–3145. https://doi.org/10.2147/DDDT.S165440

[54] H. Knopf-Marques, J. Barthes, S. Lachaal, A. Mutschler, C. Muller, F. Dufour, M. Rabineau, E.J. Courtial, J. Bystroňová, C. Marquette, P. Lavalle, N.E. Vrana, Multifunctional polymeric implant coatings based on gelatin, hyaluronic acid derivative and chain length-controlled poly(arginine), Mater. Sci. Eng. C. 104 (2019) 109898. https://doi.org/10.1016/j.msec.2019.109898

[55] M.H. Shariare, M. Rahman, S.R. Lubna, R.S. Roy, J. Abedin, A.L. Marzan, M.A. Altamimi, S.R. Ahamad, Liposomal drug delivery of Aphanamixis polystachya leaf extracts and its neurobehavioral activity in mice model, Sci. Rep. 9(2020) 1–16. https://doi.org/10.1038/s41598-020-63894-9

[56] C. Li, C. Guo, V. Fitzpatrick, A. Ibrahim, M.J. Zwierstra, P. Hanna, A. Lechtig, A. Nazarian, S.J. Lin, D.L. Kaplan, Design of biodegradable, implantable devices towards clinical translation, Nat. Rev. Mater. 5 (2020) 61–81. https://doi.org/10.1038/s41578-019-0150-z

[57] R. Makita, T. Akasaka, S. Tamagawa, Y. Yoshida, S. Miyata, H. Miyaji, T. Sugaya, Preparation of micro/nanopatterned gelatins crosslinked with genipin for biocompatible dental implants, Beilstein J. Nanotechnol. 9 (2018) 1735–1754. https://doi.org/10.3762/bjnano.9.165

[58] W.F. de Oliveira, P.B. Sales Albuquerque, N.E. Ribeiro Rodrigues, P.M. dos Santos Silva, J.F. Kennedy, M.T. dos Santos Correia, L.C. Breitenbach Barroso Coelho, Pharmaceutical applications of chitosan on medical implants: A viable alternative for construction of new biomaterials?, Carbohydr. Polym. Technol. Appl. 7 (2024) 100407. https://doi.org/10.1016/j.carpta.2023.100407

[59] E. Wu, L. Huang, Y. Shen, Z. Wei, Y. Li, J. Wang, Z. Chen, Application of gelatin-based composites in bone tissue engineering, Heliyon. 10 (2024) e36258. https://doi.org/10.1016/j.heliyon.2024.e36258

[60] E. Parsa, S. Saroukhani, F. Majlessi, H. Poorhosseini, M. Lofti-Tokaldany, A. Jalali, M. Salarifar, E. Nematipour, M. Alidoosti, H. Aghajani, A. Amirzadegan, S.E. Kassaian, Biodegradable-polymer biolimus-eluting stents versus durable-polymer everolimus-eluting stents at one-year follow-up: A registry-based cohort study, Texas Hear. Inst. J. 43 (2016) 126–130. https://doi.org/10.14503/THIJ-14-4997

[61] S.A. Stewart, J. Dom, V.J. Mcilorum, Z. Gonzalez, E. Utomo, E. Mancuso, D.A. Lamprou, R.F. Donnelly, E. Larran, Poly (caprolactone) -Based Coatings on 3D-Printed Biodegradable Implants : A Novel Strategy to Prolong Delivery of Hydrophilic Drugs, Mol. Pharm. 17 (2020) 3487–3500. https://doi.org/10.1021/acs.molpharmaceut.0c00515

[62] D. R. K. KULKARNI; K. C. PANI; C. NEUMAN, BS; AND F. LEONARD, WASHINGTON, LACTIC, Polylactic Acid for Surgical Implants, Arch Surg. 93 (1996) 840–852.

[63] D. Silva, M. Kaduri, M. Poley, O. Adir, N. Krinsky, Biocompatibility , biodegradation and excretion of polylactic acid (PLA) in medical implants and theranostic systems, Chem. Eng. J. 340 (2020) 9–14. https://doi.org/10.1016/j.cej.2018.01.010.Biocompatibility

[64] J.M. Brady, D.E. Cutright, A. Robert, G.I.N.C. Battistone, D. Biophysics, Resorption Rate , Route of Elimination , and Ultrastructure of the Implant Site of Polylactic Acid in the Abdominal Wall of the Rat *, J. Biomed Matter Res. 7 (1973) 155–166.

[65] V.N. Germanova, E. V. Karlova, L.T. Volova, A. V. Zolotarev, V. V. Rossinskaya, I.D. Zakharov, A.R. Korigodskiy, V. V. Boltovskaya, I.F. Nefedova, M. V. Radaykina, PLA-PEG Implant as a Drug Delivery System in Glaucoma Surgery: Experimental Study, Polymers (Basel). 14 (2022) 1–23. https://doi.org/10.3390/polym14163419

[66] C.C. Lin, S.J. Fu, Y.C. Lin, I.K. Yang, Y. Gu, Chitosan-coated electrospun PLA fibers for rapid mineralization of calcium phosphate, Int. J. Biol. Macromol. 68 (2014) 39–47. https://doi.org/10.1016/j.ijbiomac.2014.04.039

[67] E. Göktürk, H. Erdal, Poliglikolik Asit' in (PGA) Biyomedikal Uygulamaları, Sak. Univ. J. Sci. 169 (2017) 1–1. https://doi.org/10.16984/saufenbilder.283156

[68] N. Ashammakhi, P. Rokkanen, Absorbable polyglycolide devices in trauma and bone surgery, Biomaterials. 18 (1997) 3–9. https://doi.org/10.1016/S0142-9612(96)00107-X

[69] K.A. Athanasiou, D. Ph, J.P. Schmitz, D. Ph, Polylactic Acid-Polyglycolic Acid Implants Used in Repair of Articular Cartilage, Tissue Eng. 4 (1998) 53–63.

[70] S. Santavirta, E.S.A. Yrj, T.P. Saito, R.M. Gr, Immune Response to Polyglycolic acid implants, J Bone Jt. Surg. 72 (1990) 597–600.

[71] M.A. Naser, A.M. Sayed, W. Abdelmoez, M. Tarek, E. Wakad, M.S. Abdo, Biodegradable suture development - based albumin composites for tissue engineering applications, Sci. Rep. 14 (2024) 7912–7916. https://doi.org/10.1038/s41598-024-58194-5

[72] K. Avgoustakis, Polylactic-Co-Glycolic Acid (PLGA), Encycl. Biomater. Biomed. Eng. 4 (2008) 2259–2269. https://doi.org/10.1201/b18990-216

[73] P. Gentile, V. Chiono, I. Carmagnola, P. V. Hatton, An overview of poly(lactic-co-glycolic) Acid (PLGA)-based biomaterials for bone tissue engineering, Int. J. Mol. Sci. 15 (2014) 3640–3659. https://doi.org/10.3390/ijms15033640

[74] H.K. Makadia, S.J. Siegel, Poly Lactic-co-Glycolic Acid (PLGA) as biodegradable controlled drug delivery carrier, Polymers (Basel). 3 (2011) 1377–1397. https://doi.org/10.3390/polym3031377

[75] C.J. Destache, T. Belgum, M. Goede, A. Shibata, M.A. Belshan, Antiretroviral release from poly(DL-lactide-co-glycolide) nanoparticles in mice, J. Antimicrob. Chemother. 65 (2010) 2183–2187. https://doi.org/10.1093/jac/dkq318

[76] R.E. Eliaz, J. Kost, Characterization of a polymeric PLGA-injectable implant delivery system for the controlled release of proteins, J. Biomed. Mater. Res. 12 (1999) 1–12.

[77] M.H. S. Wachowiak, F. Danede, J.F. Willart, F. Siepmann, J. Siepmann, PLGA implants for controlled dexamethasone delivery: Impact of the polymer chemistry S., J. Drug Deliv. Sci. Technol. 86 (2023) 1–31.

[78] H.O. Alsaab, F.D. Alharbi, A.S. Alhibs, N.B. Alanazi, B.Y. Alshehri, PLGA-Based Nanomedicine : History of Advancement and Development in Clinical Applications of Multiple Diseases, Pharmaceutics. 14 (2022) 1–30.

[79] J.S. C.Bassand, J. Freitag, L. Benabed, J. Verin, F. Siepmann, PLGA implants for controlled drug release: Impact of the diameter, Eur. J. Pharm. Bio. 177 (2022) 1–39.

[80] S.R. Benhabbour, M. Kovarova, C. Jones, D.J. Copeland, R. Shrivastava, M.D. Swanson, C. Sykes, P.T. Ho, M.L. Cottrell, A. Sridharan, S.M. Fix, O. Thayer, J.M. Long, D.J. Hazuda, P.A. Dayton, R.J. Mumper, A.D.M. Kashuba, J. Victor Garcia, Ultra-long-acting tunable biodegradable and removable controlled release implants for drug delivery, Nat. Commun. 10 (2019) 1–12. https://doi.org/10.1038/s41467-019-12141-5

[81] O.O. Dosunmu, G.G. Chase, W. Kataphinan, D.H. Reneker, Electrospinning of polymer nanofibres from multiple jets on a porous tubular surface, Nanotechnology. 17 (2006) 1–14. https://doi.org/10.1088/0957-4484/17/4/046

[82] M. Doostmohammadi, H. Forootanfar, S. Ramakrishna, Regenerative medicine and drug delivery: Progress via electrospun biomaterials, Mater. Sci. Eng. C. 109 (2020) 1–22. https://doi.org/10.1016/j.msec.2019.110521

[83] J.-W. Yoo, E. Chambers, S. Mitragotri, Factors that Control the Circulation Time of Nanoparticles in Blood: Challenges, Solutions and Future Prospects, Curr. Pharm. Des. 16 (2010) 2298–2307. https://doi.org/10.2174/138161210791920496

[84] Y. Yang, H. Wu, Q. Fu, X. Xie, Y. Song, M. Xu, J. Li, Materials & Design 3D-printed polycaprolactone-chitosan based drug delivery implants for personalized administration, Mater. Des. 214 (2022) 110394. https://doi.org/10.1016/j.matdes.2022.110394

[85] G. Hatti, Polymers used as implant Biomaterials : A review, ACS Biomater. Sci. Eng. 4 (2019) 454–472.

[86] A.J.T. Teo, A. Mishra, I. Park, Y.J. Kim, W.T. Park, Y.J. Yoon, Polymeric Biomaterials for Medical Implants and Devices, ACS Biomater. Sci. Eng. 2 (2016) 454–472. https://doi.org/10.1021/acsbiomaterials.5b00429

[87] N.C. Paxton, M.C. Allenby, P.M. Lewis, M.A. Woodruff, Biomedical applications of polyethylene, Eur. Polym. J. 118 (2019) 412–428. https://doi.org/10.1016/j.eurpolymj.2019.05.037

[88] J.L. Frodel M, S. Lee, The use of high-density polyethylene implants in facial deformities, Arch. Otolaryngol. - Head Neck Surg. 124 (1998) 1219–1223. https://doi.org/10.1001/archotol.124.11.1219

[89] R. Bhattacharya, K. Mukherjee, B. Pal, Polyethylene in Orthopedic Implants: Recent Trends and Limitations, Encycl. Mater. Plast. Polym. 1–4 (2022) 777–794. https://doi.org/10.1016/B978-0-12-820352-1.00121-8

[90] B. Ramachandran, S. Chakraborty, M. Dixit, V. Muthuvijayan, A comparative study of polyethylene terephthalate surface carboxylation techniques : Characterization , in vitro haemocompatibility and endothelialization, React. Funct. Polym. 122 (2018) 22–32. https://doi.org/10.1016/j.reactfunctpolym.2017.11.001

[91] H. Cho, Root causes of post consumer High-Density Polyethylene failing in new bottles, Resour. Conserv. Re. 209 (2015) 107776.

[92] A. Gopanna, K.P. Rajan, S.P. Thomas, M. Chavali, Polyethylene and polypropylene matrix composites for biomedical applications, Elsevier Inc., 2019. https://doi.org/10.1016/B978-0-12-816874-5.00006-2

[93] G. Sternschuss, D.R. Ostergard, H. Patel, Post-implantation alterations of polypropylene in the human, J. Urol. 188 (2012) 27–32. https://doi.org/10.1016/j.juro.2012.02.2559

[94] A.A. Minea, A Review on Electrical Conductivity of Nanoparticle-Enhanced Fluids, Nanomaterials. 9 (2019) 1592–1604.

[95] L.R.R.C.J.S. Tugçe Çaykara, Maria G. Sande Nuno Azoia, Exploring the potential of polyethylene terephthalate in the design of antibacterial surfaces Tugçe, Med. Microbiol. Immunol. 209 (2020) 363–372.

[96] H.A. Anis, Gene therapy in the era of nanotechnology/a review of current data, J. Cancer Prev. Curr. Res. 10 (2019) 1–2. https://doi.org/10.15406/jcpcr.2019.10.00380

[97] J. Wang, J. Li, L. Shen, R. Ling, Z. Xu, A. Zhao, Y. Leng, N. Huang, The biomedical properties of polyethylene terephthalate surface modified by silver ion implantation, Nucl. Instruments Methods Phys. Res. Sect. B Beam Interact. with Mater. Atoms. 257 (2007) 141–145. https://doi.org/10.1016/j.nimb.2006.12.137

[98] P.R. Selvaganapathy, Exploring the surface potential of Recycled Polyethylene Terepthalate Composite Supports on the Collagen Contamination Level, Polymers (Basel). 15 (2023) 75–105. http://www.sciencedirect.com/science/article/pii/B9780444521903000045

[99] A. Victor, J. Ribeiro, F. F. Araújo, Study of PDMS characterization and its applications in biomedicine: A review, J. Mech. Eng. Biomech. 4 (2019) 1–9. https://doi.org/10.24243/jmeb/4.1.163

[100] I. Miranda, A. Souza, P. Sousa, J. Ribeiro, E.M.S. Castanheira, R. Lima, G. Minas, Properties and applications of PDMS for biomedical engineering: A review, J. Funct. Biomater. 13 (2022) 1–17. https://doi.org/10.3390/jfb13010002

[101] S. Tayyaba, M.W. Ashraf, Z. Ahmad, N. Wang, M.J. Afzal, N. Afzulpurkar, Microchannels for Biomedical Application, Processes. 9 (2021) 57–69.

[102] J.J. Jia Z, Yang C, Rhein and PDMS functionalized C/C composites as prosthetic implants for bone repair application This, Mater. Res. Express. 12 (2017) 0–24.

[103] P. Xue, Q. Li, Y. Li, L. Sun, L. Zhang, Z. Xu, Y. Kang, Surface modification of poly(dimethylsiloxane) with polydopamine and hyaluronic acid to enhance hemocompatibility for potential applications in medical implants or devices, ACS Appl. Mater. Interfaces. 9 (2017) 33632–33644. https://doi.org/10.1021/acsami.7b10260

Applications of Polymers in Surgery II
Materials Research Foundations 172 (2025) 329-357
Materials Research Forum LLC
https://doi.org/10.21741/9781644903353-14

Chapter 14

Polymers in Retinal Prosthesis

Vipul D. Prajapati[1*], P.N. Vaishnavi[1], Princy Shrivastav[2]

[1]Department of Pharmaceutics, SSR College of Pharmacy (Permanently Affiliated to Savitribai Phule Pune University), Sayli-Silvassa Road, Sayli, Silvassa, Union Territory of Dadra Nagar Haveli and Daman Diu, India

[2]Department of Pharmaceutical Chemistry, SSR College of Pharmacy (Permanently Affiliated to Savitribai Phule Pune University), Sayli-Silvassa Road, Sayli, Silvassa, Union Territory of Dadra Nagar Haveli and Daman Diu, India

*vippra2000@yahoo.com

Abstract

Retinal prostheses or prosthesis, types of medical devices, are commonly designed to restore vision in individuals suffering from the retinitis pigmentosa (RP) and age-relating macular degeneration (AMD) types of retinal degenerative diseases. These devices interface with the remaining retinal cells to stimulate visual perception through electrical signals. Polymers play a major role in the formation of retinal prostheses due to their needful major characteristics such as biocompatibility, flexibility, and conductivity. Conductive polymers, in particular, are used to create electrodes that can effectively transmit electrical impulses to the retinal cells. Moreover, the flexibility of polymeric materials allows for the fabrication of devices that conform to the curved surface of the retina, reducing mechanical stress and improving patient comfort. Recent advancements in polymeric science have led to the formulation of novel materials with enhanced properties, such as improved electrical conductivity and stability, which are essential for the long-term performance of retinal prostheses. This chapter highlights the potential of polymers to improve the functionality and durability of retinal prostheses, paving the way for more effective treatments for blindness.

Keywords

Retinal Prosthesis, Polymers, Biomedical Devices, Conductive Polymers, Biocompatible Materials, Retinal Implants, Optical Hydrogels, Surface Modification, Retinal Stimulation, Biodegradable Polymers

Contents

Materials Research Foundations 172 (2025) 329-357
https://doi.org/10.21741/9781644903353-14

1. Introduction

One of the human body's most advanced sense organs, the eye is what gives us vision. Its distinguishing qualities, such as its specific anatomy and physiology, contribute to its unique role. The six muscles that hold the two human eyes in place and facilitate their movement are responsible for their location in the eye socket. The eyeball has an estimated total capacity of 6-7 ml and an anterior posterior diameter of 24 mm. Anatomically speaking, the eye's front (anterior) and backside (posterior) regions in general house the seventy-five fraction and sixty-six fraction respectively of the ocular architecture. The posterior region is mainly composed of the sclera tissue, choroid, retina, Bruch's membrane (BM), vitreous humour, optic nerves, and retina associated blood vessels. The retina generally occupies one-third of the internal surface of posterior region of eyeball having transparent vitreous body. The front chamber of the eyeball is composed of lacrimal fluid, cornea tissues, conjunctiva segment, iris layers, ciliary body, lens, and hydrophilic fluid. Every component is distinctively created and functionally integrated, allowing for the development of a vision [1]. Light is first transmitted and refracted by eye cells, and it is subsequently transformed into electrical signals by retinal cells. The major visual pathway in the brain receives these signals after they have been refined through a sequence of synaptic contacts. A conscious experience and the activation of visual-based reflexes are produced by the interaction of several brain subregions [2–4]. Healthy vision of the eye orbital is disturbed or loss as the sever risk of retina impairment or detachment (RD) which is common in aged people. If it is not treated properly then it will lead to loss of vision and finally blindness. Hence, eye blindness surgery is the primary approach of such treatment. Since last few years the predefined polymeric retinal implants, formulated from suitable generally recognized as safe biomaterial, have been used to regain the similar normal vision of retina in severe condition of RD. This leads to understand about the complex structure of retinal and the basic needs of varieties of polymers for fabrication of suitable retinal prosthesis (RPs) types of biomedical device.

1.1 Retinal complex structural features

The retina of the eye is generally made up of nerve tissues whose layers contain the sensing elements of light called photoreceptors. It is the region of the nervous system for sensing photons, made up of several neurological processes and associated cell types, each of which plays a unique role in dealing with light visionary data. Commonly six types of neuronal region cells make the internal layered architect of the eyeballs as the laminar structure. These include photoreceptor cells into rods and cones shape, other types of the cells as the horizontal, bipolar, amacrine, ganglion, and supporting, particularly Muller glia [5]. The neural retina actually lies adjacent to the retinal pigment epithelium layer (RPE) in the posterior region of the ocular ball. The innermost nuclear layer (INL) is composed of the nuclei of amacrine, bipolar, and rod photoreceptors, whereas its outer layer (ONL) is made up of the nuclei of horizontal and cone shape cellular components. The ganglion cell layer (GCL) is the deepest nuclear layer made up of glial and ganglion cells. Bipolar cellular components located in the INL receive light impulses from photoreceptors. The impulses are then sent to the retinal ganglion cells (RGCs) and eventually go via the optic nerve to the brain [6].

The connection of suitable messages between the inner as well as outer NL, and RGCs is mediated by the cells of horizontal and amacrine located in the INL and RGC bands. The retina consists of two synaptic layers. Photoreceptors connect synaptically to bipolar and horizontal cellular components in the outermost plexiform coat, often referred to as the first main synaptic layer. In the secondary synaptic or inner plexiform layer (IPL), the ganglion cell components form synaptic interactions with bipolar and amacrine cellular components. Glial cells seen all over the retina are called Müller cells. Through radial covering in connection to the whole dense structural features of the retina, they provide support. The rod and cone photoreceptor cells' survival and proper operation depend on the RPE, which is located next to the photoreceptors [7,8].

Materials Research Foundations 172 (2025) 329-357
https://doi.org/10.21741/9781644903353-14

Retinal disorders account for around half of all occurrences of visual impairment globally. Degenerative retinal disorders, which start in the outside segments of the retina and progress in a steady way, leaving the intact inner layers until later stages of the illness, include choroideremia, retinitis pigmententosa (RP), and age-relevant macular degeneration (AMD). According to histological study, approximately ninety-three percent of RGCs in AMD patients survive, however 70% of photoreceptors are destroyed. In the extramacular area of RP, all retinal layer's experience cell loss, whereas, the macular region exhibits more preservation in the INL and GCL [9,10]. The macular cell loss in the different genetic variations of RP is the same. The extramacular area preserves a larger amount of autosomal-dominant RP cells [11,12].

1.2 Retinal prosthesis

It is a type of biomedical device, also referred to as a bionic eye or retinal implant, is intended to help people who have lost their sight as a result of specific retinal disorders or conditions. The tissue area behind the eye is known as the retina portion, photosensitive, and is mainly responsible for translating light into neural signals that the brain processes to produce vision [13,14].

Vision is significantly damaged or lost when the photosensitive receptor cells (both rods and cone types) in the retina degenerate or become non-functional, as in the case of AMD and retinitis pigmentosa. Bypassing these injured photoreceptors, retinal prostheses in common directly activating the retina's remaining healthy cells—usually the ganglion cells—to produce visual precepts [15].

1.3 Types of retinal prostheses

They are depicted in **Fig. 1**, according to reported literatures, each with unique approaches to restoring vision by targeting different portions of the retina and the photographical pathway. Epiretinal type of it is implanted in general on the internal layer of the retina, specifically in the epiretinal space. These devices target the ganglion cells and inner retinal neurons, which are responsible for transmitting visual informative signals to the brain. The common well-known examples of such type are the Argus II system. In this device, the electrode arrays are entered on the retinal layer, and visual information is captured by the periphery cameras located on the glasses worn by the patients. The camera processes visual data and transmits it without support of any wires to the implant, allowing it to stimulate retinal neurons directly. Epiretinal prostheses are advantageous because they bypass the damaged photoreceptors and interact with retinal ganglion cells, enabling functional visual perception for patients with advanced retinal degeneration [16,17].

 Materials Research Forum LLC
https://doi.org/10.21741/9781644903353-14

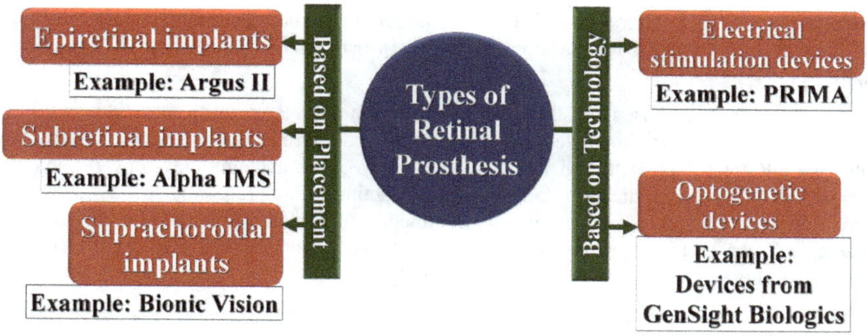

Figure 1. *Schematic illustration showing the various types of retinal prosthesis with examples based on reported literatures.*

In contrast, its subretinal type is implanted beneath the retina, between the RPE and the remaining photoreceptor cells. These prostheses in general replace the functions of the lost photo-sensitive receptors by directly stimulating the surviving bipolar cells. The Alpha-IMS type system is a prominent example of this approach. It consists of a light-sensitive microchip implanted in the subretinal space, which converts incident light into electronic power indications that activate the rest of cells of retinal region, simulating natural vision processing. The advantage of subretinal prostheses is their placement within the retina's natural light-processing layers, allowing all the patients to experience healthier and instinctive vision using the eye's optics [18].

Its third type, known as the suprachoroidal prosthesis, are located in the suprachoroidal area, situated in association between tissues of the sclera and the choroid. These prostheses are less invasive than epiretinal and subretinal devices, as they do not require extensive intraocular surgery. Suprachoroidal devices stimulate the outer layers of the retina via electrodes placed on the surface of the choroid. An example is the Bionic Eye developed by Bionic Vision of Australia. Although the visual resolution of suprachoroidal prostheses is generally lower, they offer a safer and less invasive alternative, especially for patients with severe retinal damage [19]. This positioning minimizes the risk of retinal detachment or infection and is particularly beneficial for patients who may not be candidates for more invasive retinal prostheses.

Intravitreal (IVT) prostheses, still in experimental stages, are implanted within the vitreous cavity and aim to stimulate the retina indirectly through the vitreous humour. These devices do not make direct contact with the retina, which simplifies the implantation process. However, challenges remain in efficiently stimulating the retina from this distance, which is why current research is focused on improving signal transmission and visual outcomes [20].

For patients with severe degeneration to the retina region and the optic nerves, cortical as well as optic nerve types of prostheses offer the alternative solutions. Optic nerve prostheses bypass the retina entirely, directly stimulating the optical nerves to transmit vision relevant information to the brain. Although still under development, these devices hold potential for restoring vision in patients with extensive retinal and optic nerve damage. Meanwhile, cortical prostheses, such as the Orion system, stimulate the visual cortex of the brain directly. These systems completely bypass the eye and optic nerve, offering a solution for patients with untreatable retinal or optic nerve degeneration [21]. Although cortical prostheses are still in the experimental phase, they hold significant promises for providing vision restoration in patients with widespread visual system damage.

Micro-photodiode array-based prostheses are another innovative approach, utilizing arrays of micro-photodiodes that are implanted either on or under the retina. These devices, such as the Photovoltaic Retinal implant 'PRIMA' system, are photovoltaic, meaning they convert photonic signal into electric power indications that stimulate the retina without requiring an external power source. The micro-photodiode array mimics the function of photoreceptors and provides a self-contained, efficient method of restoring vision. The lack of external power sources or cables makes these devices more user-friendly and less invasive [22].

Finally, recent advancements in flexible and wireless retinal prostheses aim to improve the biocompatibility and performance of these devices. Flexible electrode arrays conform to the curvature of the retina, providing a more stable and comfortable interface. Wireless power transmission reduces the need for external hardware and cumbersome cables. These devices, still in experimental stages, have the potential to extend the patient's overall quality of routine living life significantly by improving ease of use and reducing the risks associated with traditional implants [23].

1.3.1`Subretinal implant

Light-sensitive micro photodiode arrays make up this prosthetic retina. Since the array is located beneath the retina, it is more biologically positioned between the degenerating photonic sensitive receptor layer and the RPE layer [24]. The implant is positioned in the subretinal space via either an external region, namely scleral incision or an internal vitrectomy and retinotomy. A few thousands of tiny, photonic sensitive units placed in a silicon matrix make up the micro photodiode array. An amplifier, a microelectrode, and a diode comprise each unit [18].

Underneath the retina lies a light-sensitive micro photodiode array-based retinal prosthesis. The array is positioned between the deteriorated photonic sensitive receptor layer and the RPE in the subretinal region. The micro level photodiode array is made up of thousands of units, each with a microelectrode, amplifier, and diode inserted in a silicone matrix. The complete retinal neural network is connected to these titanium nitrite and gold microelectrodes, which are used to receive, process, amplify, and stimulate light signals. An induction coil in the retro-auricular area helps to facilitate the need for external electronics to deliver power to the implant, even while photovoltaic stimulation from natural light activates the array [25].

The physiological stimulation method of this subretinal implant and its simplicity of insertion in the sub macular area—which eliminates the requirement for an external camera or image processing unit for object localization—are among its benefits [26].

Electrodes replace absent or defective photoreceptors in a subretinal implant. In contrast to an epiretinal type of biomedical device, the electrical power activation that is generated in the innermost retinal neurons reaches the RGL via the retinal circuitry, thereby maintaining as much of a normal retinal signaling pathway as feasible. As a result, various peculiarities of human vision are maintained, such as the ability to detect edges and enhance contrast through centre-surround antagonism [17], nonlinearity and fine-scale heterogeneity of RGCs [27], and ON/OFF responses

Materials Research Foundations 172 (2025) 329-357 https://doi.org/10.21741/9781644903353-14

[28]. Adoption may be hampered by the fact that putting a device on the retinal layer is easier than the implantation or removal in the subretinal area [29].

1.3.2 Epiretinal implant

Subretinal implants differ from epiretinal implants. A micro camera on glasses is used in epiretinal implants to record visual data, which is then analysed by a movie recording processor and transformed into electrical beam signals that activate nerve fibres and ganglion cells. Since the inner retinal layers are bypassed by this direct stimulation, vision perception correction requires intricate electronic processing. Optic nerve diseases and conditions involving damaged ganglion cells are not appropriate for use with epiretinal prostheses such as Argus II. Alternatively, for such situations, research is being done on visual cortex stimulators. As of 2013 and 2011, respectively, the only FDA-approved and CE-marked epiretinal prosthesis is the Argus® II [30–32]. It is mainly used in Europe, France, Germany, USA, and Saudi region for recovering the visual performance of the patients who have AMD, and RP.

1.3.3 Argus® II epiretinal prosthesis

In 1990, this prosthesis was initially investigated. After feasibility tests of the sixteen-electrode Argus® version I was initiated in the year 2002, prospective multicentre studies employing the sixty-electrode Argus version II were initiated in year 2007. It is currently the most common useful photographic types of prosthesis in several countries because of its shown biocompatibility, reliability, and patient benefits [11]. Using an epiretinal technique, the visual microelectrode array for the Argus® II which was a man made artificial retinal implant, such system is placed into the innermost macular layer, adjacent to the nerve fibre segment. Following this, the direct stable and controlled regulated electrical activation of RGCs, bipolar cells, and/or the marginally functioning of the innermost retinal neural matrix occurs. The basic visual sense known as a phosphene is produced by these inner retinal cells when they generate an action potential that is then sent to the visual cortex in connection to the nerves and tracts of optic [13].

1.3.4 Surgical method

The eye undergoes scleral buckling preparation, just like in a pars plana vitrectomy. Once the silicone band, electronics case, and receiver coil have been inserted under the four rectus muscles, the ends are linked at the superanasal quadrant by a Watzke sleeve. The nanosized electronic casing is sutured to the superotemporal quadrant and the receiving coil is sutured to the inferotemporal quadrant under the lateral rectus using the pre-estimated limbus distances. This method was performed for a few patients as per the surgical process in which artificial retinal prosthesis of Argus® II exhibited positive results of vision. [19,33].

1.3.5 Suprachoroidal approach

The implant is in touch with the scleral wall beneath the choroid by using the suprachoroidal method. This technology may be less invasive and simpler to access for future repair of damaged vision or its replacement because it does not need intraocular surgery. There is a significant risk of bleeding due to the high vascularization of the suprachoroidal region. Unlike implants placed in other tissue microenvironments, the main objective of these implants is to promote declining peripheral vision [34–36].

2. Biomaterials in the human body

2.1 Introduction to biomaterials

A substance meant to interact with biological systems as a crucial component of a procedure intended to assess, observe, or manage bodily tissues, to add to or replace tissues, or to promote tissue regeneration [35]. The body's usage of biomaterials is conceptualized as repair, replacement, and regeneration [34,35]. Repair is restoration of nonfunctional continuity (natural process improved by biologics, technology, or active molecules in compliance with rules). Replacement involves substitutions of diseased or damaged retinal tissues or organs that is only partially functional and usually does not return the original structure (transplants, grafts, prostheses, artificial organs, etc.). Regeneration is the process of creating new tissue that is physically and functionally similar to injured or sick tissues or organs. As illustrated in the image, they can be divided into three groups according to how the tissue responds to the biomaterial. According to the reaction of the tissues with the specific functional biomaterial, in general they are classified into three possible categories as depicted in Fig. 2.

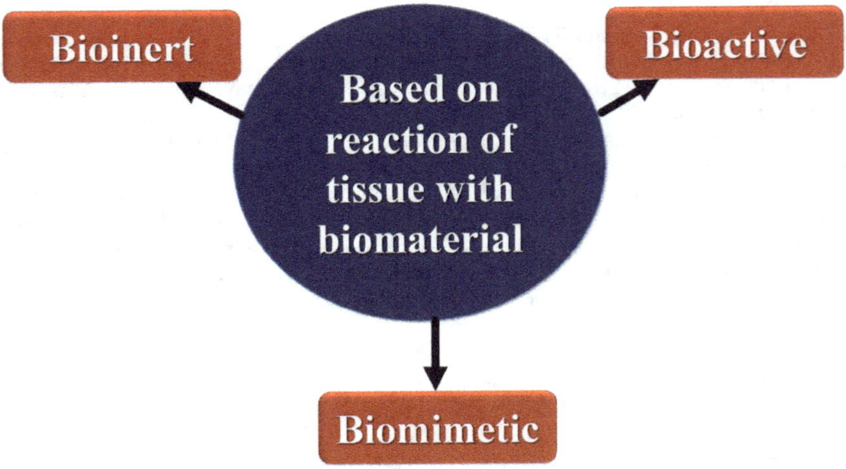

Figure 2. *Schematic image depicting the three categories of use of polymers in retinal prosthesis based on the reaction of the retinal tissues with the specific functional biomaterials.*

Chemical reactions do not occur when living tissue comes into touch with an implant. Biologically inert or naturally useful compatible polymeric substances are usually kept apart from thin bone tissues by a thin segment of fibrous tissues, while they may occasionally come into direct interaction with the surrounding tissues [37]. Biologically active components have the ability to form chemical connections with such types of tissues that are known as bioactive materials. The connection with bone tissue is called osseointegration, and it involves the direct deposition of mineral phase and collagen on the implanting layer [38]. Biomimetic substances are artificial or synthetic ones that imitate naturally available materials or that use a pattern of design inspired by nature. In order to create biomimetic materials, current regenerative medicine tactics combine

Materials Research Forum LLC
https://doi.org/10.21741/9781644903353-14

naturally origin polymeric components or peptides, which can control cellular response, with biocompatible synthetic polymers that provide mechanical and structural qualities [39].

Polymeric substances are used in a variety of usages as the biomaterials in the manufacturing of prosthetic compatible device with retinal components. Table 1 describes various uses of polymers in the realm of biomedicine.

Table 1. *Polymers used for the manufacturing of prosthetics and medical devices for various purpose.*

Category	Example of polymers	References
Utilization inside the human body	Polyamides (Nylon), polyimides, HDPE, UHMWPE, PTFE, PP, PMMA, hydrogels (e.g., PHEMA), biodegradable polyesters (PGA, PLA), polyesters (e.g., PET), polyurethanes, silicones, and PEEK	[40]
When coming into contact with human tissue	PVC, polysulfones, polyacrylonitrile (PAN), and LDPE	[41]
Utilize away from human contact	Polycarbonate (PC), copolymers (SAN, ABS), and polystyrene (PS, PST)	[42]

2.1 Biomaterials in ophthalmology

Biocompatibility, stability, degradation, sterilizability, and manufacturability are all factors that have to be taken into account, much like with materials used in other parts of the body. But particularly with materials intended for use in ophthalmic applications, optical property considerations need to be made. For materials used in contact lenses, corneal inlays or onlays, and intraocular lenses, optical qualities are very important. The materials need to have a refractive index that is at least as high as the tissue that needs to be replaced and transparent. UV-blocking chemicals are occasionally added to these materials to shield the vulnerable retina from UV light exposure [43]. Various types of polymeric materials are used to treat vision relevant function of retina. The types of polymers used for the flexible support and desired purpose in suitable types of retina relevant prosthetic implants. This can push up the working of electrical power and activate the conjugated circuits for vision in retina. For activation of photons of the prosthesis of retina, suitable organic semiconductors have been used. Various polymers for ophthalmic use are enlisted in table **2.**

Table 2. *Summary of various biomaterials used in fabrication of ophthalmic implants with specific use.*

Ophthalmic Implants	Various types of biomaterials in the fabrication of ocular implants.	Reference
CLs	PMMA, 2-hydroxyethyl methacrylate (HEMA) copolymers, silicone hydrogels	[44]
IOL	Optic: PMMA, silicone, hydrophobic and hydrophilic acrylics Haptic: PP, PMMA, polyimide, PVDF	[45]
Glaucoma shunts	Plates: silicone (impregnated with barium), PP Tubing: silicone	[46]
Inlays or onlays	Hydrogels, collagen, permeable membranes	[47]
Vitreous replacements	Silicone oil, gases	[48]
Ophthalmic viscosurgical device (OVD)	Chondroitin sulphate, sodium hyaluronic acid (NaHA), hyaluronic acid (HA), HPMC, polyacrylamide, collagen, or combinations of these materials	[49]

2.2 Biomaterials utilized in eye implants

2.2.1 Hard contact lenses (HCLs)

They are previously made of poly (methyl methacrylic acid) 'PMMA'. Nowadays they are fabricated from the copolymers of PMMA in conjugated with silicone as well as fluorine macromolecules. Methacryloxy propyl tris(trimethysiloxy) silane (TRIS) is a type of monomeric substance, potentially enhancing the $O_{2(g)}$ permeability of these polymeric substances. Silanes can enhance the $O_{2(g)}$ permeability, however because of their lipophilicity, there may be more biofouling. Although fluorinated polymers are more fouling resistant, they can also increase oxygen permeability. Finally, the majority of materials used in currently available RGP lenses are of the perfluoroalkyl-siloxyanylalkyl-methyl methacrylic acid type, which has a high diffusion and solubility coefficients and good surface qualities [50]. `

2.2.2 Soft Hydrogel Contact Lenses

The bulk of people can wear soft contact lenses because of their high level of comfort, which is a result of hydrogels' low modulus and hydrophilic content. Poly(2-hydroxymethylmethacrylate) 'PHEMA', the beginning polymeric substance used in fabrication of contact lenses (CLs), was created in 1959 by Wichterle and Lim. Its water content is about 40%, and its stability to pH, temperature, and osmolarity is comparable to that of biological circumstances, giving it advantages over previously created materials. Various different monomers have been added or utilized to these polymers to improve their water content. Particularly effective compounds include N-vinyl pyrrolidone (NVP), N, N-dimethyl acrylamide (DMA), vinyl acetate, post-hydrolysed to vinyl alcohol units, and methacrylic acid (MAA) [51].

2.2.3 Silicone hydrogel contact lenses

The CLs industry created siloxane-containing hydrogel materials to get around the drawbacks of traditional hydrogel materials without sacrificing comfort. Copolymerizing standard hydrogel monomers with innovative co-monomers, hydrophilic ally modified TRIS, or MAA or vinyl end capped polydimethylsiloxanes is how these lenses are made [52].

2.2.4 Intraocular lens (IOL) implants

In the year 1949, Sir Harold Ridley invented the tiny implant of IOL using the synthetic material 'Perspex' in general composed of a PMMA polymer. The early heavy PMMA lenses were hard to surgically fix inside the eye, necessitating a huge incision, and posing a significant risk of catastrophic complications following surgery [53]. Recent developments have resulted in modifications to the optic shape and haptics to ensure total contact of the PC-IOL to the posterior capsule, hence preventing Posterior capsule opacification (PCO) by reducing the production of secondary cataracts and epithelial migration. IOL has distinct materials for haptic and ocular components. PMMA, polypropylene (PP), polyimide, and polyvinylidene fluoride (PVDF) are examples of common haptic materials [54,55].

2.2.5 IVT implants

Due to the dynamics of the vitreous chamber and the limiting of the IVT solutions existing there, the IVT implants are designed in such a way to stay there and reduce the regularity of administration and providing long-term medical benefit. IVT implant, a type of synthetic non-erodible polymeric sustained release DDS, has an impermeable coating of ethylene vinyl acetate (EVA) that is discontinuous and from which the drug diffuses out and a semi-permeable PVA membrane. It provides continuous release for a minimum of three months, but because it is not biodegradable, a replacement surgical procedure is necessary [56–58].

2.2.6 Retisert®

The sterile IVT implant namely Retisert® is the first implant to be filled with corticosteroids. Its size is minuscule, consisting a tablet enclosed in a PVA membrane-positioned silicone elastomer cup that sits between the tablet and the release aperture.

2.2.7 Iluvien™

It is, formerly called Medidur approved by USFDA, a free floating non-erodible IVT implant mostly preferred for the therapy of diabetic macular oedema (DMO) and manufactured as well as commercialized by Alimera Sciences Company of USA. It is a polyimide cylinder-shaped tube [59]. Many biodegradable polymers are used for both the anterior and posterior chambers of DDS, as they are demonstrating their value in both contexts. Poly (lactic-co-glycolic acid) (PLGA) microspheres of triamcinolone acetonide (TA) are injected intravitreally as RETAAC to treat diabetic macular oedema (DMO). Such treatment exhibited a gradual reduction in the thickness of retina with improvement in visual performance after one year following the therapy. They're under investigation. They are shown to have no adverse effects in the first research, either from the medication or the operation itself [60].

2.2.8 PRINT® technology

Envisia Therapeutics' PRINT® Technology is a more recent method for creating intraocular implants, nanoparticles (NPs), and microparticles that can release bioactive or encapsulated drugs over an extended period of time. They consist of biodegradable materials such as Polycaprolactone

(PCL), and PLGA [61]. The biodegradable characteristics of intracameral implant namely Sorudex®, composed of PLGA and hydroxypropylmethyl cellulose (HPMC) with 60 µg dexamethasone (DEX), is intended to alleviate inflammation following cataract surgery [62].

2.2.9 Cell encapsulation technology (ECT)

ECT, AN69 copolymer (polyacrylonitrile-methyl sulfonate) microcapsules, measuring 1.5 mm in length, are mainly used to create the NT 501 and NT 503 implants. They are based on Cellular components Encapsulation Technology (ECT) and contain a genetically engineered cellular line that secretes soluble Vascular endothelial growth factor (VEGF) receptor and Ciliary Neurotrophic Factor (CNTF) [63].

2.2.10 I-vation® IVT implant

It is a titanium anchor mostly used to secure the created ECT device. It is generally composed of polyethersulfone (PES, N 99%), caprolactam as a plasticizer, and polyvinyl-pyrrolidone (PVP) as an anti-foulant, into the vitreous cavity. I-vation® is a titanium helical coil covered in 0.925 mg TA, EVA, and non-biodegradable polyamidoamine (PAMAM). The helical shape increases the surface area available for molecule's coating and enables suture-free implant attachment to the sclera [64].

2.2.11 Contact lens solutions

According to the type of lens, it is an essential requirement for efficient daily cleaning and disinfection of contact lenses for compatibility of eyes. From many functions of multipurpose cleaning solutions such as disinfection, biofouling prevention, the removal of lipid and protein deposits that have been adsorbed from the tear film is important. Majority of such systems employ aqueous soluble polymers to improve the hydration of the lens surface and the efficiency of cleaning. Some of the polymers used are Poloxamer 407 [65].

2.2.12 Intracellular iontophoresis

The scleral iontophoresis is an effective, well-tolerated, and less intrusive way to deliver medication through the sclera. It uses a little level of electric current to facilitate the drug's diffusion from the reservoir to the intended location—the retina or choroid. An electronic iontophoresis dose controller, a dispersive electrode, and a drug applicator make up the OcuPhor system. The active molecules applicator is a tiny silicone shell with a hydrogel pad to absorb the active molecule's preparation, a short flexible wires to link the conductive element to the dose controller, and a patented silver (Ag)-silver chloride (AgCl) ink conducting elements. Before using the applicator, the active molecules pad is moistened with its solution and applied to the sclera beneath the lowest eyelid. The Eyegate is made up of an annular siliconized probe with a 13 mm aperture to prevent contact with the cornea and 0.5 cm^2 of contact area. SonoEye is a novel ophthalmic gadget that is being studied. Australia's Centre for Ocular Research and Australia's Seagull Technology Pvt. Ltd. are developing it. The drug or bioactive is put into a gel and released upon the generation of an ultrasonic pulse. When tested with Avastin, it was discovered to be safe, harmless, and tolerant. The amount that is released from the gel is directly correlated with the application time and frequency, both of which are clearly higher than passive diffusion. Trans-scleral DDS is a non-invasive, successful treatment.

Materials Research Foundations 172 (2025) 329-357
https://doi.org/10.21741/9781644903353-14

3. Polymers for retinal use on the basis of obtained source

Various types of polymeric substances, classified based on their origins (naturally available, synthesized from natural resources (semi-synthetics), and synthetics), structural chain type features (linear pattern, branch or network type), paths of polymerization mechanism (types of addition, and condensation), and molecularity configurations (thermosetting, thermoplastic type, elastomeric type, and or fibres type), have been used in the fabrication of RPs. They are selected depending on the utility relevant criteria namely the behavior of polymers under processing steps 'heat and condensations', and various physicochemical characteristics. Thermal properties such as glass transition, and melting temperatures, processing heat deflection, thermal conductance of polymers during processing and utilization can be monitored under skilled supervisors. Their versatile properties can be considered in the fabrication of RPs as per the requirements of the patients recovering visual functionalities.

3.1 Natural polymers

They include polysaccharides like cellulose, chitin, and dextran, proteins like collagen, fibronectin, actin, and keratin, and polynucleotides like DNA and RNA. Their bioactive characteristics facilitate positive interactions with cells. Nevertheless, the lack of repeatable mechanical and physical qualities is a drawback for natural polymers [66,67]. Several naturally originated polymeric materials commonly referred as biomaterials, including compatible variety of albumins and varieties of hyaluronic acids, have been used to estimate the penetrability of the loaded nanomaterials generated by desolvation, nanoprecipitation, or adsorption into retinal tissue with the goal of improving the pharmacological effects of drugs. Researchers used a desolvation procedure to make negative and positive ionic human serum albumin (HAS) nanoparticles (NPs), which they then injected through intravitreal routes into the eye orbits to estimate their penetrating potentiality.

The negative charged HAS based NPs can be employed as a medication delivery carriers to the posterior portion of the ocular orbit since anionic particles diffused more readily in the eye than cationic particles did. In another work, to test for ocular toxicity, ganciclovir encapsulated albumin of originated from bovine serum (BSA), NPs were produced by coacervation using glutaraldehyde chemical cross-linker. The NPs were then injected using suitable gauze needles into the eye. Two weeks after injection, rhodopsin was identified as the desired photosensitive receptors involving in the phototransduction in retinal photoreceptors, and the particles stayed in the vitreous canula as a slimy film covering the retina. Such things didn't cause any inflammation, change the architecture of the tissue, or show any abnormalities in the development of arrestin, a protein required for visionary transduction.

3.2 Synthetic polymers

The main group of commonly used synthetic polymeric materials for the development of RPs are polyethylene (PE), polypropylene (PP), polyethylene terephthalate (PET), polystyrene (PS), polylactic acid (PLA), polycarbonate, polyvinyl chloride (PVC), PES, polyacrylic acids, polysulfones, polyethers ether ketone (PEK), elastomeric components, poly-para-xylylene (PPX), fluoropolymeric components, and varieties of silicones. Out of all these major polymeric substances, few polymerics substances that can be used in such developments are described here. When it comes to creating innovative nanoparticulate systems, these types of polymeric substances

Materials Research Forum LLC
https://doi.org/10.21741/9781644903353-14

have provided greater flexibility and versatility than natural polymers. Their physiochemical and mechanistical features are the same to those of natural originated polymers, but their elastic behavior, tensile strength, and adjustable decomposition rates are more repeatable. The United States of Food and Drug Administration (USFDA) has permitted the utility of these types of polymeric substances, including PLA, PCL, PLGA and polyglycolic acid (PGA) for utilization in several biorelevant medical usage, including tissues-engineering.

Because of their surface chemistry, variable degradation rates, adjustable physical qualities, and biocompatibility, biodegradable purely synthetic polymeric substances have been favourable features for desired usage in regenerative medicines. As a result, there has been much research done on the possibility of using different polymeric materials dependent on therapeutic as well as genetic molecules delivery systems for therapeutic approaches. The synthesis of PLGA barrier matrix and the cellular level generation of Class-III therapeutic devices also offers promise for reconstructing periodontal membranes, as evidenced by sufficient tissues creation in preclinical and clinical experiments. The biodegradable synthetic aliphatic polyester PLGA has demonstrated great promise as a drug delivery carrier among all the biomaterials. Lactide and glycolide are the two monomers that copolymerize to form the polymer known as PLGA.

The units of secondary-hydroxypropanoyl, as well as secondary-hydroxyacetyl make up the copolymer PLGA. The polymeric substance's identity and its various characteristics are estimated via its ratio of monomeric form to molecular weight, which may be adjusted to customize the characteristics of the final product. The duration of period required for maintaining the aqueous stability of the polymeric substance before breaking down in general increases with the proportionality of the lactide units. As the molecular weight of the polymeric substance grows, so does its mechanical strength. The copolymer has exceptional water solubility, is extremely soluble in water, and encapsulates molecules of almost any size. It can also be shaped and sized to any desired shape.

3.2.1 Polyacrylates

An example of a manufactured thermoplastic with an amorphous character is polyacrylates. Medical device acrylic resins are the esters of methacrylate, with methyl methacrylic acid serving as the most often utilized monomer. Acrylic acid, methacrylic acid, and cyanoacrylic acid, along with its esters, acrylonitrile, and acrylamide, serve as the foundation for acrylic resins. Methacrylic acid esters are a popular basis for acrylic resins used in medical devices. The most often used monomer is methyl methacrylic acid, which polymerizes into PMMA through free radical polymerization. Emulsion and suspension polymerization are the two processes used to polymerize methyl methacrylate.

Polymethyl methacrylate is a great material for fabrication of intraocular implants or lenses due to its remarkable optical clarity and UV transmittance, as well as its remarkable biocompatibility. It is also used in fabrication of retinal prosthesis due to its versatile compatible characteristics with electronic prosthetic to rescue vision disfunction.

The initial generation of polymeric substance contained contact lenses (CLs) were made of rigid, hard lenses made of polymethyl methacrylate. The lenses outperformed glass in terms of clarity, tough strength, and impermeable barrier to the entry of lachrymal fluid relevant film constituents because they were hydrophobic. For the purpose of enhancement of $O_{2(g)}$ permeability, stiffened gas permissible CLs were created during the year 1969 to 1970. Copolymeric substances of methyl

Materials Research Foundations 172 (2025) 329-357
https://doi.org/10.21741/9781644903353-14

methacrylic acid, namely 2,2,2-trifluoroethyl methacrylate, hexafluoroisopropyl methacrylate, and methacryloxypropyl tris (trimethylsiloxy silane), were used to attain high oxygen permeability.

3.2.2 Polyethylene

It is the most versatile polyolefin. It is composed of repeated ethylene units with similar molecular weights and has a simple structure that can be either linear or branched. Various varieties or grades of polyethylene (PE) is available such as low dense (LDPE), high dense (HDPE), linear-low dense (LLDPE), and ultra-high molecular weight (UHMWPE). All these grades of PE exhibit excellent resistance against stress crack, radiation, and having impact toughness. They have excellent corrosion-resistant property and hence, reducing the cost associated with its maintenance at home during the use by the patients. They have very little capacity of absorption of water as well as less permeability of water vapor, gases. They are chemically inert, non-toxic, biocompatible, and unsuitable for autoclaving due to low heat rebound temperature, but suitable for ethylene oxide and gamma radiation types of sterilization. HDPE and UHMWPE are the examples of linear forms, whereas LDPE and LLDPE are the branching forms. Metallocene is used as a catalyst in the production of UHMWPE, a linear PE with small, short chains along the polymeric component backbone. Compared to other polyethylene varieties, this polymer is extremely crystalline and has higher strength, stiffness, and durability. In hospitals and therapeutic environments, polyethylene grades are utilized because of their exceptional dielectric qualities and chemical resistance to various types of chemicals and disinfectants, including alcohols, alkalis, and diluted acids. UHMWPE has outstanding energy absorption, impact strength, and stress crack resistance. From the patient's point of view, take-home medical devices like prosthetics can have lower maintenance expenses due to polyethylene's resistance to corrosion. Single-use items, like syringes, that patients might need on a regular basis can also be stored at home in inexpensive polyethylene film wrapping.

HDPE is a polymer that has a density between 0.94 and 0.96 g/cm^3, is largely linear, and even has crystallinity. It is created using "stereospecific" catalysts (a technique known as Ziegler-Natta) at low pressures (between 1 and 10 atm) and temperatures (about 40±0.5°C). Compared to LDPE film, HDPE film is translucent and far less transparent (greater crystallinity reduces transparency). Compared to LDPE, it is less flexible and more stable at high temperatures. However, it can also experience photo-oxidation at ambient temperature and thermo-oxidation when exposed to oxygen.

UHMWPE is linear and has a molecular weight more than 2,000,000 Da. It is produced using a method akin to that of HDPE. UHMWPE, which is essentially a very high-level molecular weight HDPE, is often offered as powder, bars made by sintering or injection moulding, and sheets made by thermoforming. Despite having a largely linear structure, UHMWPE is less crystalline than other forms of PE variety (crystallinity ¼45%–60%) and has a comparatively lower density (0.935 g/cm^3). This is because the material's high fluid phase viscosity significantly restricts the typical crystallization methods. The material is ductile and tenacious because to these structural properties, but it is also more susceptible to creep and liquid absorption.

UHMWPE must adhere to strict guidelines set forth by standards that regulate its mechanical characteristics, and the amount of residue left over after combustion in order to be used in biomedical purposes. The main characteristics that make it suitable for use as a biomaterial are its

Materials Research Foundations 172 (2025) 329-357
https://doi.org/10.21741/9781644903353-14

low surface frictional coefficients, strong resistance to radiation, superior impact resistance compared to other plastic materials, and great resistance to abrasion.

3.2.3 Silicones

It is the main class of polymeric substances known as silicone that has an oxygen and silicon backbone as opposed to a carbon and hydrogen one. Due to their excellent characteristic with respect to development of prosthetic implants of various regions of eye especially surgical treatment approach, their use as flexible parent polymeric component for the use in ocular implant is increased. Groups that are joined to the prominent chain may be aromatic, fluorinated, or aliphatic. Silicones that are sold commercially are known as polydimethylsiloxanes because they have methyl pendant groups. Silicones are among the most significant thermoset elastomers. They are sometimes referred to as siloxanes, polyorganosiloxanes, or polysiloxanes. Not more than 3000 repeating monomer fluids, 3000–10,000 repeating monomer elastomers, somewhat crosslinking units, and cross-linked resins/adhesives are the different types of silicones. Acetoxy or chlorosilanes can be hydrolysed to create silicones. Acetoxy silanes are usually hydrolysed to yield acetic acid, which is then used to make medical grades.

Trimethyl acetoxy silane or trimethyl chlorosilane are examples of monofunctional chain transfer agents that can be used to modify the molecular weights of silicones. To create elastomers or resins, silicones with vinyl pendant functional groups can be modified via radiation or free radical initiators as the crosslinker.

They are hydrophobic, releasing no leachable or extractables, and they have minimal surface tension and water absorption. They also have exceptional electrical and dielectric qualities and exhibit outstanding resistance to chemicals, heat, and Ultra-Violet (UV) radiation. Silicone is used in second- and third-class types of biomedical devices, which include medication delivery systems, prosthetics, facial reconstruction, artificial skin, artificial organs, and catheters. Silicone fluids are used as surface modifiers and lubricants. Silicon dioxide is the most widely used dielectric material in bioresorbable electronics. Various factors relevant to polymeric substances physicochemical characteristics as well as other factors such as fabrication of prosthetic implant's similarity in shape, cost, development skill, use of prosthetic development equipment, accuracy, and repeatability are required to understand. Conjugation with all these factors, the robustness such as physical and chemical stability in all conditions, patient's wellbeing such as biocompatibility, comfort feel, microbial resistance is also required to monitor as the part of success of polymeric retinal prosthetic implants. During the use, its actual compatible comfort placement is also required to check with respect to time.

4. Polymers in nanoengineering of retinal tissue

Tissue at nanoengineering in the region of retina is a diversified field that uses varieties of functional bio-based materials, engineering, chemistry, and life care sciences to develop innovative therapeutic facts for retinal relevant rescue and its repairing. Biomaterials have been the subject of much research over the last few decades in an effort to improve treatment approaches. They have also significantly advanced retinal tissue engineering by serving as a means of delivering novel modes of therapeutic compound delivery as well as structural support for transplanted cells. Numerous biomaterials, such as peptides, inorganic materials, synthetic and naturally available polymeric substances, and many others, have been utilized for brain

Materials Research Foundations 172 (2025) 329-357 https://doi.org/10.21741/9781644903353-14

regeneration as a result of technological breakthroughs. These materials can be utilized alone or in conjunction with pharmaceuticals or cells to create nanoengineered platforms for gene/drug delivery systems intended to target brain tissue engineering or for cell-based treatments.

NPs are defined as particulates with less than 100 nm in size. The features of NPs are diversified from their original parent material due to high surface area to volume ratio. Particle size, shape, and surface properties all affect the internalization and biodistribution routes that ultimately dictate the intracellular fate of nanoparticles. The capacity of nanoparticulate systems to traverse the blood-retina barrier facilitates continuous therapy or the release of loaded suitable anti-inflammatory as well as anti-vascular molecules, which is one of the key advantages of employing them.

Various types of polymeric material as well as the GRAS approved commercially available retinal prosthetic device in which such polymeric materials used for suitable functionality for rescue of visual functionalities of damaged retina of persons is summarized in fig. 3.

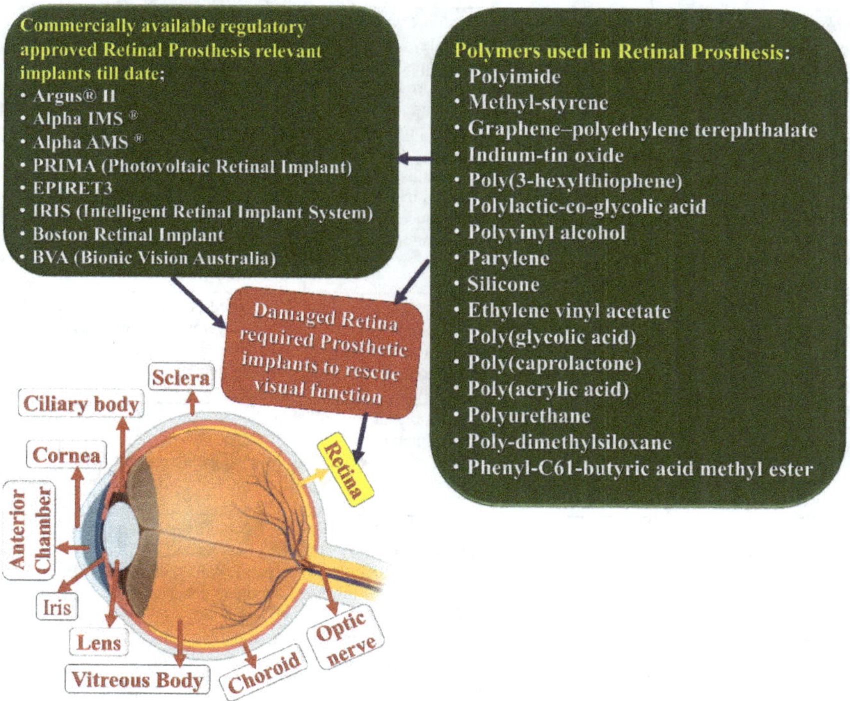

Figure 3. *An illustration depicting the types of polymeric materials and its use in GRAS approved commercial prosthetic formulation for retinal associated visual disfunction in blind patients.*

Table 3 depicts the features of various types of biomaterials or polymers along with purpose and fabrication methods for retinal nanoengineering.

Materials Research Foundations 172 (2025) 329-357 https://doi.org/10.21741/9781644903353-14

Table 3. *Summary of various biomaterials utilized for suitable purpose for retinal nanoengineering.*

Biomaterials name	Synthetic/Natural Variety	Structural features/ Fabrication lavel	Purpose/ Application	Results	Ref.
PLGA	Synthetic variety	Method of double emulsion solvent evaporation using functionalized NPs.	Use nanoparticles as a medication delivery method to treat AMD.	anti-VEGF intraceptor plasmids were successfully administered via nanoparticles, which improved gene expression in different retinal cells.	[68]
PLGA		NPs— solvent displacement.	Ocular DD device for the anti-inflammatory medication of flurbiprofen.	PLGA nanoparticles loaded with flurbiprofen enhanced the medication's effectiveness and offered steady, regulated drug administration.	[69]
PLGA		Rhodamine-6 loaded NPs—nanoprecipitation.	Examine variables that may affect RPE cells' uptake of nanoparticles intended for external retinal problems.	It is a promising vehicle for medication administration and gene transfer because it is non-toxic and rapidly internalizes, allowing for suitable genetic transfer.	[70]
Polyamidoamine/ PLGA		Hydrogel/PLGA hybrid polyamidoamine dendrimer—photopolymerization and solvent evaporation.	Delivery of brimonidine and timolol maleate, two antiglaucoma medications.	When administered, the platform improved the medication's bioavailability and bioactivity, indicating tremendous promise for a long-lasting DD method.	[71]
PLA		PEG-coated nanospheres with or without nanoprecipitation.	Acyclovir is used as an ophthalmic DD system to treat viral infections of the eyes.	PEG coated PLA nanospheres release acyclovir continuously and are far more ocular-tolerable than plain PLA nanospheres.	[72]
PCL		Nanofibers (NFr)—electrospinning (ESN).	Direct axonal development in disorders of the optic nerve.	Matched human pluripotent-derived RGCs in vitro, showcasing their capacity to direct the formation of axons.	[73]
PLCL		NFr-ESN.	Development and multiplication of human ESC-RPE cells.	PLCL nanofiber-based porous PLCL membranes facilitated human ESC-RPE cellular component adhesion, growth, and maturation in vitro.	[74]
PVA and PCL		ESN of Drug-loaded NFs.	Glaucoma medication administration.	In a rabbit with glaucoma, intraocular pressures were lower in the presence of nanofibers loaded with timolo maleate and dorzolamide than in the absence of nanofibers.	[75]
PLA		Solvent evaporation of Nano/ microparticles.	Brimonidine and timolol maleate subconjunctival route are co delivered for	Budesonide is delivered continuously and suppresses VEGF. Vascular diseases like macular degeneration have	[76]

Materials Research Foundations 172 (2025) 329-357 https://doi.org/10.21741/9781644903353-14

			vascular-related illnesses.	been linked to VEGF expression.	
Polyamide		ESN of nanofiber mats based on dendrimers.	Use nanofibers as an in vivo medication delivery method for glaucoma.	For three weeks, brimonidine tartrate, a medication for glaucoma, was given using dendrimer-based nanofiber mats. The results showed that this strategy is more effective for drug delivery than traditional eye drops.	[77]
Gelatin	Natural	Dehydrothermal and UV irradiation-loaded cross-linked gelatin NPs containing beta-fibroblast growth factor.	Benefits of bFGF in preventing photoreceptor deterioration.	NPs released bFGF selectively and continuously, which stopped photoreceptor degradation.	[78]
Albumin		Desolvation process of loaded NPs.	Medication delivery system or controlled vector for ophthalmologic effectiveness.	Non-viral gene delivery and drug- or gene-loaded nanoparticle penetration into the retina. Improved pharmacological effects of pharmaceuticals.	[79,80]
Chitosan		NPs—ionic.	Boost medication distribution to the mucosa of the eyes.	In comparison to aqueous cyclosporin A (CsA), chitosan NPs loaded with CsA exhibited therapy relevant levels of CsA in the cornea and conjunctiva.	[81]
Chitosan-PCL hybrid scaffolds	Natural and synthetic	ESN of Nanofibrous scaffolds.	RPCs should proliferate and differentiate more rapidly in order to transport damaged or impaired retinal tissue.	RPC differentiation and proliferation toward neuronal lineages were preferred.	[82]
Collagen	Natural	Nanofiber scaffolds ESN.	Create a biomimetic environment in which human ESC-RPE cells can function for the purpose of retinal tissue-engineering.	Facilitated the development, maturity, and survival of human ESC-RPE cells.	[83]

5. Retinal tissue-engineering

It is beneficial to use natural materials as scaffolds for the creation of retinal tissues because naturally occurring polymeric substances are biocompatible and resemble extracellular matrix. One of the main components of Bruch's membrane's (BM) inner collagenous membrane is collagen-I. Various natural materials have been explored for this purpose. Research has indicated that RPE cellular reattachment can occur on the inner collagenous layer [85]. It has been shown that RPE cellular proliferation, singular epithelial shape, and the passage of nutritive via the layers may all be supported by a collagen-I thin film that is custom-fabricated and less than 5 μm thick.

It has been demonstrated that gelatin, a protein-based substance produced by hydrolysing collagen, may encase retinal grafts for transplantation. Retinal sheets that were transplanted fared well and formed laminar structures [86]. Gelatin carriers cross-linked with carbodiimide were employed in a different investigation to deliver retinal sheets [87]. In one study, Shadford and associates assessed the viability of using Bombyx mori silk fibroin films for the cultivation of ARPE19 type cellular components and human RPE cell basic cultures. The establishment of the culture was slow, despite the cells retaining their tight connections and cobblestone appearance [88]. Kundu et al. showed that adult cow retinas could be decellularized upon treatment with ionic surfactants and then followed into physically stable thin membranes that might serve as a culture medium for hRPCs. Human RPC adhesion, proliferation, and differentiation were all facilitated by the decellularized matrix. On retinal films, the expression of the genes CRX, ROM1, RHODOPSIN, and NRL suggested the differentiation of photoreceptors [89].

Polymers, both synthetic and hybrid, for retinal tissue engineering when compared to their natural equivalents show that synthetic materials are more homogeneous and can be produced under regulated circumstances. Their mechanical, biological, and physicochemical characteristics can be modified to meet specific requirements. Synthetic scaffolds have a longer shelf life and are more definite and repeatable. With a specified degradation rate, the scaffolds can be built for maximum tensile toughness, porosity, and micro-level alignment [90–93].

Retinal tissue-engineering has made the utilization of diversified biodegradable polymeric substances, namely poly(lactic-lactase), PLGS, PLLA-PLGA copolymeric systems, polydimethylsiloxane (PDMS), methacrylated hydrogels, poly(glycerol-sebacate) (PGS), poly (ethylene glycol) diacrylate (PEGDA), PMMA, Parylene-C, and PCL [92]. Sorkio et al. demonstrated the development of hESC-RPE monomolecular coat in serum less cultural conditions with upper most functionalities using highly flexible, elastic, and biodegradable poly(tri-methylene) carbonate (PTMC) coats [74]. To allow for the unrestricted passage of ions and molecules through the material, Calejo et al. created a BM replacement using a close packing of higher-dimensional cells like PLA films having pore sizes ranging between 3 to 5 mm. Collagen IV was subsequently applied to the scaffolds, and pigmented hESC-RPE were planted therein. The cells multiplied, conjugated to the substrate efficiently, and showed RPE-like levels of polarization and pigmentation after six weeks [94]. RPE cells were demonstrated to develop well in vitro on a PCL-gelatin scaffolds in a different investigation [94]. Using the modernized clinically relevant grade of needle less electrospinning method, Wardke et al. created the unique ultrathin 3D nanofibrous net from collagen type-I and PLGA. Growing on this artificial membrane, human RPE cells displayed a monolayer that was appropriately orientated, had a polygonal cell shape, and had an abundance of sheet-like microvilli on its apical layers, all of which were reminiscent of native human RPE [95]. Peng et al. used a plasma-modified polydimethylsiloxane (PDMS) scaffold with a laminin coating (PDMS-PmL) to create a biorelevant mimetic scaffold that mimicked sublevel retinal Bruch's basement. It was discovered that the membrane improved adhesion, maintained viability, and promoted the functional maturation of RPE cellular components placed on it [96]. Fig. 4 depicts the different application of polymers in retinal prothesis.

 Materials Research Forum LLC
https://doi.org/10.21741/9781644903353-14

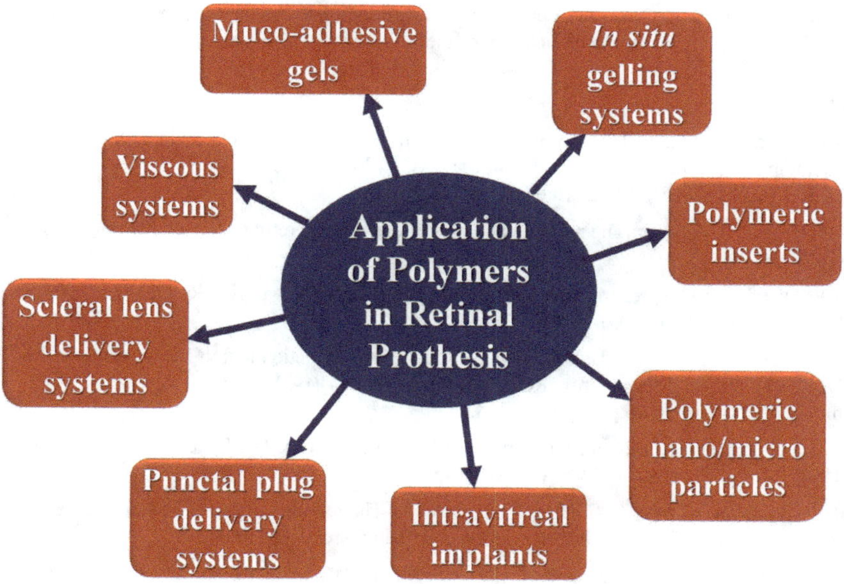

Figure 4. *Schematic illustration exhibiting the use of various polymers in retinal prosthesis in various ways.*

Conclusion

Polymers have emerged as the crucial materials in the progressive production of required retinal prostheses, primarily due to their biocompatibility, flexibility, and ability to be engineered for specific applications. The use of GRAS approved naturally originated, as well as synthetic polymeric substances in these devices offers a significant advantage in mimicking the natural retina's properties, providing a platform for integrating electronic components while ensuring minimal tissue damage. Research indicates that polymers can be effectively utilized in the fabrication of both the electrodes and the substrate of retinal prostheses. Conductive polymers, in particular, have shown promise in enhancing the interface between the device and the retina's tissue, improving signal transduction, and overall device performance. Additionally, their versatility allows for the customization of mechanical and chemical properties, which is essential for adapting the prosthesis to different patients' needs.

However, challenges remain, such as ensuring long-term stability and functionality of the polymers within the biological environment. Future advancements in polymer science, including the development of more robust and conductive components, will likely contribute a pivotal role in overcoming all types of these challenges and extending the efficacy of retinal prostheses. In conclusion, polymers hold a key position in advancing retinal prosthesis technology, offering the potentiality to restore the healthy visual functionality more effectively to the patients with retinal degenerative diseases. Ongoing research and innovation in polymer materials will be essential for optimizing the performance and longevity of these life-changing devices.

References

[1] J. Barar, A. Aghanejad, M. Fathi, Y. Omidi, Advanced drug delivery and targeting technologies for the ocular diseases, Bioimpacts. 6 (2016) 49–67. https://doi.org/10.15171/bi.2016.07

[2] H. Kolb, D. Marshak, The midget pathways of the primate retina, Doc. Ophthalmol. 106 (2003) 67–81.

[3] R. O'Rahilly, The prenatal development of the human eye, Exp. Eye Res. 21 (1975) 93–112. https://doi.org/10.1016/0014-4835(75)90075-5

[4] M. Hoon, H. Okawa, L. Della Santina, R.O.L. Wong, Functional architecture of the retina: Development and disease, Prog. Retin Eye Res. 42 (2014) 44–84. https://doi.org/10.1016/j.preteyeres.2014.06.003

[5] D.T. Hartong, E.L. Berson, T.P. Dryja, Retinitis pigmentosa, The Lancet. 368 (2006) 1795–1809. https://doi.org/10.1016/S0140-6736(06)69740-7

[6] W. Baehr, S. M. Wu, A. C. Bird, K. Palczewski, The retinoid cycle and retina disease, Vision Res. 43 (2003) 2957–2958. https://doi.org/10.1016/j.visres.2003.10.001

[7] I. J. Constable, C. M. Pierce, C.M. Lai, A.L. Magno, M.A. Degli-Esposti, M.A. French, I. L. McAllister, S. Butler, S.B. Barone, S.D. Schwartz, M.S. Blumenkranz, E.P. Rakoczy, Phase 2a Randomized Clinical Trial: Safety and Post Hoc Analysis of Subretinal rAAV.sFLT-1 for Wet Age-related Macular Degeneration, EBioMedicine. 14 (2016) 168–175. https://doi.org/10.1016/j.ebiom.2016.11.016

[8] P.A. Campochiaro, A.K. Lauer, E.H. Sohn, T.A. Mir, S. Naylor, M.C. Anderton, M. Kelleher, R. Harrop, S. Ellis, K.A. Mitrophanous, Lentiviral Vector Gene Transfer of Endostatin/Angiostatin for Macular Degeneration (GEM) Study, Hum. Gene Ther. 28 (2017) 99–111. https://doi.org/10.1089/hum.2016.117

[9] E.L. Berson, Retinitis pigmentosa. The Friedenwald Lecture, Invest. Ophthalmol. Vis. Sci. 34 (1993) 1659–1676.

[10] C. Hamel, Retinitis pigmentosa, Orphanet. J. Rare Dis. 1 (2006) 1–12. https://doi.org/10.1186/1750-1172-1-40

[11] P. Mitchell, G. Liew, B. Gopinath, T.Y. Wong, Age-related macular degeneration, The Lancet. 392 (2018) 1147–1159. https://doi.org/10.1016/S0140-6736(18)31550-2

[12] L. S. Lim, P. Mitchell, J.M. Seddon, F.G. Holz, T.Y. Wong, Age-related macular degeneration, The Lancet. 379 (2012) 1728–1738. https://doi.org/10.1016/S0140-6736(12)60282-7

[13] K. Nowik, E. Langwińska-Wośko, P. Skopiński, K.E. Nowik, J.P. Szaflik, Bionic eye review – An update, J. Clin. Neurosci. 78 (2020) 8–19. https://doi.org/10.1016/j.jocn.2020.05.041

[14] Y.H.L. Luo, L. da Cruz, A review and update on the current status of retinal prostheses (bionic eye), Br. Med. Bull. 109 (2014) 31–44. https://doi.org/10.1093/bmb/ldu002

[15] M.S. Humayun, E. de Juan, G. Dagnelie, The Bionic Eye: A Quarter Century of Retinal Prosthesis Research and Development, Ophthalmol. 123 (2016) S89–S97. https://doi.org/10.1016/j.ophtha.2016.06.044

[16] J.D. Weiland, M.S. Humayun, Retinal Prosthesis, IEEE Trans. Biomed. Eng. 61 (2014) 1412–1424. https://doi.org/10.1109/TBME.2014.2314733

[17] H. Lorach, G. Goetz, R. Smith, X. Lei, Y. Mandel, T. Kamins, K. Mathieson, P. Huie, J. Harris, A. Sher, D. Palanker, Photovoltaic restoration of sight with high visual acuity, Nat. Med. 21 (2015) 476–482. https://doi.org/10.1038/nm.3851

[18] K. Stingl, K.U. Bartz-Schmidt, D. Besch, C.K. Chee, C.L. Cottriall, F. Gekeler, M. Groppe, T.L. Jackson, R.E. MacLaren, A. Koitschev, A. Kusnyerik, J. Neffendorf, J. Nemeth, M.A.N. Naeem, T. Peters, J.D. Ramsden, H. Sachs, A. Simpson, M.S. Singh, B. Wilhelm, D. Wong, E. Zrenner, Subretinal Visual Implant Alpha IMS – Clinical trial interim report, Vision Res. 111 (2015) 149–160. https://doi.org/10.1016/j.visres.2015.03.001

[19] L. N. Ayton, P. J. Blamey, R. H. Guymer, C. D. Luu, D. A. X. Nayagam, N. C. Sinclair, M. N. Shivdasani, J. Yeoh, M. F. McCombe, R. J. Briggs, N. L. Opie, J. Villalobos, P. N. Dimitrov, M. Varsamidis, M. A. Petoe, C. D. McCarthy, J. G. Walker, N. Barnes, A. N. Burkitt, C. E. Williams, R. K. Shepherd, P. J. Allen, First-in-Human Trial of a Novel Suprachoroidal Retinal Prosthesis, PLoS One. 9 (2014) 1–27. https://doi.org/10.1371/journal.pone.0115239

[20] D. Palanker, Y. Le Mer, S. Mohand-Said, M. Muqit, J. A. Sahel, Photovoltaic Restoration of Central Vision in Atrophic Age-Related Macular Degeneration, Ophthalmol. 127 (2020) 1097–1104. https://doi.org/10.1016/j.ophtha.2020.02.024

[21] R. A. B. Fernandes, B. Diniz, R. Ribeiro, M. Humayun, Artificial vision through neuronal stimulation, Neurosci. Lett. 519 (2012) 122–128. https://doi.org/10.1016/j.neulet.2012.01.063

[22] K. Mathieson, J. Loudin, G. Goetz, P. Huie, L. Wang, T.I. Kamins, L. Galambos, R. Smith, J. S. Harris, A. Sher, D. Palanker, Photovoltaic retinal prosthesis with high pixel density, Nat. Photonics. 6 (2012) 391–397. https://doi.org/10.1038/nphoton.2012.104

[23] A. Finn, D. Grewal, L. Vajzovic, Argus II retinal prosthesis system: a review of patient selection criteria, surgical considerations, and post-operative outcomes, Clin. Ophthalmol. 12 (2018) 1089–1097. https://doi.org/10.2147/OPTH.S137525

[24] R. Daschner, A. Rothermel, R. Rudorf, S. Rudorf, A. Stett, Functionality and Performance of the Subretinal Implant Chip Alpha AMS, Sens. Mater. (2018) 179–192. https://doi.org/10.18494/SAM.2018.1726

[25] E. Özmert, U. Arslan, Retinal Prostheses and Artificial Vision, Turk. J. Ophthalmol. 49 (2019) 213–219. https://doi.org/10.4274/tjo.galenos.2019.44270

[26] D.B. Shire, S.K. Kelly, J. Chen, P. Doyle, M.D. Gingerich, S.F. Cogan, W.A. Drohan, O. Mendoza, L. Theogarajan, J.L. Wyatt, J.F. Rizzo, Development and Implantation of a Minimally Invasive Wireless Subretinal Neurostimulator, IEEE Trans. Biomed. Eng. 56 (2009) 2502–2511. https://doi.org/10.1109/TBME.2009.2021401

[27] C. H. Chou, S. Shrestha, C. D. Yang, N. W. Chang, Y. L. Lin, K. W. Liao, W. C. Huang, T. H. Sun, S. J. Tu, W. H. Lee, M. Y. Chiew, C. S. Tai, T. Y. Wei, T. R. Tsai, H. T. Huang, C.Y. Wang, H. Y. Wu, S.Y. Ho, P. R. Chen, C.H. Chuang, P.J. Hsieh, Y.S. Wu, W.L. Chen, M.J. Li, Y.C. Wu, X.Y. Huang, F. L. Ng, W. Buddhakosai, P. C. Huang, K.C. Lan, C.Y. Huang, S.L. Weng, Y.N. Cheng, C. Liang, W.L. Hsu, H.D. Huang, miRTarBase update 2018: a resource for experimentally validated microRNA-target interactions, Nucleic Acids Res. 46 (2018) D296–D302. https://doi.org/10.1093/nar/gkx1067

[28] N.S. Peachey, A.Y. Chow, Subretinal implantation of semiconductor-based photodiodes: Progress and challenges, J. Rehabil. Res. Dev. 36 (1999) 1–10.

[29] G. Roessler, T. Laube, C. Brockmann, T. Kirschkamp, B. Mazinani, M. Goertz, C. Koch, I. Krisch, B. Sellhaus, H. K. Trieu, J. Weis, N. Bornfeld, H. Ro¨thgen, A. Messner, W. Mokwa, P. Walter, Implantation and Explantation of a Wireless Epiretinal Retina Implant Device: Observations during the EPIRET3 Prospective Clinical Trial, Invest. Ophthalmol. Vis. Sci. 50 (2009) 3003–3008. https://doi.org/10.1167/iovs.08-2752

[30] C. Sekirnjak, P. Hottowy, A. Sher, W. Dabrowski, A.M. Litke, E.J. Chichilnisky, High-Resolution Electrical Stimulation of Primate Retina for Epiretinal Implant Design, J. Neurosci. 28 (2008) 4446–4456. https://doi.org/10.1523/JNEUROSCI.5138-07.2008

[31] S. Klauke, M. Goertz, S. Rein, D. Hoehl, U. Thomas, R. Eckhorn, F. Bremmer, T. Wachtler, Stimulation with a Wireless Intraocular Epiretinal Implant Elicits Visual Percepts in Blind Humans, Invest. Ophthalmol. Vis. Sci. 52 (2011) 449–455. https://doi.org/10.1167/iovs.09-4410

[32] M. Farvardin, M. Afarid, A. Attarzadeh, M.K. Johari, M. Mehryar, M.H. Nowroozzadeh, F. Rahat, H. Peyvandi, R. Farvardin, M. Nami, The Argus-II Retinal Prosthesis Implantation; From the Global to Local Successful Experience, Front. Neurosci. 12 (2018) 1–8. https://doi.org/10.3389/fnins.2018.00584

[33] I. Vázquez-Domínguez, A. Garanto, R.W.J. Collin, Molecular Therapies for Inherited Retinal Diseases—Current Standing, Opportunities and Challenges, Genes (Basel). 10 (2019) 1–30. https://doi.org/10.3390/genes10090654

[34] A.L. Saunders, C.E. Williams, W. Heriot, R. Briggs, J. Yeoh, D.A. Nayagam, M. McCombe, J. Villalobos, O. Burns, C.D. Luu, L.N. Ayton, M. McPhedran, N.L. Opie, C. McGowan, R.K. Shepherd, R. Guymer, P. J. Allen, Development of a surgical procedure for implantation of a prototype suprachoroidal retinal prosthesis, Clin. Exp. Ophthalmol. 42 (2014) 665–674. https://doi.org/10.1111/ceo.12287

[35] M.N. Shivdasani, C.D. Luu, R. Cicione, J.B. Fallon, P.J. Allen, J. Leuenberger, G.J. Suaning, N.H. Lovell, R.K. Shepherd, C.E. Williams, Evaluation of stimulus parameters and electrode geometry for an effective suprachoroidal retinal prosthesis, J. Neural Eng. 7 (2010) 1–20. https://doi.org/10.1088/1741-2560/7/3/036008

[36] L.G. Griffith, Polymeric biomaterials, Acta Mater. 48 (2000) 263–277. https://doi.org/10.1016/S1359-6454(99)00299-2

[37] J.F. Mano, R.A. Sousa, L.F. Boesel, N.M. Neves, R.L. Reis, Bioinert, biodegradable and injectable polymeric matrix composites for hard tissue replacement: state of the art and recent

developments, Compos. Sci. Technol. 64 (2004) 789–817. https://doi.org/10.1016/j.compscitech.2003.09.001

[38] W. Cao, L.L. Hench, Bioactive materials, Ceram Int. 22 (1996) 493–507. https://doi.org/10.1016/0272-8842(95)00126-3

[39] H. Shin, S. Jo, A.G. Mikos, Biomimetic materials for tissue engineering, Biomaterials. 24 (2003) 4353–4364. https://doi.org/10.1016/S0142-9612(03)00339-9

[40] N.A. Licata, A.V. Tkachenko, Self-assembling DNA-caged particles: Nanoblocks for hierarchical self-assembly, Phys. Rev. E. 79 (2009) 1–11. https://doi.org/10.1103/PhysRevE.79.011404

[41] E. A. Kamoun, X. Chen, M. S. Mohy Eldin, E. R. S. Kenawy, Crosslinked poly(vinyl alcohol) hydrogels for wound dressing applications: A review of remarkably blended polymers, Arab. J. Chem. 8 (2015) 1–14. https://doi.org/10.1016/j.arabjc.2014.07.005

[42] L. L. Hench, R. J. Splinter, W. C. Allen, T. K. Greenlee, Bonding mechanisms at the interface of ceramic prosthetic materials, J. Biomed. Mater. Res. 5 (1971) 117–141. https://doi.org/10.1002/jbm.820050611

[43] S. M. Kurtz, J. N. Devine, PEEK biomaterials in trauma, orthopedic, and spinal implants, Biomaterials. 28 (2007) 4845–4869. https://doi.org/10.1016/j.biomaterials.2007.07.013

[44] T. Goda, K. Ishihara, Soft contact lens biomaterials from bioinspired phospholipid polymers, Expert Rev. Med. Devices. 3 (2006) 167–174. https://doi.org/10.1586/17434440.3.2.167

[45] M. Chehade, M.J. Elder, Intraocular lens materials and styles: A review, Aust. N Z J. Ophthalmol. 25 (1997) 255–263. https://doi.org/10.1111/j.1442-9071.1997.tb01512.x

[46] L. Pinchuk, I. Riss, J. F. Batlle, Y. P. Kato, J. B. Martin, E. Arrieta, P. Palmberg, R. K. Parrish, B. A. Weber, Y. Kwon, J. Parel, The development of a micro-shunt made from poly(styrene- block -isobutylene- block -styrene) to treat glaucoma, J. Biomed. Mater. Res. B Appl. Biomater. 105 (2017) 211–221. https://doi.org/10.1002/jbm.b.33525

[47] S. Morimoto, F. B. W. Rebello de Sampaio, M. M. Braga, N. Sesma, M. Özcan, Survival Rate of Resin and Ceramic Inlays, Onlays, and Overlays, J. Dent. Res. 95 (2016) 985–994. https://doi.org/10.1177/0022034516652848

[48] K.E. Swindle, N. Ravi, Recent advances in polymeric vitreous substitutes, Expert Rev. Ophthalmol. 2 (2007) 255–265. https://doi.org/10.1586/17469899.2.2.255

[49] D. Tognetto, P. Cecchini, R. D'Aloisio, R. Lapasin, Mixed polymeric systems: New ophthalmic viscosurgical device created by mixing commercially available devices, J. Cataract Refract. Surg. 43 (2017) 109–114. https://doi.org/10.1016/j.jcrs.2016.11.035

[50] J. Singh, K.K. Agrawal, Polymeric Materials for Contact Lenses, J. Macromol. Sci. Part C Polym. Rev. 32 (1992) 521–534. https://doi.org/10.1080/15321799208021431

[51] C.S.A. Musgrave, F. Fang, Contact Lens Materials: A Materials Science Perspective, Mater. 12 (2019) 1–10. https://doi.org/10.3390/ma12020261

[52] K. Ishihara, X. Shi, K. Fukazawa, T. Yamaoka, G. Yao, J. Y. Wu, Biomimetic-Engineered Silicone Hydrogel Contact Lens Materials, ACS Appl. Bio. Mater. 6 (2023) 3600–3616. https://doi.org/10.1021/acsabm.3c00296

[53] D. Bozukova, C. Pagnoulle, R. Jérôme, C. Jérôme, Polymers in modern ophthalmic implants—Historical background and recent advances, Mater. Sci. Eng. R Rep. 69 (2010) 63–83. https://doi.org/10.1016/j.mser.2010.05.002

[54] M. Karayilan, L. Clamen, M.L. Becker, Polymeric Materials for Eye Surface and Intraocular Applications, Biomacromolecules. 22 (2021) 223–261. https://doi.org/10.1021/acs.biomac.0c01525

[55] J. M. Legeais, L. P. Werner, G. Legeay, B. Briat, G. Renard, In vivo study of a fluorocarbon polymer-coated intraocular lens in a rabbit model, J. Cataract Refract. Surg. 24 (1998) 371–379. https://doi.org/10.1016/S0886-3350(98)80326-X

[56] Y. E. Choonara, V. Pillay, M. P. Danckwerts, T. R. Carmichael, L. C. du Toit, A review of implantable intravitreal drug delivery technologies for the treatment of posterior segment eye diseases, J. Pharm. Sci. 99 (2010) 2219–2239. https://doi.org/10.1002/jps.21987

[57] J. B. Christoforidis, S. Chang, A. Jiang, J. Wang, C. M. Cebulla, Intravitreal Devices for the Treatment of Vitreous Inflammation, Mediators Inflamm. 2012 (2012) 1–8. https://doi.org/10.1155/2012/126463

[58] A. Pandhare, P. Bhatt, H.S. Saluja, Y.V. Pathak, Biodegradable Polymeric Implants for Retina and Posterior Segment Disease, Drug Delivery for the Retina and Posterior Segment Disease, in: J.K. Patel, V. Sutariya, J.R. Kanvar, Y.V.Pathak (Eds.), Springer Cham, Springer Nature, Switzerland, 2018, pp. 273–291. https://doi.org/10.1007/978-3-319-95807-1_15

[59] C. A. Arcinue, O. M. Cerón, C. S. Foster, A Comparison Between the Fluocinolone Acetonide (Retisert) and Dexamethasone (Ozurdex) Intravitreal Implants in Uveitis, J. Ocul. Pharmacol. 29 (2013) 501–507. https://doi.org/10.1089/jop.2012.0180

[60] S. Leinonen, I. Immonen, K. Kotaniemi, Fluocinolone acetonide intravitreal implant (Retisert®) in the treatment of sight threatening macular oedema of juvenile idiopathic arthritis-related uveitis, Acta Ophthalmol. 96 (2018) 648–651. https://doi.org/10.1111/aos.13744

[61] J.M. DeSimone, Co-opting Moore's law: Therapeutics, vaccines and interfacially active particles manufactured via PRINT®, J. Control. Release. 240 (2016) 541–543. https://doi.org/10.1016/j.jconrel.2016.07.019

[62] J.R. Seal, M.R. Robinson, J. Burke, M. Bejanian, M. Coote, M. Attar, Intracameral Sustained-Release Bimatoprost Implant Delivers Bimatoprost to Target Tissues with Reduced Drug Exposure to Off-Target Tissues, J. Ocul. Pharmacol. Ther. 35(1) (2019) 50-57.

[63] W. Tao, Application of encapsulated cell technology for retinal degenerative diseases, Expert Opin. Biol. Ther. 6 (2006) 717–726. https://doi.org/10.1517/14712598.6.7.717

[64] K. Nayak, M. Misra, A review on recent drug delivery systems for posterior segment of eye, Biomed. Pharmacother. 107 (2018) 1564-1582.

[65] F. Siedenbiedel, J. C. Tiller, Antimicrobial Polymers in Solution and on Surfaces: Overview and Functional Principles, Polymers (Basel). 4 (2012) 46–71. https://doi.org/10.3390/polym4010046

[66] T. Jiang, W. I. Abdel-Fattah, C. T. Laurencin, In vitro evaluation of chitosan/poly(lactic acid-glycolic acid) sintered microsphere scaffolds for bone tissue engineering, Biomater. 27 (2006) 4894–4903. https://doi.org/10.1016/j.biomaterials.2006.05.025

[67] B. Ongpipattanakul, T. Nguyen, T. F. Zioncheck, R. Wong, G. Osaka, L. DeGuzman, W. P. Lee, L. S. Beck, Development of tricalcium phosphate/amylopectin paste combined with recombinant human transforming growth factor beta 1 as a bone defect filler, J. Biomed. Mater. Res. 36 (1997) 295–305. https://doi.org/10.1002/(SICI)1097-4636(19970905)36:3<295::AID-JBM4>3.0.CO;2-9

[68] S. R. Singh, H. E. Grossniklaus, S. J. Kang, H. F. Edelhauser, B. K. Ambati, U. B. Kompella, Intravenous transferrin, RGD peptide and dual-targeted nanoparticles enhance anti-VEGF intraceptor gene delivery to laser-induced CNV, Gene Ther. 16 (2009) 645–659. https://doi.org/10.1038/gt.2008.185

[69] E. Vega, M. A. Egea, O. Valls, M. Espina, M. L. García, Flurbiprofen Loaded Biodegradable Nanoparticles for Ophtalmic Administration, J. Pharm. Sci. 95 (2006) 2393–2405. https://doi.org/10.1002/jps.20685

[70] R. A. Bejjani, D. BenEzra, H. Cohen, R. Jutta, A. Charlotte, J. C. Jeanny, G. Golomb, F. F. Behar-Cohen, Nanoparticles for gene delivery to retinal pigment epithelial cells, Mol. Vis. 11 (2005) 124–132.

[71] H. Yang, P. Tyagi, R. S. Kadam, C. A. Holden, U. B. Kompella, Hybrid Dendrimer Hydrogel/PLGA Nanoparticle Platform Sustains Drug Delivery for One Week and Antiglaucoma Effects for Four Days Following One-Time Topical Administration, ACS Nano. 6 (2012) 7595–7606. https://doi.org/10.1021/nn301873v

[72] C. Giannavola, Influence of Preparation Conditions on Acyclovir-Loaded Poly-d,l-Lactic Acid Nanospheres and Effect of PEG Coating on Ocular Drug Bioavailability, Pharm. Res. 20 (2003) 584–590. https://doi.org/10.1023/A:1023290514575

[73] V. M. Sluch, C. ha O. Davis, V. Ranganathan, J. M. Kerr, K. Krick, R. Martin, C. A. Berlinicke, N. Marsh-Armstrong, J. S. Diamond, H. Q. Mao, D. J. Zack, Differentiation of human ESCs to retinal ganglion cells using a CRISPR engineered reporter cell line, Sci. Rep. 5 (2015) 1–17. https://doi.org/10.1038/srep16595

[74] A. Sorkio, S. Haimi, V. Verdoold, K. Juuti-Uusitalo, D. Grijpma, H. Skottman, Poly(trimethylene carbonate) as an elastic biodegradable film for human embryonic stem cell-derived retinal pigment epithelial cells, J. Tissue Eng. Regen Med. 11 (2017) 3134–3144. https://doi.org/10.1002/term.2221

[75] Gagandeep, T. Garg, B. Malik, G. Rath, A. K. Goyal, Development and characterization of nano-fiber patch for the treatment of glaucoma, Eur. J. Pharm. Sci. 53 (2014) 10–16. https://doi.org/10.1016/j.ejps.2013.11.016

[76] U. B. Kompella, N. Bandi, S. P. Ayalasomayajula, Subconjunctival Nano- and Microparticles Sustain Retinal Delivery of Budesonide, a Corticosteroid Capable of Inhibiting

VEGF Expression, Invest. Ophthalmol. Vis. Sci. 44 (2003) 1192–1201. https://doi.org/10.1167/iovs.02-0791

[77] M. G. Lancina, S. Singh, U. B. Kompella, S. Husain, H. Yang, Fast Dissolving Dendrimer Nanofiber Mats as Alternative to Eye Drops for More Efficient Antiglaucoma Drug Delivery, ACS Biomater. Sci. Eng. 3 (2017) 1861–1868. https://doi.org/10.1021/acsbiomaterials.7b00319

[78] T. Sakai, N. Kuno, F. Takamatsu, E. Kimura, H. Kohno, K. Okano, K. Kitahara, Prolonged Protective Effect of Basic Fibroblast Growth Factor–Impregnated Nanoparticles in Royal College of Surgeons Rats, Invest. Ophthalmol. Vis. Sci. 48 (2007) 3381–3387. https://doi.org/10.1167/iovs.06-1242

[79] A. K. Zimmer, P. Chetoni, M. F. Saettone, H. Zerbe, J. Kreuter, Evaluation of pilocarpine-loaded albumin particles as controlled drug delivery systems for the eye. II. Co-administration with bioadhesive and viscous polymers, J. Contr. Release. 33 (1995) 31–46. https://doi.org/10.1016/0168-3659(94)00059-4

[80] Y. Mo, M. E. Barnett, D. Takemoto, H. Davidson, U. B. Kompella, Human serum albumin nanoparticles for efficient delivery of Cu, Zn superoxide dismutase gene, Mol. Vis. 13 (2007) 746–757.

[81] A. M. De Campos, A. Sánchez, M. J. Alonso, Chitosan nanoparticles: a new vehicle for the improvement of the delivery of drugs to the ocular surface. Application to cyclosporin A, Int. J. Pharm. 224 (2001) 159–168. https://doi.org/10.1016/S0378-5173(01)00760-8

[82] B. Noorani, F. Tabandeh, F. Yazdian, Z. S. Soheili, M. Shakibaie, S. Rahmani, Thin natural gelatin/chitosan nanofibrous scaffolds for retinal pigment epithelium cells, Int. J. Polym. Mater. 67 (2018) 754–763. https://doi.org/10.1080/00914037.2017.1362639

[83] A. Sorkio, P. J. Porter, K. Juuti-Uusitalo, B. J. Meenan, H. Skottman, G. A. Burke, Surface Modified Biodegradable Electrospun Membranes as a Carrier for Human Embryonic Stem Cell-Derived Retinal Pigment Epithelial Cells, Tissue Eng. Part A. 21 (2015) 2301–2314. https://doi.org/10.1089/ten.tea.2014.0640

[84] C. R. Wittmer, T. Claudepierre, M. Reber, P. Wiedemann, J. A. Garlick, D. Kaplan, C. Egles, Multifunctionalized Electrospun Silk Fibers Promote Axon Regeneration in the Central Nervous System, Adv. Funct. Mater. 21(22) (2011) 4232–4242. https://doi.org/10.1002/adfm.201100755

[85] N. S. Bhatt, D. A. Newsome, T. Fenech, T. P. Hessburg, J. G. Diamond, M. V. Miceli, K. E. Kratz, P. D. Oliver, Experimental Transplantation of Human Retinal Pigment Epithelial Cells on Collagen Substrates, Am. J. Ophthalmol. 117 (1994) 214–221. https://doi.org/10.1016/S0002-9394(14)73079-X

[86] J. T. Lu, C. J. Lee, S. F. Bent, H. A. Fishman, E. E. Sabelman, Thin collagen film scaffolds for retinal epithelial cell culture, Biomater. 28 (2007) 1486–1494. https://doi.org/10.1016/j.biomaterials.2006.11.023

[87] J. Y. Lai, Y. T. Li, Evaluation of cross-linked gelatin membranes as delivery carriers for retinal sheets, Mater. Sci. Eng. C. 30(5) (2010) 677–685. https://doi.org/10.1016/j.msec.2010.02.024

[88] A. M. A. Shadforth, K. A. George, A. S. Kwan, T. V. Chirila, D. G. Harkin, The cultivation of human retinal pigment epithelial cells on Bombyx mori silk fibroin, Biomater. 33 (2012) 4110–4117. https://doi.org/10.1016/j.biomaterials.2012.02.040

[89] J. Kundu, A. Michaelson, K. Talbot, P. Baranov, M. J. Young, R. L. Carrier, Decellularized retinal matrix: Natural platforms for human retinal progenitor cell culture, Acta Biomater. 31 (2016) 61–70. https://doi.org/10.1016/j.actbio.2015.11.028

[90] D. A. Bernards, R. B. Bhisitkul, P. Wynn, M. R. Steedman, O. T. Lee, F. Wong, S. Thoongsuwan, T. A. Desai, Ocular Biocompatibility and Structural Integrity of Micro- and Nanostructured Poly(caprolactone) Films, J. Ocul. Pharmacol. 29 (2013) 249–257. https://doi.org/10.1089/jop.2012.0152

[91] L. Lu, M. J. Yaszemski, A. G. Mikos, Retinal pigment epithelium engineering using synthetic biodegradable polymers, Biomater. 22(24) (2001) 3345–3355. https://doi.org/10.1016/S0142-9612(01)00172-7

[92] Lichun Lu, C. A. Garcia, A. G. Mikos, Retinal pigment epithelium cell culture on thin biodegradable poly(DL-lactic-co-glycolic acid) films, J. Biomater. Sci. Polym. Ed. 9(11) (1998) 1187–1205. https://doi.org/10.1163/156856298X00721

[93] T. Hadlock, S. Singh, J. P. Vacanti, B. J. McLaughlin, Ocular Cell Monolayers Cultured on Biodegradable Substrates, Tissue Eng. 5(3) (1999) 187–196. https://doi.org/10.1089/ten.1999.5.187

[94] S. Rahmani, F. Tabandeh, S. Faghihi, G. Amoabediny, M. Shakibaie, B. Noorani, F. Yazdian, Fabrication and characterization of poly(ε-caprolactone)/gelatin nanofibrous scaffolds for retinal tissue engineering, International Journal of Polymeric Materials and Polymeric Biomaterials. 67 (2018) 27–35. https://doi.org/10.1080/00914037.2017.1297939

[95] P. H. Warnke, M. Alamein, S. Skabo, S. Stephens, R. Bourke, P. Heiner, Q. Liu, Primordium of an artificial Bruch's membrane made of nanofibers for engineering of retinal pigment epithelium cell monolayers, Acta Biomater. 9 (2013) 9414–9422. https://doi.org/10.1016/j.actbio.2013.07.029

[96] P. Peng, J. Namkung, M. Barnes, C. Sun, A Meta-Analysis of Mathematics and Working Memory: Moderating Effects of Working Memory Domain, Type of Mathematics Skill, and Sample Characteristics, J Educ Psychol. 108(4) (2016) 455-473. https://doi.org/10.1037/edu0000079.supp

Materials Research Foundations 172 (2025) 358-380
https://doi.org/10.21741/9781644903353-15

Chapter 15

Polymers for Vascular Applications

Satish Polshettiwar[1*], Nilesh Mahajan[2], Dishank Purandare[1], Rakesh Mishra[3], Arpana Patil[4]

[1]School of Health Sciences and Technology, Dr Vishwanath Karad MIT World Peace University, Pune, India- 411038

[2]Department of Pharmaceutics, Dadasaheb Balpande College of Pharmacy, Besa, Nagpur, 440037 (M.S.) India

[3]Dr DY Patil Institute of Pharmaceutical Sciences & Research, Pimpri-Pune 411018

[4]STES's Smt. Kashibai Navale College of Pharmacy, Kondhwa, Pune

Abstract

Polymers, both natural and synthetic, play a critical role in the development of vascular applications due to their biocompatibility, mechanical properties, and versatility. This chapter explores the use of natural polysaccharides such as collagen, chitosan, and alginate, highlighting their ability to support tissue regeneration and wound healing in vascular contexts. It also examines synthetic polymers like polyethylene glycol (PEG), polylactic acid (PLA), polycaprolactone (PCL), and polyurethane, which offer tunable properties for vascular grafts, stents, and drug delivery systems. The chapter delves into key considerations for biocompatibility, mechanical strength, flexibility, and the development of drug delivery systems with antithrombogenic properties. Additionally, it discusses biodegradable and synthetic vascular grafts, stents, and vascular patches, emphasizing innovations like coatings and modifications to enhance graft performance. The future of vascular applications is closely tied to advances in smart polymers, nanotechnology, and tissue engineering, with emerging fields like 3D bioprinting offering customized, patient-specific solutions. The regulatory, ethical, and economic considerations surrounding the adoption of these polymers are also addressed. By integrating polymers into vascular medicine, this chapter underscores the potential of both natural and synthetic materials to revolutionize the treatment of vascular diseases, driving sustainable and innovative advancements in healthcare.

Keywords

Polymers, Vascular Applications, Biocompatibility, Vascular Grafts, Nanocomposites, Stents Polymer

Contents

Materials Research Forum LLC
https://doi.org/10.21741/9781644903353-15

1. Introduction

Vascular health ensures that the blood vessels function properly, delivering oxygen and nutrients while removing waste. Good vascular health prevents diseases like hypertension, stroke, and atherosclerosis, lowering the risk of heart attacks and strokes. [1]. Additionally, adequate vascular function is vital for cognitive health, as disorders affecting the arteries supplying the brain are

Materials Research Forum LLC
https://doi.org/10.21741/9781644903353-15

strongly linked to an elevated risk of dementia and cognitive decline [2]. Organs such as the kidneys and liver also depend on efficient blood flow for optimal function, and poor vascular health can lead to organ damage, contributing to conditions like chronic kidney disease. Furthermore, healthy blood vessels are essential for effective wound healing and recovery from injuries by ensuring the delivery of critical nutrients and immune cells to damaged tissues. Maintaining vascular integrity can also prevent PAD, a condition that leads to poor circulation, particularly in the legs, causing pain, numbness, and a heightened risk of infections. By preventing PAD, vascular health helps preserve mobility and overall quality of life. Atherosclerosis is a long-term condition where plaques form within the walls of arteries, leading to a narrowing and stiffening of these blood vessels. This process can impede blood circulation and elevate the risk of serious incidents like strokes and heart attacks. Plaque formation starts off with destruction to the endothelium. Plaque, consists of fats, cholesterol, calcium, and other components, begins with damage to the endothelium, the artery's innermost layer. This damage is often a consequence of high blood pressure, tobacco use, or elevated cholesterol levels. Atherosclerosis begins when cholesterol infiltrates the artery wall after endothelial injury [1,3-5]. Cholesterol oxidizes and is absorbed by macrophages, forming foam cells that create fatty streaks. These streaks can develop into fibrous plaques, which may rupture, causing blood clots and artery blockages. Risk factors include family history, high blood pressure, elevated cholesterol, diabetes, and obesity. While often asymptomatic, symptoms arise when arteries narrow significantly or block [2,6,7]. Diagnosing atherosclerosis often involves techniques such as ultrasound, angiography which help visualize blockages within the arteries. Blood tests are frequently used to assess cholesterol levels. Treatment typically includes medications like statins to reduce cholesterol, antihypertensive drugs to manage blood pressure, and antiplatelet agents to reduce blood clot risk. In advanced stages, surgical interventions like angioplasty, stent insertion, or bypass surgery may be required to restore normal flow of blood. Coronary artery disease (CAD), a prevalent cardiovascular condition, involves plaque buildup in the coronary arteries, reducing blood flow to the heart. This can result in symptoms such as chest pain, fatigue, and angina [8]. CAD results from complex interactions between genetic, environmental, and behavioural factors. Smoking, diabetes, obesity, hypertension, and hyperlipidemia are risk factors. Plaque accumulation in coronary arteries may lead to partial or total blockage, diminishing oxygen delivery to the heart muscle. Chronic angina may develop under physical or mental stress, while acute coronary syndromes, such as myocardial infarction, occur when a plaque ruptures and forms a blood clot that obstructs blood flow. Diagnostic tools such as electrocardiograms (ECGs), stress testing, echocardiography, and coronary angiography are used to treat CAD [9]. CAD management involves lifestyle changes, medications, and interventional procedures. Heart-healthy diets, regular exercise, quitting smoking, and maintaining a healthy weight are critical for both preventing CAD and managing existing conditions. Medications such as ACE inhibitors, beta-blockers, statins, and antiplatelet drugs help control risk factors and prevent complications [10]. When arterial blockages are severe, procedures like coronary artery bypass grafting (CABG) might be needed to restore blood flow. Deep vein thrombosis (DVT), a condition where blood clots form in deep veins, mostly in the legs, can cause severe complications like pulmonary embolism (PE) if untreated. PE occurs when a clot travels to the lungs, leading to sudden shortness of breath, chest pain, and increased heart rate. The risk factors include advanced age, obesity, smoking, and family history [11]. Symptoms include redness, swelling, pain, and a warm sensation in the affected limb, but DVT can sometimes occur without noticeable symptoms, making it particularly dangerous. Diagnosing DVT often involves imaging techniques like ultrasound, D-dimer blood tests, and sometimes venography or MRI. DVT

Materials Research Foundations 172 (2025) 358-380 https://doi.org/10.21741/9781644903353-15

treatment and prevention focus on managing existing clots and preventing new ones from forming. Preventative measures include maintaining a healthy weight, avoiding long periods of immobility, engaging in regular physical activity, and using compression stockings. High-risk individuals might receive anticoagulants like warfarin or heparin to prevent clots, while thrombolytic therapy can dissolve existing ones. Severe cases may need surgery. Lymphedema, involves significant swelling in limbs due to lymphatic system compromise. Primary lymphedema is often inherited, while secondary lymphedema arises from damage or obstruction of the lymphatic system due to cancer treatments, infections, or surgery [12,13]. Symptoms range from mild discomfort and swelling to severe edema that impairs mobility and quality of life. Diagnosis involves physical examinations and imaging tests such as Doppler ultrasonography, CT scans, MRI lymphangiography, and lymphoscintigraphy. Treatments include physical activity, manual lymph drainage, compression therapy, and, in severe cases, surgical interventions [14]. Managing lymphedema requires lifestyle changes, such as avoiding strenuous activities, protecting the skin from injury, and maintaining a healthy weight [15,16]. Support groups and educational resources help individuals cope with the physical and psychological challenges of lymphedema. Hypertension, is a chronic condition where arterial blood pressure is consistently high leading to increased risks of heart attacks, strokes, and other cardiovascular issues [17-19]. Often called the "silent killer," hypertension may cause symptoms such as headaches, shortness of breath, or nosebleeds, though many people experience no symptoms. The pathophysiology involves complex interactions between the heart, blood vessels, kidneys, and nervous system, leading to increased blood volume and vascular resistance. Management includes lifestyle changes, such as a heart-healthy diet, regular exercise, weight management, and reducing alcohol intake. Pharmacological treatments include diuretics, beta-blockers, ACE inhibitors, and angiotensin II receptor blockers (ARBs). Despite effective treatments, hypertension remains a significant global cause of morbidity and mortality. In vascular medicine, polymers play a critical role in the development of medical devices such as vascular grafts, stents, and drug delivery systems [20,21]. Polymers are macromolecules composed of repeating monomer units. Their biocompatibility, mechanical strength, flexibility, permeability, surface properties, thermal stability [22] and sterilizability are important characteristics for medical applications. Biocompatibility ensures that polymers do not trigger adverse immune responses when implanted in the body. Surface modifications can enhance biocompatibility by promoting cell adhesion and proliferation. Mechanical strength and flexibility allow polymers to withstand the dynamic forces of blood flow and pressure. Permeability is crucial for drug delivery systems, allowing the controlled release of medications over time. Hydrophilicity and hydrophobicity affect the interaction between polymers and biological tissues, influencing cell adhesion [23] and immune responses. Polymers have been used in medicine since the 1940s, with advancements in biodegradable polymers and tissue engineering [24].

2. Synthetic and Natural Polymers for Vascular Applications

Polymers have emerged as key materials in the development of advanced solutions for vascular applications due to their unique properties, which range from biocompatibility and biodegradability to mechanical strength and versatility.

2.1 Natural polymers

Collagen: Collagen, a key component of the native extracellular matrix (ECM), is widely used in vascular engineering due to its biocompatibility, biodegradability, and support for cell growth.

Materials Research Foundations 172 (2025) 358-380 https://doi.org/10.21741/9781644903353-15

However, its limitations, including weak mechanical strength and fast degradation, pose challenges for its use in vascular scaffolds [25]. To overcome its inherent mechanical limitations, collagen is often combined with materials like electrochemically aligned collagen filaments, electrospun fibers, and synthetic polymers such as polycaprolactone. These combinations enhance the mechanical properties, such as rupture force and dynamic compliance, making them comparable to or better than human arteries and veins. Additionally, collagen-based materials promote endothelial cell viability, wound healing, and tissue repair, with applications as hemostatic agents, drug delivery systems, vascularization. Hybrid materials that blend collagen with synthetic polymers further improve performance and durability, making them ideal for variety of vascular applications, like vascular grafts and stents, where collagen coatings enhance biocompatibility, reduces thrombosis and promote endothelialization [26].

Chitosan: It is a biopolymer obtained from chitin, is increasingly utilized in VTE due to its low toxicity, biocompatibility and biodegradability [27]. It exhibits properties such as anticoagulant effects, hemostatic capabilities, and a structure similar to ECM components, making it effective for vascular regeneration by inhibiting vascular smooth muscle cells. Chitosan-based scaffolds are valuable for healing of wound, regeneration of tissue and drug delivery, promoting angiogenesis and controlled release of drugs. Additionally, its antimicrobial properties help prevent infections in vascular applications. However, chitosan's relatively low mechanical strength has led to its combination with other materials, such as collagen or synthetic polymers, to enhance its properties for use in vascular grafts and scaffolds [28]. Recent developments include a novel formulation using different gelation processes to produce chitosan gels with improved mechanical properties suitable for in vivo vascular implants [29].

Hyaluronic acid: This, is a naturally occurring polysaccharide that is majorly used for its vascular applications due to its property of excellent biocompatibility and its property to modulate cell behaviour. Also the application in vascular engineering, HA promotes adhesion of cells, their proliferation, migration which are crucial for endothelialization and angiogenesis, making it essential for bioengineered blood vessels. Its hydrophilic nature retains moisture, creating an optimal environment for tissue regeneration and wound healing, while its anti-inflammatory properties reduce the risk of adverse immune responses [30,31]. HA can chemically be modified to form nanoparticles enabling controlled drug delivery, further enhancing vascular repair and regeneration. Additionally, HA-coated vascular devices, such as stents, exhibit reduced thrombogenicity, improving hemocompatibility and reducing the risk of complications like blood clots. These attributes make HA a versatile and valuable material in developing vascular grafts, stent coatings, contributing to, reduced thrombosis, and enhanced healing in vascular tissues [32,33].

2.2 Synthetic Polymers

Polyethylene glycol (PEG): This polymer has a range of uses in medicine and surgery, owing to its biocompatibility, hydrophilicity, and its potential to modify the attributes of molecule surfaces. Within pharmaceutical frameworks, PEGylation that involves the attachment of PEG to therapeutic molecules, such as drugs and proteins, to enhance their solubility, stability, and circulation time in the body. This modification reduces the immunogenicity of these molecules and prolongs their half-life, making them more effective in treating vascular diseases such as thrombosis and atherosclerosis. PEG is also used to coat nanoparticles in drug delivery systems, ensuring that they evade the immune system and target vascular tissues more effectively [34,35].

Materials Research Forum LLC
https://doi.org/10.21741/9781644903353-15

PEG-based hydrogels are employed in tissue regeneration to form scaffolds that resemble the extracellular matrix, encouraging blood vessel development and tissue healing. PEG is also used in synthetic grafts to improve biocompatibility and hemocompatibility, resulting in better integration with host tissues [36,37]. PEG's role in hemostasis and wound healing is also significant. It is used in the formulation of surgical sealants and adhesives that help control bleeding and seal blood vessels after surgery, thereby reducing the risk of leaks and promoting healing. PEG is also incorporated into wound dressings designed for vascular injuries, where it helps maintain a moist environment, promotes healing, and reduces the risk of infection. Furthermore, PEG's anti-fouling properties are harnessed in the coating of vascular medical devices, such as stents and catheters and helps prevent blood clot formation and enhances the performance of the devices [38-40].

Polylactic acid (PLA): This is a biocompatible polymer gaining prominence in vascular medicine and surgery, particularly in applications where biodegradability is critical. It is widely used in bioresorbable vascular stents, which provide temporary support to blood vessels after angioplasty. These stents maintain vessel patency and gradually degrade into lactic acid, reducing the need for long-term implants and minimizing the risk of late stent thrombosis [41]. PLA is also integral to drug-eluting stents, acting as a carrier for anti-proliferative drugs, preventing restenosis, and safely degrading after drug release [42]. In tissue engineering and vascular grafts, PLA scaffolds support new blood vessel growth and eventually degrade as tissues regenerate, reducing reliance on permanent materials [43,44]. PLA is also used in absorbable sutures for vascular surgeries, providing support during wound healing and eliminating the need for suture removal. PLA-based hemostatic agents promote clot formation and gradually degrade post-surgery [45,46].

Polycaprolactone (PCL): Polycaprolactone (PCL) is commonly utilized in the creation of synthetic vascular grafts, offering extended mechanical support, allowing the body sufficient time to regenerate its own tissue. As these grafts degrade, they are gradually taken over by natural tissue, reducing the need for permanent synthetic implants [47,48]. In vascular tissue engineering, PCL is utilized to develop scaffolds that temporarily support new blood vessel formation. Its mechanical properties can be customized to fit the requirements of vascular tissues, and its slow degradation allows the scaffold to remain effective throughout the tissue regeneration process [49,50]. PCL plays a critical function in drug-eluting stents, where it helps release therapeutic agents slowly over time to prevent restenosis after stent placement. PCL controls drug release and gradually degrades after delivery. It's also used in nanoparticulate drug delivery for vascular treatments, enhancing effectiveness and reducing side effects. Moreover, PCL is instrumental in haemostasis and wound healing, where it is used in the as a haemostatic agent to control bleeding during and after vascular surgeries. These agents promote clot formation and degrade gradually as healing progresses [51].

Polyurethane (PU): PU a versatile polymer that is utilized in vascular applications due to its excellent mechanical properties and flexibility. Its adaptability allows for engineering tailored to specific functions, making it valuable for the making of vascular devices. PU is commonly used in synthetic vascular grafts, which replace or bypass damaged blood vessels. These grafts combine strength, elasticity, and durability, closely mimicking natural blood vessels and providing long-term stability, essential for withstanding dynamic blood flow conditions such as plasticity and pressure changes. PU is also used in vascular stents, where it provides a biocompatible and flexible coating that reduces restenosis and minimizes tissue growth and blood clot formation. Its elasticity enables the stent to conform to vessel walls without compromising blood flow [52]. In catheter

Materials Research Foundations 172 (2025) 358-380 https://doi.org/10.21741/9781644903353-15

manufacturing, PU's flexibility, biocompatibility, and resistance to thrombosis make it ideal for extended vascular access during medical procedures [53]. PU is integrated into drug-eluting stents and catheters to control the release of therapeutic agents, reducing inflammation and restenosis while limiting systemic exposure. It also aids in haemostasis by controlling bleeding during vascular surgeries and promoting clot formation. Additionally, PU-based vascular patches repair or close blood vessels, providing either biodegradable or long-lasting support based on the application [54].

3. Biocompatibility and Mechanical Properties of Polymers in Vascular Applications:

Biocompatibility is crucial for preventing adverse reactions such as immune responses, inflammation, or rejection, which can lead to complications like infection, device failure, or additional surgeries. For example, non-biocompatible materials in vascular grafts can trigger thrombosis or hyperplasia, causing graft occlusion. Enhancing tissue integration is another key aspect of biocompatibility, as materials like titanium in orthopaedic implants bond well with bone, improving implant stability and reducing loosening risks. Additionally, biocompatible materials contribute to the long-term performance of medical devices by reducing degradation or complications, ensuring effectiveness over time. For instance, biocompatible polymers in drug delivery systems allow for controlled, sustained medication release, improving therapeutic outcomes [55]. Moreover, biocompatibility reduces complications in implantable devices such as pacemakers and stents; materials like Nitinol, used in stents, demonstrate excellent biocompatibility, minimizing risks like thrombosis and restenosis [56].

3.1 Techniques for assessing biocompatibility:

Cytotoxicity: Biocompatibility is evaluated using cell culture assays. Qualitative tests include direct contact, agar diffusion, and MEM elution, while quantitative tests like the MTT assay measure cell viability, commonly used to assess toxicity in compounds, pollutants, and antineoplastic drugs.

Irritation test: The local irritation of materials and devices is assessed on animal skin or mucous membranes, reflecting the expected clinical use and slightly exaggerating conditions for safety. The Intracutaneous Test involves intradermal injections to evaluate erythema and edema, recommended for devices with external or internal contact. The Primary Skin Irritation Test applies test materials to rabbit skin, while Mucous Membrane Irritation Tests assess irritation using extracts, focusing on areas like the vagina, cheek pouch, and eyes.

Acute Systemic Toxicity and Sub chronic toxicity: The Acute Systemic Toxicity test evaluates leachable from devices that may cause systemic toxicity by injecting extracts into mice, monitoring for toxic signs post-injection. This test is essential for all blood-contact devices and those interacting with internal tissues. The Material-Mediated Pyrogen test assesses a material's potential to induce fever, the rabbit pyrogen test is preferred for biocompatibility. Sub-chronic toxicity tests evaluate harmful effects from prolonged exposure, typically lasting up to 90 days in rats, and are mandatory for permanent devices.

Genotoxicity: Genotoxicity evaluations detect mutagens that can cause genetic damage, classified into point mutations, DNA damage, and chromosome damage. The Ames test is commonly used for point mutations, while Mouse Lymphoma and HGPRT assays assess both point and

Materials Research Foundations 172 (2025) 358-380
https://doi.org/10.21741/9781644903353-15

chromosome damage. ISO 10993-1 mandates genotoxicity testing for permanent and long-term contact devices.

Implantation Tests: Biocompatibility tests for medical devices and biomaterials that are with direct contact of living tissue, such as sutures, assess safety for short- or long-term implantation. These studies evaluate biochemical exchanges, cellular interactions, and immune responses, with histopathological analysis offering insights into the tissue response at the implant sites.

Hemocompatibility: Blood-contacting devices, such as catheters and vascular prostheses, must be evaluated for blood compatibility. The Haemolysis Assay measures red blood cell damage, while Coagulation Assays (PT and PTT) assess blood clotting pathways. Thrombogenicity testing is typically performed in vivo, and Complement Activation Testing evaluates inflammatory responses in plasma for devices in contact with blood.

3.2 Mechanical properties: strength and flexibility, durability under physiological conditions

In vascular applications, the mechanical properties of polymers, including strength, flexibility, and durability under physiological conditions, are crucial for ensuring functionality and longevity. Polymers must exhibit high tensile strength to withstand the mechanical stresses and pressures from blood flow, preventing structural failure. Flexibility is equally important, allowing the material to accommodate the dynamic movements and pulsations within blood vessels without compromising integrity. Polyethylene terephthalate (PET, or Dacron) is commonly used in vascular grafts due to its impressive strength and flexibility, enabling it to withstand pulsatile blood flow and conform to the natural movements of blood vessels. PET grafts are often woven or knitted to enhance compliance, thereby mimicking the natural arteries. Similarly, polytetrafluoroethylene (PTFE, or Teflon) is utilized in bypass grafts for its excellent chemical resistance, low friction, and durability, ensuring flexibility and preventing kinking in small-diameter applications. Durability is critical for maintaining performance over prolonged exposure to the complex conditions of the human body, including resistance to degradation from bodily fluids and mechanical wear. Polyurethane (PU) is favoured for blood-contacting devices due to its elasticity and biocompatibility, while poly(L-lactic acid) (PLLA) offers biodegradability for vascular stents and drug delivery systems, allowing for customizable durability. Together, these properties ensure that polymers effectively serve as scaffolds or components in vascular applications, enhancing the overall success and safety of medical devices and implants [57].

3.3 Bio-functionality, drug delivery capabilities, antithrombogenic properties, degradability, stability, and their impact on long-term use

In vascular applications, the choice of polymers is crucial due to their roles in bio-functionality, drug delivery capabilities, and antithrombogenic properties. These materials must mimic the properties of blood vessels while fulfilling specific requirements such as biocompatibility, durability, and resistance to thrombosis. Bio-functionality describes a polymer's capacity to interact safely and effectively with biological systems. Biocompatible polymers are widely used due to their ability to support cell adhesion and degrade into non-toxic components. These materials also improve vascular devices by providing controlled drug release, aiding in the prevention of restenosis and promoting healing. PLGA allows for adjustable degradation rates, while hyaluronic acid hydrogels can target specific tissues for enhanced therapeutic effects.

Materials Research Foundations 172 (2025) 358-380 https://doi.org/10.21741/9781644903353-15

Antithrombogenic properties are vital for preventing blood clot formation, which can lead to severe complications; polymers like PTFE and polyurethanes can be modified with anticoagulants or endothelial cell-promoting factors, while hydrophilic surfaces, such as those modified with PEG, minimize protein adsorption and platelet adhesion, thereby improving overall vascular health. Polymers used in vascular applications must strike a balance between degradability and stability to ensure immediate functionality and long-term safety, which is crucial for the performance of implants such as stents, grafts, and scaffolds. Degradable polymers are vital for temporary implants, as controlled degradation enhances patient safety by eliminating the need for removal procedures, while stable polymers are required for permanent devices to withstand the physiological environment without losing functionality or causing adverse effects. The degradation behaviour of polymers in vivo is influenced by factors like enzymatic activity, pH levels, and mechanical stresses; for example, poly(L-lactic acid) (PLLA) in biodegradable stents undergoes hydrolytic degradation into lactic acid, preventing any additional surgeries. In vitro, laboratory studies simulate the body's environment to evaluate degradation rates and stability, employing accelerated aging tests to mimic oxidative conditions and providing insights into long-term material performance before clinical application. Ongoing research continues to enhance the degradation and stability of these materials, thereby improving their safety and effectiveness in vascular applications. While degradability is advantageous for temporary implants, such as bioresorbable vascular scaffolds, it must be carefully controlled to ensure the material provides sufficient support during the critical healing period before it degrades [58-60]. The degradation products must also be biocompatible to avoid inflammatory responses or other adverse effects. Polymers like PGA and PLGA are generally used for their predictable degradation profiles [61]. For permanent vascular applications, such as non-degradable stents or grafts, stability is crucial.

4. Polymer applications in vascular applications

Polymers are essential in advancing drug delivery systems offering customizable, versatile, and biocompatible solutions for grafts, stents, tissue engineering, and targeted therapies [62].

4.1 Synthetic vascular grafts

Synthetic vascular grafts, made from materials such as Polyethylene Terephthalate (PET), Polytetrafluoroethylene (PTFE), and Polyurethane (PU), play a crucial role in vascular surgery. PET is strong, durable, biocompatible, resistant to degradation, and has a low infection risk, making it ideal for large-diameter grafts like those used for the aorta and large arteries (e.g., Dacron grafts). PTFE is chemically inert, non-reactive, and biofouling-resistant, suitable for small-diameter grafts, arteriovenous (AV) access for hemodialysis, and peripheral vascular grafts (e.g., Gore-Tex grafts). PU's elasticity, flexibility, and hemocompatibility make it particularly useful for small-diameter grafts like those used in coronary artery bypass grafting (CABG). Key design aspects include porosity, where micro-porous materials promote tissue integration, and macro-porous materials allow for faster endothelialization; compliance, which must balance stiffness and elasticity to match native vessels; and surface modifications, such as heparin coating to reduce clot formation or endothelial cell seeding to mimic natural vessel lining. Synthetic grafts are commonly applied in aortic aneurysm repair using PET or PTFE for their strength and durability, and in peripheral artery disease (PAD) with PTFE for its flexibility and thrombosis resistance. PTFE is also used in AV fistulas for dialysis due to its biocompatibility and non-reactive nature. Despite advantages like availability in various sizes, customizability, and durability, synthetic grafts face challenges such as thrombogenicity, infection risks, compliance mismatch leading to intimal hyperplasia, and limited success in small-diameter applications like coronary arteries, where lower patency affects performance.

4.2 Biodegradable vascular grafts

Biodegradable vascular grafts represent a significant advancement in vascular surgery, providing temporary support to damaged or diseased blood vessels while promoting natural tissue regeneration. These grafts degrade gradually, eliminating the need for permanent implants. PLA offers good mechanical strength and an adjustable degradation rate, while PGA has a faster degradation rate but is typically combined with PLA to balance its mechanical strength and degradation. PLGA, with tunable degradation rates [63], is extensively used in controlled delivery of drugs and temporary scaffolds, while polymers of natural origin like gelatin and collagen, when combined with synthetic polymers, improve tissue regeneration. The design of biodegradable grafts involves key factors such as degradation rate, which must align with tissue regeneration to offer temporary support, and mechanical properties, which are enhanced by blending polymers or using reinforcement techniques. Biocompatibility, essential for minimizing adverse reactions, can be tailored through surface modifications, and porosity facilitates cell infiltration and tissue integration, developed through techniques like electrospinning and 3D printing. Biodegradable grafts hold promise in small-diameter vascular grafts, where synthetic grafts often fail due to thrombosis and intimal hyperplasia, and in peripheral artery disease (PAD), where they help bypass occluded arteries and promote healing. In coronary artery bypass grafting (CABG), they offer an alternative to autologous veins, and in pediatric applications, biodegradable grafts accommodate growth by being replaced with natural tissue, avoiding repeat surgeries. The advantages of these grafts include reduced long-term complications, promotion of natural healing, elimination of the need for removal, and customizable properties. However, challenges remain, such as thrombogenicity, mechanical stability, controlled degradation, and biocompatibility.

4.3 Coatings and modifications to improve graft performance

Coatings and surface modifications are essential for enhancing the performance of grafts by addressing issues such as thrombogenicity, infection, and tissue integration. Anti-thrombogenic coatings, using substances like heparin, nitric oxide (NO), and hydrophilic materials, prevent blood clot formation. Heparin-bonded Dacron grafts inhibit thrombin and clotting factors, while NO-releasing PTFE grafts maintain an anti-thrombogenic environment. Hydrophilic surfaces, such as polyethylene glycol (PEG), resist protein adsorption and platelet adhesion, enhancing hemocompatibility. Anti-infective coatings, including antibiotics like rifampicin or gentamicin, reduce infection risk, and silver-impregnated grafts provide broad-spectrum antimicrobial activity. Antimicrobial peptide (AMP) coatings also disrupt bacterial membranes without fostering antibiotic resistance. Endothelialization-enhancing coatings, such as growth factors like VEGF and bFGF, promote endothelial cell proliferation, while autologous cell-seeding mimics natural blood vessel lining to improve biocompatibility. Bioactive coatings using extracellular matrix (ECM) proteins like collagen and fibronectin support tissue integration, and polydopamine creates bioactive layers for further functionalization. Smart coatings, including pH- and temperature-responsive systems, adapt to physiological changes, releasing drugs in response to triggers like inflammation or infection for targeted, controlled delivery. These advanced coatings enhance the functionality of vascular grafts, leading to better patient outcomes by mitigating thrombosis,

Materials Research Foundations 172 (2025) 358-380
https://doi.org/10.21741/9781644903353-15

preventing infection, and promoting tissue integration. Continuous innovations in materials science and biomedical engineering are driving the development of more versatile and effective graft coatings, offering new therapeutic possibilities for vascular disease treatment.

4.4 Stents for vascular use

Stents play a crucial role in treating vascular diseases, with advancements in polymer science significantly enhancing their performance and applications [64]. Polymer-coated metal stents combine the mechanical strength of metals with the biocompatibility of polymers, improving biocompatibility, reducing thrombogenicity, and allowing for drug-eluting capabilities, while providing strong mechanical support. However, if the polymer is not biocompatible or if drug release is poorly controlled, there may be a risk of delayed Endothelialization; examples include stainless steel or cobalt-chromium stents coated with biocompatible polymers like Polyethylene Glycol (PEG) or Polyurethane (PU). Fully polymeric stents, made entirely from biodegradable materials, offer temporary support with gradual degradation, making them suitable for coronary artery disease and areas requiring minimal long-term foreign material. They promote natural vessel healing and have fewer long-term complications, but require careful management of mechanical integrity and degradation rates, such as Polylactic Acid (PLA) and Polyglycolic Acid (PGA) stents. Drug-eluting stents (DES) release medications to prevent restenosis while providing mechanical support, commonly used for coronary and peripheral artery diseases. DES reduce the incidence of restenosis through targeted drug delivery, minimizing systemic side effects, and maintaining prolonged therapeutic effects at the stent site. Despite their advantages, considerations include the potential for late stent thrombosis and the necessity for long-term antiplatelet therapy, along with the challenge of balancing drug release kinetics; examples include Sirolimus-eluting, Paclitaxel-eluting, and Everolimus-eluting stents. Continuous innovations aim to improve patient outcomes by addressing the limitations of traditional stent technologies.

4.5 Vascular patches and their use in surgical repair

Vascular defects associated with cardiovascular diseases frequently necessitate surgical intervention, often employing vascular patches that play a crucial role in sealing blood vessels post-surgery. These patches must exhibit high biocompatibility to minimize complications such as thrombosis, inflammation, calcification, or neointimal hyperplasia, which can lead to surgical failure. Typically made from materials like polyester-urethane or knitted polyester, sometimes impregnated with collagen, vascular patches are designed to be thin, elastic, and maintain blood tightness, biocompatibility, and mechanical stability over time. Many patches feature a microporous structure to enhance tissue integration or alignment guidelines for precise implantation. Their applications are particularly significant in procedures such as carotid artery endarterectomy (CEA), where they address small or irregular arteriotomies or repair distal internal carotid arteries. Vascular patches can be developed from both synthetic and biological materials. Synthetic materials include polytetrafluoroethylene (PTFE), which is biocompatible, non-thrombogenic, and available as expanded PTFE (ePTFE) with a microporous structure, making it suitable for arterial and venous reconstruction; polyethylene terephthalate (PET), commercially known as Dacron, is widely used due to its high tensile strength, durability, and biocompatibility, particularly for large vessel reconstruction and aortic aneurysm repairs; and polyurethane (PU), a flexible and elastic material used for peripheral artery repair and small-diameter applications, often enhanced by combining with other materials [65]. Biological materials, on the other hand, include

Materials Research Foundations 172 (2025) 358-380
https://doi.org/10.21741/9781644903353-15

bovine pericardium, noted for its excellent biocompatibility and tissue integration properties; human pericardium, both autologous and allogeneic, valued for its superior compatibility and effectiveness in repairing delicate structures, especially in pediatric surgery; porcine small intestinal submucosa (SIS), a biodegradable material effective in promoting tissue regeneration; and decellularized human or animal vessels, which serve as natural scaffolds to support cellular repopulation and integration. Vascular patches are utilized in various surgical repairs, including arterial and venous reconstruction to reinforce or replace damaged vessels, aneurysm repair, where patches stabilize the affected area following resection, and congenital heart defect repair, addressing anomalies like septal defects with patches made from bovine pericardium or SIS. They are also critical in carotid endarterectomy to widen arteries after plaque excision, with PTFE and bovine pericardium being common materials, and in peripheral artery disease (PAD) repair, where patches restore blood flow by widening or reconstructing occluded vessels. Furthermore, vascular patches are vital in haemodialysis access, where they reinforce arteriovenous (AV) fistulas, ensuring reliable vascular access. The choice of synthetic or biological materials for these patches depends on the specific clinical requirements, with synthetic options offering durability and biological alternatives providing compatibility. Continuous advancements in materials and construction techniques are enhancing the performance and application scope of vascular patches, thus playing a significant role in improving patient outcomes in cardiovascular surgeries.

4.6 Polymer applications in vascular medicine

Polymers play a key role in vascular medicine. Localized drug delivery targets specific sites, such as particular blood vessels or tissues, ensuring high local concentrations of drugs while reducing exposure to the rest of the body. Various polymer-based delivery vehicles are utilized for this purpose. Microspheres and nanoparticles offer customizable release profiles and are instrumental in delivering chemotherapeutic agents and anti-inflammatory drugs to treat vascular diseases and tumours. Hydrogels, composed of materials such as alginate, are hydrophilic and capable of retaining large amounts of water, making them ideal for injectable applications that provide controlled drug release and support angiogenesis in ischemic tissues. Polymeric coatings, applied to medical devices like stents and vascular grafts, allow for sustained drug release to prevent restenosis and reduce thrombogenicity. Additionally, polymeric films and patches, made from materials like chitosan and PLGA, facilitate localized drug delivery through diffusion or degradation-controlled mechanisms, addressing post-operative adhesions and preventing restenosis. Lastly, polymeric implants designed for long-term drug release are particularly beneficial for chronic vascular conditions such as atherosclerosis and peripheral artery disease, providing sustained therapeutic intervention at the site of the disease. Collectively, these polymer-based systems significantly improve patient compliance and treatment outcomes in vascular medicine.

5. Innovations and future directions

Polymers have revolutionized vascular applications by offering biocompatibility, mechanical strength, and adaptability, essential for addressing cardiovascular diseases. Advanced polymers are used in stents, grafts, and catheters, designed to mimic the natural elasticity of blood vessels, reducing risks like thrombosis and restenosis. Recent innovations include bioresorbable scaffolds that degrade over time, eliminating the need for additional surgeries. Nanotechnology further

Materials Research Foundations 172 (2025) 358-380
https://doi.org/10.21741/9781644903353-15

enhances targeted drug delivery. As research advances, personalized vascular therapies are expected to transform cardiovascular care [66].

5.1 Smart polymers temperature-sensitive polymers ph-sensitive polymers

Smart polymers have revolutionized vascular applications due to their ability to respond to environmental stimuli, enabling controlled and targeted therapeutic effects [67]. Notably, temperature-sensitive and pH-sensitive polymers play key roles in drug delivery, tissue engineering, and vascular interventions. Temperature-sensitive polymers like poly(N-isopropylacrylamide) (PNIPAM) undergo a phase transition near body temperature, allowing for controlled drug release, particularly in inflamed tissues. PNIPAM-based hydrogels, for example, enable localized anti-inflammatory drug delivery by responding to temperature changes. Additionally, polymers like PEG-PLA have been used in embolization therapy to block abnormal blood flow. pH-sensitive polymers, such as poly(acrylic acid) (PAA), respond to pH variations in environments like tumors, releasing drugs precisely at the target site to improve efficacy and reduce side effects [68-70]. Another example is poly(methacrylic acid) (PMAA), used in drug-eluting stents, where the polymer coating responds to the acidic environment of atherosclerotic plaques, swelling and releasing anti-proliferative drugs to prevent restenosis [71]. This targeted drug delivery approach reduces the risk of adverse effects in non-target areas. Additionally, pH-sensitive polymers are being explored for vascular grafts and tissue engineering, where polymer scaffolds can release growth factors in response to the slightly acidic environment of regenerating tissues, thus promoting healing, new blood vessel growth, and better integration of the graft with surrounding tissues.

5.2 Nanotechnology in polymers nanoparticles for targeted therapy nanofibers and their applications

Nanotechnology has greatly advanced vascular medicine by integrating nanoparticles and nanofibers into polymer-based materials, enhancing targeted therapy and improving treatment outcomes for vascular diseases. Polymeric nanoparticles, typically 1 to 100 nanometers in size, offer precise drug delivery by binding to specific cellular markers, reducing systemic side effects, and providing controlled, sustained release. For instance, biodegradable PLGA nanoparticles are used to treat atherosclerosis by delivering anti-inflammatory drugs directly to plaque sites. Nanofibers, made from polymers like PCL and collagen, mimic the extracellular matrix and are crucial in tissue engineering scaffolds and wound dressings due to their high surface area and strength [72, 73]. In vascular tissue engineering, nanofiber scaffolds support endothelial cell growth, promoting new blood vessel formation and biodegrade as tissue forms, eliminating the need for removal. Functionalized with bioactive molecules like vascular endothelial growth factor (VEGF), nanofibers accelerate tissue regeneration. They are also used in drug-eluting stents, reducing restenosis rates by delivering therapeutic agents directly to vascular sites [74].

5.3 Tissue engineering and regenerative medicine-scaffolds

There has been a significant development especially in vascular tissue engineering, where developing artificial blood vessels is essential for treating cardiovascular diseases [75]. These vessels are constructed using scaffolds that provide a 3D framework to support cell attachment, growth, and differentiation, mimicking the natural extracellular matrix (ECM) of blood vessels. Choosing the right scaffold material is crucial, requiring biocompatibility, biodegradability, and

Materials Research Forum LLC
https://doi.org/10.21741/9781644903353-15

appropriate mechanical properties to support vascular cells. Natural polymers like collagen, elastin, and fibrin are biocompatible and promote cell adhesion due to their similarity to native ECM, but often lack sufficient mechanical strength. Synthetic polymers such as polyglycolic acid (PGA), polylactic acid (PLA), and polycaprolactone (PCL) provide more control over scaffold properties like degradation rates and mechanical strength, though they may require surface modifications to enhance cell interactions. Advanced fabrication techniques like electrospinning, 3D printing, and bioprinting enable the creation of scaffolds with architectures that closely mimic natural blood vessel structures. Adding bioactive molecules, such as peptides and proteins, boosts the bioactivity of scaffolds, encouraging vascular cell growth and organization. Stem cells are integral to vascular tissue engineering because they can differentiate into endothelial cells, smooth muscle cells, and pericytes, all of which are essential for creating functional blood vessels. Stem cell sources include embryonic stem cells (ESCs), induced pluripotent stem cells (iPSCs), and adult stem cells such as mesenchymal stem cells (MSCs) and endothelial progenitor cells (EPCs). While ESCs and iPSCs are highly versatile, concerns around ethics and tumorigenesis limit their application. Adult stem cells, including MSCs and EPCs, are safer and have demonstrated potential in vascular regeneration. Scaffold properties like elasticity and stiffness affect stem cell differentiation, while growth factors like VEGF, PDGF, FGF, and TGF-β regulate the formation of functional blood vessels. Controlled release of these factors through microspheres or hydrogel matrices ensures localized, sustained delivery to support effective vascular tissue formation.

5.4 Advances in bioprinting3D printing of vascular structures customization and patient-specific solutions

Advances in 3D bioprinting have revolutionized the creation of vascular structures, offering solutions to the limitations of traditional grafts, such as size mismatch, immunogenicity, and limited growth potential. This technology enables the precise fabrication of complex vascular networks, enhancing graft integration and functionality. The process starts with advanced imaging techniques like MRI or CT scans, which produce detailed models of a patient's vascular anatomy, serving as blueprints for bioprinting grafts that fit seamlessly into the vascular system. This customization is particularly important for pediatric patients, where growth potential and anatomical changes must be considered to ensure long-term success. Recent advancements in materials have been crucial in developing effective vascular structures. Modern bio-inks, composed of living cells and biocompatible polymers, such as biodegradable polyesters like poly(propylene fumarate), have shown promise in maintaining graft patency and functionality over time. The choice of materials is critical, as they must possess suitable mechanical properties to support vascular tissue growth. Integrating vascular networks into 3D-printed tissues is vital for nutrient transport and waste removal, which are critical for the viability of larger constructs. Besides this, challenges remain, like safeguarding the long-term viability also the functionality of bioprinted grafts, particularly regarding vascular network integrity and thrombosis prevention. Additionally, the evolving regulatory landscape for bioprinted devices may affect the speed of clinical adoption.

6. Challenges and Considerations

The use of polymers in vascular applications, such as stents and grafts, presents a range of challenges and considerations [76] crucial for ensuring these medical devices' safety, efficacy, and longevity. A primary concern is biocompatibility, where polymers must not trigger adverse

Materials Research Foundations 172 (2025) 358-380
https://doi.org/10.21741/9781644903353-15

immune responses or thrombosis while promoting endothelialisation to reduce the risk of clot formation. Achieving the right balance between flexibility and mechanical strength is vital, as these materials must withstand the dynamic forces of blood flow without degrading prematurely. Thrombogenicity, or the material's potential to cause clotting, is a significant issue, often addressed through surface modifications and smooth surfaces to minimize platelet activation. In drug-eluting devices, polymers must enable precise, sustained drug release, ensuring therapeutic efficacy at the target site without systemic side effects. Degradation and stability are also critical; polymers must degrade at a controlled rate, with non-toxic by-products. Sterilization methods pose another challenge, as polymers must endure processes like gamma irradiation without compromising their properties. Additionally, the consistency of polymer properties during large-scale manufacturing is essential to ensure reliable device performance. Regulatory and ethical considerations further complicate the selection and use of polymers, as they must meet stringent standards and, in some cases, avoid materials that raise ethical concerns, such as those derived from animals. Finally, the cost and accessibility of polymers are significant factors, particularly in resource-limited settings, where the choice of material can impact the affordability and availability of vascular devices. Balancing these challenges requires ongoing research and innovation to advance the field and improve patient outcomes in vascular medicine.

6.1 Overview of regulatory requirements case studies of successful approvals

Navigating the regulatory and approval processes for polymer-based medical devices in vascular applications [77] involves rigorous compliance with standards set by bodies like the U.S. FDA and the European Medicines Agency (EMA). In the U.S., the approval pathway depends on the device's risk level. The 510(k) pathway is used for devices similar to existing ones, requiring proof of equivalent safety and effectiveness. For higher-risk or novel devices, the Premarket Approval (PMA) process demands extensive clinical data and manufacturing scrutiny. The De Novo pathway allows for risk-based assessment of low to moderate-risk devices without predicates. In the EU, obtaining CE marking under the Medical Devices Regulation (MDR) involves a conformity assessment by Notified Bodies, particularly for high-risk devices, and is based on robust clinical evidence. Compliance with international standards for Quality Management Systems and risk management is essential. Manufacturers must ensure effective risk identification and mitigation strategies. Post-market surveillance is critical for monitoring the device's performance, ensuring its continued safety and effectiveness in real-world settings. This ongoing oversight is crucial for maintaining regulatory compliance and patient safety over time.

Case Study 1: Cypher Drug-Eluting Stent: This stent [78] developed by Cordis Corporation, was the first of its kind to receive FDA approval. This stent is coated with a polymer that slowly releases an anti-proliferative drug to prevent the re-narrowing of arteries, a common issue with traditional bare-metal stents. The regulatory journey for the Cypher stent involved the PMA pathway due to its novel combination of a drug and a polymer-coated stent. Extensive preclinical testing and large-scale clinical trials, including randomized controlled trials with over 1,300 patients, were conducted to demonstrate the stent's safety and efficacy. The results showed a significant reduction in major adverse cardiac events (MACE) compared to bare-metal stents, ultimately leading to their approval. Post-market surveillance was critical in monitoring the long-term safety of the Cypher stent, including the risk of late stent thrombosis, which has since informed the design and regulation of subsequent generations of drug-eluting stents.

Materials Research Forum LLC
https://doi.org/10.21741/9781644903353-15

Case Study 2: Absorb GT1 Bioresorbable Vascular Scaffold: It [79] represented a groundbreaking innovation in vascular therapy by providing temporary support to the blood vessel before gradually dissolving, leaving no permanent implant behind. The Absorb GT1 was approved through the PMA pathway, reflecting its novel design and the need for substantial clinical data. Clinical trials involving over 2,000 patients demonstrated that the Absorb scaffold was non-inferior to traditional drug-eluting stents in terms of safety and efficacy, leading to its approval. However, the post-market experience revealed challenges, particularly an increased risk of late scaffold thrombosis, which eventually led to the device being withdrawn from the market. This case study underscores the critical importance of long-term clinical data and post-market surveillance in the regulatory process, highlighting how even approved devices must be continuously monitored to ensure they meet safety standards in real-world settings.

6.2 Clinical trials and evidence importance of clinical trials review of major clinical trials involving vascular polymers

Clinical trials are paramount in the development and approval of vascular polymers, as they provide the necessary evidence to demonstrate safety, efficacy, and performance in real-world applications. These trials are designed to assess the clinical outcomes, potential risks, and long-term effects of polymer-based devices in the vascular system. The importance of clinical trials lies in their ability to generate robust data that informs regulatory decisions and guides clinical practice. A review of major clinical trials involving vascular polymers highlights the critical role these studies play in advancing medical technology. For example, trials involving drug-eluting stents, such as the SPIRIT and ENDEAVOR series, have shown significant reductions in restenosis rates compared to bare-metal stents, leading to widespread adoption and regulatory approval. Similarly, trials for bioresorbable vascular scaffolds, like the ABSORB trials, have provided insights into the benefits and challenges of using temporary scaffolds in vascular interventions, although long-term safety concerns have tempered initial enthusiasm. These studies underscore the essential nature of clinical trials in validating the clinical utility of vascular polymers, ensuring that these devices meet the rigorous standards required for patient safety and therapeutic effectiveness.

6.3 Ethical and economic considerations cost-effectiveness of polymer-based solutions ethical issues in polymer research and application

When considering the use of polymers in vascular applications, both ethical and economic aspects play crucial roles in their adoption and development. The cost-effectiveness of polymer-based solutions is often highlighted due to their ability to offer durable, biocompatible, and versatile materials that can reduce long-term healthcare costs by improving patient outcomes and minimizing the need for repeated interventions. For instance, polymer-coated stents and grafts can significantly decrease the incidence of restenosis, leading to fewer surgeries and hospitalizations, thus proving economically beneficial over time [80]. However, these innovations also raise ethical issues, particularly in research and application. The development and testing of new polymer-based devices must adhere to strict ethical standards, ensuring that clinical trials are conducted with informed consent, transparency, and a clear risk-benefit analysis to protect patient safety. Additionally, there is an ethical obligation to ensure equitable access to these advanced treatments, avoiding a scenario where only affluent populations benefit from cutting-edge technologies. Balancing the economic advantages of polymer-based vascular solutions with ethical

Materials Research Foundations 172 (2025) 358-380
https://doi.org/10.21741/9781644903353-15

considerations is essential to ensure that these innovations contribute positively to healthcare while maintaining trust and fairness in medical practice.

7. Real-world Applications and outcomes.

In recent years, polymers have become crucial in vascular applications, leading to significant advancements in patient care. For example, bioresorbable stents made from polymers like polylactic acid (PLA) have been successfully used to treat coronary artery disease (CAD). Clinical trials have shown that bioresorbable stents maintain arterial patience and support better patient outcomes, spurring further research into advanced materials. Similarly, polymer-coated vascular grafts used in bypass surgery offer advantages over traditional synthetic grafts made from materials like Dacron and ePTFE. The polymer coatings, often treated with bioactive agents like heparin, provide antithrombotic properties. Patients with polymer-coated grafts typically experience fewer complications and better outcomes. Drug-eluting balloons (DEBs) are another successful application of polymers in vascular treatments. These devices use a polymer coating to deliver therapeutic agents directly to vascular lesions, proving effective in treating conditions like peripheral artery disease (PAD) and in-stent restenosis. The controlled release of drugs such as paclitaxel inhibits cell proliferation, significantly lowering restenosis rates compared to plain balloons and some drug-eluting stents. The success of polymer-based vascular products highlights the importance of selecting appropriate polymer materials with balanced biocompatibility, mechanical strength, and degradation rates. Integrating bioactive agents can enhance performance, but their release must be carefully managed. Long-term studies are essential to fully assess these products' safety and effectiveness across diverse patient populations. Overall, polymer-based solutions offer distinct advantages over traditional methods, including reduced long-term complications and better patient-specific outcomes, making them increasingly favourable in clinical practice.

Conclusion

Polysaccharides and synthetic polymers have demonstrated significant potential in advancing both sustainable and innovative technologies, particularly in the field of vascular applications. Natural polysaccharides such as collagen, chitosan, and alginate provide excellent biocompatibility and biodegradability, making them ideal for tissue engineering, drug delivery systems, and wound healing. Their ability to support cell adhesion, proliferation, and controlled drug release has enhanced their utility in biomedical fields, while their biodegradability addresses environmental concerns. Synthetic polymers, including polyethylene glycol (PEG), polylactic acid (PLA), polycaprolactone (PCL), and polyurethane, offer mechanical strength, flexibility, and durability, which are critical for long-term vascular applications. These materials are widely used in the development of vascular grafts, stents, and patches, providing structural support and promoting tissue regeneration. Recent innovations in smart polymers, nanotechnology, and bioprinting have further expanded the possibilities for personalized and targeted therapies, offering patient-specific solutions in regenerative medicine. Despite their immense potential, challenges remain in scaling up production, ensuring cost-effectiveness, and navigating regulatory hurdles. The future of polysaccharides and synthetic polymers in vascular applications will rely on overcoming these barriers through continued research, clinical trials, and advances in material science. As the demand for sustainable, biocompatible materials grows, these polymers are poised to play a crucial role in improving patient outcomes, driving innovation in medical technologies, and supporting eco-friendly practices across industries.

Acknowledgment

The authors are thankful to the School of Health Sciences and Technology, Department of Pharmaceutical Sciences, Dr. Vishwanath Karad MIT World Peace University, Department of Pharmaceutics, Dadasaheb Balpande College of Pharmacy; Dr DY Patil Institute of Pharmaceutical Sciences & Research and STES's Smt. Kashibai Navale College of Pharmacy for providing the necessary facilities to carry out the work for this article.

References

[1] P. Libby, P.M. Ridker, G.K. Hansson, Progress and challenges in translating the biology of atherosclerosis, Nature. 473 (2011) 317-325. https://doi.org/10.1038/nature10146

[2] P.B. Gorelick, A. Scuteri, S.E. Black, C. Decarli, S.M. Greenberg, C. Iadecola, L.J. Launer, S. Laurent, O.L. Lopez, D. Nyenhuis, R.C. Petersen, J.A. Schneider, C. Tzourio, D.K. Arnett, D.A. Bennett, H.C. Chui, R.T. Higashida, R. Lindquist, P.M. Nilsson, G.C. Roman, F.W. Sellke, S. Seshadri, Vascular contributions to cognitive impairment and dementia: a statement for healthcare professionals from the american heart association/american stroke association, Stroke. 42 (2011) 2672-2713. https://doi.org/10.1161/STR.0b013e3182299496

[3] A.J. Lusis, Atherosclerosis, Nature. 407 (2000) 233-241. https://doi.org/10.1038/35025203

[4] G.K. Hansson, Inflammation, atherosclerosis, and coronary artery disease, N Engl J Med. 352 (2005) 1685-1695. https://doi.org/10.1056/NEJMra043430

[5] R. Ross, Atherosclerosis--an inflammatory disease, N Engl J Med. 340 (1999) 115-126. https://doi.org/10.1056/NEJM199901143400207

[6] S. Jebari-Benslaiman, U. Galicia-García, A. Larrea-Sebal, J.R. Olaetxea, I. Alloza, K. Vandenbroeck, A. Benito-Vicente, C. Martín, Pathophysiology of Atherosclerosis, Int J Mol Sci. 23 (2022) 3346-3384. https://doi.org/10.3390/ijms23063346

[7] V. Fuster, L. Badimon, J.J. Badimon, J.H. Chesebro, The pathogenesis of coronary artery disease and the acute coronary syndromes (1), N Engl J Med. 326 (1992) 242-250. https://doi.org/10.1056/NEJM199201233260406

[8] A.K. Malakar, D. Choudhury, B. Halder, P. Paul, A. Uddin, S. Chakraborty, A review on coronary artery disease, its risk factors, and therapeutics, J Cell Physiol. 234 (2019) 16812-16823. https://doi.org/10.1002/jcp.28350

[9] V. Smith, D. Devane, C.M. Begley, M. Clarke, Methodology in conducting a systematic review of systematic reviews of healthcare interventions, BMC Med Res Methodol. 11 (2011) 15-21 https://doi.org/10.1186/1471-2288-11-15

[10] U. Ralapanawa, R. Sivakanesan, Epidemiology and the Magnitude of Coronary Artery Disease and Acute Coronary Syndrome: A Narrative Review, J Epidemiol Glob Health. 11 (2021) 169-177. https://doi.org/10.2991/jegh.k.201217.001

[11] C. Kearon, E.A. Akl, J. Ornelas, A. Blaivas, D. Jimenez, H. Bounameaux, M. Huisman, C.S. King, T.A. Morris, N. Sood, S.M. Stevens, J.R.E. Vintch, P. Wells, S.C. Woller, L. Moores, Antithrombotic Therapy for VTE Disease: CHEST Guideline and Expert Panel Report, Chest. 149 (2016) 315-352. https://doi.org/10.1016/j.chest.2015.11.026

[12] S.G. Rockson, Lymphedema, Am J Med. 110 (2001) 288-295. https://doi.org/10.1016/S0002-9343(00)00727-0

[13] S.J. Greene, G.C. Fonarow, M. Vaduganathan, S.S. Khan, J. Butler, M. Gheorghiade, The vulnerable phase after hospitalization for heart failure, Nat Rev Cardiol. 12 (2015) 220-229. https://doi.org/10.1038/nrcardio.2015.14

[14] O. Kayıran, C. De La Cruz, K. Tane, A. Soran, Lymphedema: From diagnosis to treatment, Turk J Surg. 33 (2017) 51-57. https://doi.org/10.5152/turkjsurg.2017.3870

[15] R. Mortimer, M. Privopoulos, S. Kumar, The effectiveness of hydrotherapy in the treatment of social and behavioral aspects of children with autism spectrum disorders: a systematic review, J Multidiscip Healthc. 7 (2014) 93-104. https://doi.org/10.2147/JMDH.S55345

[16] G. Peppas, V.G. Alexiou, E. Mourtzoukou, M.E. Falagas, Epidemiology of constipation in Europe and Oceania: a systematic review, BMC Gastroenterol. 8 (2008) 5-12. https://doi.org/10.1186/1471-230X-8-5

[17] P.K. Whelton, R.M. Carey, The 2017 Clinical Practice Guideline for High Blood Pressure, JAMA. 318 (2017) 2073-2074. https://doi.org/10.1001/jama.2017.18209

[18] O.A. Carretero, S. Oparil, Essential hypertension. Part I: definition and etiology, Circulation. 101 (2000) 329-35. https://doi.org/10.1161/01.CIR.101.3.329

[19] R.P. Lifton, A.G. Gharavi, D.S. Geller, Molecular mechanisms of human hypertension, Cell. 104 (2001) 545-56. https://doi.org/10.1016/S0092-8674(01)00241-0

[20] L. Claesson-Welsh, Vascular permeability--the essentials, Ups J Med Sci. 120 (2015) 135-143. https://doi.org/10.3109/03009734.2015.1064501

[21] S. Fathi-Karkan, B. Banimohamad-Shotorbani, S. Saghati, R. Rahbarghazi, S. Davaran, A critical review of fibrous polyurethane-based vascular tissue engineering scaffolds, J Biol Eng. 16 (2022) 6-24. https://doi.org/10.1186/s13036-022-00286-9

[22] A. Premkumar, D.A. Kolin, K.X. Farley, J.M. Wilson, A.S. McLawhorn, M.B. Cross, P.K. Sculco, Projected Economic Burden of Periprosthetic Joint Infection of the Hip and Knee in the United States, J Arthroplasty. 36 (2021) 1484-1489. https://doi.org/10.1016/j.arth.2020.12.005

[23] L. Claesson-Welsh, Vascular permeability--the essentials, Ups J Med Sci. 120 (2015) 135-143. https://doi.org/10.3109/03009734.2015.1064501

[24] R. Parenteau-Bareil, R. Gauvin, F. Berthod, Collagen-Based Biomaterials for Tissue Engineering Applications, Materials (Basel). 3 (2010) 1863-1887. https://doi.org/10.3390/ma3031863

[25] F. Copes, N. Pien, S. Van Vlierberghe, F. Boccafoschi, D. Mantovani, Collagen-Based Tissue Engineering Strategies for Vascular Medicine, Front Bioeng Biotechnol. 166 (2019). https://doi.org/10.3389/fbioe.2019.00166

[26] S. Ravi, E.L. Chaikof, Biomaterials for vascular tissue engineering, Regen Med. 5 (2010) 107-120. https://doi.org/10.2217/rme.09.77

[27] E. Khor, L.Y. Lim, Implantable applications of chitin and chitosan, Biomaterials. 24 (2003) 2339-2349. https://doi.org/10.1016/S0142-9612(03)00026-7

[28] M. NV. Ravi. Kumar, A Review of Chitin and Chitosan Applications, Reactive and Functional Polymers. 46 (2000) 1-27. https://doi.org/10.1016/S1381-5148(00)00038-9

[29] S. Verma, C.D. Mazer, A.T. Yan, T. Mason, V. Garg, H. Teoh, F. Zuo, A. Quan, M.E. Farkouh, D.H. Fitchett, S.G. Goodman, R.M. Goldenberg, M. Al-Omran, R.E. Gilbert, D.L. Bhatt, L.A. Leiter, P. Jüni, B. Zinman, K.A. Connelly, Effect of Empagliflozin on Left Ventricular Mass in Patients With Type 2 Diabetes Mellitus and Coronary Artery Disease: The EMPA-HEART CardioLink-6 Randomized Clinical Trial, Circulation. 140 (2019) 1693-1702. https://doi.org/10.1161/CIRCULATIONAHA.119.042375

[30] J.A. Burdick, G.D. Prestwich, Hyaluronic acid hydrogels for biomedical applications, Adv Mater. 23 (2011) H41-56. https://doi.org/10.1002/adma.201003963

[31] L. Ouyang, C.B. Highley, C.B. Rodell, W. Sun, J.A. Burdick, 3D Printing of Shear-Thinning Hyaluronic Acid Hydrogels with Secondary Cross-Linking, ACS Biomater Sci Eng. 2 (2016) 1743-1751. https://doi.org/10.1021/acsbiomaterials.6b00158

[32] S.F. Grieco, E. Castrén, G.M. Knudsen, A.C. Kwan, D.E. Olson, Y. Zuo, T.C. Holmes, X. Xu, Psychedelics and Neural Plasticity: Therapeutic Implications, J Neurosci. 42 (2022) 8439-8449. https://doi.org/10.1523/JNEUROSCI.1121-22.2022

[33] R.T. Hoare, D.S. Kohane, Hydrogels in drug delivery: Progress and challenges, Polymer. 49 (2008) 1993-2007. https://doi.org/10.1016/j.polymer.2008.01.027

[34] S. Zalipsky, Chemistry of polyethylene glycol conjugates with biologically active molecules, Advanced Drug Delivery Reviews. 16 (1995) 157-182. https://doi.org/10.1016/0169-409X(95)00023-Z

[35] F.M. Veronese, A. Mero, The impact of PEGylation on biological therapies, BioDrugs. 22 (2008) 315-329. https://doi.org/10.2165/00063030-200822050-00004

[36] C.C. Lin, K.S. Anseth, PEG hydrogels for the controlled release of biomolecules in regenerative medicine, Pharm Res. 26 (2009) 631-643. https://doi.org/10.1007/s11095-008-9801-2

[37] I.M. El-Sherbiny, M.H. Yacoub, Hydrogel scaffolds for tissue engineering: Progress and challenges, Glob Cardiol Sci Pract. 2013 (2013) 316-342. https://doi.org/10.5339/gcsp.2013.38

[38] L. Sanders, J. Nagatomi, Clinical applications of surgical adhesives and sealants, Crit Rev Biomed Eng. 42 (2014) 271-292. https://doi.org/10.1615/CritRevBiomedEng.2014011676

[39] D.F. Torchiana DF, Polyethylene glycol based synthetic sealants: potential uses in cardiac surgery, J Card Surg. 18 (2003) 504-506. https://doi.org/10.1046/j.0886-0440.2003.00305.x

[40] S.L. Yeh, P. Deval, W.B. Tsai, Fabrication of Transparent PEGylated Antifouling Coatings via One-Step Pyrogallol Deposition, Polymers (Basel). 15 (2023) 2731-2742. https://doi.org/10.3390/polym15122731

[41] V. Kočka, P. Toušek, P. Widimský, Absorb bioresorbable stents for the treatment of coronary artery disease, Expert Rev Med Devices. 12 (2015) 545-557. https://doi.org/10.1586/17434440.2015.1080119

[42] R.X. Yin, D.Z. Yang, J.Z. Wu, Nanoparticle drug- and gene-eluting stents for the prevention and treatment of coronary restenosis, Theranostics. 4 (2014) 175-200. https://doi.org/10.7150/thno.7210

[43] M. Santoro, S.R. Shah, J.L. Walker, A.G. Mikos, Poly(lactic acid) nanofibrous scaffolds for tissue engineering, Adv Drug Deliv Rev. 107 (2016) 206-212. https://doi.org/10.1016/j.addr.2016.04.019

[44] G. Matsumura, N. Hibino, Y. Ikada, H. Kurosawa, T. Shin'oka, Successful application of tissue engineered vascular autografts:clinical experience, Biomater. 24 (2003) 2303-2308. https://doi.org/10.1016/S0142-9612(03)00043-7

[45] M.A. Naser, A.M. Sayed, W. Abdelmoez, M.T. El-Wakad, M.S. Abdo, Biodegradable suture development-based albumin composites for tissue engineering applications, Sci Rep. 14 (2014) 7912-7926. https://doi.org/10.1038/s41598-024-58194-5

[46] Z. Alhulaybi, I. Dubdub, M. Al-Yaari, A. Almithn, A.F. Al-Naim, H. Aljanubi, Pyrolysis Kinetic Study of Polylactic Acid, Polymers (Basel). 15 (2022) 12-27. https://doi.org/10.3390/polym15010012

[47] D. Di Francesco, A. Pigliafreddo, S. Casarella, L. Di Nunno, D. Mantovani, F. Boccafoschi, Biological Materials for Tissue-Engineered Vascular Grafts: Overview of Recent Advancements Biomolecules. 13 (2023) 1389-1414. https://doi.org/10.3390/biom13091389

[48] J.P. Cuenca, A. Padalhin, B.T. Lee, Small-Diameter Decellularized Vascular Graft with Electrospun Polycaprolactone, Materials Letters. 284 (2021) 128973. https://doi.org/10.1016/j.matlet.2020.128973

[49] N. Muzzio, S. Moya, G. Romero, Multifunctional Scaffolds and Synergistic Strategies in Tissue Engineering and Regenerative Medicine, Pharmaceutics. 13 (2021) 792-830 https://doi.org/10.3390/pharmaceutics13060792

[50] R. Dwivedi, S. Kumar, R. Pandey, A. Mahajan, D. Nandana, D.S. Katti, D. Mehrotra, Polycaprolactone as biomaterial for bone scaffolds: Review of literature, J Oral Biol Craniofac Res. 10 (2020) 381-388. https://doi.org/10.1016/j.jobcr.2019.10.003

[51] A.S. Udriște, A.C. Burdușel, A.G. Niculescu, M. Rădulescu, A.M. Grumezescu AM, Coatings for Cardiovascular Stents-An Up-to-Date Review, Int J Mol Sci. 25 (2024) 1078-1102. https://doi.org/10.3390/ijms25021078

[52] M. Cui, Z. Chai, Y. Lu, J. Zhu, L. Chen, Developments of polyurethane in biomedical applications: A review, Resources Chemicals and Materials. 2 (2023) 262-276. https://doi.org/10.1016/j.recm.2023.07.004

[53] S. Wendels, L. Avérous, Biobased polyurethanes for biomedical applications, Bioact. Mater. 6 (2020) 1083-1106. https://doi.org/10.1016/j.bioactmat.2020.10.002

[54] G. Dang, J. Gu, J. Song, Z. Li, J. Hao, Y. Wang, C. Wang, T. Ye, F. Zhao, Y. Zhang, F.R. Tay, L. Niu, L. Xia, Multifunctional polyurethane materials in regenerative medicine and

tissue engineering, Cell Reports Physical Science. 5 (2024) 102053. https://doi.org/10.1016/j.xcrp.2024.102053

[55] V.R. Sinha, A. Trehan, Biodegradable microspheres for protein delivery, J Control Release. 90 (2003) 261-280. https://doi.org/10.1016/S0168-3659(03)00194-9

[56] R. Langer, J.P. Vacanti, Tissue engineering, Science. 260 (1993) 920-926. https://doi.org/10.1126/science.8493529

[57] M.S. Shive, J.M. Anderson, Biodegradation and biocompatibility of PLA and PLGA microspheres, Adv Drug Deliv Rev. 28 (1997) 5-24. https://doi.org/10.1016/S0169-409X(97)00048-3

[58] J.M. Anderson, A. Rodriguez, D.T. Chang, Foreign body reaction to biomaterials, Semin Immunol. 20 (2008) 86-100. https://doi.org/10.1016/j.smim.2007.11.004

[59] H.K. Makadia, S.J. Siegel, Poly Lactic-co-Glycolic Acid (PLGA) as Biodegradable Controlled Drug Delivery Carrier, Polymers (Basel). 3 (2011) 1377-1397. https://doi.org/10.3390/polym3031377

[60] M. Vert, Degradable and bioresorbable polymers in surgery and in pharmacology: beliefs and facts, J Mater Sci Mater Med. 20 (2009) 437-446. https://doi.org/10.1007/s10856-008-3581-4

[61] J.C. Middleton, A.J. Tipton, Synthetic biodegradable polymers as orthopedic devices, Biomaterials. 21 (2000) 2335-2346. https://doi.org/10.1016/S0142-9612(00)00101-0

[62] D. Durán-Rey, V. Crisóstomo, J.A. Sánchez-Margallo, F.M. Sánchez-Margallo, Systematic Review of Tissue-Engineered Vascular Grafts, Front Bioeng Biotechnol. 9 (2021) 771400. https://doi.org/10.3389/fbioe.2021.771400

[63] I.E. Yalcin, T.G. Sadikoglu, Design parameters for electrospun biodegradable vascular grafts, Journal of Industrial Textiles. 47 (8) 2205-2227. https://doi.org/10.1177/1528083716654470

[64] A.S. Udriște, A.G. Niculescu, A.M. Grumezescu, E.Bădilă, Cardiovascular Stents: A Review of Past, Current, and Emerging Devices, Materials. 14 (2021) 2498-2520. https://doi.org/10.3390/ma14102498

[65] S. Wendels, L. Avérous, Biobased polyurethanes for biomedical applications, Bioactive Materials. 6 (2021) 1083-1106. https://doi.org/10.1016/j.bioactmat.2020.10.002

[66] S.H. Im, Y. Jung, S.H. Kim, Current status and future direction of biodegradable metallic and polymeric vascular scaffolds for next-generation stents, Acta Biomater. 60 (2017) 3-22. https://doi.org/10.1016/j.actbio.2017.07.019

[67] B. Hilmi, Z.A. Abdul Hamid, H. Md Akil, B.H. Yahaya, The Characteristics of the Smart Polymeras Temperature or pH-responsive Hydrogel, Procedia Chemistry. 19 (2016) 406-409. https://doi.org/10.1016/j.proche.2016.03.031

[68] A. Fattah-alhosseini, R. Chaharmahali, S. Alizad, M. Kaseem, B. Dikici, A review of smart polymeric materials: Recent developments and prospects for medicine applications, Hybrid Advances. 5 (2024) 100178. https://doi.org/10.1016/j.hybadv.2024.100178

[69] J. Tan, Q. Yang, G. Hu, H. Zhang, L. Pei, J. Wang, Experimental study on the temperature-sensitive behavior of poly-n-isopropylacrylamide/graphene oxide composites and the flexible conductive cotton fabrics, Polymer Testing. 110 (2022) 107563. https://doi.org/10.1016/j.polymertesting.2022.107563

[70] M. Santhamoorthy, T.T. Vy Phan, V. Ramkumar, C.J. Raorane, K. Thirupathi, S.C. Kim, Thermo-Sensitive Poly (N-isopropylacrylamide-co-polyacrylamide) Hydrogel for pH-Responsive Therapeutic Delivery, Polymers (Basel). 14 (2022) 4128-4142. https://doi.org/10.3390/polym14194128

[71] J. Singh, P. Nayak, pH-responsive polymers for drug delivery: Trends and opportunities, Journal of Polymer Science. 61 (2023) 2828-2851. https://doi.org/10.1002/pol.20230403

[72] A.A. Yetisgin, S. Cetinel, M. Zuvin, A. Kosar, O. Kutlu, Therapeutic Nanoparticles and Their Targeted Delivery Applications, Molecules. 25 (2020) 2193-2224. https://doi.org/10.3390/molecules25092193

[73] M. Zhang, W. Song, Y. Tang, X. Xu, Y. Huang, D. Yu, Polymer-Based Nanofiber-Nanoparticle Hybrids and Their Medical Applications, Polymers. 14 (2022) 351-378. https://doi.org/10.3390/polym14020351

[74] Z. Yuan, D. Sheng, L. Jiang, M. hafiq, A.R. Khan, R. Hashim, Y. Chen, B. Li, X. Xie, J. Chen, Y. Morsi, X. Mo, S. Chen, Vascular Endothelial Growth Factor-Capturing Aligned Electrospun Polycaprolactone/Gelatin Nanofibers Promote Patellar Ligament Regeneration, Acta Biomaterialia. 140 (2022) 233-246. https://doi.org/10.1016/j.actbio.2021.11.040

[75] M. Hosseinkhani, D. Mehrabani, M.H. Karimfar, S. Bakhtiyari, A. Manafi, R. Shirazi, Tissue engineered scaffolds in regenerative medicine, World J Plast Surg. 3 (2014) 3-7.

[76] M.X. Li, Q.Q. Wei, H.L. Mo, Y. Ren, H.J. Lu, Y.K. Joung, Challenges and advances in materials and fabrication technologies of small-diameter vascular grafts, Biomater Res. 27 (2023) 91. https://doi.org/10.1186/s40824-023-00424-4

[77] S. Ravi, E.L. Chaikof, Biomaterials for vascular tissue engineering, Regen Med. 5 (2010) 107-120. https://doi.org/10.2217/rme.09.77

[78] S.B. Choi, CYPHER coronary stents and risk of thrombosis, CMAJ. 169 (2003) 218.

[79] M. Nakamura, N. Suzuki, K. Fujii, J. Furuya, T. Kawasaki, T. Kimura, T. Sakamoto, K. Tanabe, H. Kusano, K.A. Stockelman, K. Kozuma, The Absorb GT1 Bioresorbable Vascular Scaffold System- 5-Year Post-Market Surveillance Study in Japan, Circ J. 88 (2024) 863-872. https://doi.org/10.1253/circj.CJ-23-0877

[80] R. Gaspar, R. Duncan, Polymeric carriers: Preclinical safety and the regulatory implications for design and development of polymer therapeutics, Advanced Drug Delivery Reviews. 61 (2009) 1220-1231. https://doi.org/10.1016/j.addr.2009.06.003

Applications of Polymers in Surgery II
Materials Research Forum LLC
Materials Research Foundations 172 (2025) 381-396
https://doi.org/10.21741/9781644903353-16

Chapter 16

Role of Polymers in Cosmetics and Personal Care Harvests: Precarious Impact on Human Health and Environment

Venkatramana Losetty[1], M. Dhanalakshmi[1], C. Hazarathaiah Yadav[1], K. Sivakumar[2*]

[1]Department of Chemistry, Vel Tech Rangarajan Dr. Sagunthala R&D Institute of Science and Technology, Chennai, India

[2]Department of Chemistry, S.V. Arts College (TTD's), Tirupati, India

*sivakumarkasi64@gmail.com

Abstract

This chapter discusses the application of polymers in cosmetics and personal care products and also explores how they are hazardous to harm human health and the environment. Cosmetics are used globally for beauty and contain many chemical substances, especially different types of polymers added during formulation. The quality, functionality, and shelf life are increased by the presence of these polymer substances which act as additives, fragrances, preservatives, stabilizers, surfactants, dyes, and shine enhancers. However, many toxic and hazardous polymer-based substances are present in cosmetics beyond acceptable levels. Cosmetic formulations become more attractive and marketable to consumers when bonded with polymers. The polymers are mixed with cosmetic products as modifiers and stabilizers including makeup, skincare, and hair care products. Some polymers are responsible for serious skin damage and even organ damage and can lead to cancer. This chapter will discuss the usage of polymers in cosmetics, and determine their types, like natural, semi-synthetic, and synthetic polymers present in cosmetics, and how their unique properties enhance the need in cosmetic applications. Many of these polymers are bioactive and tend to bio-accumulate and environmental impact which gives significant risk to human health and ecosystems. Therefore, natural polymers are highly valued in cosmetic formulations for their biocompatibility, safety, and environmental friendliness.

Keywords

Cosmetic, Polymer, Human health, Environment, Formulation, Stabilizer, Biocompatibility, Ecosystem

Contents

1. Introduction

Cosmetics are products applied to the body to improve how we look, feel, or smell. As shown in Figure 1, creams, lotions, powders, perfumes, lipsticks, nail polishes, and beauty products are some of examples cosmetics and personal care products. Towards innovation, consumer trends, and the need for self-expression and personal care the cosmetics industry has become a multi-billion dollar global market.

***Figure 1**: Different types of cosmetics and personal care products are available in the market. The pictures were downloaded from Google using the keyword "cosmetics".*

Personal care products often offer more benefits beyond cleaning and these benefits are often seen after just one use. They wear perfumes that smell good and also offer defenses like deodorants or

 Materials Research Forum LLC
https://doi.org/10.21741/9781644903353-16

antiperspirants [1]. Besides the visual benefits, these products can boost the self-esteem of people with medical conditions that makeup can hide. However, different chemicals are used since this class of consumer products has many properties. Globally cosmetics are used to improve appearance and formulas are getting more and more chemical substances, especially polymers [2]. These polymers increase the quality, functionality, and shelf life of cosmetic and personal care products by being stabilizers, perfumes, preservatives, surfactants, colors, and shine enhancers. Modern cosmetics and personal hygiene products are mostly made of polymers. In skincare, hair care, and makeup goods their versatility makes the product performance, texture, stability, and lifespan better. They offer desired properties like smooth application, moisture retention, and long-lasting effects. So they can be used as thickeners, film formers, emulsifiers, and stabilizers [3].

The growing demand for cosmetic products in recent years has led to the usage of both bio-polymers and synthetic polymers in the formulations of cosmetics [4]. Moreover, some polymers, particularly those synthesized from synthetic sources, have raised issues about possible health and environmental implications. Since many synthetic polymers may not be bio-degradable, concerns have been raised over their ability to accumulate with living organisms, long-term stability in the environment, and also drastic effects on all ecosystems. Some polymers, which can enter into the skin and interact with internal organs, are linked to skin irritation and other health hazards [5]. Polymers are normally observed in skincare products such as body oils, moisturizers, sunscreens, liquid soaps, shampoos, conditioners, split-end treatments, hair dye, styling gels, nail care, makeup, fragrances, *etc*,. A miscellaneous range of polymers is needed for many applications and is used in cosmetic products, which may differ in terms of composition, preparation methods, and physicochemical properties [6]. Some polymers are crucial raw materials in cosmetic formulations that are needed to make high-performing products. These polymers are categorized as natural polymers, synthetic polymers, and semi-synthetic polymers consisting of unit monomers as described in Figure 2. These polymers are widely used in cosmetics because of their structural diversity, which can be used in cosmetic formulations to adjust rheology, thicken, stabilize or destabilize foam, emulsify, provide hold, condition, and form films, among other functions.

Materials Research Forum LLC
https://doi.org/10.21741/9781644903353-16

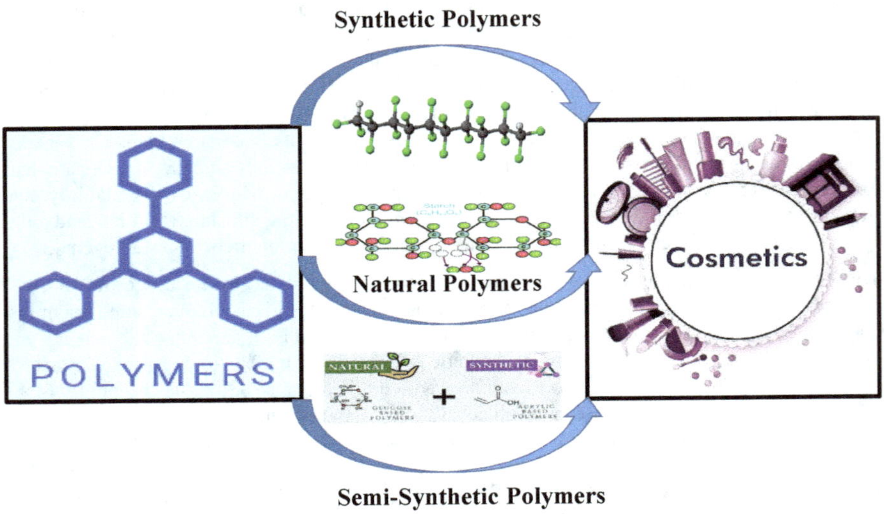

***Figure 2**: Representation of various polymers (synthetic, natural, and semi-synthetic) used in cosmetics manufacturing.*

2. Polymers in cosmetics-data analysis

Based on the analysis, the worldwide demand for polymer ingredients in the personal care and cosmetic sector is anticipated to increase at a compound annual growth rate (CAGR) of 3.8% from 2021 to 2026. Many personal care products, and several cosmetics utilize polymers, because of their versatile qualities that enhance formulations along with various advantages [7]. Natural, synthetic, and semi-synthetic polymers are among the various types of polymers that are used in products for a variety of functions that include film formation, emulsification, rheology, modification, providing protective layers, serving as emollients, and enhancing aesthetics. The polymers that are usually used for various purposes are caprylic triglycerides, polyacrylates, polyvinyl amides, alkylene oxide-based homopolymers and copolymers, silicones, and polyvinyl acetate [8]. These are applied in different areas such as hair care, skincare, color cosmetics, and sun protection.

2.1 Polymers in the cosmetics-Asia Pacific field

Due to the lifestyle changes of the existing consumer and the rapid rise in population, the Asia Pacific region is expected to lead the polymers in cosmetics market growth. Furthermore, it is estimated that the polymer industry would have significant development in North America and European countries.

Materials Research Forum LLC
https://doi.org/10.21741/9781644903353-16

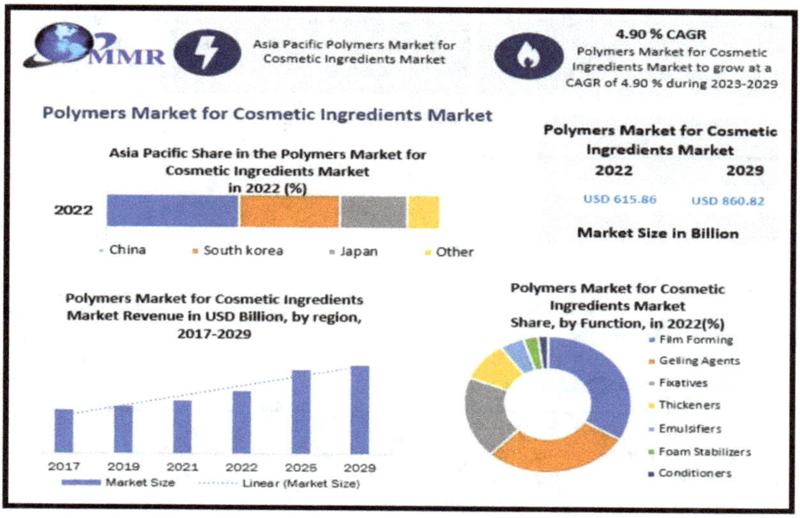

Figure 3: *Asian-specific polymers market for cosmetic ingredients market. The picture was downloaded from Google using the keyword "polymers in cosmetics-data analysis".*

Figure 3 describes the Asian-specific polymers market for cosmetic ingredients, the picture was downloaded from Google using the keyword "polymers in cosmetics-data analysis". The market size linearly increases from 2017 to 2029, among all Asian Pacific countries, China has the highest share. Additionally, Figure 3 describes the various polymers that are utilized in cosmetics and health care products. In order to ensure the safe use of polymers in cosmetic products, Regulatory organizations like U.S. Food and Drug Administration (FDA) and the European Union's Cosmetics Regulation (EC No 1223/2009) have developed standards. However, increasing concerns about the environmental effects of synthetic polymers have led to discrimination in their examination. For instance, the European Union has taken action to outlaw the usage of plastic materials within the cosmetics and formulation of personal care products, and the remaining countries are considering endorsing laws along similar lines [9].

2.2 Polymers in cosmetics-India market

India has qualified linear growth in its personal care and cosmetics division, with more shelf space being allocated in boutiques and retail outlets nationwide for the past decade. Numerous multinational brands have entered in the Indian market, due to the infrastructure facilities and low pricing strategies [10]. This growth will result in India accounting for 5% of the global cosmetics market, placing it among the top five markets worldwide in terms of revenue by the year 2025. Furthermore, the cosmetic market is expected to expand owing to the increase in consumer's preference for good cosmetic products, including organic, herbal, and Ayurveda items. Some important sectors such as color cosmetics, perfumes, specialized skincare lotions, hair care, and makeup products are projected to see significant growth in the market. The Indian beauty market remains one of the fastest-growing globally, with a population beyond a billion and increasing

Materials Research Foundations 172 (2025) 381-396
https://doi.org/10.21741/9781644903353-16

disposable incomes. India is a highly striking market for multinational corporations looking to expand their market share. Figure 4 describes the country-wide share of the polymer market in cosmetic ingredients and regional insights. The country, China has the largest proportion of about 58.3% of plastic materials present in beauty and health care products. Moreover, Johnson and Johnson used the highest polymer ingredients in cosmetic products.

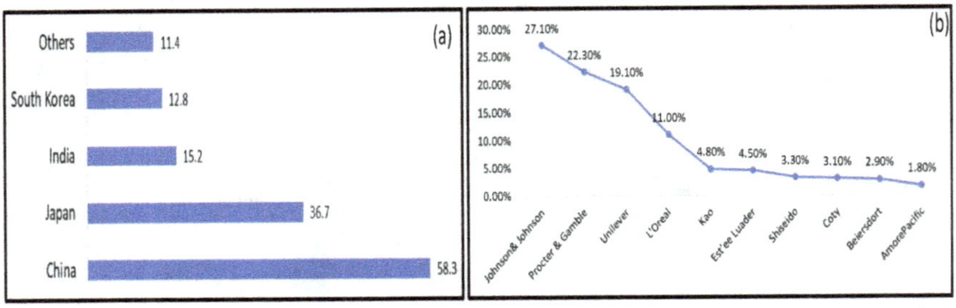

***Figure 4**: The polymers market in cosmetic ingredients analysis, (a) country-wise polymers in cosmetic products, and (b) polymers market in cosmetic regional insights. The keyword "cosmetic market" was used to download the pictures from Google.*

Polymers are extensively used in the cosmetic industry during the manufacturing of cosmetic products. Generally, the polymers which are used in cosmetics and personal care products are categorized into natural polymers, synthetic polymers, and semi-synthetic polymers as described in Figure 5. Proteins and polysaccharides are the best examples of natural polymers that are generated from plant and animal sources. Some natural polymers namely hyaluronic acid, collagen, and alginate are a few examples of substances that are highly esteemed for their moisturizing, film-forming, and skin-conditioning properties [11]. Natural polymers have been widely used in cosmetic applications because of their great consumer appeal, safety, bio-compatibility nature, and environmental friendliness. They are adjustable and suitable for many functions such as stabilizing and modifying makeup and skin and hair care products. Similarly, the frequently utilized natural polymers in cosmetic applications are polysaccharides, starch, xanthan gum, guar gum, carrageenan, alginate, pectin, gelatin, agar, collagen, and hyaluronic acid. Starch, which is a naturally occurring polysaccharide is of two types that include granule starch and soluble starch which is found in a variety of applications. When the starch is heated during extraction, soluble starch dries and becomes moisture-resistant. The moisture-free stretch can give skin and hair a smooth appearance. On the other hand, the hydrogen bond present in the granule starch has been altered rendering it unbreakable, resulting in skin and hair that are less greasy and softer [12]. Starches can be derived from various sources, including non-ionic polymeric compounds like potato, corn, wheat, and cassava root, because of the presence of varied amylose concentrations, each of which gives cosmetic compositions with different qualities. Furthermore, starch is frequently mixed with other naturally occurring polymers, like chitosan, to strengthen its antioxidant qualities and provide the skin with further advantages.

3. Types of polymers in cosmetics

Materials Research Foundations 172 (2025) 381-396
https://doi.org/10.21741/9781644903353-16

Polymers are extensively used in the cosmetic industry during the manufacturing of cosmetic products. Generally, the polymers which are used in cosmetics and personal care products are categorized into natural polymers, synthetic polymers, and semi-synthetic polymers as described in Figure 5.

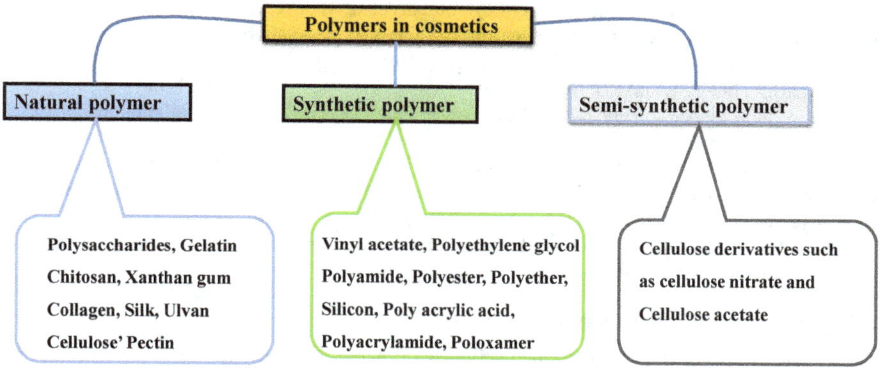

***Figure 5:** Different types of polymeric materials utilized in beauty and personal care products*

3.1 Natural polymers in cosmetics

Proteins and polysaccharides are the best examples of natural polymers generated from plant and animal sources. Some natural polymers namely hyaluronic acid, collagen, and alginate are a few examples of substances that are highly esteemed for their moisturizing, film-forming, and skin-conditioning properties[13]. Natural polymers have been widely used in cosmetic applications because of their great consumer appeal, safety, bio-compatibility nature, and environmental friendliness. They are adjustable and suitable for many functions such as stabilizing and modifying makeup and skin and hair care products. Similarly, the frequently utilized natural polymers in cosmetic applications are polysaccharides, starch, xanthan gum, guar gum, carrageenan, alginate, pectin, gelatin, agar, collagen, and hyaluronic acid. Starch, which is a naturally occurring polysaccharide is of two types that include granule starch and soluble starch which is found in a variety of applications. When the starch is heated during extraction, soluble starch dries and becomes resistant to moisture. The moisture-free stretch can give skin and hair a smooth appearance [14]. On the other hand, the hydrogen bond present in the granule starch has been altered and renders it unbreakable, resulting in skin and hair that are less greasy and softer. Starches can be derived from various sources, including non-ionic polymeric compounds like potato, corn, wheat, and cassava root, because of the presence of varied amylose concentrations, each of which gives cosmetic compositions with different qualities. Furthermore, starch is frequently mixed with other naturally occurring polymers, like chitosan, to strengthen its antioxidant qualities and provide the skin with further advantages [15].

3.2 Synthetic polymers in cosmetics

On the other hand, some synthetic polymers such as silicones, polyacrylates, and polyethylene glycols (PEGs) are wholly synthesized in laboratories. These polymers are highly preferred in

Materials Research Foundations 172 (2025) 381-396 https://doi.org/10.21741/9781644903353-16

cosmetic formulations because they may be personalized to attain certain features like water resistance, enhanced durability, film formation, and emulsifiers. Silicon materials have had more attention in the cosmetics and personal care sector in recent years [16]. Some cosmetic products, particularly deodorants, shampoos, antiperspirants, and lotions have silicon as a major ingredient, and its residents include bio-methane and cyclomethycaine. Those ingredients play an important role in cosmetics as suspending agents, emulsification, and associative thickening. The anionic or non-ionic variants of synthetic polymers like polyethylene glycols (PEGs) find considerable use in cosmetics. They function as emulsifiers, which make it possible for oily and watery components to blend well, as dispersion enhancers. Furthermore, they also function as emollients, which aid in softening and lubricating the skin and also in delivering other compounds deeper into the skin. Various common beauty products such as skincare, cosmetics, bath and shaving products, skin cleansers, shampoos, hair conditioners, and deodorants contain these polymeric ingredients [17]. Moreover, apart from the looks and beauty of the person, some drawbacks regarding the effects of synthetic polymers on the environment and possible health dangers when synthetic polymers in cosmetics have been particularly vocal.

3.3. Semi-synthetic polymers in cosmetics

Semi-synthetic polymers, which include cellulose derivatives, are natural polymers that undergo chemical modification and are commonly utilized in creams and lotions as stabilizers and thickeners. The two main classes of cellulose derivatives are cellulose ethers and cellulose esters, each of which has unique mechanical and physicochemical characteristics. These cellulose derivatives have exclusive properties such as resistance to heat, hydrolysis, oxidation, bio-degradation, and surface activity [17]. They also can form thermoplastic films and exhibit viscosity properties in solution. Moreover, cellulose esters with good film-forming properties such as cellulose acetate, cellulose acetate phthalate, cellulose acetate butyrate, cellulose acetate trimellitate, and hydroxypropyl methylcellulose phthalate are generally water-insoluble substances. In addition, other high water-soluble substances are cellulose ethers which include methylcellulose, ethyl cellulose, hydroxyethyl cellulose, hydroxypropyl methylcellulose, carboxymethyl cellulose, and sodium carboxymethyl cellulose. The above said polymers are frequently applied in cosmetics and personal health care products such as creams, shampoos, lotions, and gels as thickening, stabilizing, gelling, and bio-adhesive agents[18]. Some kinds of semi-synthetic polymers including agar, pectin, cellulose derivatives, starches, acacia, sodium alginate, and gelatin are less possible to become contaminated by microorganisms than cellulose derivatives.

4. Functions of polymers in cosmetics

The cosmetics and personal care industries utilize several polymers to modify the functionality of substances. The polymer showed a dramatic impact on the rheological properties including viscosity and surface tension during the formulation. These polymers may change the applications of products, water sensitivity, active ingredient distribution, and viscosity of the final product [19]. For water-based products, thickeners (polymers) made of both natural and artificial materials are employed. One or more fatty structuring agents such as long-chain fatty alcohols, triglycerides, and natural or synthetic waxes are added to the anhydrous systems which on gradual contracting upon cooling provide structural stability and facilitates mold release. Synthetic waxes which are made from long-chain hydrocarbons that are obtained from petroleum or from some monomers

Materials Research Foundations 172 (2025) 381-396
https://doi.org/10.21741/9781644903353-16

Nitrocellulose and Acrylate copolymers	Film formation nature, skin protection, and enhanced product aesthetics
Hydroxyethyl cellulose, Hydroxypropyl methylcellulose, and polyacrylic acid	Product shelf life and rheological properties
Cellulose	Film formation, Rheological properties, and well-hydrated property
Chitosan	Moisturizer, long-term color properties, and antimicrobial properties
Hyaluronic acid	Skin protector, moisturizer, and antiaging properties
Carrageenan	pH controller and gelling properties
Collagen	Antioxidant, antihypertensive, and lipid-lowering properties
Agar	Stabilizer, Thickener, and gelling properties
Xanthan gum	Rheology property, and emulsifier
Chitin	Filmmaker and antibacterial agent
Alginate	Stabilizing and thickness properties
Pullulan	Anti-wrinkle property and film formation property
Nylon-12	Viscosity controller, and bulking
Teflon	Binding agent, skin conditioner, and slip modifier
Nylon-6	Bulking substance and viscosity controller
Allyl stearate	Film formation
Poly(propylene terephthalate)	Skin conditioner and emulsion stabilizer
Styrene acrylate	Coloring agent and aesthetics
Poly(methyl methacrylate)	Function as sorbent for ingredient delivery

5. Polymers in cosmetics-new trend

Customized beauty is indeed projected to be a major trend in the future of cosmetic and beauty care products. Several customers are searching for products matched to their skin types, tones, and preferences. National and international companies offer custom-made advice, formulations, and even specialized products through different technologies such as Artificial Intelligence, online quizzes, and skincare apps. The attractiveness of clean beauty is still developing, highlighting non-toxic, cruelty-free, and ecologically friendly materials. This satisfies consumer requirements for eco-friendly packaging composed of biodegradable materials and reusable bins or storage containers. Another trend in cosmetics is called "*skinification*", which involves using skincare

Materials Research Foundations 172 (2025) 381-396
https://doi.org/10.21741/9781644903353-16

elements such as ceramides, vitamin C, and hyaluronic acid to nurture the skin while applying makeup [24]. These ingredients have some advantages for the environment, waterless beauty, which uses less water to generate concentrated formulae, thus becoming more and more popular. Finally, multipurpose items such as skincare cosmetics or beauty products with SPF are attractive and more widespread as a way to satisfy consumers who value productivity. The physical, chemical, toxicological, and biological characteristics of polymer compounds as well as their interactions with the enzyme active sites can be predicted by using some computer-based techniques such as molecular modeling, molecular docking, DFT, and quantitative structure-activity relationship studies.

Cosmetic flow powders or compacted powders have been used as an alternative to the polymers in cosmetics for decades. This is because of the unique properties of the compacted powders such as adhesiveness, slipperiness, absorbency, and a sleek, efflorescent effect that imparts a healthier, more vibrant, and younger appearance to the skin. Nowadays, aerated powders with micro or nanosized structures or active carriers disseminated in emulsions, gels, or liquids are used to create the suspensions. These are the common kinds of solid cosmetic dosage forms that are essential to maintain body appearance and condition. The polymer Polymethylsilsesquioxane (PSA) is commonly utilized as an antiaging agent in skincare products. The biocompatible polymer polymethyl methacrylate (PMMA) is used in various cosmetics such as foundation for face and eyes, and also in hair and nail care products [25]. In addition to orthodontic devices and dental prosthetics, it is utilized in aesthetic treatments to soften tissue fillers, which reduces wrinkles permanently and in treating atrophic acne scars. The methyl methacrylate cross polymer is also utilized as a component to increase the viscosity of non-aqueous materials.

In addition, natural and biodegradable compounds that are alternatives to synthetic polymers are being actively applied in beauty care products. Polysaccharide-related micro-carriers and nano-carriers offer capable strategies to overcome physical and metabolic barriers, thereby improving cosmetic effectiveness. The most versatile biopolymers utilized in cosmetic and personal care applications are the alginates, and the biopolymers which are derived from sea algae. These alginates are linear anionic polysaccharides that belong to the class of copolymers consisting of α-L-guluronic acid and β-D-mannuronic acid units linked by glycosidic bonds. Since alginate salts are non-toxic in nature and eco-friendly benign, they are frequently utilized in cosmetic products.

Additionally, methylmethacrylate cross-polymer functions as a non-aqueous viscosity enhancer in cosmetic products. Dimer acids such as polyamides like Nylon 6 or Nylon 12 powder are chemically neutral, which can be prepared on heating the vegetable-derived fatty acids over clay [26]. The resulting substances are highly viscous, lipophilic, and non-crystalline oils. A variety of polyamide products are produced by forming amide bonds between diamine and mono-functional terminators. These polyamides possess some characteristic features like flexibility, excellent absorption tendency, increased usability in various industries, and high spreadability. The polyamide layer is produced through the common technique such as interfacial polymerization reaction between a poly-functional amine such as m-phenylenediamine and a polyfunctional acyl chloride like trimesoyl chloride. Depending on the concentration of the polyamide utilized in the polymerization technique, these polymers can form shear-thinning oil-based gels and also the modulus of the Nylon-12 gel varies from a deformable gel to a clear hard stick. The synthesis mechanism is believed to result from a supramolecular hydrogen-bonded network.

Materials Research Foundations 172 (2025) 381-396
https://doi.org/10.21741/9781644903353-16

6. Environmental and health concerns

Polymers have extensive properties and benefits when blended with cosmetics and personal care products. Even though, their presence has raised concerns about their environmental and health impacts. Particularly, some synthetic polymers have been analyzed for their potential risks to human beings and ecosystems when they accumulate with microorganisms.

6.1 Non-biodegradability and environmental impact

Some synthetic polymers that are used in cosmetics may not biodegradable are inert chemically and do not decompose biologically. It is very hard for bacteria to break down into small pieces because they are generated by lengthy chains of hydrogen and carbon atoms. The non-biodegradable polymers that are widely used in the manufacture of cosmetics are linear high-density polyethylene (HDPE), branched low-density polyethylene (LDPE), polyethylene terephthalate (PET), polyethylene (PE), polypropylene (PP), polystyrene (PS), and polyvinyl chloride (PVC) [27]. Moreover, some synthetic polymers such as microplastics and nanoplastics are used in exfoliating scrubs in cosmetic products. Synthetic polymers, the certain concentration of polyethylene used in film-forming products, they are not biodegradable, once these products are washed off the skin or hair they may enter into the wastewater systems or water resources eventually affect the environment and ecosystems. Recent studies show that microplastics accumulate in oceans, rivers, and lakes, where they pose a threat to aquatic life. For instance, marine organisms can ingest microplastics, which can lead to physical harm or exposure to toxic substances that stick to the plastic's surface.

6.2 Bioaccumulation

The usage of synthetic polymers in cosmetics and personal care products increases additional concern, because they have the potential to bioaccumulate with living organisms. Some synthetic polymers break down into smaller particles such as microplastics, which may be absorbed by living organisms such as plants, animals, and humans. The longstanding effects of bioaccumulated polymers in the human body and ecosystems are not yet fully understood. However, there is growing confirmation that they may interject biological processes and contribute to toxicity in wildlife.

6.3 Health risks

In addition to environmental apprehensions, some synthetic polymers and semisynthetic polymers have been connected to health issues. For instance, certain concentrations of PEG compounds can be contaminated with ethylene oxide, a known carcinogen. Furthermore, some synthetic polymers can penetrate the skin, which raises disquiet about their potential to interact with internal organs or disrupt the endocrine systems of humans. Even though the concentrations of polymer substances in cosmetics are normally low, repeated acquaintance over time may pose risks, particularly for individuals with sensitive skin or underlying health conditions.

Summary and conclusion

This chapter discussed the applications of the natural, semi-synthetic, synthetic, and other modified derivatives of polymers which possess distinct physical, chemical, biological, and pharmaceutical properties particularly in cosmetics and personal care products such as skin care, hair care, and nail care products. Furthermore, their remarkable qualities help in promoting active ingredients in the personal care and in the cosmetic sectors. The interrelated properties of multi-functional polymers, such as physiochemical characteristics, biodegradability, functional performance, biological and environmental safety, necessitate through characterization to tailor them for specific applications. Polymers must not only exhibit strength, flexibility, and chemical resistance, but also ensure environmental sustainability, biocompatibility and safety. Comprehensive testing including spectroscopic, thermal, mechanical and biodegradability assessment is crucial to optimize these polymers for diverse and advanced applications. Some synthetic polymers break down into smaller particles such as microplastics, which may be absorbed by living organisms such as plants, animals, and humans. The longstanding effects of bioaccumulated polymers in the human body and ecosystems are not yet fully understood. However, there is growing confirmation that they may interject biological processes and contribute to toxicity in wildlife. Cosmetic flow powders or compacted powders have been used as an alternative to the polymers in cosmetics for decades. This is because of the unique properties of the compacted powders such as adhesiveness, slipperiness, absorbency, and a sleek, efflorescent effect that imparts a healthier, more vibrant, and younger appearance to the skin.

Acknowledgment

The authors express sincere thanks to Department of Chemistry, Vel Tech Rangarajan Dr. Sagunthala R&D Institute of Science and Technology, Chennai, India and

S.V. Arts College (TTD's), Tirupati, India for the constant support and encouragement.

References

[1] I. Aranaz, N. Acosta, C. Civera, B. Elorza, J. Mingo, C. Castro, M. Gandía, A. Heras Caballero, Cosmetics and Cosmeceutical Applications of Chitin, Chitosan and Their Derivatives, Polymers 10 (2018) 213. https://doi.org/10.3390/polym10020213

[2] A. Okunola A, O. Kehinde I, A. Oluwaseun, A. Olufiropo E, Public and Environmental Health Effects of Plastic Wastes Disposal: A Review, J. Toxicol. Risk Assess. 5 (2019). https://doi.org/10.23937/2572-4061.1510021

[3] A.L.V. Cubas, R.T. Bianchet, I.M.A.S.D. Reis, I.C. Gouveia, Plastics and Microplastic in the Cosmetic Industry: Aggregating Sustainable Actions Aimed at Alignment and Interaction with UN Sustainable Development Goals, Polymers 14 (2022) 4576. https://doi.org/10.3390/polym14214576

[4] T.F.R. Alves, M. Morsink, F. Batain, M.V. Chaud, T. Almeida, D.A. Fernandes, C.F. Da Silva, E.B. Souto, P. Severino, Applications of Natural, Semi-Synthetic, and Synthetic Polymers in Cosmetic Formulations, Cosmetics 7 (2020) 75. https://doi.org/10.3390/cosmetics7040075

[5] C. Campanale, C. Massarelli, I. Savino, V. Locaputo, V.F. Uricchio, A Detailed Review Study on Potential Effects of Microplastics and Additives of Concern on Human Health, Int. J. Environ. Res. Public. Health 17 (2020) 1212. https://doi.org/10.3390/ijerph17041212

[6] C.J. Moore, Synthetic polymers in the marine environment: A rapidly increasing, long-term threat, Environ. Res. 108 (2008) 131–139. https://doi.org/10.1016/j.envres.2008.07.025

[7] R.P. Babu, K. O'Connor, R. Seeram, Current progress on bio-based polymers and their future trends, Prog. Biomater. 2 (2013) 8. https://doi.org/10.1186/2194-0517-2-8

[8] T.J. Oliveira, T.C.M. Segato, G.P. Machado, D. Grotto, A.F. Jozala, Evolution of Bacterial Cellulose in Cosmetic Applications: An Updated Systematic Review, Molecules 27 (2022) 8341. https://doi.org/10.3390/molecules27238341

[9] R.H. Waring, R.M. Harris, S.C. Mitchell, Plastic contamination of the food chain: A threat to human health?, Maturitas 115 (2018) 64–68. https://doi.org/10.1016/j.maturitas.2018.06.010

[10] A. Sionkowska, Current research on the blends of natural and synthetic polymers as new biomaterials: Review, Prog. Polym. Sci. 36 (2011) 1254–1276. https://doi.org/10.1016/j.progpolymsci.2011.05.003

[11] H. Aliyar, G. Schalau, Recent Developments in Silicones for Topical and Transdermal Drug Delivery, Ther. Deliv. 6 (2015) 827–839. https://doi.org/10.4155/tde.15.39

[12] H.C. Nam, W.H. Park, Eco-friendly poly(lactic acid) microbeads for cosmetics via melt electrospraying, Int. J. Biol. Macromol. 157 (2020) 734–742. https://doi.org/10.1016/j.ijbiomac.2019.11.240

[13] P. Morganti, G. Morganti, M.-B. Coltelli, Natural Polymers and Cosmeceuticals for a Healthy and Circular Life: The Examples of Chitin, Chitosan, and Lignin, Cosmetics 10 (2023) 42. https://doi.org/10.3390/cosmetics10020042

[14] L. Panariello, A. Vannozzi, P. Morganti, M.-B. Coltelli, A. Lazzeri, Biobased and Eco-Compatible Beauty Films Coated with Chitin Nanofibrils, Nanolignin and Vitamin E, Cosmetics 8 (2021) 27. https://doi.org/10.3390/cosmetics8020027

[15] S. Usman, A.F. Abdull Razis, K. Shaari, M.N.A. Amal, M.Z. Saad, N. Mat Isa, M.F. Nazarudin, S.Z. Zulkifli, J. Sutra, M.A. Ibrahim, Microplastics Pollution as an Invisible Potential Threat to Food Safety and Security, Policy Challenges and the Way Forward, Int. J. Environ. Res. Public. Health 17 (2020) 9591. https://doi.org/10.3390/ijerph17249591

[16] R.P. Gawade, S.L. Chinke, P.S. Alegaonkar, Polymers in cosmetics, in: Polym. Sci. Innov. Appl., Elsevier, 2020: pp. 545–565. https://doi.org/10.1016/B978-0-12-816808-0.00017-2

[17] A.L. Jones, R.S.S. Kramer, Facial Cosmetics and Attractiveness: Comparing the Effect Sizes of Professionally-Applied Cosmetics and Identity, PLOS ONE 11 (2016) e0164218. https://doi.org/10.1371/journal.pone.0164218

[18] J. Kozlowska, N. Stachowiak, A. Sionkowska, Collagen/Gelatin/Hydroxyethyl Cellulose Composites Containing Microspheres Based on Collagen and Gelatin: Design and Evaluation, Polymers 10 (2018) 456. https://doi.org/10.3390/polym10040456

[19] H.K. Ageel, S. Harrad, M.A.-E. Abdallah, Occurrence, human exposure, and risk of microplastics in the indoor environment, Environ. Sci. Process. Impacts 24 (2022) 17–31. https://doi.org/10.1039/D1EM00301A

[20] I.E. Napper, R.C. Thompson, Plastic Debris in the Marine Environment: History and Future Challenges, Glob. Chall. 4 (2020) 1900081. https://doi.org/10.1002/gch2.201900081

Materials Research Forum LLC
https://doi.org/10.21741/9781644903353-16

[21] H. Aliyar, G. Schalau, Recent Developments in Silicones for Topical and Transdermal Drug Delivery, Ther. Deliv. 6 (2015) 827–839. https://doi.org/10.4155/tde.15.39

[22] M.F. Maitz, Applications of synthetic polymers in clinical medicine, Biosurface Biotribology 1 (2015) 161–176. https://doi.org/10.1016/j.bsbt.2015.08.002

[23] V. Losetty, G.N. Reddy, C.H. Yadav, K. Sivakumar, Study on Microplastic Content in Cosmetic Products and Their Detrimental Effect on Human Health, in: Inamuddin, T. Altalhi, V.C. Fernandes (Eds.), Toxic Eff. Micro- Nanoplastics, 1st ed., Wiley, 2024: pp. 253–264. https://doi.org/10.1002/9781394238163.ch11

[24] A. Zielińska, C. Martins-Gomes, N.R. Ferreira, A.M. Silva, I. Nowak, E.B. Souto, Anti-inflammatory and anti-cancer activity of citral: Optimization of citral-loaded solid lipid nanoparticles (SLN) using experimental factorial design and LUMiSizer®, Int. J. Pharm. 553 (2018) 428–440. https://doi.org/10.1016/j.ijpharm.2018.10.065

[25] S. Gupta, R. Bansal, S. Gupta, N. Jindal, A. Jindal, Nanocarriers and nanoparticles for skin care and dermatological treatments, Indian Dermatol. Online J. 4 (2013) 267. https://doi.org/10.4103/2229-5178.120635

[26] T.D. Fornes, D.R. Paul, Structure and Properties of Nanocomposites Based on Nylon-11 and -12 Compared with Those Based on Nylon-6, Macromolecules 37 (2004) 7698–7709. https://doi.org/10.1021/ma048757o

[27] A. Okunola A, O. Kehinde I, A. Oluwaseun, A. Olufiropo E, Public and Environmental Health Effects of Plastic Wastes Disposal: A Review, J. Toxicol. Risk Assess. 5 (2019). https://doi.org/10.23937/2572-4061.1510021

Applications of Polymers in Surgery II
Materials Research Foundations 172 (2025) 397-412

Materials Research Forum LLC
https://doi.org/10.21741/9781644903353-17

Chapter 17

Polymers in Dentistry

Huma Iftekhar[1*], Adeela Majid[1]
[1]Department of Conservative Dentistry & Endodontics, AMU
* huma.iftekhar@yahoo.com

Abstract

Dentistry is notable for its diverse array of materials, which encompasses polymers, metal alloys, ceramics, composite materials and inorganic salts. Foundational understanding of polymer characteristics used in dentistry is beneficial because it offers insights pertinent for healthcare practice. Polymeric materials have found increased use in biological applications because of their many beneficial features and broad range of applicability. Significant advancement in the domains of engineering and nanotechnology has made these polymers possible to be customized for particular dental needs. In the present chapter, utilization of polymers in different branches of dentistry is discussed.

Keywords

Dentistry, Polymers, Biofilm, Dental Caries, Denture Base

Contents

Materials Research Foundations 172 (2025) 397-412
https://doi.org/10.21741/9781644903353-17

1. Introduction

Dental treatment includes a wide range of procedures such as cavity fillings, tooth replacement, and cavity lining. Traditionally, metallic and polymeric biomaterials have been utilized for these purposes [1]. The selection of biomaterials is based on their mechanical and physical characteristics and their lack of toxicity. Polymers, because of their outstanding physics and mechanical properties, as well as their biocompatibility, play a vital role in dentistry [2]. These of a wide range of polymers in these procedures has been extensive, playing a key role in enabling major advancements in clinical and reconstructive oral care. [3].

Polymers are macromolecules characterized by their high molecular weight made up of repetitive structural components derived from their individual monomers. In dentistry, widely employed polymers include polycarbonate (PC), polyethylene (PE), polylactic acid (PLLA), polymethyl methacrylate (PMMA), hexamethyldisilazane (HMDC), polydimethylsiloxane, polyethylene glycol (PEG), polyurethane (PUR), polycaprolactone, and polypyrrole (PPy) [4].

The tooth is made up of a complex arrangement of biopolymers. It includes enamel (topmost covering of the crown of the tooth), dentin (layer directly below the outer surface), pulp (contains sensory nerves and blood vessels), cementum (layer covering the root), periodontal ligament (tissue that anchors the tooth), and alveolar bone (structure that anchors the tooth in the mouth). These components consist of various biopolymeric building blocks made up of proteins composed of repeated monomer components that have carbon as their structural framework, creating distinct organic structures. [5]. Since the main components of a tooth are made up of polymers, dental biomaterials aim to mimic the natural tooth's polymeric structure [6].

These days, polymers are commonly employed for restorative purposes as a, prevention of biofilm formation, management of tooth decay with substances utilized in fabrication of partial and full dental prosthesesand lately for cellular engineering and restoration of tooth minerals, drug delivery, implants among various additional purposes [7].

2. Applications of polymers in different branches of dentistry are as follows:

Prosthodontics	Denture bases, artificial teeth, delicate liner materials, temporary restoratives, core build-up, impression and luting materials, and maxillofacial prostheses.
Restorative Dentistry	Biofilm and dental caries prevention, Pit and fissure sealants, cavity fillings, resin and glass-ionomer cements, Dentin bonding agents.
Orthodontics	Invisible aligners, brackets, adhesives, elastic chains
Endodontics	Gutta-percha, cements sealers, and elastic rubber dam sheets.
Equipment	Mouth guards (athletic gear), defensive eyewear, Mixing bowls and spatulas.

Periodontics	Drug delivery
Oral surgery	Implants & grafts

2.1 Role of polymer in biofilm prevention and dental caries development

The human oral cavity is home to more than 700 different bacteria. Biofilms comprise the bulk of microorganisms in oral environment that attaches to hard and soft tissues. Dental caries, periodontal disorders, and infections from oral implants are just a few of the oral illnesses that can result from disruptions in the oral microbiome. To address the issues caused by harmful bacteria due to microbial imbalance in the mouth, a variety of materials have been developed, some of which are listed below.

Cationic polymers- They interact with and damage the negatively polarized cell wall of bacteria, leading to their destruction. They are less likely to face resistance from bacteria and can also penetrate biofilms effectively. A variety of them utilized in designing of biomaterial, such as amino-containing, quaternary ammonium polymer sand guanidine-based polymers, [8].

Polymer conjugated with antibiotics-Polymers can be conjugated with antibiotics through ether, ester or amide bond, to enhance theirstability and solubility. To inhibit the formation of biofilms and bacterial growth, various antibacterial agents, such as fluoride varnish or inorganic nanoparticles (NPs) containing quaternary ammonium compounds can be added with natural products into dental materials. For example, the antimicrobial ciprofloxacin was initially modified with the polymerizable monomer 2-hydroxyethyl methacrylate (HEMA), resulting in the development of a monomer containing Ciprofloxacin. It was then combined with HEMA to produce an antibacterial copolymer. After incorporation into resin, it showed sustained inhibition of Streptococcus mutans. The use of nanoparticles alongside antimicrobials boosts the effectiveness against biofilms by delivering drugs effectively and disrupting biofilm structures with antimicrobial agents by releasing metal ions and antimicrobial agents and prevents microbial colonization [9].

Polymers in caries prevention

Polymers play an important role in caries prevention by improving the delivery of fluoride and other antimicrobial agents, thereby reducing bacterial growth and demineralization of tooth enamel. They can be incorporated into dental materials such as sealants and varnishes to provide a sustained release of protective agent [10]. For example, fluoride-releasing polymers have been shown to effectively restrain the formation of dental caries by continuously releasing fluoride ions that strengthen enamel. Various fluoride-containing dental restoratives can be found on the market, including amalgams, glass ionomers cements (GIC), resin-modified GIC, compomers and composites.

With the advancement in nanotechnology, nanomaterials acting against caries emerged as a significant breakthrough in caries study. Nanoparticles are integrated with polymers or applied as coatings on various biomaterials. Various mechanisms have been suggested for how nanoparticles work, such as the release of metal ions, oxidative stress, and other non-oxidative processes. A common theory is that Nanoparticles with a positive charge are attracted to the bacterial cell membranes that are negatively polarized through electrostatic interactions. This attraction disrupts

Materials Research Foundations 172 (2025) 397-412
https://doi.org/10.21741/9781644903353-17

the cell membrane, causing it to become more permeable, which can result in the rupture of the membrane and leakage of cellular contents. [10].

Nano anti-cavity materials with remineralizing properties are resins or adhesives enhanced with biologically active nanoparticles that help restore the collagen that is exposed in the dental interface and stop it from deteriorating, thereby enhancing durability of bonding. These modified resins can release Ca2+ ions and PO43- ions in acidic conditions, aiding in the dental tissue remineralisation to protect against cavities [11]. Such as Nanosized Calcium Fluoride (NCaF2), Nano Particulate Hydroxyapatite (NHAP), Bioactive Glass Nanoparticle (NBG) and Nanosized Amorphous Calcium-phosphate Particle (NACP).

Chitosan, a natural polymer derived from deacetylated chitin, is used in toothpaste with fluoride and stannous ions to prevent enamel erosion and abrasion. Research by Carvalho and Lussii [12] demonstrated that this fluoride-stannous-chitosan toothpaste effectively reduces tooth surface loss. Chitosan's high positive zeta-potential allows it to adhere to enamel, forming a protective, acid-resistant layer when combined with mucin from saliva. Additionally, stannous ions contribute to enamel protection by forming amorphous deposits and integrating into eroded enamel and dentin [13].

Propolis is a resin-like material sourced from beehives, known for its antibacterial, antifungal, antiviral, anti-inflammatory properties, antioxidant and antitumor. It inhibits the activity of glucosyltransferase enzymes in S.mutans bacteria and prevents caries. It can also be used in mouthwashes to support oral health and prevent cavities because of its anti-plaque and antibacterial effects [14].

2.2 Polymers in restorative dentistry

Composite restorative materials represent a significant advancement in modern biomaterials, successfully replicating the look and function of natural teeth. These materials are made up of three main components the polymer-based resin, the filler, and the interface between them. The resin phase includes monomers that, when exposed to visible light, undergo polymerization, turning a liquid into a polymer with strong cross-linking. This process is initiated by the formation of active centers, often radicals, which drive polymerization [15]. Composite restorations have several drawbacks, including polymerization shrinkage, potential for material fractures and the risk of secondary caries. Additionally, composites and resins tend to retain greater microbial film compared to materials like amalgam and dental cement.

Nano-silver nanocomposites have been made using polymerswith improved characteristics and increased antibacterial effectiveness. Travan et al. [16] created a nanocomposite with antimicrobial properties by incorporating chitosan modified with lactose with silver nanoparticles into heat-polymerized PMMA, which is used in dental applications.

Resin-modified glass ionomer cements are a combination of glass ionomer cements and composite resin, featuring both polymerizable components and acid-base reaction. These materials include fluoro-aluminosilicate glasses, polyacrylic acid, photo-initiators, water and a water-soluble methacrylate monomer like hydroxyethyl methacrylate (HEMA), which may or may not be attached to the polyacrylic acid chemically [17].In RMGIC, the presence of polymerizable functional groups enables faster curing when exposed to light or chemicals, while permitting the acid-base reaction to continue even after the polymerization process.

Pit and fissure sealant- Pit and fissure sealant are made generally resin based which can be polymerized by UV light, auto-polymerized or chemically cured. As the occlusal surfaces of posterior teeth, especially molars vary greatly in shape, featuring deep enamel fissures that accumulate food and encourage plaque buildup which make them more susceptible to caries than smoother surfaces. These sealants create a physical barrier that prevents microorganism and food particle accumulation, helping to stop the onset and progression of caries.

2.3 Polymers in endodontics

Root canal treatment involves debridement of the canals followed by obturation of these canals. The primary objective of root canal obturation is to create a hermetic seal. The most common obturating material is the Gutta Percha. The latest development in endodontic filling materials involves the use of a polymer which is hydrophilic in nature within the canal, known as the SmartSeal system. This system features obturation points called Pro-points, which have a polyamide core wrapped in a hydrophilic polymer coating which expands on absorbing water from the canal and provide a tight [18].

Resin based sealers-Resin-based sealers effectively create a solid monoblock structure within the root canal. Gutta percha coated with polybutadiene-di-isocyanate-methacrylate adhesive has both hydrophilic and hydrophobic properties which bond well with gutta percha and methacrylate resin. This leads to a strong, unified seal within the canal. Additionally, a resin blend containing polycaprolactone and dimethacrylate, known as Resilon, offers an alternative to gutta percha for filling root canals [19].

2.4 Polymers in orthodontics

Polymers have been widely employed in orthodontics because of their favorable mechanical properties and aesthetic qualities. Polymers in orthodontics are classified into brackets, adhesives, elastic chains and elastomeric modules.

Various types of orthodontic polymers available are polyurethane, polycarbonate, polyamides, polyethylene and polymethyl methacrylate [20]. The thermoplastic material significantly decreases in viscosity when heated, making it easy to mold and shape. Once the heat is removed, it returns to a rigid state and holds the new shape. They are mostly used for the manufacturing of invisible orthodontic appliances [21].

Polyethylene terephthalate glycol (PET-G) -It is a resin that is non-crystalline having properties as aesthetically transparent and shiny, high impact resistance, remarkable ductility, excellent chemical inertness which makes it suitable for the manufacturing of clear aligners.

Polypropylene is a lightweight thermoplastic polymer with a partially crystalline structure, comparable to high-density polyethylene. It possesses several notable properties, including resistance to chemicals, humidity, and heat, as well as good surface hardness and dimensional stability. Additionally, it exhibits excellent resistance to flexural fatigue, load, tension, impact, and abrasion, and offers insulating properties.

Polymer brackets- as the demand for the esthetics has increased during the treatment in fixed orthodontics, polymers bracketshas been introduced in the market. These polymers brackets are made up of urethane-dimethacrylate, high-density polyethylene, Polycarbonate and polyoxymethylene brackets. They provide better esthetics, but the main drawback is decreased wear resistance and hardness [22].

Shape-memory polymers in orthodontics offer significant capacity for both functionality and aesthetics. These materials are particularly useful for creating transparent flexible polymeric wires, which can be transformed into arch wires with a specific elasticity modulus of elasticity when exposed to stimuli like heat or light. Shape-memory polyurethane (PU) combined with poly (ε-caprolactone) (PCL) has been used to create such PU wires, used to correct misaligned teeth [20].

Biocompatible films made up of polymers are applied to Nickel -titanium wires to enhance their mechanical properties and corrosion resistance. Various polymeric coatings effective for preventing corrosion on stainless steel, NiTi and other materials include polypyrrole/hydroxyapatite, polyetheretherketone, Polyurethane polyamide, hexamethyldisilazane, polytetrafluoroethylene, graphene oxide/Hydroxyapatite as well as nanoparticles of fullerene-like tungsten disulfide [23].

2.5 Polymers in prosthodontics

Denture base material-

Polymethyl methacrylate (PMMA) is mostly used for manufacturing denture bases and acrylic teeth. PMMA has acceptable aesthetic for denture base materials, with simple processing technique, being tissue tolerant, color stable, low cost, with remarkable polishing ability and marginal adaptation [24,25].Synthetic polymers along with PMMA including bakelite (phenol-formaldehyde) cellulose nitrate, epoxy resins, nylon, polystyrene and vinyl polymers (polyvinyl chloride and polyvinyl acetate)have been used as denture base material however, these materials are more challenging to mold compared to acrylics[26]. Exothermic polymerization, high polymerization shrinkage, tissue irritation linked to excess free monomer and low strength and wear resistance are main disadvantages of PMMA along with it has relatively low impact and flexural strengths, resulting in a higher likelihood of fractures. Furthermore, the uneven surfaces of polymer post-fabrication encourage the attachment of microorganisms to the denture areas near abutment teeth, which could negatively affect oral health and hygiene. Nanoparticles of zirconium, silver nanoparticles or platinum nanoparticles [27,28] added to PMMA to decrease colonization of bacteria or fungus. Polymeric coatings are utilized as antibiotics on prostheses to inhibit development of biofilm. Shibata et al. [29] studied the impact of a polymer poly (2-methacryloyloxyethyl phosphorylcholine-co-n-butyl methacrylate) (PMB), on PMMA's ability to prevent biofilm development. Their findings showed that the PMB polymer film effectively inhibited sucrose-dependent biofilm formation by S. mutans on PMMA denture bases. This study suggested that this biocompatible PMB film could help to reduce biofilm accumulation on PMMA denture surfaces.

2.6 Polymers in implantology

Osseointegration relies on an optimal environment, conducive to its success, leading to various surface modifications of dental implants. These procedures include reinforced fibers having high strength and additives such as antimicrobial agents or hydroxyapatite incorporated through thermoset polymers, that cures at low temperatures [30].Composite made from a polymer and carbon fibers achieved effective osseointegration. The thermoset polymer matrix, along with the carbon fibers, forms covalent bonds that offer robust support for bone structure

 Materials Research Forum LLC
https://doi.org/10.21741/9781644903353-17

while ensuring excellent osseointegration. [31]. Biomimetic coated Titanium surfaces with poly(lactic-co-glycolic acid) (PLGA)/collagen nanofibers and nano-hydroxyapatite (n-HA) are investigated for bone and dental implant surfaces to enhance osseointegration [32].The coating enhanced initial cell adhesion, differentiation and proliferation, as well as mineral deposition on the implant interface.

3. Recent advancements of polymers in dentistry and their applications

3.1 Polyether ketone Ketone (PEKK)

PAEK (polyaryletherketone) is a thermoplastic polymer which consist of 2 more recognized molecule PEKK and polyetheretherketone (PEEK). The main differences between the two is PEKK contains an additional ketone group, which enhances its polarity and backbone stiffness, resulted inelevated glass transition and melting temperatures.[33]

PEKK is a high-performance thermoplastic polymer introduced by Bonner in 1962, initially used in various industrial and military contexts. Recently, it has gained attention as a biomaterial for dental and medical applications due to its superior properties. PEKK is extensively used in restorative, prosthetic, and implant dentistry, showing great promise for cranial and orthopedic implants (Fig.1). Its mechanical strength and second ketone group allow for extensive surface modifications, making it suitable for various biomedical applications [34].

Properties of PEKK (Schwitalla et al. 2015; Alqurashi et al. 2021) [35]

Physical and mechanical properties	**Biological properties**
1. Has high melting point and compressive strength as compared to other polymers.	**1.** It provides metal free restorations beneficial for patients with hypersensitivity issues.
2. It exhibits comparable compressive strength but has a lower elasticity modulus than dentin.	2. PEKK shows excellent biocompatibility and has received FDA approval for use in spinal surgeries and oro-maxillofacial surgeries.
3. PEKK's shock absorption capabilities, combined with its adequate strength and fracture resistance, enhance its potential as a restorative material.	3. The ketone group in PEKK enhances the potential for surface chemical modifications, resulting in a more complex surface topography, a micro-rough texture and increased surface area. These features influence cell behavior and osteointegration on the PEKK surface.
4. PEKK has the fatigue limit (754 N) which is reported remarkably higher	

Materials Research Foundations 172 (2025) 397-412
https://doi.org/10.21741/9781644903353-17

compared to nickel-chromium (586 N) and zirconia (422 N).	4. PEKK shows less bacterial attachment.

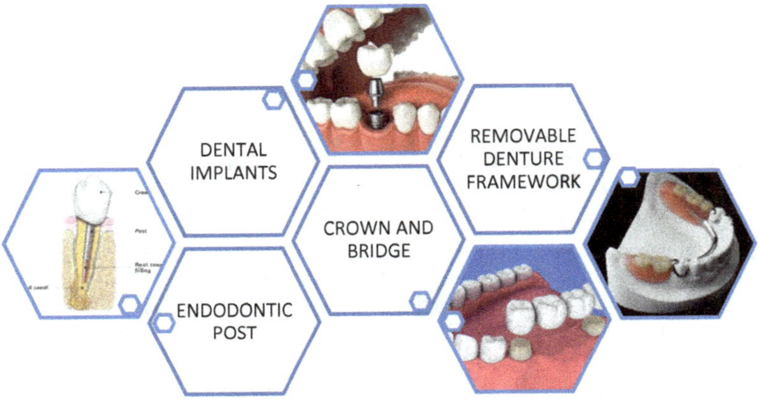

Fig 1- Showing application of PEKK in dentistry.

3.2 Stimuli-responsive restorative dental polymers (smart polymers)

Stimuli-responsive polymers can detect environmental changes, process this information, and initiate specific actions in a controlled manner. While the development of these materials is still in its early stages, they present numerous opportunities for dental applications. The primary challenges lie in the nature of polymeric dental networks, designed for mechanical strength, lacking the capability for the flexible restructuring of chemical bonds [36].

These stimuli can be physical, biological or chemical and materials can be single-responsive, dual-responsive, or multi-responsive based on their response to one, two, or multiple stimuli respectively.

Mechanical Stimuli:

These polymers use mechanical energy to induce chemical transformations, enabling self-healing capabilities. Elongation and compression trigger conformational changes, alignment of polymer chains, and disruption of excimers/disruption of supramolecular interactions. In dentistry, self-healing systems that utilize microcapsules typically consist of poly(urea-formaldehyde) shells that encase triethylene glycol dimethacrylate (TEGDMA) [37].The mechanism of crack filling by self healing polymers is depicted in the form of flow chart(Fig 2).

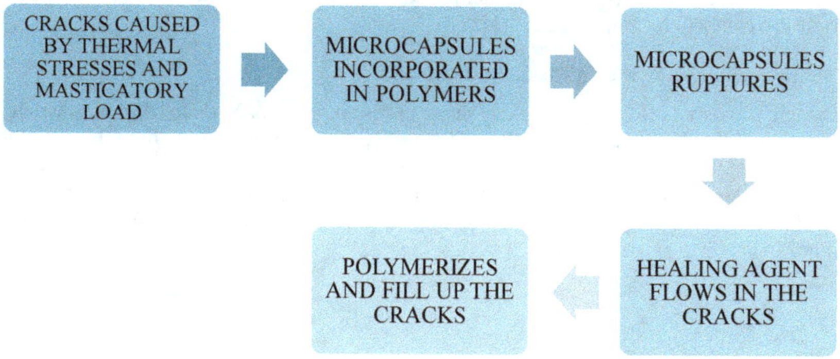

Fig.2 Flowchart showing mechanism of filling of microcracks through self-healing polymers

Glucose-Responsive Polymers:

These polymers detect elevated glucose levels and undergo conformational changes through glucose oxidation by glucose oxidase (GOx), reversible covalent bond or glucose-binding protein interactionsformation with boronic acids. GOx-based systems respond to pH changes from glucose oxidation by-products, not glucose itself. In dental materials, glucose-responsive polymers could help prevent or disrupt bacterial biofilm formation by sensing glucose levels or diol groups in glucans [38].

pH- Responsive Polymers:

These polymers have ionizable acidic or basic groups that change their structure and properties in response to pH fluctuations. They combat bacterial biofilms by responding to pH changes caused by acidogenic bacteria. pH-sensitive materials, like dental resins with pH-responsive monomers such as do-decylmethylaminoethyl methacrylate (DMAEM), show significant antimicrobial effects against bacteria likeStreptococcus mutans. Polymers like Poly-1A switch from neutral to positively charged states in acidic conditions, enhancing antimicrobial activity [39].

Temperature-Responsive Polymers:

Thermo-responsive polymers undergo a phase transition at specific temperatures. The main challenge in developing these composites is maintaining required properties in the oral environment with high glass transition temperatures (Tg). Strategies include grafting low-Tg stimuli-responsive segments onto the polymer backbone, enhancing mobility, and creating favorable conditions for polymer chain rearrangements at lower temperatures [40].

Multi-Responsive Polymers:

Dual- or multi-responsive systems have distinct functional moieties that respond to different stimuli in various ways: parallel, serial, or causal. These systems offer enhanced precision and broader response ranges but require sophisticated design and synthesis to ensure compatibility and balance. In dentistry, multi-responsive systems are used in thermoplastic networks for orthodontic aligners and hydrogels, such as poly (ether urethane). Several of them materials, like shape

Materials Research Foundations 172 (2025) 397-412
https://doi.org/10.21741/9781644903353-17

memory polymers, respond to aqueous and thermal stimuli, facilitating conformational changes at physiologically compatible temperatures [41].

3.3 Shape-memory polymers (SMPs) in dentistry

Shape-changing materials (SCMs) alter their shape in response to appropriate stimuli, gradually changing shape while exposed to the stimulus and returning to their original form once the stimulus is removed. SMPs, also known as actively moving polymers (AMPs), react to various external stimuli such as magnetism, temperature, moisture, specific wavelengths, electricity, pH and certain chemicals_[42].The various types of shape memory polymers and their properties are depicted below (Fig.3, 4).

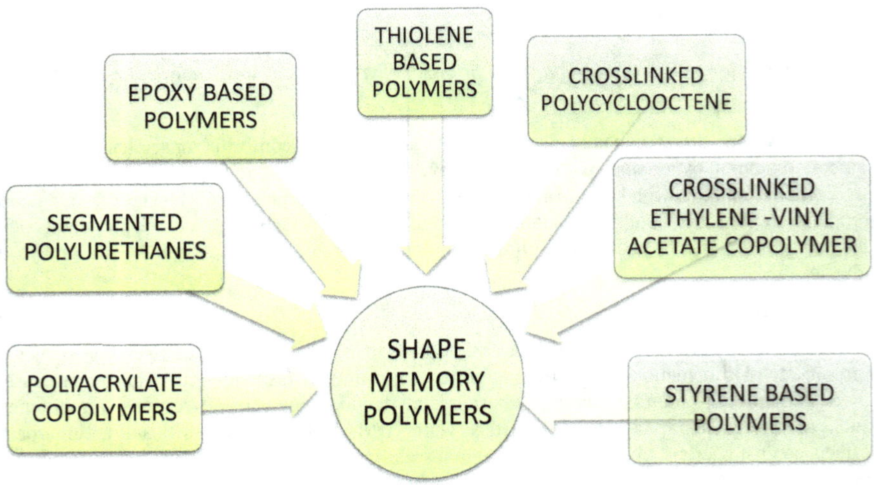

Fig 3- Types of shape memory polymers.

https://doi.org/10.21741/9781644903353-17

-

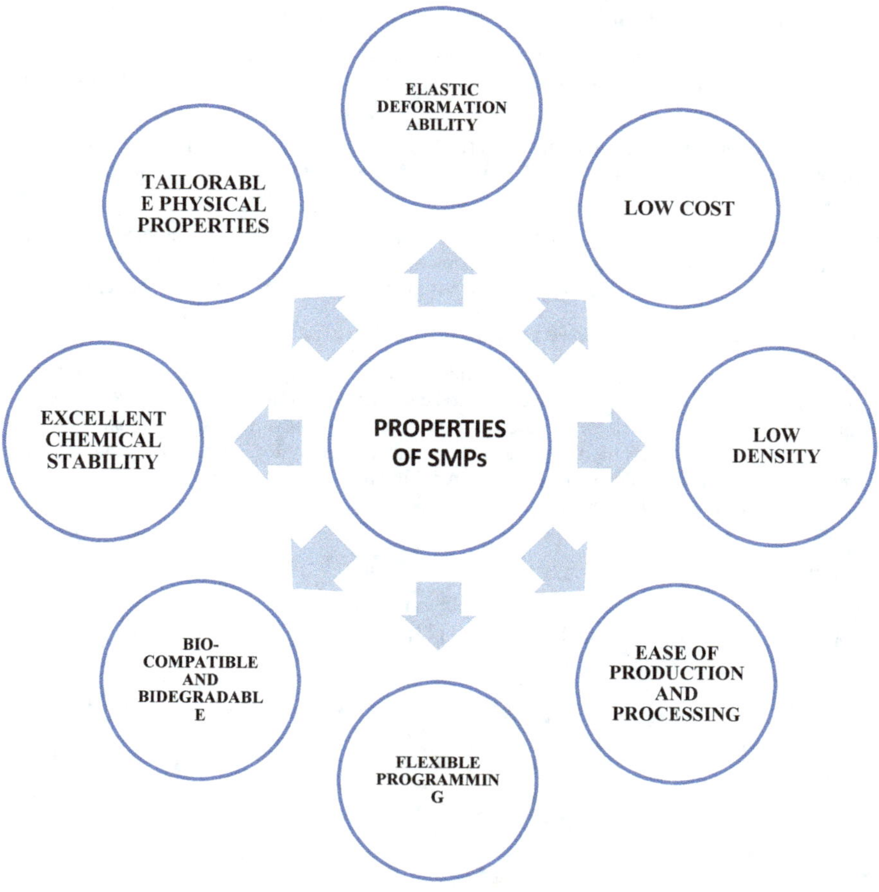

Fig.4 Properties of Shape Memory Polymers (SMPs)

Applications of SMPs in Various Specialties of Dentistry

(i). Orthodontics: SMPs are used in arch wires and clear aligners, gaining popularity due to aesthetic demand. Elastic modules made from SMPs show less force degradation at oral temperature over time.

Materials Research Forum LLC
https://doi.org/10.21741/9781644903353-17

(ii). Endodontics: Self-adjustable filling materials developed from SMPs effectively seal root canals under thermal stimulation. SMPs are also used in endodontic instrument extractors that heat to revert back to their initial form, facilitating the removal of broken instruments.

(iii). Prosthodontics: Shape-memory resin crowns adapt perfectly to abutment surfaces without cement. SMP-lined dentures adapt gradually to the alveolar ridge, ensuring stability and reducing adjustments.

(iv). Restoration: Innovative chromatic materials combining SMPs with optical change elements adapt in response to stimuli, simplifying color selection and improving aesthetic outcomes in restorations.

(v). Oral Surgery and Implantology: SMPs serve as active substrates for cell culture, with potential for tissue engineering and scaffold development. SMP-based implants expand and fit tightly within cavities, providing instant fixation and rapid osseointegration with the alveolar bone.

3.4 Polymers in drug delivery systems

Polymeric frameworks for the administration of antimicrobial agents, biologically active substances and various cell types have garnered considerable global attention. These systems offer innovative solutions for targeted therapy, enhancing efficacy and reducing side effects of treatments. By utilizing polymers, researchers can create controlled-release formulations that improve drug stability and bioavailability. Additionally, the versatility of polymers allows for customization in drug delivery applications, making them suitable for a wide range of medical conditions (Fig.5).

Examples include:

(i). Chlorhexidine dihydrochloride- Chitosan nanoparticles (CSNPs) combined with peptidomimetic derivative glutathione (GSH) [43].

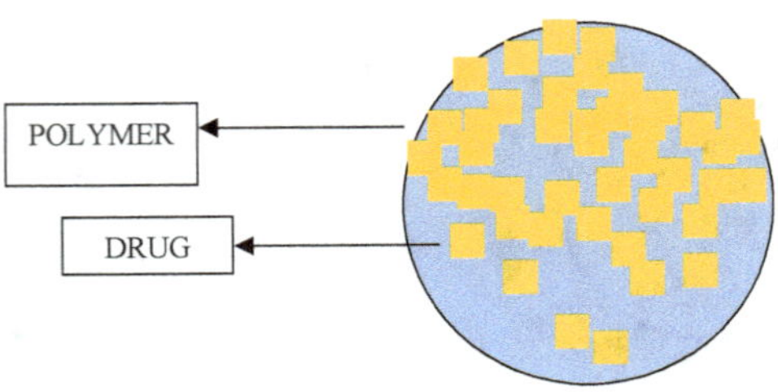

Fig. 5- polymer-carrying drug

(ii). Polylactic acid (PLA) and polyglycolic acid (PLGA) can deliver several medications or bioactive substances to treat oral diseases and enhance tissue regeneration. These formulations can be shaped into scaffolds, nanoparticles, microspheresand fiber mats, loaded with single or dual drugs for controlled release.

(iii). Antibacterial nanofibrous poly(L-lactic acid) (PLLA) scaffolds with well-interconnected macroporous structures and PLGA nanospheres loaded with doxycycline [44].

(iv). pH-responsive nanoparticles for targeted topical delivery of farnesol, made from diblock copolymers, release farnesol upon core destabilization in acidic conditions, compromising the mechanical stability of biofilms for easier disruption [44].

3.5 Bone grafts

Bone grafts are tissues transplanted to repair or rebuild damaged or missing bones. They can be sourced from the patient's body (autograft), a donor (allograft), or be synthetic (alloplastic). Synthetic bone substitutes closely replicate the biological characteristics of natural bone, addressing immunogenicity and donor site complications. Commonly used polymers include polylactic acid, polyglycolic acid, and polycaprolactone, known as aliphatic polyesters. Hard tissue replacement (HTR) Synthetic Bone™ composed of Polymethyl methacrylate, polyhydroxyethylmethacrylate, and calcium hydroxide is used in the treatment of periodontal intrabony and furcation defect.[45]

Advantages of Polymer Grafts:

Polymer bone grafts are advantageous for their versatility and biocompatibility. However, challenges include acidic degradation products affecting local pH, osteoconductivity, and cell adhesion. Modifications such as adding hydroxyapatite (HA) or tricalcium phosphate (TCP) or using coatings like silk fibroin with VEGF have shown potential to enhance bone regeneration. Additionally, 3D-printed polylactic acid-based biopolymers with specific pore sizes have improved cell proliferation and differentiation,[46]

References

[1] Roach M, Base metal alloys used for dental restorations and implants, Dent Clin North Am. 51(2007)603-627. https://doi.org/10.1016/j.cden.2007.04.001

[2] Deb S., PolymersIn Dentistry, Proc Inst MechEng H.21 (1998) 453-64. https://doi.org/10.1243/0954411981534213

[3] Bertassoni, L.E., Uskokovic, V., Nanotechnology in dental sciences: moving towards a finer way of doing dentistry, Materials. 3(3) (2010) 1674-1691. https://doi.org/10.3390/ma3031674

[4] Maitz M.F, Applications of synthetic polymers in clinical medicine, Biosurf Biotribol.1(2015)1 161-176. https://doi.org/10.1016/j.bsbt.2015.08.002

[5] Lai WF, Oka K, Jung HS, Advanced functional polymers for regenerative and therapeutic dentistry, Oral Dis.21(5) (2015)550-557. https://doi.org/10.1111/odi.12281

[6] Ferracane, J.L.,Current trends in dental composites. Crit. Rev, Oral Biol. Med. 6(1995) 302-318. https://doi.org/10.1177/10454411950060040301

[7] Roach M. Base metal alloys used for dental restorations and implants. Dent Clin North Am.51(2007) 603-624.Deb S. Polymers in dentistry. Proc Inst Mech Eng H. 21(1998)453-464. https://doi.org/10.1016/j.cden.2007.04.001

[8] Jia B, Zhang B, Li J, Qin J, Huang Y, Huang M, et al., Emerging polymeric materials for treatment of oral diseases: design strategy towards a unique oral environment, Chem Soc Rev.53 (2024)3273-301. https://doi.org/10.1039/D3CS01039B

Materials Research Foundations 172 (2025) 397-412
https://doi.org/10.21741/9781644903353-17

[9] Dinesh Rokaya, Viritpon Srimaneepong, Polymeric materials and films in dentistry: An overview ,J Adv Res.14 (2018 Nov) 25-34. https://doi.org/10.1016/j.jare.2018.05.001

[10] Chen H, Gu L, Liao B, Zhou X, Cheng L, Ren B., Advances of Anti-Caries Nanomaterials, Molecules.25(2020)5047. https://doi.org/10.3390/molecules25215047

[11] Vyavhare S, Sharma DS, Kulkarni VK, Effect of three different pastes on remineralization of initial enamel lesion: an in vitro study, J Clin Pediatr Dent.39(2015)149-60. https://doi.org/10.17796/jcpd.39.2.yn2r54nw24l03741

[12] Carvalho T.S., Lussi A, Combined effect of a fluoride-, stannous- and chitosan-containing toothpaste and stannous-containing rinse on the prevention of initial enamel erosion-abrasion, J Dent.42 (2014)450-459. https://doi.org/10.1016/j.jdent.2014.01.004

[13] Dinesh Rokaya, ViritponSrimaneepong,Polymeric materials and films in dentistry: An overview, J Adv Res.14 (2018) 25-34. https://doi.org/10.1016/j.jare.2018.05.001

[14] Farnoush Mohammadi, Applications of Propolis in Dentistry: A Review, Ethiop J Health Sci.28(2018) 505-512.

[15] Cramer NB, Stansbury JW, Bowman CN, Recent advances and developments in composite dental restorative materials, J Dent Res. 90(2011)402-16. https://doi.org/10.1177/0022034510381263

[16] Travan A., Marsich E., Donati I., Benincasa M., Giazzon M., Felisari L, Silver-polysaccharide nanocomposite antimicrobial coatings for methacrylic thermosets,Acta Biomater.7(2011)337-346. https://doi.org/10.1016/j.actbio.2010.07.024

[17] Coutinho E, Yoshida Y, Inoue S, Fukuda R, Snauwaert J, Nakayama Y, et al., Gel phase formation at resin-modified glass-ionomer/tooth interfaces, J Dent Res. 86(2007)656-61. https://doi.org/10.1177/154405910708600714

[18] Shashank Arora, Vibha Hegde, Comparative evaluation of a novel smart-seal obturating system and its homogeneity of using cone beam computed tomography: In vitro simulated lateral canal study ,J Conserv Dent.17(4)(2014) 364-368. https://doi.org/10.4103/0972-0707.136512

[19] Sandeep Thakur, Mithra N Hegde, Darshana Devadiga, Resin Based Root Canal Sealers: An Overview,Research Journal of Pharmaceutical, Biological and Chemical Sciences. 4(2013)1501-1509.

[20] Hassan, R., Aslam Khan, M.U., Abdullah, A.M., Abd Razak, S.I, A Review on Current Trends of Polymers in Orthodontics: BPA-Free and Smart Materials,Polymers. 13 (2021)1409. https://doi.org/10.3390/polym13091409

[21] Bichu YM, Alwafi A, Liu X, Andrews J, Ludwig B, Bichu AY, Zou B,Advances in orthodontic clear aligner materials,Bioact Mater. 22 (2022) 384-403. https://doi.org/10.1016/j.bioactmat.2022.10.006

[22] Julia Krauss, Andreas Faltermeier, Michael Behr, Peter Proff, Evaluation of alternative polymer bracket materials, Am J Orthod Dentofacial Orthop. 137(3) (2010)362-7. https://doi.org/10.1016/j.ajodo.2008.12.020

Materials Research Forum LLC
https://doi.org/10.21741/9781644903353-17

[23] Torabi M., Sadrnezhaad S.K., Corrosion behavior of polypyrrole/hydroxyapatite nanocomposite thin films electropolymerized on NiTi substrates in simulated body fluid,Mater Corros.62(2011)252-257. https://doi.org/10.1002/maco.200905399

[24] Cunningham, J., Bond strength of denture teeth to acrylic bases,J. Dentistry. 21(1993) 274-280. https://doi.org/10.1016/0300-5712(93)90106-Z

[25] Jo, L.J., Shenoy, K.K., Shetty, S., Flexural strength and hardness of resins for interim fixed partial dentures, Indian J. Dent. Res. 22 (2011) 71-76. https://doi.org/10.4103/0970-9290.79992

[26] Khindria, S., Mittal, S., Sukhija, U., Evolution of denture base materials, J Indian Prosthodontic Soc.9(2)(2009)64-69. https://doi.org/10.4103/0972-4052.55246

[27] Gad M.M., Al-Thobity A.M., Shahin S.Y., Alsaqer B.T., Ali A.A, Inhibitory effect of zirconium oxide nanoparticles on Candida albicans adhesion to repaired polymethyl methacrylate denture bases and interim removable prostheses: a new approach for denture stomatitis prevention, Int J Nanomed.12(2017)5409-5419. https://doi.org/10.2147/IJN.S142857

[28] Li Z, Sun J, Lan J, Qi Q,Effect of a denture base acrylic resin containing silver nanoparticles on Candida albicans adhesion and biofilm formation, Gerodontology. 3(2016) 3209-16.

[29] Shibata Y., Yamashita Y., Tsuru K., Ishihara K., Fukazawa K., Ishikawa K, Preventive effects of a phospholipid polymer coating on PMMA on biofilm formation by oral streptococci,Appl Surf Sci. 390(2016)602-607. https://doi.org/10.1016/j.apsusc.2016.08.108

[30] Petersen R.C. Titanium implant osseointegration problems with alternate solutions using epoxy/carbon-fiber-reinforced composite. Metals (Basel).4(2014)549-569. https://doi.org/10.3390/met4040549

[31] Petersen R.C.,Bisphenyl-polymer/carbon-fiber-reinforced composite compared to titanium alloy bone implant,Int J Polym Sci. (2011)168924. https://doi.org/10.1155/2011/168924

[32] Ravichandran R., Ng C.C., Liao S., Pliszka D., Raghunath M., Ramakrishna S., Biomimetic surface modification of titanium surfaces for early cell capture by advanced electrospinning,Biomed Mater. 7(1)(2012)015001. https://doi.org/10.1088/1748-6041/7/1/015001

[33] Maloo LM, Toshniwal SH, Reche A, Paul P, Wanjari MB, A Sneak Peek Toward Polyaryletherketone (PAEK) Polymer: A Review, Cureus.14(11) (2022) 31042. https://doi.org/10.7759/cureus.31042

[34] Alqurashi H, Khurshid Z, Syed AUY, Rashid Habib S, Rokaya D, Zafar MS, Polyetherketoneketone (PEKK): An emerging biomaterial for oral implants and dental prostheses,J Advert Res. 28(2021)87-95. https://doi.org/10.1016/j.jare.2020.09.004

[35] Schwitalla AD, Spintig T, Kallage I, Müller WD,Flexural behavior of PEEK materials for dental application,Dent Mater.31(11)(2015)1377-84. https://doi.org/10.1016/j.dental.2015.08.151

[36] Fugolin, A.P.P.; Huynh, B.; Rajasekaran, S.P,Innovations in the Design and Application of Stimuli-Responsive Restorative Dental Polymers, Polymers.15(2023)3346. https://doi.org/10.3390/polym15163346

[37] Wiggins KM, Brantley JN, Bielawski CW,Methods for activating and characterizing mechanically responsive polymers, Chem Soc Rev.42(2013) 130-47. https://doi.org/10.1039/c3cs35493h

[38] Wang J, Yu J, Zhang Y, Zhang X, Kahkoska AR, Chen G, etal., Charge-switchable polymeric complex for glucose-responsive insulin delivery in mice and pigs,Sci Adv. 5(2019)4357. https://doi.org/10.1126/sciadv.aaw4357

[39] Ofridam F, Tarhini M, Lebaz N, Gagnière É, Mangin D, Elaissari A, Ph -sensitive polymers: Classification and some fine potential applications, Polym Adv Technol. 32(4)(2021)1455-84. https://doi.org/10.1002/pat.5230

[40] Urban, M.W. Stimuli-Responsive Materials: From Molecules to Nature Mimicking Materials Design; Royal Society of Chemistry2019.

[41] Schattling P, Jochum FD, Theato P, Multi-stimuli responsive polymers - the all-in-one talents,Polym Chem.5(1)(2014)25-36. https://doi.org/10.1039/C3PY00880K

[42] Bruni A, Serra FG, Deregibus A, Castroflorio T,Shape-Memory Polymers in Dentistry: Systematic Review and Patent Landscape Report,Materials.12(14) (2019) https://doi.org/10.3390/ma12142216

[43] Marks R. Bioengineering and the Skin: Based on the Proceedings of the European Society for Dermatological Research Symposium, held at the Welsh National School of Medicine, Cardiff, 19-21 July 1979. Springer Science & Business Media; (2012). 327 p.

[44] Xu X, He L, Zhu B, Li J, Li J,Advances in polymeric materials for dental applications,Polym Chem. 8 (2017)807-23. https://doi.org/10.1039/C6PY01957A

[45] Zhao R, Yang R, Cooper PR, Khurshid Z, Shavandi A, Ratnayake J, Bone Grafts and Substitutes in Dentistry: A Review of Current Trends and Developments, Molecules. 26 (2021) 3007. https://doi.org/10.3390/molecules26103007

[46] Parani M, Lokhande G, Singh A, Gaharwar AK, Engineered Nanomaterials for Infection Control and Healing Acute and Chronic Wounds,ACS Appl Mater Interfaces. 8(2016)10049-69. https://doi.org/10.1021/acsami.6b00291

Applications of Polymers in Surgery II — Materials Research Forum LLC
Materials Research Foundations 172 (2025) 413-460 — https://doi.org/10.21741/9781644903353-18

Chapter 18

Polymers in Blood Vessels Reconstruction

Joy Hoskeri H[1,*], Vidyashree Suryavanshi[1], Preeti Gadyal[1], Nivedita Pujari S[2], Arun K. Shettar[3]

[1]Department of Bioinformatics, Karnataka State Akkamahadevi Women University, Vijayapura-586109, Karnataka, India

[2]Albert Einstein College of Medicine, Bronx-10461, New York USA

[3]Cytxon Biosolutions Pvt Ltd, Huballi, 580031, Karnataka, India

*joybioinformatics@gmail.com, joybioinfo@gmail.com

ORCID ID: 0000-0001-5763-7115

Abstract

Vascular tissue architectural knowledge is the key for repairing vascular abnormalities and also offers opportunities for targeted therapies for vascular diseases. Despite the advancements in the field of biomaterials vascular implantation is still a challenge. Importance of natural polymers like collagen, elastin, chitosan, alginate, fibrin, gelatin, and synthetic polymers like polyethylene-glycol, poly(L-lactic-co-glycolic-acid), polyvinyl-alcohol, polypropylene, polycaprolactone, poly(phenylene-vinylene) in blood vessel reconstruction, primarily due to their biocompatibility and ability to support cellular activities essential for vascular repair is covered in this chapter. Polymeric aortic vessels using advanced 3D printing technique demonstrate the potential of polymer scaffolds in creating customized vascular replacements.

Keywords

Cardiovascular Diseases, Blood Vessel Reconstruction, Polymers, Haemostatics, Sealants

Contents

Materials Research Foundations 172 (2025) 413-460
https://doi.org/10.21741/9781644903353-18

Materials Research Foundations 172 (2025) 413-460
https://doi.org/10.21741/9781644903353-18

1. Introduction

The circulatory system, including lymphatic and cardiovascular vascular systems, is an essential part of the organism. The cardiovascular system, which is made up of the heart, arteries, veins, and capillaries, is in charge of pumping blood throughout the body [1]. In vertebrates, interstitial fluids are the channels for transporting hormones, nutrients, and oxygen into cells and are also used to eliminate waste, such as carbon dioxide [2]. The determination of congenital heart abnormalities, which are often identified by fetal heart ultrasound and MRI screening, requires an understanding of the development of the circulatory system during the embryonic stage [3]. In addition, new circulation systems, such as those used in fuel cell vehicles, show improvements in the flexibility and efficiency of temperature control and improve the overall performance of the system [4]. Furthermore, to improve patient care, a circulatory support system with implanted blood pumps and advanced sensors is essential to monitor the cardiac characteristics and operation parameters of pumps [5].

Blood vessels are different from other types of body tissue in that they are located exclusively in the vascular system of the body. All tissue and organs in the body are supplied mainly with blood,

Materials Research Foundations 172 (2025) 413-460 https://doi.org/10.21741/9781644903353-18

bones, and other metabolites by the blood vessels that form the vascular tree. Blood vasculature also plays a role in thermoregulation in response to environmental changes [6]. The body's blood arteries are so large that they come together to form the circulatory system. The arteries, arteries, capillaries, veins, and veins that make up this cardiovascular system are connected lines that function independently [7]. Long blood vessels called arteries and veins run throughout the body; the terms "capillaries," "veins," and "arteries" refer to the smaller blood vessels that join the arteries and veins. Blood leaves the heart and flows to every region of the body via the aorta, capillaries, systemic arteries, veins, and systemic veins.

Elastic arteries form a very small proportion of peripheral arteries, which are primarily known as systemic arteries. Tables and economics are the main branches of the system's arterial branches [8]. The arteries function as a convertible tube and the primary veins separates from the larger veins to form smaller veins, then disperse into different tissues as capillaries. The diameter of the capillary is smaller than the diameter of human hair. The internal wall of an internal volume tube consists of a single layer of flat and expandable fibrils and endothelial cells with their basic membranes [9]. The transport of metabolites from tissues into the blood and nutrients from the blood to tissues takes place through the capillary walls. Venules form when capillaries are united. The two small veins eventually combine to form two important systemic veins, which provide blood to the heart. They also continue to merge into small veins [10].

Vascular health problems include many diseases, including atherosclerosis, inflammatory vasculitis, genetic diseases affecting blood vessels, endocrine diseases such as diabetes, and infectious agents that cause vascular diseases [11]. Age and lifestyle factors aggravate atherosclerosis, which is defined by lipid buildup on the artery wall and causes cardiovascular disorders such as myocardial infarction and stroke [12]. Additionally, there are significant variations in vascular health between racial, ethnic, and socioeconomic groups, and social health variables have a significant impact on illness presentation and outcome [13]. Understanding different pathologies that affect vasculature is essential to the development of accurate diagnosis, effective treatment, and targeted interventions to address these complex health problems.

According to specific conditions, vascular disorders can manifest different symptoms. Common symptoms include mild injuries, petechiae, contusions, mucous membranes, and spontaneous blood from small vessels characteristic of vascular bleeding disorders [14]. Headaches are common symptoms of many vascular diseases, particularly subarachnoid bleeding, and arteritis, where they are the most obvious features. Patients with vascular dementia often experience cognitive decline, mood disorders, emotional instability, and focal neurological disorders due to damage to brain cells affecting the cerebrovascular system [15]. In addition, common behavioral and psychological symptoms such as apathy, depression symptoms, irritation, agitation, aggression, and aggression in vascular dementia and the symptoms profile of small vessels and large vessels differ [16]. Understanding these different symptoms is crucial to identifying and treating vascular disorders accurately.

Numerous risk factors, including genetic predisposition, obesity, hypertension, diabetes, and smoking, have an impact on vascular disorders. Diseases such as peripheral artery disease, cerebral vascular disease, renal artery stenosis, and cardiovascular disease are associated with these risk factors [17]. Furthermore, diet and lifestyle choices have a significant impact on an individual's susceptibility to circulatory system diseases. The advancement of cardiovascular disease,

Materials Research Forum LLC
https://doi.org/10.21741/9781644903353-18

including dementia linked to vascular pathology, can be prevented or slowed down with the help of effective management of vascular risk factors such as smoking, obesity, diabetes, dyslipidemia, hypertension, diabetes, and aging [15]. Furthermore, sedentary habits, obesity, irregular sleep patterns, and depression all worsen small vessel diseases, a primary cause of neurological disorders like kidney failure, blindness, cerebral infarction, and dementia.

The etiology, diagnosis, and treatment of illnesses such as cerebral vascular diseases (CSVDs), cardiovascular diseases, and vascular conditions are all influenced by genetics. Genetics also plays a significant role in vascular diseases. Research has highlighted the genetic factors associated with these diseases, including genetic mutations such as ELMO2, which cause intraskeletal vascular malformations [18], as well as new genetic factors identified in CSVD, which contribute to the identification of subtypes and better clinical management. Genome-wide association studies have revealed DNA variants that cause coronary diseases, strokes, and hypertension, highlighting the genetic composition of common cardiovascular diseases [19]. Understanding the genetic basis of vascular abnormalities not only helps diagnosis but also provides opportunities for targeted therapies and preventive interventions that ultimately improve the management and forecasting of people with vascular diseases.

Vascular diseases are widespread and require a large-scale reconstruction of blood vessels. Various types of blood vessel reconstruction, including autografts, allografts, and biomaterials [20] have been developed. One crucial surgical approach for treating individuals with tumor vessels and atherosclerosis is blood vascular replacement. Although autografts are considered gold standards because of their excellent patency, their use is limited due to poor availability, disease, and deformity [21]. Allograft is usually an alternative to autograft, but its use and infection risk have moral and ethical problems. Biomaterials can solve these problems, and major progress is expected in this field. This review examines the use of autografts, allografts, and biomaterials, focusing on the advantages and disadvantages of biomaterials [22].

Given the surgical time needed for the vascular bypass and the post-operative re-endothelialization after many vascular reconstruction procedures, implants must be developed with superior results over autogenous vessels [23]. Even though self-contained vessels are thought to be the ideal materials for artery bypass surgery, there aren't as many of these procedures performed because of equipment shortages and poor quality. Because of the lack of a suitable donor and the significant sacrifice of healthy arterial segments, the correct choice of autogenous vessels as materials is difficult [24]. Allografts are considered an alternative to autogenous vessels but are not recommended for the risk of infection and the use of heavy immunosuppressants. Since the 1950s, many implant materials have been used to reconstruct vascular organs. However, the requirements for vascular implants are very strict. Although several attempts have been made to produce biomaterials that exhibit such benign properties, no fully successful implants have yet been manufactured [25]. The healing of host tissue is a key issue in artery reconstruction. Researchers are exploring new techniques to control the biological reactions involved in the healing process. Biomaterials with exceptional biomechanical qualities can be crucial to the outcome of vascular bypass surgery [26].

Blood vessels are essential to the supply of oxygen and the nutrition of the organs. In plastic surgery, blood vessels are reconstructed not only to cure certain tissue necrosis but also to repair defects in skin and muscle tissues with free flaps [27]. In addition, the reconstruction of blood vessels can solve problems of bone transplantation or fractures that cannot be healed [28].

Materials Research Foundations 172 (2025) 413-460
https://doi.org/10.21741/9781644903353-18

Anastomosis of the transplanted intestines is made possible by the repair of blood vessels, which is advantageous for the excretory system's circulation [29]. In general, adequate blood supply increases the possibility of wound healing and transplantation. Previously, blood vessel reconstruction was divided into arterial and vascular surgery. Since the mechanical strength and blood pressure of the veins are lower than those of the arteries, the reconstruction of the veins may be more difficult [30,31]. Currently, basic research, surgical technology, and materials for the reconstruction of blood vessels have been developed and improved, and some of the very difficult operations and procedures to save limbs have been successfully applied to clinical practice.

2. Types of blood vessel reconstruction

Blood vessels are essential for oxygen and organ nutrition. In plastic surgery, blood vessels are rebuilt not only to cure specific tissue necrosis but also to repair skin and muscle tissue defects using free flaps [27]. Furthermore, repairing blood vessels can solve bone transplantation and repair broken bones [28]. In the case of the digestive system, the reconstruction of blood vessels provides anastomosis of transplanted intestines and helps to circulate excrements [29]. In general, adequate blood supply increases the likelihood of wound healing and transplantation. Previously, the reconstruction of blood vessels was divided into vascular and arterial surgery. Because the mechanical resistance and blood pressure of the vein are lower than those of the arteries, the reconstruction of the vein may be more difficult [30,31]. Currently, basic research, surgical technology, and materials for rehabilitating blood vessels have been developed and improved, and some of the very difficult operations and procedures to save limbs have been successfully implemented in clinical practice.

2.1 Grafting techniques

Blood vessel graft plays avital role in cardiac tissue engineering and surgery. Various approaches have been used, including synthetic materials such as Teflon chips for arterial reconstruction [32], Electrospinning techniques for producing biodegradable polymer transplants with biologically active molecules [33], and the use of engineering and cell techniques to fabricate small-diameter vascular grafts for the external arteries [34]. Coronary bypass transplantation (CABG) often uses stem vein transplantation and internal thoracic arteries, and emerging evidence supports the superiority of multiple artery transplantation to improve long-term patency [35]. These techniques are aimed at creating biocompatible and mechanically robust grafts that seamlessly integrate with host blood vessels, promote tissue reconstruction, and ensure optimal exchange of nutrients and waste after implantation [34]. Further research is needed to validate and optimize these approaches to improve the results of blood vessel reconstruction.

2.2 Autologous grafts

The use of internal iliac vessels to reconstruct vascular organs in intestinal transplants reduces vascular complications, facilitates transplantation, and reduces the risk of thrombosis [36]. Self-grafts have greater mechanical qualities, smooth muscle differentiation, and collagen/elastin expression, which improves their biocompatibility and prognosis for blood vessel restoration [37]. Autologous transplants offer a higher quality of life and better results in breast reconstruction than implants. They also show ongoing progress in reducing donor site morbidity and improving reconstruction methods. Autologous grafts offer advantages in vascular reconstruction due to low infection risks, high patency rates, and the ease of adapting to suitable channel preparations, as

Materials Research Foundations 172 (2025) 413-460
https://doi.org/10.21741/9781644903353-18

research [63]. All things considered, xenografting has a great deal of promise for vascular applications, but more study is required to resolve its drawbacks and enhance clinical outcomes.

2.9 Allografting

Allograft in blood vessels involves transplanting vascular tissue from a donor to a recipient, which has shown promise in several studies. Research shows that allografts can effectively restore blood flow and improve vascular function, especially in the case of severe arteries or injuries [64]. However, the success of allografting is often limited by issues such as immune rejection and the long-term viability of the transplantation [65]. Recent advances have focused on improving graft integration and reducing rejection rates through techniques such as the preconditioning of grafts and the use of immunosuppressive therapies [66]. Furthermore, bioengineering approaches, such as decellularized vascular scaffolds, have been studied to create more compatible transplants [67]. Despite these innovations, the challenges of achieving optimal results remain and further research is needed to improve technology and improve patient prognosis of vascular allograft.

2.10 Endovascular techniques

It has been demonstrated that endovascular technologies play a significant role in the management of vascular disorders, particularly when it comes to the treatment of ailments like aneurysms and peripheral artery diseases. The success of these minimally invasive techniques, which use catheters and imaging guides to navigate the vascular system and shorten recovery times and associated difficulties with open surgery, has been demonstrated by recent research [68,69]. Studies show that endovascular procedures, such as endovascular aneurysm repair (EVAR), perform better than traditional approaches in terms of mortality and morbidity rates [70]. Some studies do, however, also point out drawbacks, such as the possibility of end leaks and the requirement for ongoing monitoring following the procedure [71]. Overall, although endovascular technology represents significant progress in vascular surgery, ongoing research is essential to address its limitations and optimize patient outcomes. The integration of these methods into clinical practice continues to evolve, reflecting their increasing importance in modern medicine.

2.11 Balloon angioplasty

A common procedure for recovering blood vessels, balloon angioplasty is generally used to treat stenosis or blockages. According to research, this technique effectively dilates closed arteries, increasing blood flow and reducing symptoms of vascular disease [72]. To compress the plaque against the vessel wall, a balloon catheter is introduced into the vessel and inflated at the location of obstruction [73]. Recent research emphasizes the importance of auxiliary therapies such as balloons that eliminate drugs, which have the potential to reduce restenosis rates compared to conventional angioplasty [74]. However, there are some disadvantages, such as the possibility of re-retracing the ship and the requirement for subsequent procedures [75]. In addition, the effectiveness of surgery depends critically on the selection of the patient and the specificities of the vascular injury [76]. Given all this, balloon angioplasty remains the gold standard of vascular interventions, but continuous development and cautious treatment of patients are required to maximize results.

2.12 Stent placement

Materials Research Foundations 172 (2025) 413-460
https://doi.org/10.21741/9781644903353-18

The placement of needles in the reconstruction of blood vessels is a crucial intervention to restore vascular patency and improve blood flow. Research shows that stenting can significantly improve the outcome of various vascular conditions, including coronary and peripheral artery diseases. For example, studies have shown the effectiveness of drug-soluble stents in reducing restless rates compared to conventional metal stents and thus improving long-term patency rates [77,78]. Furthermore, advances in stent design, such as bioresorbable scaffolds, have shown promise in reducing the complications associated with permanent implants [79,80]. There are still difficulties, though, such as the possibility of thrombosis and the requirement for two rounds of antiplatelet therapy following surgery [81]. Overall, although the placement of the stent is a valuable technique for vascular reconstruction, ongoing research is essential to optimize the materials and designs of the stent to improve patient outcomes and reduce associated risks.

3. Revolution in blood vessel reconstruction through 3D printing technologies

3D printing technology has become a transformative approach to blood vessel reconstruction, providing significant advances in tissue engineering. Research indicates that 3D printing can create complex vascular structures that closely imitate natural blood vessels and increase the possibility of successful integration and function *in vivo* [82]. Various materials, including biocompatible hydrogels and polymers, have been used to manufacture these structures, allowing individualization based on specific patient needs [83]. Furthermore, studies emphasize the importance of incorporating endothelial cells and elastin muscles into printed structures to promote vascularization and improve mechanical properties [84]. However, there are still challenges, such as ensuring the long-term reliability and functionality of printed packaging and addressing the scaling of production for clinical applications [85]. Overall, 3D printing is a promising possibility for the reconstruction of blood vessels, but continued research is crucial to overcome existing limitations and improve clinical results [86].

3.1 Cell-free scaffolds

Free cell scaffolding has appeared as a positive approach to the reconstruction of blood vessels, using its ability to support tissue regeneration without the complications associated with cell-based therapy. Research indicates that these scaffolds can be constructed from various biomaterials and provide a favorable environment for vascular tissue formation. For example, Itoh highlights the potential of biodegradable polymers in creating scaffolds that mimic extracellular matrix and facilitate cell adhesion and proliferation required for vascularization [87]. Fernández-Pérez et al. also emphasize the importance of scaffold architecture and mechanical properties that have a major impact on cell behavior and tissue integration [88]. Saito and others discussed the role of bioactive molecules in scaffolds, which improves angiogenesis and improves the overall healing process [89]. However, challenges remain such as ensuring sufficient nutrient distribution and the long-term stability of scaffolds in vivo. Overall, cell-free scaffolds represent a versatile and effective strategy for reconstructing blood vessels and are being studied to optimize their design and functionality [90].

3.2 Cell-based scaffolds

Cell-based scaffolds are emerging as a promising approach to rebuilding blood vessels, combining cell regeneration potential with scaffolding structure support. According to studies, these scaffolds can enhance endothelial cell growth and function, which is critical for the proper vascularization

Materials Research Foundations 172 (2025) 413-460
https://doi.org/10.21741/9781644903353-18

and integration of the host tissues [91]. Sassi et al., for instance, stress the significance of biocompatible materials that encourage cell adhesion and proliferation and aid in the development of functioning blood arteries. Furthermore, Yudintceva et al. highlighted how the scaffold's structure affects how cells behave, indicating that particular designs might maximize waste elimination and nutrition flow, enhancing cell survival and function [92]. However, there are drawbacks, like the difficulty of maintaining patience over time and the possibility of thrombosis, which necessitates more investigation into scaffold composition and surface modifications [93]. Overall, although cell-based scaffolds have significant potential, ongoing studies are essential to address these challenges and improve their clinical application in vascular tissue engineering.

3.3 Surgery involved in blood vessel reconstruction

Surgical repair of blood vessels usually involves various techniques aimed at restoring blood flow and vascular integrity. A common approach is vascular graft, in which synthetic or biological grafts are used to bypass occluded vessels, as pointed out by Zholborsov et al., which discusses the effectiveness of different graft materials in improving patient outcomes [94]. Furthermore, Song et al. emphasizes the importance of minimally invasive techniques to reduce recovery time and complications associated with conventional open surgeries [95]. Tádler et al. further elaborated on the role of non-invasive vascular procedures such as angioscopy and stenting under local anesthesia, thus improving patient safety and comfort [96]. Ohtsuka et al. also noted advances in imaging technology to support preoperative planning and intraoperative navigation and to improve the accuracy of these reconstruction surgeries. Together, these studies highlight the evolution of surgical techniques for the reconstruction of blood vessels and focus on patient-focused results and technological integration.

3.4 Polymers used in vessel reconstruction

Many polymer kinds are employed in the field of blood vessel regeneration, each with specific benefits. Because biodegradable polymers like polycaprolactone (PCL) and polylactic glycollic acid (PLGA) are biocompatible and can promote tissue regeneration over time, they are frequently utilized. Furthermore, the biological characteristics of natural polymers like collagen and fiber that encourage cell adhesion and proliferation are coming to light. The creation of artificial hydrogels that can replicate the mechanical characteristics of real tissues and provide an environment that is favorable to cell activity is another recent development that is worth mentioning [97]. Furthermore, it has been demonstrated that vascularization and healing are enhanced when bioactive compounds are added to polymer matrices. To maximize these materials' performance for clinical applications, more study is necessary as it is still challenging to determine their mechanical strength and long-term stability [98]. All things considered, the incorporation of various polymer kinds in blood vessel restoration signifies a comprehensive strategy meant to enhance patient results. Polymers are crucial to medicine, particularly when it comes to blood vessel regeneration. Since polymers are big molecules made up of repeating structural units, they have been used in a variety of medicinal applications because they are controllable, versatile, and biocompatible. Polymers have transformed the treatment of cardiovascular disorders and offered new options for individuals in need of vascular intervention when it comes to the repair of blood vessels [98].

Historically, the use of polymers in the reconstruction of blood vessels dates back to the early 20th century, when researchers began exploring synthetic materials for medical applications. One of the main figures in this field was the well-known cardiovascular surgeon Dr. Michael DeBakey,

Materials Research Foundations 172 (2025) 413-460
https://doi.org/10.21741/9781644903353-18

who pioneered the development of synthetic vascular implants made of Dacron polyester. His work set the stage for later developments in this area by laying the groundwork for the use of polymers in blood channel repair [99]. The polymers utilized in blood artery restoration have a substantial influence and provide novel therapeutic options for individuals suffering from vascular conditions such as peripheral arteries, aneurysms, and atherosclerosis. In addition to serving as a scaffold for tissue regeneration and stimulating the growth of new blood vessels, polymers can be utilized to replicate the characteristics of native blood vessels. This has decreased the chance of problems, enhanced the long-term results, and improved the outcome for patients undergoing vascular treatments [100]. The creation of bioresorbable polymers is one of the most notable developments in the field of polymers used in blood vessel restoration. These polymers are designed to degrade over time and allow the body to gradually replace synthetic materials with natural tissues [101]. Bioresorbent polymers are used for the manufacture of vascular stents, which are small mesh tubes inserted into narrow or blocked blood vessels to restore blood flow. These stents provide temporary support to the vessel and gradually disintegrate as the vessel heals and regains its function.

Prominent individuals who have made significant contributions to the field of polymers utilized in blood vessel repair include renowned bioengineer and biomaterials pioneer Dr. Robert Langer. The creation of novel polymers for use in tissue engineering, vascular prosthesis, and drug delivery systems is the main area of study for Dr. Langer. His studies have had a significant influence on the field and have sparked the creation of novel materials and techniques for blood vessel repair [102]. Blood artery repair using polymers is viewed differently by patients, physicians, and researchers. Some believe that polymers are a cutting-edge technology that can be used to treat vascular disorders in novel ways that enhance patient outcomes. Others emphasized the need for more study and patient monitoring during vascular surgeries, voicing concerns about the long-term safety and efficacy of synthetic materials in the body.

4. Synthetic polymers

4.1 Polyethylene glycol

Because polyethylene glycol (PEG) polymers are biocompatible and can enhance the mechanical characteristics of vascular implants, they have become a viable material for blood vessel repair. Studies have demonstrated that PEG may be successfully utilized to create hydrogels that resemble extracellular matrix and encourage cell adhesion and proliferation, two essential processes for the proper integration of veins [103]. Furthermore, PEG-based scaffolds have shown improved durability in animal models, suggesting their potential for long-term functionality in vascular applications. Furthermore, studies emphasize the versatility of PEG in changing surface properties to reduce thrombogenesis, thus improving the performance of grafts in a physiological environment [104]. However, there are some limitations, such as the need to optimize the formulation of PEG to balance mechanical resistance and degradation rates. Overall polymers are a significant advancement in the field of vascular tissue engineering, and research is still being done to address current issues. Using various methods, polyethylene glycol (PEG) enhances blood vessel repair, hence promoting angiogenesis and improving biocompatibility. Studies have demonstrated that PEG can alter the surface characteristics of scaffolds, enhancing cell adhesion and proliferation, two critical processes for the engineering of vascular tissue [105]. Furthermore, PEG's hydrophilic properties contribute to the preservation of humidity and the inhibition of

Materials Research Foundations 172 (2025) 413-460
https://doi.org/10.21741/9781644903353-18

thrombogenesis, which in turn fosters the proliferation of endothelial cells. Furthermore, PEG can be utilized to make hydrogels that replicate extracellular matrix, which promotes the migration and organization of vascular cells [106]. Research has demonstrated that PEG-based materials can regulate growth factor release, which enhances the integration of modified tissues into the host vasculature and promotes angiogenesis even further. Though PEG has bright futures, there are several drawbacks as well, such as the possibility of immunological reactions and the requirement to further refine its mechanical characteristics for certain uses. PEG is an effective instrument for developing vascular reconstruction techniques all around.

4.2 Polyglycolic acid

Because polyglycolic acid (PGA) is biocompatible and biodegradable, its potential for blood artery regeneration is becoming more widely acknowledged. Studies demonstrate that PGA is a useful tool for promoting cell adhesion and proliferation in vascular transplants, two processes that are essential for the proper integration of host tissues [107]. Additionally, research has demonstrated that PGA scaffolds can promote endothelial cell development, enhance the healing process, and lower the risk of thrombosis. Furthermore, it is possible to modify the mechanical characteristics of PGA to resemble those of real blood arteries, which are necessary to preserve vascular function in physiological settings [108]. There are several restrictions, though, like the requirement to modify the surface to increase its hydrophilicity and enhance cell interaction. Although using PGA to repair the vascular system is a potential way to enhance surgical outcomes, more study is required to fully realize its clinical advantages.

4.3 Polylactic acid

Polylactic acid (PLA) has become a promising biomaterial for reconstructing blood vessels due to its biocompatibility and biodegradability. Research suggests that PLA supports endothelial cell growth and promotes vascularization, making it suitable for the creation of vascular transplants [109]. Furthermore, studies have shown that PLA-based scaffolds can be modified to improve mechanical properties and degradation rates, which are crucial to maintaining vessel integrity under physiological conditions [110,111]. However, challenges remain, such as the need to improve the surface properties to promote adhesion and cell proliferation [112]. Recent developments in the addition of bioactive chemicals to PLA have demonstrated the possibility of getting around these restrictions and enhancing the material's in vivo performance [113]. All things considered, PLA offers several benefits for vascular applications; nonetheless, further research is necessary to maximize its therapeutic application for vessel regeneration. Furthermore, PLA's mechanical qualities can be modified to satisfy certain needs, which are helpful for applications like scaffolds for tissue engineering, drug delivery systems, and sutures. According to research, PLA can promote tissue regeneration and cell proliferation, which increases its applicability in regenerative medicine. But there are drawbacks as well, like comparatively slow degradation rates and certain flaws that could impair performance in particular applications [114].

4.4 Poly(L-lactic-co-glycolic acid)

Because poly(L-lactic-co-glycolic acid) (PLGA) is biocompatible and biodegradable, it has grown more and more well-known for its use in blood vessel restoration. Research suggests that PLGA-based scaffolds, which have good mechanical qualities and encourage endothelial cell proliferation, can successfully assist vascular regeneration. For example, a study showed that

Materials Research Foundations 172 (2025) 413-460
https://doi.org/10.21741/9781644903353-18

when combined with poly(isosorbide sebacate) fibers, the proangiogenic expression profile of human cervical vein endothelial cells (HUVECs) improved their regenerative potential [115]. In addition, the PLGA scaffold was optimized to mimic the mechanical properties of the interior architectures of the arteries, thereby facilitating endothelial cell survival and proliferation [110]. In addition, Research has demonstrated that the incorporation of growth factors and nanomaterials into polylactic acid scaffolds enhances their anti-infective characteristics and facilitates the integration of tissue [109]. Overall, polylactic acid (PLGA) is a promising material for blood vessel repair due to its versatility and effectiveness in fostering settings that are favorable for vascular tissue creation [116].

4.5 Polyurethane

Polyurethane polymers (PUs) are important materials in the reconstruction of blood vessels, especially for small-diameter vascular grafts. Research shows that the mechanical properties of PU, such as elasticity and molecular design, enable it to mimic natural blood vessels. For example, a hydrophilic PU elastomer developed through cross-linking with polyethylene glycol (PEG) exhibits multi-biological functions similar to native endothelium, including nitrogen oxide release and anti-thrombotic properties that are essential to the prevention of thrombosis and intimate hyperplasia [117]. In addition, PU-based scaffolds have demonstrated promising hemocompatibility and performance in regenerative medicine, although challenges such as thrombus formation and restlessness persist. Recent advances include the development of orthogonally functional PUs that can be modified with anti-thrombogenic agents and endothelial inducer peptides, increasing both biocompatibility and cell proliferation [118]. These findings collectively underline the potential of PU in the field of vascular tissue engineering and highlight the need for ongoing research to address its limitations.

4.6 Polycaprolactone

The importance of polycaprolactone (PCL) in blood vessel rebuilding is becoming more widely acknowledged, particularly in the context of small-diameter vascular transplant development. Because of its advantageous biomechanical characteristics, moderate rate of degradation, and biocompatibility, PCL is a great option for tissue engineering vascular grafts (TEGs) [119]. Research has indicated that PCL can be enhanced through many changes, like coating with polyvinyl alcohol (PVA), which increases water resistance and decreases platelet absorption, leading to greater in vivo performance. Moreover, it has been demonstrated that adding bioactive molecules to PCL patches, such as vascular endothelial growth factor (VEGF), speeds up cell infiltration and endothelial layer development, enhancing vascular regeneration [120]. In addition, PCL grafts with nitrate function have been developed to release nitrate oxides, promote the alignment of endothelial cells, and reduce calcification, which are essential to the success of long-term grafts [121]. Together, these findings highlight the versatility and effectiveness of PCL in vascular reconstruction applications.

4.7 Hyaluranic acid

Because it is a component of the extracellular matrix (ECM) of the vascular tissue, hyaluronic acid (HA) is crucial to the regeneration of blood vessels. HA is a crucial part of the extracellular matrix (ECM) that supports blood vessel flexibility and structural integrity as well as the body's ability to respond to mechanical force in physiological settings [122]. It is synthesized by HA synthases and

Materials Research Foundations 172 (2025) 413-460
https://doi.org/10.21741/9781644903353-18

can influence vascular cell behavior, promoting smooth muscle cell differentiation, migration, and proliferation, which are crucial processes in vascular remodeling and pathology. In addition, the incorporation of HA into biomaterials such as polyurethane improves their biological properties for vascular applications. Studies have shown that bulk incorporation of HA into these materials improves their stability and performance, especially in fluid conditions, which is essential to maintain the function of endothelial cells and minimize adverse reactions such as protein adsorption and platelet adhesion [123]. Thus, HA not only supports structural aspects of the blood vessel but also improves the function of vascular transplantation, highlighting the multifaceted role of vascular reconstruction [124].

4.8 Polydioxanone

Since polydioxanone (PDO) is biodegradable and has mechanical qualities, it is becoming better acknowledged for its use in blood vessel regeneration. According to research, PDO fibers can be utilized to create vascular grafts that mimic the mechanical properties of native arteries, accelerating the process of endothelialization and soft muscle cell regeneration [125]. In hybrid grafts, PDO is combined with polycaprolactone (PCL) to improve mechanical resistance and hydrophilicity, thereby enabling improved vascular regeneration and in vitro patency. In addition, PDO-based multi-block copolymers have improved elasticity compared to conventional vascular prostheses and are suitable for temporary tissue replacement in regenerative therapy [126]. Overall, the mechanical properties of PDO, combined with its biodegradability, provide it with promising material for the development of effective vascular transplants, although further research is needed to optimize its integration and performance in clinical conditions.

4.9 Polypropylene

Polypropylene (PP) has become a vital substance in the development of vascular grafts and blood vessel regeneration. Research demonstrating an 81% late patency rate for polypropylene screws compared to lower rates for other materials like Dacron and ePTFE indicates that it is a suitable choice for an artery replacement due to its high tensile strength and inert qualities [127]. In addition, the inclusion of coatings, such as a methacrylic acid copolymer, enhances the vascular response by promoting vascularization and reducing inflammation, which is essential for successful integration and healing. Furthermore, innovative approaches using biodegradable variants such as polypropylene fumarate (PPF) are being studied, in which PPF scaffolds are enhanced with elastin to mimic natural vascular structures, responding to the need for bioactive and absorbable materials [128]. In general, polypropylene and its derivatives have promising potential for vascular applications, balancing mechanical properties and biological compatibility.

4.10 Polyvinyl alcohol

Because of its advantageous qualities, such as its low thrombogenicity and biocompatibility, polyvinyl alcohol (PVA) has become a potential material for blood artery restoration. Studies reveal that the application of PVA coating onto other materials, such as polycaprolactone (PCL), can improve the mechanical characteristics of vascular grafts, including their resilience and stiffness [129]. Furthermore, polyelectrolyte multilayer-modified PVA-based vascular patches have demonstrated enhanced hemocompatibility and quick endothelialization, two critical factors in the effective healing of damaged arteries. Furthermore, PVA hydrogels have non-thrombogenic profiles and encourage positive blood-biomaterial interactions, which make them appropriate for vascular graft applications, especially when copolymerized with dextran [130]. Further research supports the use of PVA/gelatin blends in the creation of vascular scaffolds with long-term patency by highlighting their capacity to improve endothelial cell adhesion and proliferation. Overall, PVA's versatility and compatibility with biological systems position it as a valuable candidate in vascular reconstruction strategies [131].

4.11 Polycarbonate

The biocompatibility and mechanical qualities of polycarbonate polymers are gaining recognition for their potential in blood vessel restoration. Comparing PMDO to standard poly(trimethylene carbonate) (PTMC), a study on a glycerol-based aliphatic polycarbonate showed considerable improvements in blood compatibility, with PMDO decreasing platelet adhesion by 33% and fibrinogen denaturation by 23% [132]. This enhanced performance is attributed to the polymer's unique hydration properties, which also promote hydrolytic biodegradability. Additionally, polycarbonate urethane films incorporating hydroxyl-terminated polydimethylsiloxane (PDMS) have shown improved biocompatibility while maintaining mechanical stability, making them suitable for artificial vessels [133]. These results imply that polycarbonate urethanes treated with PMDO and PDMS are both attractive alternatives for vascular tissue engineering. However, more investigation is required to completely comprehend their long-term performance and integration into biological systems.

5. Natural polymers

5.1 Collagen

In particular, collagen-based polymers are essential for vascular tissue engineering and blood vessel regeneration. Numerous studies demonstrate how well collagen works to create scaffolds that resemble the extracellular matrix (ECM), which is crucial for vascular regeneration. Promising possibilities for vascular grafts include human-like collagen/chitosan scaffolds, which have shown favorable mechanical characteristics and biocompatibility, boosting cell adhesion and extracellular matrix formation [134]. Additionally, collagen-blended biodegradable polymer nanofibers have shown potential in supporting endothelial cell viability and maintaining their phenotype, which is vital for successful graft integration. Furthermore, the manipulation of collagen microstructure through cross-linking can influence vessel network formation, with specific configurations promoting vessel stabilization and growth [135]. However, challenges such as the balance between mechanical strength and biological functionality remain, as excessive cross-linking can inhibit smooth muscle cell activity. Overall, collagen polymers are integral to developing effective vascular grafts, with ongoing research aimed at optimizing their properties for clinical applications [97].

5.2 Elastin

Blood vessel regeneration greatly benefits from the use of elastin polymers, especially in the creation of synthetic tissues and vascular grafts. According to research, elastin-like recombinases (ELRs) improve cell-mediated remodeling and viscoelasticity, two mechanical qualities of collagen-based scaffolds that are essential for simulating the vascular extracellular matrix (ECM) [136]. Furthermore, vascular grafts made of elastin-silk fibroin have been created; these grafts exhibit good biocompatibility and encourage endothelial cell adhesion, both of which are critical for the patency and functionality of the graft. Additionally, the potential for vascular repair applications of synthetic hydrogels comprising elastin-like proteins has been suggested by their ability to sustain endothelial cell survival and angiogenic behavior [137]. Collectively, these studies highlight the importance of elastin polymers in enhancing the structural and functional properties of vascular constructs, thereby addressing challenges in vascular tissue engineering and graft performance.

5.3 Chitosan

Chitosan, a biopolymer derived from chitin, has shown significant promise in blood vessel reconstruction through various innovative applications. Studies show that chitosan can improve the mechanical characteristics and biocompatibility of vascular grafts. For example, a study showed that an electrospun tube made of poly-L-lactic acid (PLLA), chitosan, and collagen had good hemocompatibility and tensile strength, which made it appropriate for use in vascular graft applications [138]. Furthermore, the potential of bacterial cellulose modified by chitosan as a scaffold in tissue-engineered blood arteries was assessed; it demonstrated good biocompatibility and no unfavorable reactions in vivo. PEGylated chitosan has also been used to make bilayered vascular grafts that resemble natural tissue, demonstrating its adaptability in vascular restoration [139]. Overall, these findings underscore chitosan's role in enhancing the performance of vascular grafts, contributing to improved outcomes in vascular reconstruction procedures.

5.4 Alginate

Blood vessel regeneration benefits significantly from the use of alginate polymers, especially when it comes to the creation of hydrogels that closely resemble natural vascular architecture. Research indicates that alginate-based hydrogels, often combined with other materials like collagen and gelatin, can be engineered to create vascular grafts with tunable properties such as wall thickness and mechanical strength [140]. For instance, a hydrogel composed of sodium alginate, gellan gum, and gelatin demonstrated excellent biocompatibility and structural stability, making it suitable for bioengineered vascular grafts. Additionally, alginate di-aldehyde combined with gelatin has been successfully utilized to fabricate vessel-like structures, supporting cell proliferation and migration, which are crucial for vascular integration [141]. However, while these alginate-based solutions show promise, challenges such as biodegradability and long-term functionality remain areas for further investigation. Overall, alginate polymers are pivotal in advancing vascular tissue engineering, offering innovative approaches to address vascular diseases.

5.5 Fibrin

Fibrin polymers play a crucial role in blood vessel reconstruction, particularly in vascular tissue engineering (VTE). Fibrin from autologous blood is an excellent option for vascular graft creation because it functions as a biocompatible scaffold that encourages cell adhesion, migration, and angiogenesis. Its high porosity and biodegradability make it easier for growth agents to be released under control, which promotes tissue regeneration [142]. Fibrin's poor strength and stiffness, among other mechanical limitations, make it difficult to use in therapeutic settings. Recent developments, such as the use of transluminal compression techniques, have demonstrated promise in improving the mechanical characteristics of constructions made of fibrin, leading to increases in their strength and capacity for endothelialization [143]. Additionally, alternative polymerization methods that do not rely on exogenous thrombin have demonstrated superior mechanical

Materials Research Forum LLC
https://doi.org/10.21741/9781644903353-18

properties and reduced thrombogenicity, further supporting the use of fibrin in vascular applications. Overall, fibrin's unique properties and ongoing innovations in its application underscore its potential in blood vessel reconstruction[144].

5.6 Silk fibroin

Silk fibroin has emerged as a promising polymer for blood vessel reconstruction due to its favorable mechanical and biological properties. According to research, silk fibroin can be used to successfully generate small-diameter vascular grafts (SDVGs), showing good patency rates and encouraging endothelialization, an essential step for the proper integration and function of the graft [145]. According to research conducted in vivo, scaffolds made of silk can help generate collagen and elastic fibers and help deposit extracellular matrix (ECM), simulating the shape of genuine blood arteries. Furthermore, silk fibroin is a better option than synthetic materials because of its low chemical reactivity, which lowers the chance of graft rejection [146]. Furthermore, silk fibroin can be combined with other materials, such as polyethylene glycol, to enhance the creation of functional vascular networks, which are essential for tissue and organ reconstruction [147]. Overall, silk fibroin's unique properties position it as a valuable candidate for vascular tissue engineering.

5.7 Gelatin

Gelatin polymers play a significant role in blood vessel reconstruction, primarily due to their biocompatibility and ability to support cellular activities essential for vascular repair. For instance, According to a study, poly(lactic acid)-gelatin scaffolds efficiently boosted macrophage polarisation and enhanced the recruitment of endothelial cells and fibroblasts, which are essential for vascular regeneration, when mixed with Ag-Fe_3O_4 nanowires and basic fibroblast growth factor (bFGF)[109]. Furthermore, superior anti-thrombogenic qualities and increased endothelial cell affinity were demonstrated by gelatin-based electrospun nanofibers, especially those mixed with chondroitin sulfate and polycaprolactone [148]. These attributes made them potential candidates for vascular tissue engineering. Moreover, angiogenesis and wound healing were markedly enhanced by gelatin methacryloyl hydrogels functionalized with endothelial progenitor cells, demonstrating the versatility of gelatin in stimulating the development of new blood vessels[149]. Taken together, these investigations highlight the potential of gelatin polymers in creating efficient scaffolds for tissue engineering and vascular restoration.

6. Conjugated and doped polymers in blood vessel reconstruction

6.1 Poly(3, 4-ethylene dioxythiophene)

Poly(3,4-ethylenedioxythiophene) (PEDOT) polymers are emerging as promising materials for blood vessel reconstruction due to their unique properties. The discovery of methacrylate-terminated PEDOT, which improves conductivity and processability and is appropriate for biomedical applications, such as vascular grafts, is highlighted in recent publications. For PEDOT-based materials to successfully integrate and function, they must have mechanical qualities that resemble real blood vessels as well as biocompatibility [150]. Additionally, the fabrication of polymeric aortic vessels using advanced techniques like 3D printing demonstrates the potential of these materials in creating customized vascular replacements that can address specific medical conditions, such as connective tissue disorders. However, while PEDOT shows promise,

Materials Research Foundations 172 (2025) 413-460
https://doi.org/10.21741/9781644903353-18

challenges remain regarding its solubility and processing, which ongoing research aims to overcome. Overall, Vascular graft technology has advanced significantly using PEDOT polymers, which combine electrical conductivity and advantageous mechanical qualities for successful blood vessel repair [151].

6.2 Polyaniline (PANI)

Polyaniline (PANI) polymers have shown significant promise in blood vessel reconstruction, particularly in the development of vascular scaffolds. Research indicates that PANI-coated polyurethane (PANI-PU) electrospun fibers enhance endothelial cell adhesion and proliferation, demonstrating improved anticoagulation properties compared to uncoated fibers. The PANI-PU scaffolds exhibited a hemolysis rate of only 0.14%, indicating good blood compatibility, and a plasma recalcification time of 123 seconds, which suggests effective anticoagulant properties [152]. Additionally, PANI's intrinsic electrical activity has been linked to increased tissue regeneration rates, making it a favorable choice for tissue engineering applications, including vascular repair. While other materials like chitosan and PCL have been explored for artificial blood vessels, PANI's unique properties may offer superior mechanical and biological performance for small-diameter vascular grafts [150]. Overall, PANI-based scaffolds represent a promising avenue for enhancing vascular reconstruction strategies.

6.3 Poly(phenylene vinylene) (PPV)

The potential of poly(phenylene vinylene) (PPV) polymers in vascular tissue engineering is becoming more widely acknowledged, especially in the creation of scaffolds for blood artery repair. Due to the increasing prevalence of cardiovascular disorders and the shortcomings of currently available synthetic materials, which frequently fail in smaller applications due to poor patency rates, the research emphasizes the necessity for small-diameter vascular grafts [150,153]. To replicate natural blood vessels, PPV-based scaffolds can be developed to have the necessary mechanical qualities and biocompatibility. Research has, for example, shown how to create polymeric scaffolds with microstructural characteristics and regulated porosity to improve tissue ingrowth and cell seeding. Additionally, customized vascular grafts that can adjust to physiological conditions can be made possible by the application of cutting-edge methods like 3D printing. Overall, the integration of PPV polymers in vascular grafts presents a promising avenue for improving outcomes in vascular reconstruction, addressing both mechanical and biological challenges in current synthetic options [151].

6.4 Polyfluorene

Applications of polyfluorene polymers, especially polytetrafluoroethylene (PTFE), in blood vessel repair have been thoroughly investigated. Because of their mechanical qualities and biocompatibility, PTFE grafts are preferred for use in vascular prostheses. However, problems including poor hemocompatibility and thrombosis have been reported, which contributes to a high explantation rate in clinical settings [154]. Researchers have looked into surface changes to increase graft function and endothelialization, such as micropatterning with cell adhesion peptides, to solve these problems [154]. Furthermore, by releasing antiplatelet medicines in reaction to oxidative stress, the integration of reactive oxygen species (ROS)--responsive anti-thrombogenic medications into polymer-thermal fibrillation (PTFE) grafts has demonstrated potential in preventing thrombosis. The use of PTFE in infected wounds is still debatable despite these developments, with conflicting results being documented [155]. Overall, while polyfluorene polymers like PTFE present significant potential for vascular reconstruction, ongoing research is essential to optimize their performance and address existing limitations.

6.5 Poly(thiophene)

Due to their special structural and functional characteristics, poly(thiophene) polymers, in particular thiophene copolymers, show promise in the restoration of blood vessels. After curing, these copolymers—which contain hydroxyl and long-chain alkane groups—show good bonding characteristics and electric conductivity, which qualifies them for usage in tissue engineering applications [156]. Scaffold design plays an important role in blood vascular tissue creation because pore size affects cell infiltration and interaction. Although the majority of the research focuses on poly(e-caprolactone) (PCL) scaffolds, thiophene-based materials can also benefit from the scaffold design principles, which implies that they could be efficiently built to support endothelial and smooth muscle cell co-cultures [157]. Moreover, the use of biological materials, including harvested blood vessels, highlights the importance of integrating synthetic polymers like poly(thiophene) with biological scaffolds for enhanced structural support and functionality in vascular applications. Thus, while direct studies on poly(thiophene) in blood vessel reconstruction are limited, the existing research indicates a potential for these materials in future applications [158].

6.6 Poly(p-phenylene sulphide)

In the field of vascular tissue engineering, poly(p-phenylene sulfide) (PPS) is showing promise, especially for blood vessel repair. Because of its excellent mechanical and thermal stability, it can be used as a scaffold to create vascular grafts, which are necessary for patients with heart conditions who need vascular access. The long-term success of vascular implants is hampered by conditions including restenosis and graft thrombosis, which are being addressed by the development of biodegradable polymeric scaffolds, such as PPS [159]. While PPS has shown excellent water pressure cracking strength, its application in vascular engineering is still under investigation, with ongoing research focusing on optimizing its properties for better hemocompatibility and integration with biological tissues. However, compared to more commonly used materials like polyurethanes, the specific advantages and limitations of PPS in vascular applications require further exploration to fully establish its efficacy and safety in clinical settings [160].

6.7 Poly(3,3-diaalkylqualerthiophene)

Poly(3,3-diaalkylqualerthiophene) is not explicitly mentioned in the provided abstracts; however, the research highlights advancements in materials for blood vessel reconstruction. For instance, the development of biocompatible photopolymer formulations, such as polyacrylate elastomers, demonstrates promising mechanical properties and cytocompatibility for artificial vascular grafts, particularly in small-diameter applications where traditional materials fail to maintain patency rates[150]. These materials can be fabricated using stereolithography, allowing for precise control over structure and porosity, which is crucial for tissue ingrowth and endothelialization. Additionally, the integration of biofunctionalization techniques, such as applying heparin and RGDC-peptides, enhances endothelial cell adhesion, further supporting the viability of these synthetic constructs [161]. While the specific role of poly(3,3-diaalkylqualerthiophene) in this

 Materials Research Forum LLC
https://doi.org/10.21741/9781644903353-18

context remains unclear, the ongoing research into alternative polymers and fabrication methods is vital for improving vascular graft outcomes.

7. Polymers in hemostatics

7.1 Chitosan

The natural polysaccharide chitosan, which is derived from chitin, has become an important component in the creation of hemostatic medicines because of their distinct biological characteristics and biocompatibility. Research indicates that chitosan-based materials, including sponges, hydrogels, and dressings, effectively promote hemostasis by enhancing red blood cell aggregation and platelet activation, which are critical for clot formation [162,163]. Recent advancements have focused on modifying chitosan chemically to improve its hemostatic performance, addressing limitations such as rigidity in pure chitosan sponges. For instance, combining chitosan with hydroxypropylmethylcellulose has resulted in sponges that exhibit superior flexibility and faster hemostatic capabilities [164]. Additionally, chitosan's ability to interact with plasma proteins further enhances its efficacy in various surgical and trauma settings. Overall, Because of its flexibility and adaptability, chitosan is a good option for creating hemostatic polymers that work well in medical applications [165,166].

7.2 Collagen

Collagen polymers' special qualities make them essential for hemostatics and wound healing. According to research, collagen improves hemostatic effects by encouraging platelet aggregation and making it easier for a stable fibrin clot to form—both of which are necessary for halting bleeding. A study on collagen-containing polyester dressings demonstrated that these materials significantly reduced bleeding time compared to traditional dressings, highlighting collagen's effectiveness in hemostasis while maintaining non-adhering properties that prevent further injury during dressing changes [167]. Additionally, bioactive polymers, including collagen, contribute to wound healing by providing a moist environment and delivering growth factors, which are vital for tissue repair. The integration of collagen with other materials, such as tissue factors and silica nanoparticles, has shown promise in enhancing coagulation pathways, suggesting that collagen-based systems can be optimized for improved hemostatic performance [168]. Overall, collagen polymers are integral to developing advanced hemostatic agents and wound dressings.

7.3 Gelatin

Because of their biocompatibility and capacity to stimulate coagulation, gelatin polymers are becoming more widely acknowledged for their use in hemostatic applications. Numerous gelatin-based products intended to improve hemostatic performance have been highlighted in recent investigations. For example, a Janus gelatin sponge embedded with zein nanoparticles outperformed conventional gelatin sponges in vitro and in vivo, exhibiting improved hemostatic characteristics under both coagulation-dependent and coagulopathic conditions [169]. A gelatin/sodium polyacrylate composite sponge was presented in a different study. It demonstrated notable blood-clotting properties and good biocompatibility when loaded with thrombin. Additionally, a polyphenol-functionalized chitosan/gelatin sponge showed enhanced liquid absorption and antioxidant properties, further supporting hemostatic performance. Lastly, a gelatin-based adhesive hydrogel was developed, showcasing excellent adhesion to wet tissues and

Materials Research Forum LLC
https://doi.org/10.21741/9781644903353-18

rapid hemostatic action, indicating its potential to replace conventional agents [170]. Collectively, these findings underscore the versatility and effectiveness of gelatin polymers in developing advanced hemostatic materials.

7.4 Alginate

Alginate polymers have emerged as significant materials in hemostatic applications due to their biocompatibility and effectiveness in promoting blood clotting. Clinical studies demonstrate that alginate polymer polysaccharide hydrogels can significantly reduce bleeding time and enhance healing in patients with gastroduodenal ulcers, effectively preventing recurrent bleeding and improving treatment outcomes [171]. Additionally, alginate-based hemostatic materials exhibit high absorption capacity and cell viability, making them suitable for emergencies where rapid hemostasis is critical. Innovations such as macroporous alginate hydrogels have further improved hemostatic efficacy by enhancing blood absorption and coagulation factor delivery, leading to substantial reductions in blood loss [172]. Moreover, composite scaffolds incorporating alginate with other natural polymers have shown promising results in terms of antimicrobial properties and wound healing, indicating their potential as versatile hemostatic agents. Overall, alginate polymers represent a promising avenue for developing effective hemostatic materials in clinical settings [173].

7.5 Fibrin

As they function as a natural framework for clot formation and tissue repair, fibrin polymers are essential to hemostasis. Their polymerization involves a spontaneous self-assembly process, where fibrin monomers aggregate to form a stable network, essential for effective hemostatic function[174].

7.5.1 Fibrin Polymerization and Hemostasis

Self-Assembly Process: Thrombin activity transforms fibrogen into fibrin, resulting in the creation of fibrin polymers that aggregate side to side and end to end [175]. Stabilization: Covalent bonds are added to the fibrin network during this phase, improving its structural integrity and resistance to deterioration.

Hemostatic Glue: Fibrin is used as a hemostatic agent during surgery to minimize blood loss and promote quick clot formation [174]. Composite Hydrogels: Advances in fibrin-based hydrogels aim to increase mechanical characteristics and stability, making them appropriate for many biomedical applications.

While fibrin's role in hemostasis is well-established, challenges remain in optimizing its properties for broader clinical applications, particularly regarding its mechanical strength and degradation rates.

7.6. Hyaluronic acid

Hyaluronic acid (HA) plays a crucial role in hemostatic applications because of its biocompatibility, biodegradability, and ability to create hydrogels. Numerous novel HA-based materials intended for efficient hemorrhage control are highlighted in recent investigations.

Hemostatic Characteristics of HA Sponges: Thermal cross-linking produces HA sponges, which have superior mechanical qualities and a large fluid absorption capacity. When compared to conventional gelatin sponges, in vitro studies demonstrate a considerable reduction in coagulation time and blood loss, with up to 73% less bleeding time observed in in vivo models [176].

Hydrogel Innovations: HA is utilized in hydrogen bond-crosslinked hydrogels, which demonstrate self-healing and adhesive properties, enhancing their effectiveness in managing uncontrolled bleeding. These hydrogels can be prepared rapidly and show good biocompatibility [177].

Combination with Other Polymers: The integration of HA with poly(lactide-co-glycolide) (PLGA) enhances hemostatic performance while providing anti-infection properties. The PLGA-HA fibrous fragments achieve rapid hemostasis within 28 seconds, showcasing their potential in clinical applications [178].While HA-based materials show promise in hemostatic applications, challenges remain in optimizing their mechanical properties and ensuring consistent performance across various clinical scenarios. Further exploration is required to address these limitations and enhance their efficacy.

7.7 PEG

Polyethylene glycol (PEG) has emerged as a significant component in the development of innovative hemostatic agents, enhancing their efficacy and biocompatibility. Various studies highlight PEG's role in creating effective hemostatic powders and gels that facilitate rapid clot formation and wound closure.

Biomimetic Hemostatic Powder: A study introduced a hemostatic powder combining PEG with tannic acid and a triblock copolymer, demonstrating rapid gelation and effective sealing on wet surfaces, independent of the natural clotting process [179].

CA-PEG-CA Hemostatic Powder: Another research developed a poly(ethylene glycol)-di(cyanoacrylate) powder that forms an in situ gel upon contact with tissue, enhancing adhesion and promoting platelet aggregation for effective hemostasis[180].

Chitosan-PEG Gels: Gels combining chitosan and PEG, loaded with natural extracts, showed significant hemostatic action in vivo, outperforming traditional products [181].

Injectable PEG Hydrogels: Novel hydrogels based on PEG exhibited rapid gelation and strong adhesive properties, significantly reducing hemostasis time in animal models [182].

While PEG-based hemostatic agents show promise, challenges remain in ensuring their long-term stability and biodegradability in clinical applications.

7.8 Polyurethane

Because of their mechanical qualities, biocompatibility, and capacity to encourage quick clot formation, polyurethane (PU) polymers have become important materials in hemostatic applications. Current research demonstrates that different PU formulations and architectures improve hemostatic effectiveness.

Compared to conventional hydrogels, the PAA-PU Janus hydrogel dramatically lowers the hemostatic time and blood loss due to its asymmetric interfacial toughness. Its distinct structure increases coagulation cascade events and encourages tissue attachment [183].

Materials Research Foundations 172 (2025) 413-460
https://doi.org/10.21741/9781644903353-18

PU-TA/Gel sponges demonstrate high porosity and water absorption, leading to rapid clot formation and effective skin regeneration in animal models. These sponges showed the shortest clotting times and minimal bleeding [184].

A PU dressing incorporating chitosan and usnic acid not only provides hemostatic functions but also exhibits broad-spectrum antibacterial properties, making it suitable for wound care [185,186].

Thermoplastic PU films, combined with various biocompatible materials, offer flexibility and comfort while maintaining effective hemostatic capabilities, indicating a promising future for PU in clinical settings [185,186].

While PU polymers show great promise in hemostatic applications, challenges remain in optimizing their properties for specific clinical scenarios, particularly regarding long-term biocompatibility and integration with biological tissues.

7.9 Polyvinyl alcohol

PVA, or polyvinyl alcohol, has become a prominent polymer in hemostatic applications because of its efficient blood-clotting capabilities and biocompatibility. Many PVA formulations and alterations that improve its hemostatic properties are highlighted in recent studies.

Electrospun Membranes: Following plasma treatment, PVA membranes coupled with polylactic acid (PLA) exhibited a 79% improvement in hemostatic performance, exceeding untreated PVA membranes by a substantial margin and exhibiting a coagulation time of 258 seconds [187].

Applications for Hydrogels: PVA hydrogels produced by freeze-thaw techniques considerably decreased bleeding time and volume and raised survival rates in rat models from 75% to 91.7% when compared to conventional therapies [188].

Composite Bandages: PVA integrated with silk fibroin and graphene oxide resulted in a hemostatic dressing with a blood clotting ability of 89% within 15 seconds, showcasing high absorption and biocompatibility [189].

While PVA demonstrates promising hemostatic properties, alternative materials like poly(acrylic acid) and polyvinylpyrrolidone composites also show potential in clinical settings, indicating a diverse landscape of hemostatic agents that may complement or compete with PVA-based solutions[190].

8. Polymers in flowable agents

The development of flowable agents for blood vessel reconstruction involves innovative polymer formulations that enhance biocompatibility and mechanical properties. Recent studies highlight various polymer systems that can be utilized in this context.

8.1 Bovine thrombin (Active)

Bovine thrombin has been widely utilized as a flowable hemostatic agent in surgical settings, particularly for its ability to promote localized hemostasis. However, its application is accompanied by significant safety concerns and evolving alternatives.Bovine thrombin is effective in controlling bleeding during surgeries, especially in complex procedures like spinal surgery, where traditional methods may fail[191].In a rodent model of hepatic injury, bovine thrombin was tested as a topical agent; however, it did not significantly improve survival or hemodynamic

Materials Research Foundations 172 (2025) 413-460
https://doi.org/10.21741/9781644903353-18

stability compared to saline or suture ligation[192].The use of bovine thrombin has been linked to antibody-mediated coagulopathy, which can lead to severe complications, including life-threatening bleeding and anaphylaxis. Due to these risks, the FDA has mandated warnings for products containing bovine thrombin, prompting a shift towards human-derived alternatives that exhibit lower immunogenicity[193].In summary, while bovine thrombin serves as a useful hemostatic agent, its associated risks have led to increased interest in safer alternatives, such as human plasma-derived and recombinant thrombin products.

8.2 Bovine gelatin

Bovine gelatin has emerged as a significant component in flowable hemostatic agents, enhancing surgical outcomes by effectively controlling bleeding. Recent studies highlight its efficacy, safety, and potential complications associated with its use. Flowable hemostatic agents made from bovine gelatin have shown superior performance in stopping bleeding, particularly in challenging surgical scenarios like uneven or deep wounds. In a rabbit model, these agents resulted in shorter bleeding times and higher success rates in achieving hemostasis compared to traditional gelatin sponges[194].The combination of gelatin with thrombin enhances clot formation, making it a vital tool in surgeries where rapid hemostatic action is required[195].Despite their effectiveness, bovine gelatin-based agents can trigger allergic reactions, particularly in patients with a history of animal allergies. Reports of intraoperative anaphylaxis highlight the need for careful patient screening before administration[195,196].The safety profile of these agents has been evaluated, showing no significant adverse effects in controlled studies, but awareness of potential allergic responses remains crucial[194].While bovine gelatin serves as an effective flowable hemostatic agent, its use must be balanced with the risk of allergic reactions, necessitating thorough patient assessments before surgical procedures.

9. Sealants

9.1 Fibrin sealant

Fibrin sealants like TISSEEL and EVICEL, have shown significant efficacy in promoting hemostasis during vascular reconstruction procedures. These agents enhance surgical outcomes by effectively controlling bleeding at suture lines, thereby reducing the need for additional interventions.In a research contrasting TISSEEL Lyo with manual compression, at 4 minutes after application, 84.3% of patients had reached hemostasis, compared to only 20.8% in the control group ($p < 0.001$)[197]. Comparably, EVICEL outperformed manual compression at 39% (odds ratio 11.34, $p < 0.001$) in terms of success rate in reaching hemostasis, with an 85% success rate at 4 minutes[198].Minimal occurrence of severe side effects, with the TISSEEL group seeing no intraoperative bleeding and the rates of adverse events for the fibrin sealant and manual compression groups being similar. Fibrin sealants have a good safety profile because they do not raise the risk of bleeding-related complications [197,198]. Despite the effectiveness of thrombin sealants, there are still some worries about the possibility of side effects, which were marginally more common in the groups that received fibrin sealants. Nevertheless, they are an important tool in vascular surgery due to their advantages in producing quick hemostasis.

9.2 PEG sealant

Materials Research Foundations 172 (2025) 413-460
https://doi.org/10.21741/9781644903353-18

The use of polyethylene glycol (PEG) sealants in blood vessel reconstruction has shown promising results in enhancing hemostasis and reducing complications. Various studies highlight the effectiveness and safety of PEG-based sealants in surgical applications.

A study on CoSeal, a PEG sealant, demonstrated significant reductions in intraoperative and postoperative transfusion requirements during aortic surgeries, with a notable decrease in postoperative drainage loss (985 ml vs. 1709 ml) and a trend towards fewer thoracotomies [199]. The ArterX Vascular Sealant (AVS) showed a 60.5% immediate hemostasis rate compared to 39.6% with Gelfoam Plus, indicating superior performance in arterial surgeries[200].The novel PGSA-based SETALIUM™ sealant exhibited excellent biocompatibility and achieved hemostasis in 84% of clinical cases without adverse events, supporting its use in vascular reconstructions[201]. The Chol-apGltn/4S-PEG sealant demonstrated high burst strength and minimal swelling, suggesting its potential for cardiovascular applications [202]. While PEG sealants show significant benefits in vascular reconstruction, ongoing research is essential to further validate their long-term safety and effectiveness in diverse surgical contexts.

9.3 Cyanoacrylate sealant

Cyanoacrylate sealants have emerged as a promising adjunct in vascular reconstruction, enhancing hemostasis and potentially reducing complications. Their application spans various surgical contexts, demonstrating efficacy in managing bleeding and preventing infections. In comparison to conventional techniques, cyanoacrylate dramatically shortened the time to hemostasis in a trial involving 151 patients, with mean durations of 119.3 seconds versus 403.8 seconds for controls [203]. In a different trial, 93.6% of anastomotic sites achieved rapid hemostasis within one minute, accounting for 71.3% of cases [204].The use of cyanoacrylate as a skin sealant in vascular surgeries was hypothesized to lower surgical site infections by two-thirds, although the trial was discontinued before conclusive results were reached[205].Cyanoacrylate has been successfully utilized in kidney transplantation, where it facilitated vascular anastomoses without significant bleeding, showcasing its potential in complex cases[206].Additionally, it has been applied in endovenous treatments for venous insufficiency, indicating its broad applicability in vascular disorders[207]. While cyanoacrylate sealants show promising applications in vascular reconstruction. To completely comprehend their long-term effects and potential consequences, more research is required.

9.4 Glutaraldehyde sealant

The use of glutaraldehyde sealants in blood vessel reconstruction has shown promising results, particularly in the context of grafting techniques. Research indicates that glutaraldehyde-treated grafts can serve as effective substitutes for autologous veins, demonstrating comparable patency and morphological characteristics.A study involving rabbits found that glutaraldehyde-treated homologous vein grafts exhibited similar patency rates to autologous veins, with no significant differences in thrombus formation or intimal hyperplasia over time[208].Long-term research on umbilical vein grafts stabilized with glutaraldehyde for limb salvage reported cumulative patency rates of 76.4% for popliteal reconstructions at 36 months, indicating their reliability as alternatives to saphenous veins[209,210].Glutaraldehyde stabilization preserves the original architecture of the graft, enhancing thromboresistance and biocompatibility, which is crucial for successful vascular reconstruction. The absence of significant biodegradation or aneurysmal deterioration in long-term follow-ups supports the graft's durability [209]. While glutaraldehyde-treated grafts show

Materials Research Foundations 172 (2025) 413-460
https://doi.org/10.21741/9781644903353-18

potential, concerns regarding the inflammatory response and long-term outcomes necessitate further research to optimize their application in clinical settings.

10. New horizons in polymer science and blood vessel reconstruction

Recent advancements in polymer science are significantly enhancing blood vessel reconstruction techniques, particularly through innovative imaging and fabrication methods. These developments are crucial for improving surgical outcomes and tissue engineering applications.

10.1 3D Reconstruction Techniques

A semiparametric Gaussian mixture model has been developed for automated 3D reconstruction of blood vessels from CT data, improving accuracy and efficiency compared to traditional methods[212].Another method utilizes a multi-scale vessel enhancement algorithm to preprocess images, enabling the generation of accurate blood vessel tree images through shortest path calculations[213].Studies on drug interactions within vascular networks are made easier by the CLARITY-to-REALITY (C2R) approach, which enables the transcription of in vivo blood vessel networks onto perfusable microfluidic devices using readily available materials [214].A novel 3D-printed sacrificial material, Polyvinyl Alcohol (PVA), has been employed to create intricate blood vessel structures, enhancing nutrient perfusion and cell viability in engineered tissues[215].While these advancements present promising solutions, challenges remain in scaling these technologies for widespread clinical use, particularly regarding the accessibility of sophisticated equipment and materials.

Polymers are significantly transforming blood vessel reconstruction techniques through innovative scaffold designs and biocompatible materials. These advancements enhance the integration and functionality of vascular grafts, addressing critical challenges in tissue engineering.

10.2 Advanced Scaffold Materials

Hybrid Scaffolds: The use of polymer-grafted graphene scaffolds has shown promise in microvascular tissue engineering, combining biocompatibility with mechanical strength, essential for vascular applications [216].

Formulations for Photopolymers: (Meth)acrylate-based photoelastomers are being developed for artificial vascular grafts; these photoelastomers have customized mechanical properties that resemble real blood vessels, which are essential for reconstructions with small diameters [103].

Multiblock Co-polymers: Research on elastic multiblock co-polymers indicates improved elasticity and shape-memory effects, enhancing the mechanical performance of vascular prostheses [217].Cellularized Alginate Sheets: Utilizing alginate membranes for blood vessel reconstruction has demonstrated effective cell proliferation and orientation, crucial for functional tissue integration[218].

Biodegradable Polymers: The development of scaffolds made of biodegradable polymers enables the production of neovessels that can adapt to injury and remodel, thereby resolving problems such as thrombosis and restenosis [219].

Blood artery regeneration has greatly advanced thanks to polymers, especially in the development of tissue engineering techniques and small-diameter vascular grafts (SDBVs). Critical issues like thrombosis and intimal hyperplasia, which have traditionally restricted the viability of synthetic grafts, are addressed by these improvements.

Recent innovations in fabrication technologies, including 3D printing and micro stereolithography, allow for the creation of biocompatible polymer scaffolds that mimic natural blood vessels' mechanical properties. These techniques enable precise control over scaffold architecture, promoting tissue ingrowth and vascularization [220,221].

Scaffolds made of biodegradable polymers are being used to assist cell growth and, over time, break down to allow for natural tissue integration. The incorporation of growth factors and endothelial cells into these scaffolds enhances neovascularization, crucial for successful graft integration [221,222].

11. Clinical applications and future directions

The ongoing development of polymer-based graft shows promise for clinical applications, with the potential to improve long-term patency rates in vascular surgeries [219,220]. Future research is expected to focus on optimizing polymer formulations and fabrication methods to further enhance the performance of engineered blood vessels. While these advancements are promising, challenges such as restenosis and graft thrombosis remain prevalent, indicating a need for continued innovation in polymeric materials and techniques for vascular reconstruction.Coating the decellularized arteries using poly(ethylmethacrylate-co-diethylaminoethyl acrylate) significantly improved tensile strength and load-bearing capacity, essential for vascular grafts under physiological conditions[224]. The development of electrospun polymer scaffolds from polyhydroxybutyrate/valerate and polycaprolactone demonstrated the ability to maintain endothelial cell viability under pulsatile flow, crucial for graft functionality [225]. Human umbilical vein endothelial cells exhibited outstanding compatibility with the optimized pre-polyglycerol sebacate-poly lactic acid scaffold, which encouraged cell proliferation and sustained structural integrity over time [226]. Polymer scaffolds can be engineered to enhance neovascularization, which is vital for the integration and functionality of grafts in tissue engineering applications [222]. The polymer coating on decellularized vessels not only facilitated endothelial cell binding but also reduced platelet adhesion, thereby minimizing thrombosis risk [224]. While polymers significantly enhance vascular reconstruction techniques, challenges remain in achieving optimal integration and long-term functionality in vivo, necessitating ongoing research in this field.

References

[1] Raisa, Cristina, Teodoro, da, Silva., Patric, Emerson, Oliveira, Gonçalves., Ligia, de, Loiola, Cisneros. (2017). Anatomical Principles of the Circulatory System. doi: 10.1007/978-3-319-46059-8_2

[2] Marc, J., Klowden. (2013). Chapter 7 – Circulatory Systems. doi: 10.1016/B978-0-12-415819-1.00007-6

[3] Julia, Büchele. (2023). Circulatory System. doi: 10.1007/978-981-19-9538-5_7

[4] Michael, Sonnekalb., Sebastian, Fink. (2019). Circulatory system for fuel cell vehicle.

[5] Harjes, Daniel, I., Mercier, Nichole. (2021). Circulatory support systems including controller and plurality of sensors and methods of operating same.

[6] Tara, B., M., Smith. (2022). Architecture of the Blood Vessels. doi: 10.1007/978-981-19-7122-8_1

[7] Martin, Caon. (2024). Lecture 20: Anatomy, Structure and Function of Blood Vessels. doi: 10.1007/978-3-031-56296-9_1

[8] Phat, N., Vuong., Colin, Berry. (2002). Histology of vessels. doi: 10.1007/978-2-8178-0786-7_1

[9] J., Gordon, Betts., Kelly, A., Young., James, A., Wise., Eddie, Johnson., Brandon, Poe., Dean, H., Kruse., Oksana, Korol., Jody, E., Johnson., Mark, Womble., Peter, DeSaix. (2013). Structure and Function of Blood Vessels.

[10] Linda, M., Ernst., Michael, K., Fritsch. (2019). Blood Vessels and Lymphatic Vessels. doi: 10.1007/978-3-030-11425-1_2

[11] Michael, H., Birnbaum. (2022). Diseases of small and medium-sized blood vessels. doi: 10.1016/b978-0-12-822224-9.00020-7

[12] Meghan, Mott., Katherine, Pahigiannis., Walter, J., Koroshetz. (2014). Small Blood Vessels: Big Health Problems National Institute of Neurological Disorders and Stroke Update. Stroke. https://doi.org/10.1161/STROKEAHA.114.007113

[13] Mary, N., Sheppard. (2012). Diseases of blood vessels. Surgery (oxford). https://doi.org/10.1016/J.MPSUR.2015.04.006

[14] Michael, A., Briones. (2009). Vascular Bleeding Disorders. Transfusion Medicine and HemostasisClinical and Laboratory Aspects. https://doi.org/10.1016/B978-0-12-374432-6.00104-4

[15] Barbara, Schildkrout. (2014). Disease #68: Vascular Dementia. doi: 10.1002/9781394260447.ch96

[16] Salka, S., Staekenborg., Tanja, Su., Elisabeth, C.W., van, Straaten., Roger, Lane., Philip, Scheltens., Frederik, Barkhof., Wiesje, M., van, der, Flier. (2010). Behavioural and psychological symptoms in vascular dementia; differences between small- and large-vessel disease. Journal of Neurology, Neurosurgery, and Psychiatry. https://doi.org/10.1136/JNNP.2009.187500

[17] Quazi, S., Capuzzo, Am., Malik, Ja. (2021). Genomic Mapping in Detection of Vascular Disorders. doi: 10.20944/PREPRINTS202103.0209.V1

[18] Riwaj, Bhagat., Sandro, Marini., Jose, R., Romero. (2023). Genetic considerations in cerebral small vessel diseases. Frontiers in Neurology. https://doi.org/10.3389/fneur.2023.1080168

[19] Takayuki, Morisaki., Hiroko, Morisaki. (2017). Genetics of Hereditary Large Vessel Diseases. doi: 10.1002/9780470015902.A0027328

[20] GOLDBERG, V. M., & Stevenson, S. (1987). Natural history of autografts and allografts. *Clinical Orthopaedics and Related Research®*, *225*, 7-16.

[21] Matsumura, G., Hibino, N., Ikada, Y., Kurosawa, H., & Shin'oka, T. (2003). Successful application of tissue engineered vascular autografts: clinical experience. *Biomaterials*, *24*(13), 2303-2308.

[22] Teebken, O. E., Pichlmaier, M. A., Brand, S., & Haverich, A. (2004). Cryopreserved arterial allografts for in situ reconstruction of infected arterial vessels. *European journal of vascular and endovascular surgery*, *27*(6), 597-602.

[23] Gutowski, P., Guziewicz, M., Ilzecki, M., Kazimierczak, A., Lawson, J. H., Prichard, H. L., ... & Niklason, L. E. (2023). Six-year outcomes of a phase II study of human-tissue engineered blood vessels for peripheral arterial bypass. *JVS-Vascular Science*, *4*, 100092.

[24] Hu, K., Li, Y., Ke, Z., Yang, H., Lu, C., Li, Y., ... & Wang, W. (2022). History, progress and future challenges of artificial blood vessels: A narrative review. *Biomaterials Translational*, *3*(1), 81.

[25] Weiss, S., Bachofen, B., Widmer, M. K., Makaloski, V., Schmidli, J., & Wyss, T. R. (2021). Long-term results of cryopreserved allografts in aortoiliac graft infections. *Journal of vascular surgery*, *74*(1), 268-275.

[26] Deng, X., Gould, M., & Ali, M. A. (2022). A review of current advancements for wound healing: Biomaterial applications and medical devices. *Journal of Biomedical Materials Research Part B: Applied Biomaterials*, *110*(11), 2542-2573.

[27] Adigbli, G., Alshomer, F., Maksimcuka, J., & Ghali, S. (2016). Principles of plastic surgery, wound healing, skin grafts and flaps. *Textbook of plastic and reconstructive surgery*, *1*, 3-37.

[28] Stegen, S., Van Gastel, N., & Carmeliet, G. (2015). Bringing new life to damaged bone: the importance of angiogenesis in bone repair and regeneration. *Bone*, *70*, 19-27.

[29] Cunha-Melo, J. R., & Costa, G. (2014). Intestinal transplantation: evolution and current status. *MedicalExpress*, *1*, 307-322.

[30] Sourav, Sadhukhan., Atanu, Pal., Aniruddha, Datta. (2023). Outcome of Central Venoplasty and AVFistuloplasty in a Tertiary Care Centre in Eastern India. Journal of the American Society of Nephrology. https://doi.org/10.1681/asn.20233411s1540c

[31] Anna, Olds., W., Hampton, Gray., Markian, Bojko., Carly, Weaver., Jeanette, N., Cleveland., Michael, E., Bowdish., Winfield, J., Wells., Vaughn, A., Starnes., Sandeep, Kumar. (2024). Surgical Pulmonary Arterioplasty at Bidirectional Cavopulmonary Anastomosis Leads to Favorable Pulmonary Hemodynamics at Final Stage Palliation. JTCVS open. https://doi.org/10.1016/j.xjon.2024.02.006

[32] Francesco, Giovanardi., Francesco, Nudo., Quirino, Lai., M., Garofalo., Adriano, Consolo., E., Choppin, de, Janvry., G.A., Arroyo, Murillo., P., Ursi., D., Stabile., Fabio, Melandro., P.B., Berloco., Renzo, Pretagostini., Luca, Poli. (2019). Surgical Technique Notes of Arterial Vascular Reconstruction During Kidney Transplantation: Personal Experience and Literature Review. doi: 10.1016/J.TRANSPROCEED.2018.04.072

[33] Leonid, S., Barbarash., Ehl, Gudin, Jakov, L, Vovich., V., V., Sevostyanova., A., S., Golovkin. (2013). Tissue-Engineered Vascular Graft and Its Fabrication Approach.

[34] Ziyu, Wang., Suzanne, M., Mithieux., Anthony, S., Weiss. (2019). Fabrication Techniques for Vascular and Vascularized Tissue Engineering. Advanced Healthcare Materials. https://doi.org/10.1002/ADHM.201900742

[35] D., Vervoort., Malak, Elbatarny., Rodolfo, V., Rocha., Stephen, E., Fremes. (2023). Reconstruction Technique Options for Achieving Total Arterial Revascularization and Multiple Arterial Grafting. Stomatology. https://doi.org/10.3390/jcm12062275

[36] Guosheng, Wu., Yinglun, Wu., Mian, Wang., Wentong, Zhang., Chaoxu, Liu., Tingbo, Liang. (2021). Vascular reconstruction of segmental intestinal grafts using autologous internal iliac vessels.. Gastroenterology Report. https://doi.org/10.1093/GASTRO/GOAB016

[37] Chao, Lin, Chen., How, Ran, Guo., Ying, Jan, Wang., Hong, Tai, Chang., Chui, Yi, Pan., Ho, Yi, Tuan-Mu., Hsiu, Chuan, Lin., Chao, Yi, Chen., Jin, Jia, Hu. (2019). Combination of inductive effect of lipopolysaccharide and in situ mechanical conditioning for forming an autologous vascular graft in vivo. Scientific Reports. https://doi.org/10.1038/S41598-019-47054-2

[38] Serkan, Ketenciler., Kamil, Boyacıoğlu., İlknur, Akdemir., Gürkan, Kömürcü., Adil, Polat. (2018). Autologous Saphenous Vein Panel Graft for Vascular Reconstruction.. Annals of Vascular Surgery. https://doi.org/10.1016/J.AVSG.2018.05.037

[39] Alfonso, Lapergola., Alfonso, Lapergola., Emanuele, Felli., Thomas, Rebiere., Didier, Mutter., Patrick, Pessaux. (2020). Autologous peritoneal graft for venous vascular reconstruction after tumor resection in abdominal surgery: a systematic review.. Updates in Surgery. https://doi.org/10.1007/S13304-020-00730-9

[40] Wouter, J., Geelhoed., K.E.A., van, der, Bogt., T.C., Rothuizen., Febriyani, Damanik., Jaap, F., Hamming., Carlos, Mota., M.S., van, Agen., H.C., de, Boer., M., Tobón, Restrepo., Boris, Hinz., A., Kislaya., Christian, Poelma., A.J., van, Zonneveld., Ton, J., Rabelink., Lorenzo, Moroni., Joris, I., Rotmans. (2020). A novel method for engineering autologous non-thrombogenic in situ tissue-engineered blood vessels for arteriovenous grafting.. Biomaterials. https://doi.org/10.1016/J.BIOMATERIALS.2019.119577

[41] Santh, Prakash, Lanka., Guruprasad, Rai., Revanth, Maramreddy., Arvind, Kumar, Bishnoi., V., M., Ruvin, Kumara., Ganesh, Kamath, Sevagur. (2023). Saphenous Vein Tributaries in Sequential Vein Grafting for Coronary Artery Disease. doi: 10.21203/rs.3.rs-3507838/v1

[42] Rasit Dinc (2023). In Situ Saphenous Vein Bypass Grafting in Peripheral Arterial Disease of the Lower Extremities. Cardio Open, doi: 10.33140/coa.08.03.04

[43] Lars, Saemann., L., Wernstedt., Sabine, Pohl., M., Stiller., Jan, Willsch., Britt, Hofmann., Gábor, Veres., Andreas, Simm., Gábor, Szabó. (2023). Impact of Age on Endothelial Function of Saphenous Vein Grafts in Coronary Artery Bypass Grafting. Stomatology. https://doi.org/10.3390/jcm12175454

[44] Ryo, Nakamura., Kentaro, Honda., Hideki, Kunimoto., Takahiro, Fujimoto., Kota, Agematsu., Yoshiharu, Nishimura. (2023). Impact of Graft Velocity on Saphenous Vein Graft Atherosclerosis after Coronary Artery Bypass Grafting.. Annals of Thoracic and Cardiovascular Surgery. https://doi.org/10.5761/atcs.oa.23-00066

[45] Huiru, Chen., Zilan, Wang., Ke, Si., Xiaoxiao, Wu., Hanyu, Ni., Yan-Chi, Tang., Wei, Li., Zheng, Wang. (2023). External stenting for saphenous vein grafts in coronary artery bypass grafting: A meta-analysis.. European Journal of Clinical Investigation. https://doi.org/10.1111/eci.14046

[46] C., A., Dietl. (2006). Angiographic Studies of the Radial Artery Graft. doi: 10.1007/3-540-30084-8_19

[47] Francesco, Cabrucci., Beatrice, Bacchi., Aleksander, Dokollari., Massimo, Bonacchi. (2023). Radial artery harvesting with harmonic scalpel: fully no-touch technique.. Multimedia Manual of Cardiothoracic Surgery. https://doi.org/10.1510/mmcts.2023.018

[48] Junjiro, Kobayashi. (2009). Radial artery as a graft for coronary artery bypass grafting.. Japanese Circulation Journal-english Edition. https://doi.org/10.1253/CIRCJ.CJ-09-0322

[49] Sun, Zong-quan. (2007). Role of Radial Artery Graft in Coronary Artery Bypass Grafting. Chinese Journal of Clinical Thoracic and Cardiovascular Surgery,

[50] Prechtel, E. (2003). *Design, Synthesis, and Characterization of a Tissue Engineered Small Caliber Vascular Graft*. Drexel University.

[51] Moore, W. R., Graves, S. E., & Bain, G. I. (2001). Synthetic bone graft substitutes. *ANZ journal of surgery*, *71*(6), 354-361.

[52] Edward, Chen., Lih-Sheng, Turng. (2024). A double-expanded polytetrafluoroethylene fabrication method for increased mechanical compliance in tubular vascular graft applications. doi: 10.1002/pen.26652

[53] Roberta, Cassano., P., Perri., Antonio, Esposito., F, Intrieri., Roberta, Sole., Federica, Curcio., Sonia, Trombino. (2023). Expanded Polytetrafluoroethylene Membranes for Vascular Stent Coating: Manufacturing, Biomedical and Surgical Applications, Innovations and Case Reports. Membranes. https://doi.org/10.3390/membranes13020240

[54] Dongfang, Wang., Yiyang, Xu., Lixia, Wang., Xiaofeng, Wang., Cuihong, Ren., Bo, Zhang., Qian, Li., James, A., Thomson., Lih-Sheng, Turng. (2020). Expanded Poly(tetrafluoroethylene) Blood Vessel Grafts with Embedded Reactive Oxygen Species (ROS)-Responsive Antithrombogenic Drug for Elimination of Thrombosis. ACS Applied Materials & Interfaces. https://doi.org/10.1021/ACSAMI.0C07868

[55] Marco, Amabili., Prabakaran, Balasubramanian., Giovanni, Ferrari., Giulio, Franchini., Francesco, Giovanniello., Eleonora, Tubaldi. (2020). Identification of viscoelastic properties of Dacron aortic grafts subjected to physiological pulsatile flow.. Journal of The Mechanical Behavior of Biomedical Materials. https://doi.org/10.1016/J.JMBBM.2020.103804

[56] Cristiano, Spadaccio., Alberto, Rainer., Raffaele, Barbato., Marcella, Trombetta., Massimo, Chello., Bart, Meyns. (2019). The long-term follow up of large-diameter Dacron® vascular grafts in surgical practice: a review.. Journal of Cardiovascular Surgery. https://doi.org/10.23736/S0021-9509.16.08061-7

[57] Turkmen, Arikan., Elmar, Mammadov., Ertan, Emek., Birkan, Bozkurt., N., Inan, Gurcan., Pinar, Yazici., Tolga, Sahin., Ayfer, Serin., Unal, Aydin., Yaman, Tokat. (2019). Utility of Polyethylene Terephthalate (Dacron) Vascular Grafts for Venous Outflow

Reconstruction in Living-Donor Liver Transplantations.. doi: 10.1016/J.TRANSPROCEED.2019.02.040

[58] Asik, Ali, Mohamed, Ali., Praveen, Sharma., Rujuta, Rege., Saveetha, Rajesh., Kulasekaran, Nadhamuni. (2016). Dacron graft aneurysm with dissection.. Indian Journal of Radiology and Imaging. https://doi.org/10.4103/0971-3026.195783

[59] John, M., Trahanas., Omar, A., Jarral., G., Chad, Hughes. (2022). A dissection … of the Dacron?. JTCVS techniques. https://doi.org/10.1016/j.xjtc.2022.03.022

[60] Gavin, R., Meehan., Hannah, E., Scales., Rowland, Osii., Mariana, De, Niz., Mariana, De, Niz., Jennifer, C., Lawton., Matthias, Marti., Paul, Garside., Alister, Craig., James, M., Brewer. (2020). Developing a xenograft model of human vasculature in the mouse ear pinna.. Scientific Reports. https://doi.org/10.1038/S41598-020-58650-Y

[61] Zhihong, Dong., Atsushi, Imai., Sudha, Krishnamurthy., Zhaocheng, Zhang., Benjamin, David, Zeitlin., Jacques, E., Nör. (2013). Xenograft Tumors Vascularized with Murine Blood Vessels May Overestimate the Effect of Anti-Tumor Drugs: A Pilot Study. PLOS ONE. https://doi.org/10.1371/JOURNAL.PONE.0084236

[62] Stefan, Schlatt., Birgit, Westernströer., Kathrin, Gassei., Jens, Ehmcke. (2010). Short title: Testis xenografting Summary sentence: The circulatory connection between donor and host in testis xenografting is established by outgrowing donor and ingrowing host blood vessels.

[63] Marius, Harpa., Sânziana, Flămând, Oltean., Alexandra, Iulia, Puscas., Raluca, Truta., David, Emanuel, Anitei., Claudiu, Ghiragosian., Cosmin, Banceu., I., Movileanu., Ovidiu, Simion, Cotoi., Raluca, Niculescu., Horatiu, Suciu., Dan, Simionescu., Hussam, Al, Hussein. (2023). Bioengineered Small-Diameter Vascular Xenografts as an Alternative to Autologous Vascular Grafting for Emergency Revascularization – a Preliminary Study. Journal of Cardiovascular Emergencies. https://doi.org/10.2478/jce-2023-0021

[64] Kerbi, Alejandro, Guevara-Noriega., Albert, Martinez, Toiran., Bruno, Alvarez-Concejo., Jose, Luis, Pomar. (2018). Historical Overview of Vascular Allografts Transplantation.

[65] Kerbi, Alejandro, Guevara-Noriega., Albert, Martinez, Toiran., Bruno, Alvarez-Concejo., Jose, Luis, Pomar. (2019). Historical Overview of Vascular Allograft Transplantation. doi: 10.15420/VER.2018.15.1

[66] Holger, Konrad., Anja, Wahle., Wolfgang, Altermann., Gerald, Schlaf. (2017). Strong Humoral Anti-HLA Immune Response Upon Arbitrarily Chosen Allogeneic Arterial Vessel Grafts. Journal of clinical & cellular immunology. https://doi.org/10.4172/2155-9899.1000525

[67] Paola, Tracanelli., F., Romani. (2016). Vascular Homograft Procurement. doi: 10.1007/978-3-319-28416-3_17

[68] José, Ignacio, Bilbao., A., Martínez., P., Domínguez, Echavarri., Octavio, Cosín., B., Zudaire. (2005). Técnicas terapéuticas endovasculares Endovascular therapeutic techniques.

[69] José, Ignacio, Bilbao., Martínez, de, la, Cuesta, A., Domínguez, Echavarri, P., Cosín, O., L., Desloques., B., Zudaire. (2005). Endovascular therapeutic techniques. Anales Del Sistema Sanitario De Navarra,

[70] Roger, M., Greenhalgh., Jean-Pierre, Becquemin. (2001). Vascular and endovascular surgical techniques.

[71] Jan, Brunkwall., Klaus, Lackner. (2004). The influence of endovascular techniques on our surgical practice: a 10-year view.. Journal of Endovascular Therapy. https://doi.org/10.1583/04-1313.1

[72] Jie-Long, Lin., Chen, Haiming., Feng-Cheng, Lin., Jieying, Li., Cheng-Xin, Xie., Wen-Liang, Guo., Xiu-Fen, Huang., Cheng, Hong. (2020). Application of DynaCT angiographic reconstruction in balloon pulmonary angioplasty. European Radiology. https://doi.org/10.1007/S00330-020-07028-9

[73] Blaine, Andrew, Schneider. (2016). Balloon angioplasty catheter coating to encourage vessel repair and further reduce restenosis.

[74] Pengbin, Tang., Dongjin, Huang., Yin, Wang., Gong, Ruobin., Wen, Tang., Youdong, Ding. (2016). Position based balloon angioplasty. doi: 10.1145/3013971.3013996

[75] Alistair, J., McCleary., Michael, J., Gough. (2001). Vascular surgeon's experience with intraoperative angioplasty.. Anz Journal of Surgery. https://doi.org/10.1046/J.1445-2197.2001.02212.X

[76] S., Marlene, Grenon. (2012). Angioplasty Balloon, Stents, and Stent Grafts. doi: 10.1007/978-1-4614-0839-0_4

[77] Sarah, Pradhan., Rachel, D., Vanderlaan., Lee, Benson. (2024). Extravascular Central Venous Line Removal and Endovascular Covered Stent Implantation Guided by Three-Dimensional Computed Tomographic Reconstruction. CJC pediatric and congenital heart disease. https://doi.org/10.1016/j.cjcpc.2023.12.005

[78] Georgia, S., Karanasiou., Claire, Conway., Michail, I., Papafaklis., Augusto, C., Lopes., Kostas, A., Stefanou., Lambros, S., Athanasiou., Lampros, K., Michalis., Elazer, R., Edelman., Dimitrios, I., Fotiadis. (2014). Finite element analysis of stent implantation in a three-dimensional reconstructed arterial segment.. doi: 10.1109/EMBC.2014.6944902

[79] Wei, Jichang., Gong, Xiaoyan., Zhou, Weichao., Xiaoming, Zhang., Liu, Hao. (2018). Stent for implantation into blood vessel.

[80] Huanming, Xu., Yuqian, Mei., Xiaofeng, Han., Jianyong, Wei., Paul, N., Watton., Wan, Jia., Anqiang, Li., Duanduan, Chen., Jiang, Xiong. (2019). Optimization schemes for endovascular repair with parallel technique based on hemodynamic analyses.. International Journal for Numerical Methods in Biomedical Engineering. https://doi.org/10.1002/CNM.3197

[81] Liu, Mulong., Xu, Wenchao. (2017). Blood vessel stent and manufacturing method thereof.

[82] Bing, li., Yan, Shu., Hailin, Ma., Kun, Cao., Yuen, Yee, Cheng., Zhilin, Jia., Xiao, Ma., Hongfei, Wang., Kedong, Song. (2024). Three-dimensional printing and decellularized-extracellular-matrix based methods for advances in artificial blood vessel fabrication: A review.. Tissue & Cell. https://doi.org/10.1016/j.tice.2024.102304

[83] Jaewoo, Choi., Eun, Ji, Lee., Woong, Bi, Jang., Sang-Mo, Kwon. (2023). Development of Biocompatible 3D-Printed Artificial Blood Vessels through Multidimensional Approaches. Journal of Functional Biomaterials. https://doi.org/10.3390/jfb14100497

[84] Hector F. Salazar, Gillian M. O'Connell, A.B., George S. Corpuz, Sophia Salingaros, Abby H. Chopoorrian, Kassandra Carrion, Celia Aboaf, Xue Dong, and Jason A. Spector (2023). A Novel 3D Printed Sacrificial Material for the Development of Perfused Intricate Blood Vessel Structures. Plastic and reconstructive surgery. Global open. https://doi.org/10.1097/01.gox.0000938168.04392.a0

[85] Jun, Hua, Zhu., Xinwang, Wang., Lin, Lin., Wen, Zeng. (2023). 3D bioprinting for vascular grafts and microvasculature. International Journal of bioprinting. https://doi.org/10.36922/ijb.0012

[86] Lingtong, Zhou., Yuanchang, Li., Qin, Tu., Jinyi, Wang. (2023). A 3D Printing Mold Method for Rapid Fabrication of Artificial Blood Vessels. Colloids and Surfaces A: Physicochemical and Engineering Aspects. https://doi.org/10.1016/j.colsurfa.2023.130952

[87] Manabu, Itoh. (2021). Scaffold-Free Autologous Cell-Based Vascular Graft for Clinical Application. doi: 10.1007/978-3-030-58688-1_9

[88] Julia, Fernández-Pérez., Kenny, A., van, Kampen., Carlos, Mota., Matthew, B., Baker., Lorenzo, Moroni. (2023). Flexible, Suturable, and Leak-free Scaffolds for Vascular Tissue Engineering Using Melt Spinning. ACS Biomaterials Science & Engineering. https://doi.org/10.1021/acsbiomaterials.3c00535

[89] Junichi, Saito., Junichi, Saito., Utako, Yokoyama., Utako, Yokoyama., Takashi, Nakamura., Tomomitsu, Kanaya., Takayoshi, Ueno., Yuji, Naito., Toshio, Takayama., Makoto, Kaneko., Shigeru, Miyagawa., Yoshiki, Sawa., Yoshihiro, Ishikawa. (2021). Scaffold-free tissue-engineered arterial grafts derived from human skeletal myoblasts.. Artificial Organs. https://doi.org/10.1111/AOR.13930

[90] Mehdi, Atari., Abbas, Saroukhani., Maziar, Manshaei., Peiman, Bateni., Anousheh, Zargar, Kharazi., Elham, Vatankhah., Shaghayegh, Haghjooy, Javanmard. (2023). Preclinical in vivo assessment of a cell-free multi-layered scaffold prepared by 3D printing and electrospinning for small-diameter blood vessel tissue engineering in a canine model.. Biomaterials Science. https://doi.org/10.1039/d3bm00642e

[91] Salha, Sassi., Tatsuya, Watanabe., Toshiharu, Shinoka. (2023). Scaffold and Cell-Based Tissue Engineering Approaches as Alternative Therapy for Blood Vessel Disease. Cardiology research and cardiovascular medicine. https://doi.org/10.29011/2575-7083.100097

[92] Natalia, M., Yudintceva., Yu., A., Nashchekina., Maxim, Shevtsov., V., B., Karpovich., G., I., Popov., I., A., Samusenko., N., A., Mikhailova. (2021). Small-Diameter Vessels Reconstruction Using Cell Tissue-Engineering Graft Based on the Polycaprolactone. Cell and Tissue Biology. https://doi.org/10.1134/S1990519X21060110

[93] Hidekazu, Sekine., Teruo, Okano. (2020). Capillary Networks for Bio-Artificial Three-Dimensional Tissues Fabricated Using Cell Sheet Based Tissue Engineering.. International Journal of Molecular Sciences. https://doi.org/10.3390/IJMS22010092

 Materials Research Forum LLC
https://doi.org/10.21741/9781644903353-18

[94] A.A., Zholborsov., Talaibek, Baisekeev., Ahmad, Niyazov., A., Mamyshov., A.A., Niyazov., B., Zhynzhyrov., N., Osmonbekova. (2023). History of Reconstructive Vascular Surgery. Бюллетень науки и практики. https://doi.org/10.33619/2414-2948/96/33

[95] Tae, K., Song., E., John, Harris., Shyam, S., Raghavan., Jeffrey, A., Norton. (2009). Major Blood Vessel Reconstruction During Sarcoma Surgery. Archives of Surgery. https://doi.org/10.1001/ARCHSURG.2009.149

[96] Petr, Štádler., Miroslav, Špaček., Matous, P., Vitásek, P., Korisková, Z., Michálek, P. (2004). [Laparoscopic vascular reconstructions--initial experience].. Rozhledy v chirurgii : měsíčník Československé chirurgické společnosti,

[97] Bin, Jiang., Rachel, Suen., Jiao, Jing, Wang., Zheng, Zhang., Jason, A., Wertheim., Guillermo, A., Ameer. (2016). (1) Mechanocompatible Polymer-Extracellular-Matrix Composites for Vascular Tissue Engineering.. Advanced Healthcare Materials. https://doi.org/10.1002/ADHM.201501003

[98] Qianhan, Zeng., Jing, Zhou., Ying, Ji., Hansheng, Wang. (2024). A semiparametric Gaussian mixture model for chest CT-based 3D blood vessel reconstruction.. Biostatistics. https://doi.org/10.1093/biostatistics/kxae013

[99] Singh, C., Wong, C. S., & Wang, X. (2015). Medical textiles as vascular implants and their success to mimic natural arteries. Journal of functional biomaterials, 6(3), 500-525.

[100] Stowell, C. E., & Wang, Y. (2018). Quickening: Translational design of resorbable synthetic vascular grafts. Biomaterials, 173, 71-86.

[101] Oberhauser, J. P., Hossainy, S., & Rapoza, R. J. (2009). Design principles and performance of bioresorbable polymeric vascular scaffolds. EuroIntervention, 5(Suppl F), F15-F22.

[102] Pankajakshan, D., & Agrawal, D. K. (2010). Scaffolds in tissue engineering of blood vessels. Canadian journal of physiology and pharmacology, 88(9), 855-873.

[103] Stefan, Baudis., Christian, Heller., Robert, Liska., Juergen, Stampfl., Helga, Bergmeister., Guenter, Weigel. (2009). (4) (Meth)acrylate-based photoelastomers as tailored biomaterials for artificial vascular grafts. Journal of Polymer Science Part A. https://doi.org/10.1002/POLA.23352

[104] Zhu, T., Gu, H., Zhang, H., Wang, H., Xia, H., Mo, X., & Wu, J. (2021). Covalent grafting of PEG and heparin improves biological performance of electrospun vascular grafts for carotid artery replacement. Acta Biomaterialia, 119, 211-224.

[105] Rosellini, E., Giordano, C., Guidi, L., & Cascone, M. G. (2024). Biomimetic Approaches in Scaffold-Based Blood Vessel Tissue Engineering. Biomimetics, 9(7), 377.

[106] Song, J., & Gerecht, S. (2023). Hydrogels to Recapture Extracellular Matrix Cues That Regulate Vascularization. Arteriosclerosis, thrombosis, and vascular biology, 43(8), e291-e302.

[107] Soodabeh, Gorgani., Anousheh, Zargar, Kharazi., Shaghayegh, Haghjooy, Javanmard., Mohammad, Rafiinia. (2020). Improvement of Endothelial Cell Performance in an Optimized Electrospun Pre-polyglycerol Sebacate-Poly Lactic Acid Scaffold for Reconstruction of Intima

in Coronary Arteries. Journal of Polymers and The Environment. https://doi.org/10.1007/S10924-020-01749-0

[108] Carrabba, M., Fagnano, M., Ghorbel, M. T., Rapetto, F., Su, B., De Maria, C., ... & Madeddu, P. (2024). Development of a Novel Hierarchically Biofabricated Blood Vessel Mimic Decorated with Three Vascular Cell Populations for the Reconstruction of Small-Diameter Arteries. Advanced Functional Materials, 34(7), 2300621.

[109] Guoxing, Liao. (2024). Customized Vascular Repair Microenvironment: Poly(lactic acid)-Gelatin Nanofibrous Scaffold Decorated with bFGF and Ag@Fe3O4 Core-Shell Nanowires.. ACS Applied Materials & Interfaces. https://doi.org/10.1021/acsami.4c09269

[110] Soodabeh, Gorgani., Anousheh, Zargar, Kharazi., Shaghayegh, Haghjooy, Javanmard., Mohammad, Rafiinia. (2020). Improvement of Endothelial Cell Performance in an Optimized Electrospun Pre-polyglycerol Sebacate-Poly Lactic Acid Scaffold for Reconstruction of Intima in Coronary Arteries. Journal of Polymers and The Environment. https://doi.org/10.1007/S10924-020-01749-0

[111] Jun, Zhang., He, Lei., Guo, Wei., Jiang, Xuefeng., Fu, Lei., Zhao, Yue., Luxia, Zhang., Lutao, Yang., Yajuan, Li., Yutong, Wang., Mo, Hong., Jian, Shen. (2019). Zwitterionic Polymer-Grafted Polylactic Acid Vascular Patches Based on a Decellularized Scaffold for Tissue Engineering.. ACS Biomaterials Science & Engineering. https://doi.org/10.1021/ACSBIOMATERIALS.9B00684

[112] Iffa, A., Fiqrianti., Prihartini, Widiyanti., Muhammad, A., Manaf., Claudia, Y., Savira., Nadia, Rifqi, Cahyani., Fitria, Renata, Bella. (2018). Poly-L-lactic Acid (PLLA)-Chitosan-Collagen Electrospun Tube for Vascular Graft Application.. Journal of Functional Biomaterials. https://doi.org/10.3390/JFB9020032

[113] Salvatore, Buscemi., Vincenzo, Davide, Palumbo., A., Maffongelli., S., Fazzotta., Fabio, Salvatore, Palumbo., Mariano, Licciardi., Calogero, Fiorica., Roberto, Puleio., Giovanni, Cassata., L., Fiorello., Giuseppe, Buscemi., A.I., Lo, Monte. (2017). Electrospun PHEA-PLA/PCL Scaffold for Vascular Regeneration: A Preliminary in Vivo Evaluation.. doi: 10.1016/J.TRANSPROCEED.2017.02.017

[114] Lalit, Ranakoti., Brijesh, Gangil., Prabhakar, Bhandari., Tej, Singh., Shubham, Sharma., Jujhar, Singh., Sunpreet, Singh. (2023). Promising Role of Polylactic Acid as an Ingenious Biomaterial in Scaffolds, Drug Delivery, Tissue Engineering, and Medical Implants: Research Developments, and Prospective Applications. Molecules. https://doi.org/10.3390/molecules28020485

[115] Monika, Smiga-Matuszowicz., Jakub, Wlodarczyk., Małgorzata, Skorupa., Dominika, Czerwińska-Główka., Kaja, Fołta., Małgorzata, Pastusiak., Małgorzata, Adamiec-Organiściok., Magdalena, Skonieczna., Roman, Turczyn., Michał, Sobota., Katarzyna, Krukiewicz. (2023). Biodegradable Scaffolds for Vascular Regeneration Based on Electrospun Poly(L-Lactide-co-Glycolide)/Poly(Isosorbide Sebacate) Fibers. International Journal of Molecular Sciences. https://doi.org/10.3390/ijms24021190

[116] Gülce, Taşkor, Önel. (2023). Synthesis and Characterization of Poly(lactic-co-glycolic acid) Derived with LGlutamic Acid and L-Aspartic Acid. Erzincan University Journal of Science and Technology. https://doi.org/10.18185/erzifbed.1235522

[117] Shuo, Li., Lei, Yang., Zijian, Zhao., Xiaoniu, Yang., Hongying, Lv. (2024). (1) A Polyurethane-based Hydrophilic Elastomer with Multi-biological Functions for Small-diameter Vascular Grafts.. Acta Biomaterialia. https://doi.org/10.1016/j.actbio.2024.01.006

[118] Jun, Fang., Jialing, Zhang., Jun, Du., Yanjun, Pan., Jing, Shi., Yongxuan, Peng., Weiming, Chen., Liu, Yuan., Sang-Ho, Ye., William, R., Wagner., Meng, Yin., Xiumei, Mo. (2016). (5) Orthogonally Functionalizable Polyurethane with Subsequent Modification with Heparin and Endothelium-Inducing Peptide Aiming for Vascular Reconstruction. ACS Applied Materials & Interfaces. https://doi.org/10.1021/ACSAMI.6B04289

[119] Natalia, M., Yudintceva., Yu., A., Nashchekina., Maxim, Shevtsov., V., B., Karpovich., G., I., Popov., I., A., Samusenko., N., A., Mikhailova. (2021). (1) Small-Diameter Vessels Reconstruction Using Cell Tissue-Engineering Graft Based on the Polycaprolactone. Cell and Tissue Biology. https://doi.org/10.1134/S1990519X21060110

[120] V., V., Sevostianova., A., V., Mironov., E., O., Krivkina., M., V., Khanova., E., A., Velikanova., Vera, G., Matveeva., T., V., Glushkova., L., V., Antonova., Leonid, S., Barbarash. (2019). (3) Biodegradable poly(ε-caprolactone) VEGF-containing vascular patches for angioplasty. doi: 10.1063/1.5132188

[121] Sen, Yang., Xueni, Zheng., Meng, Qian., He, Wang., Fei, Wang., Yongzhen, Wei., Adam, C., Midgley., Ju, He., Hongyan, Tian., Qiang, Zhao. (2021). (4) Nitrate-Functionalized poly(ε-Caprolactone) Small-Diameter Vascular Grafts Enhance Vascular Regeneration via Sustained Release of Nitric Oxide. Frontiers in Bioengineering and Biotechnology. https://doi.org/10.3389/FBIOE.2021.770121

[122] Arianna, Parnigoni., Manuela, Viola., Evgenia, Karousou., Simona, Rovera., Cristina, Giaroni., Alberto, Passi., Davide, Vigetti. (2022). (1) Role of hyaluronan in pathophysiology of vascular1 endothelial and smooth muscle cells.. American Journal of Physiology-cell Physiology. https://doi.org/10.1152/ajpcell.00061.2022

[123] Amaliris, Ruiz., Kashmila, R., Rathnam., Kristyn, S., Masters. (2014). (3) Effect of Hyaluronic Acid Incorporation Method on the Stability and Biological Properties of Polyurethane-Hyaluronic Acid Biomaterials. Journal of Materials Science: Materials in Medicine. https://doi.org/10.1007/S10856-013-5092-1

[124] Davide, Vigetti., Manuela, Viola., Evgenia, Karousou., Anna, Genasetti., Manuela, Rizzi., Moira, Clerici., Barbara, Bartolini., Paola, Moretto., Giancarlo, De, Luca., Alberto, Passi. (2008). (5) Vascular pathology and the role of hyaluronan.. The Scientific World Journal. https://doi.org/10.1100/TSW.2008.145

[125] Soo, kyeong, Yang., Muhammad, Shafiq., Muhammad, Shafiq., Daeheum, Kim., Chulhwan, Park., Youngmee, Jung., Youngmee, Jung., Soo, Hyun., Soo, Hyun., Soo, Hyun. (2016). (1) Characterization and preparation of bioinspired resorbable conduits for vascular reconstruction. Macromolecular Research. https://doi.org/10.1007/S13233-016-4042-4

[126] Rumiana, Tzoneva., Claudia, Weckwerth., Barbara, Seifert., Marc, Behl., Matthias, Heuchel., Iana, Tsoneva., Andreas, Lendlein. (2011). (3) In vitro evaluation of elastic multiblock co-polymers as a scaffold material for reconstruction of blood vessels.. Journal of Biomaterials Science-polymer Edition. https://doi.org/10.1163/092050610X537147

[127] Howard, P., Greisler., Charles, W., Tattersall., Scott, C., Henderson., Emma, A., Cabusao., Jacqueline, D., Garfield., Dae, Un, Kim. (1992). (2) Polypropylene small-diameter vascular grafts. Journal of Biomedical Materials Research. https://doi.org/10.1002/JBM.820261009

[128] Silvia, Scaglione., Rossella, Barenghi., Szabolcs, Beke., Luca, Ceseracciu., Ilaria, Romano., Francesca, Sbrana., Paola, Stagnaro., Fernando, Brandi., Massimo, Vassalli. (2013). (4) Characterization of a bioinspired elastin-polypropylene fumarate material for vascular prostheses applications. doi: 10.1117/12.2021754

[129] Naohiro, Wakabayashi., Takumi, Yoshida., Kyohei, Oyama., Daisuke, Naruse., Masahiro, Tsutsui., Yuta, Kikuchi., Daisuke, Koga., Hiroyuki, Kamiya. (2022). (1) Polyvinyl alcohol coating prevents platelet adsorption and improves mechanical property of polycaprolactone-based small-caliber vascular graft. Frontiers in Cardiovascular Medicine. https://doi.org/10.3389/fcvm.2022.946899

[130] Nuno, Alexandre., Elísio, Costa., Susana, Coimbra., Susana, Coimbra., A., Silva., A., Silva., A, Lopes., Miguel, Rodrigues., Marta, Santos., Ana, Colette, Maurício., José, D., Santos., Ana, Lúcia, Luís. (2015). (4) In vitro and in vivo evaluation of blood coagulation activation of polyvinyl alcohol hydrogel plus dextran-based vascular grafts. Journal of Biomedical Materials Research Part A. https://doi.org/10.1002/JBM.A.35275

[131] Julia, M., Ino., Ervi, Sju., Véronique, Ollivier., Evelyn, K.F., Yim., Didier, Letourneur., Catherine, Le, Visage. (2013). (5) Evaluation of hemocompatibility and endothelialization of hybrid poly(vinyl alcohol) (PVA)/gelatin polymer films.. Journal of Biomedical Materials Research Part B. https://doi.org/10.1002/JBM.B.32977

[132] Valentina, Montagna., Valentina, Montagna., Junko, Takahashi., Meng, Yu, Tsai., Takayuki, Ota., Nicolas, Zivic., Seigou, Kawaguchi., Takashi, Kato., Masaru, Tanaka., Haritz, Sardon., Kazuki, Fukushima., Kazuki, Fukushima. (2021). (1) Methoxy-Functionalized Glycerol-Based Aliphatic Polycarbonate: Organocatalytic Synthesis, Blood Compatibility, and Hydrolytic Property.. ACS Biomaterials Science & Engineering. https://doi.org/10.1021/ACSBIOMATERIALS.0C01460

[133] Rong, Zhu., Xinyu, Wang., Jing, Yang., Yiyu, Wang., Zongrui, Zhang., Yuanjing, Hou., Fei, Lin. (2016). (3) Influence of hydroxyl-terminated polydimethylsiloxane on high-strength biocompatible polycarbonate urethane films. Biomedical Materials. https://doi.org/10.1088/1748-605X/12/1/015011

[134] Chenhui, Zhu., Daidi, Fan., Zhiguang, Duan., Wenjiao, Xue., Longan, Shang., Fulin, Chen., Yane, Luo. (2009). (3) Initial investigation of novel human-like collagen/chitosan scaffold for vascular tissue engineering.. Journal of Biomedical Materials Research Part A. https://doi.org/10.1002/JBM.A.32256

[135] Catherine, F., Whittington., Mervin, C., Yoder., Sherry, L., Voytik-Harbin. (2013). (5) Collagen-polymer guidance of vessel network formation and stabilization by endothelial colony forming cells in vitro.. Macromolecular Bioscience. https://doi.org/10.1002/MABI.201300128

[136] Dimitria, Bonizol, Camasão., Miguel, González-Pérez., Sara, Palladino., Matilde, Alonso., José, Carlos, Rodríguez-Cabello., Diego, Mantovani. (2020). (1) Elastin-like

recombinamers in collagen-based tubular gels improve cell-mediated remodeling and viscoelastic properties.. Biomaterials Science. https://doi.org/10.1039/D0BM00292E

[137] Vivek, A., Kumar., Jeffrey, M., Caves., Jeffrey, M., Caves., Carolyn, A., Haller., Carolyn, A., Haller., Erbin, Dai., Liying, Liu., Stephanie, J., Grainger., Stephanie, J., Grainger., Elliot, L., Chaikof., Elliot, L., Chaikof., Elliot, L., Chaikof. (2013). (3) Acellular vascular grafts generated from collagen and elastin analogs.. Acta Biomaterialia. https://doi.org/10.1016/J.ACTBIO.2013.05.024

[138] Iffa, A., Fiqrianti., Prihartini, Widiyanti., Muhammad, A., Manaf., Claudia, Y., Savira., Nadia, Rifqi, Cahyani., Fitria, Renata, Bella. (2018). (3) Poly-L-lactic Acid (PLLA)-Chitosan-Collagen Electrospun Tube for Vascular Graft Application.. Journal of Functional Biomaterials. https://doi.org/10.3390/JFB9020032

[139] Anlin, Yin., Xiaorong, Lan., Weihua, Zhuang., Zhonglan, Tang., Yan, Li., Yunbing, Wang. (2020). (5) PEGylated chitosan and PEGylated PLCL for blood vessel repair: An in vitro study.. Journal of Biomaterials Applications. https://doi.org/10.1177/0885328219875937

[140] Antunes, M., Bonani, W., Reis, R. L., Migliaresi, C., Ferreira, H., Motta, A., & Neves, N. M. (2022). Development of alginate-based hydrogels for blood vessel engineering. Biomaterials Advances, 134, 112588.

[141] Florian, Ruther., Thomas, Distler., Aldo, R., Boccaccini., Rainer, Detsch. (2019). (5) Biofabrication of vessel-like structures with alginate di-aldehyde—gelatin (ADA-GEL) bioink. Journal of Materials Science: Materials in Medicine. https://doi.org/10.1007/S10856-018-6205-7

[142] Vera, G., Matveeva., M., U., Khanova., Larisa, V., Antonova., Leonid, S., Barbarash. (2020). (1) Fibrin – a promising material for vascular tissue engineering. Russian Journal of Transplantology and Artificial Organs. https://doi.org/10.15825/1995-1191-2020-1-196-208

[143] Florian, Helms., Axel, Haverich., Ulrike, Böer., Mathias, Wilhelmi. (2021). (3) Transluminal compression increases mechanical stability, stiffness and endothelialization capacity of fibrin-based bioartificial blood vessels.. Journal of The Mechanical Behavior of Biomedical Materials. https://doi.org/10.1016/J.JMBBM.2021.104835

[144] Vera, G., Matveeva., Evgenia, A., Senokosova., V., V., Sevostianova., Mariam, Yu., Khanova., Tatiana, V., Glushkova., Tatiana, N., Akentieva., Larisa, V., Antonova., Leonid, S., Barbarash. (2022). (5) Advantages of Fibrin Polymerization Method without the Use of Exogenous Thrombin for Vascular Tissue Engineering Applications. Advances in Cardiovascular Diseases. https://doi.org/10.3390/biomedicines10040789

[145] Alberto, M., Settembrini., Gianluca, Buongiovanni., Piergiorgio, Settembrini., Antonio, Alessandrino., Giuliano, Freddi., Giulia, Vettor., Eugenio, Martelli. (2023). (4) In-vivo evaluation of silk fibroin small-diameter vascular grafts: state of art of preclinical studies and animal models. Frontiers in Surgery. https://doi.org/10.3389/fsurg.2023.1090565

[146] Piotr, Koczoń., Alicja, Dąbrowska., Ewa, Laskowska., Małgorzata, Łabuz., Katarzyna, Maj., Jakub, Masztakowski., Bartłomiej, J., Bartyzel., Andrzej, Bryś., Joanna, Bryś., Eliza, Gruczyńska-Sękowska. (2023). (1) Applications of Silk Fibroin in Human and Veterinary Medicine. Materials. https://doi.org/10.3390/ma16227128

[147] Hong, Chen., Hong, Chen., Hong, Chen., Yajie, Zhang., Tianyu, Ni., Pi, Ding., Yue, Zan., Yue, Zan., Xue, Cai., Xue, Cai., Yiwei, Zhang., Min, Liu., Renjun, Pei. (2021). (2) Construction of a Silk Fibroin/Polyethylene Glycol Double Network Hydrogel with Co-Culture of HUVECs and UCMSCs for a Functional Vascular Network. doi: 10.1021/ACSABM.0C00353

[148] Xiangqian, Kong., He, Yuxiang., Hua, Zhou., Peixian, Gao., Lei, Xu., Zonglin, Han., Le, Yang., Mo, Wang. (2021). (3) Chondroitin Sulfate/Polycaprolactone/Gelatin Electrospun Nanofibers with Antithrombogenicity and Enhanced Endothelial Cell Affinity as a Potential Scaffold for Blood Vessel Tissue Engineering.. Nanoscale Research Letters. https://doi.org/10.1186/S11671-021-03518-X

[149] Yixi, Chen., Mao, Ye., Xiaofei, Wang., Wenqiang, Li., Weijian, Chen. (2022). (5) Functionalized gelatin/strontium hydrogel bearing endothelial progenitor cells for accelerating angiogenesis in wound tissue regeneration.. Biomaterials advances. https://doi.org/10.1016/j.bioadv.2022.212803

[150] Stefan, Baudis., Christian, Heller., Robert, Liska., Juergen, Stampfl., Helga, Bergmeister., Guenter, Weigel. (2009). (5) (Meth)acrylate-based photoelastomers as tailored biomaterials for artificial vascular grafts. Journal of Polymer Science Part A. https://doi.org/10.1002/POLA.23352

[151] Anna, Effenberger., Jade, Greenberg., Vidhya, Dhar., Anusha, Aggarwal., Tyler, Lee., N., Sanjeeva, Murthy., Thomas, Emge. (2018). (3) Fabrication and Characterization of Polymeric Aortic Vessels. doi: 10.1109/URTC45901.2018.9244812

[152] Yumei, Li., Yumei, Li., Rui, Zhao., Xiang, Li., Chuying, Wang., Huiwei, Bao., Shudan, Wang., Jing, Fang., Jinqiu, Huang., Ce, Wang. (2019). (2) Blood-compatible Polyaniline Coated Electrospun Polyurethane Fiber Scaffolds for Enhanced Adhesion and Proliferation of Human Umbilical Vein Endothelial Cells. Fibers and Polymers. https://doi.org/10.1007/S12221-019-8735-0

[153] Alida, Abruzzo., Calogero, Fiorica., Vincenzo, Davide, Palumbo., Roberta, Altomare., Giuseppe, Damiano., Maria, Concetta, Gioviale., Giovanni, Tomasello., Mariano, Licciardi., Fabio, Salvatore, Palumbo., Gaetano, Giammona., Attilio, Ignazio, Lo, Monte. (2014). (1) Using Polymeric Scaffolds for Vascular Tissue Engineering. International Journal of Polymer Science. https://doi.org/10.1155/2014/689390

[154] Marie, Claude, Boivin., Pascale, Chevallier., Stéphane, Turgeon., Jean, Lagueux., Gaétan, Laroche. (2011). (2) Micropatterning Polymer Materials to Improve Endothelialization. Advanced Materials Research. https://doi.org/10.4028/WWW.SCIENTIFIC.NET/AMR.409.777

[155] Dhiraj, M., Shah., Robert, P., Leather., John, D., Corson., Allastair, M., Karmody. (1984). (5) Polytetrafluoroethylene grafts in the rapid reconstruction of acute contaminated peripheral vascular injuries.. American Journal of Surgery. https://doi.org/10.1016/0002-9610(84)90227-7

[156] Yuan, Meirong., Xu, Yongjin., Chen, Yuqi. (2019). (1) Thiophene copolymer, preparation method and applications thereof.

[157] Nazely, Diban., Dimitrios, Stamatialis. (2011). (2) Functional Polymer Scaffolds for Blood Vessel Tissue Engineering. Macromolecular Symposia. https://doi.org/10.1002/MASY.201100038

[158] M'hamed, Chahma. (2005). (5) Synthesis and characterization of poly(thiophene sulfides) prepared via chemically initiated oxidative polymerization methods. Synthetic Metals. https://doi.org/10.1016/J.SYNTHMET.2005.06.023

[159] Yamanoue, Hisashi., Noguchi, Gen., Saito, Kei. (2019). (1) Poly(phenylene sulfide) resin composition, production method therefor, and molded article.

[160] Peiyuan, Zuo., Abbas, Tcharkhtchi., Mohammadali, Shirinbayan., Joseph, Fitoussi., Farid, Bakir. (2019). (4) Overall Investigation of Poly (Phenylene Sulfide) from Synthesis and Process to Applications—A Review. Macromolecular Materials and Engineering. https://doi.org/10.1002/MAME.201800686

[161] Birgit, Huber., Sascha, Engelhardt., Wolfdietrich, Meyer., Hartmut, Krüger., Annika, Wenz., Veronika, Schönhaar., Günter, E., M., Tovar., Günter, E., M., Tovar., Petra, J., Kluger., Petra, J., Kluger., Kirsten, Borchers., Kirsten, Borchers. (2016). (3) Blood-Vessel Mimicking Structures by Stereolithographic Fabrication of Small Porous Tubes Using Cytocompatible Polyacrylate Elastomers, Biofunctionalization and Endothelialization.. Journal of Functional Biomaterials. https://doi.org/10.3390/JFB7020011

[162] A., Zubareva., V., P., Varlamov., Anna, M., Nechaeva., N., N., Drozd. (2023). (1) Chemical modification of chitosan for developing of new hemostatic materials: A review.. International Journal of Biological Macromolecules. https://doi.org/10.1016/j.ijbiomac.2023.127608

[163] Xiuxue, Lei. (2023). (4) Recent advances of chitosan as a hemostatic material: Hemostatic mechanism, material design and prospective application. Carbohydrate Polymers. https://doi.org/10.1016/j.carbpol.2023.121673

[164] Chunyan, Yu., Yanju, Lu., Jinhui, Pang., lu, Li. (2023). (2) A hemostatic sponge derived from chitosan and hydroxypropylmethylcellulose. Journal of The Mechanical Behavior of Biomedical Materials. https://doi.org/10.1016/j.jmbbm.2023.106240

[165] Peng, Fan., Yanbo, Zeng., Dionisio, Zaldivar-Silva., Lissette, Agüero., Shige, Wang. (2023). (5) Chitosan-Based Hemostatic Hydrogels: The Concept, Mechanism, Application, and Prospects. Molecules. https://doi.org/10.3390/molecules28031473

[166] Roberta, Cassano., Paolo, Perri., Edoardo, Scarcello., Paolo, Piro., Roberta, Del, Sole., Federica, Curcio., Sonia, Trombino. (2024). (3) Chitosan Hemostatic Dressings: Properties and Surgical Applications. Polymers. https://doi.org/10.3390/polym16131770

[167] Chih-Tsung, Chang. (2023). (1) The hemostatic effect and wound healing of Novel collagen-containing polyester dressing.. Journal of Biomaterials Science-polymer Edition. https://doi.org/10.1080/09205063.2023.2230842

[168] Zhuang, Shi., Chengcheng, Shi., Chengkun, Liu., Haiyan, Sun., Si-Yuan, Ai., Xiaodan, Liu., Haoyu, Wang., Yunsong, Gan., Hua-zhen, Dai., Xiaoqiang, Wang., Fang, Huang. (2022). (5) Incorporation of tissue factor-integrated liposome and silica nanoparticle into collagen

hydrogel as a promising hemostatic system. Journal of Biomaterials Science-polymer Edition. https://doi.org/10.1080/09205063.2022.2156769

[169] Yu, Wang., Jie, Lin., Hao, Fu., Bingran, Yu., Guochao, Zhang., Yang, Hu., Fu-Jian, Xu. (2023). (1) A Janus Gelatin Sponge with a Procoagulant Nanoparticle-Embedded Surface for Coagulopathic Hemostasis.. ACS Applied Materials & Interfaces. https://doi.org/10.1021/acsami.3c15517

[170] Haihua, Wang., Mengxi, Wang., Jingheng, Wu., Shilei, Zhu., Yanan, Ye., Yushan, Liu., Ke, Li., Ruyue, Li., Yuhang, Zhang., Meng, Wei., Xin, Yang., Leixin, Meng. (2024). (5) Nature-Inspired Gelatin-Based Adhesive Hydrogel: A Rapid andUser-Friendly Solution for Hemostatic Applications.. Advanced Healthcare Materials. https://doi.org/10.1002/adhm.202304444

[171] Sergey, Barannikov., Evgeniy, Cherednikov., Galina, Polubkova., A., K., Vorontsov., Yu., V., Maleev., Alexey, Evgenievich, Bolkhovitinov., G., V., Prokhorov. (2024). (1) First experience of using alginate polymer polysaccharide hemostatic hydrogel in complex endoscopic treatment of unstable gastroduodenal ulcer bleeding: Clinical cases. Kubanskij naučnyj medicinskij vestnik. https://doi.org/10.25207/1608-6228-2024-31-3-73-92

[172] Xiaoqiang, Wang., Chang, Liu., Chengkun, Liu., Zhuang, Shi., Fang, Huang. (2024). (3) Development of alginate macroporous hydrogels using sacrificial CaCO3 particles for enhanced hemostasis.. International Journal of Biological Macromolecules. https://doi.org/10.1016/j.ijbiomac.2023.129141

[173] Jayavardhini, Bhoopathy., Weslen, Vedakumari, Sathyaraj., Beryl, Vedha, Yesudhason., Selvarajan, Rajendran., Sankari, Dharmalingam., Jayashri, Seetharaman., Ranjitha, Muthu., Ramachandran, Murugesan., Subramanian, Raghunandhakumar., Suresh, Kumar, Anandasadagopan. (2023). (5) Haemostatic potency of sodium alginate/aloe vera/sericin composite scaffolds - preparation, characterisation, and evaluation.. Artificial Cells Nanomedicine and Biotechnology. https://doi.org/10.1080/21691401.2023.2293784

[174] Raúl, Sanz-Horta., Ana, I., Matesanz., Alberto, Gallardo., Helmut, Reinecke., José, L., Jorcano., Pablo, Martínez, Acedo., Diego, Velasco., Carlos, Elvira. (2023). (1) Technological advances in fibrin for tissue engineering. Journal of Tissue Engineering. https://doi.org/10.1177/20417314231190288

[175] Tímea, Feller., Simon, D., Connell., Robert, A.S., Ariëns. (2021). (3) Why fibrin biomechanical properties matter for hemostasis and thrombosis.. Journal of Thrombosis and Haemostasis. https://doi.org/10.1111/JTH.15531

[176] Chengkun, Liu., Zi, Li., Zhuang, Shi., Zhidong, Ma., Shihai, Liu., Xiaoqiang, Wang., Fang, Huang. (2024). (1) Thermo-assisted fabrication of a novel shape-memory hyaluronic acid sponge for non-compressible hemorrhage control. International Journal of Biological Macromolecules. https://doi.org/10.1016/j.ijbiomac.2024.133657

[177] Yue, Xie., Pan, Gao., Fang-Fang, He., Chun-Yun, Zhang. (2022). (3) Application of Alginate-Based Hydrogels in Hemostasis. Gels. https://doi.org/10.3390/gels8020109

[178] Han, Yu., Qiaohong, Xiao., Guilin, Qi., Feixiang, Chen., Biyue, Tu., Suo, Zhang., Yinping, Li., Yun-Ru, Chen., Peng, Duan. (2022). (4) A Hydrogen Bonds-Crosslinked

Hydrogels With Self-Healing and Adhesive Properties for Hemostatic. Frontiers in Bioengineering and Biotechnology. https://doi.org/10.3389/fbioe.2022.855013

[179] Kanaparedu, P., C., Sekhar., Xunhui, Zhang., Huimin, Geng., Qun, Yu., Peiyu, Zhang., Jiwei, Cui. (2023). (1) Biomimetic Hemostatic Powder Derived from Coacervate-Immobilized Thermogelling Copolymers.. Biomacromolecules. https://doi.org/10.1021/acs.biomac.3c00840

[180] Jia, Bing, Wang., Cheng-xin, Li., Wei, Zhang., Weimin, Huang., Zhiqing, Liu., Rui, Shi., Shiyuan, Wang., Shan, Liu., Weiguo, Shi., Yunpeng, Li., Liang-Fa, Xu. (2023). (2) A contact-polymerizable hemostatic powder for rapid hemostasis.. Biomaterials Science. https://doi.org/10.1039/d3bm00075c

[181] Yoo, Eun, Lee., Yeonjeong, Kim., Kyung, Min, Park. (2023). (3) Actively cross-linking hemostatic sealant enables rapid hemostasis and wound closure.. Biotechnology Journal. https://doi.org/10.1002/biot.202200517

[182] Huijun, Ye., Yiwen, Xian., Shurong, Li., Chong, Zhang., Decheng, Wu. (2022). (5) In situ forming injectable γ-poly(glutamic acid)/PEG adhesive hydrogels for hemorrhage control.. Biomaterials Science. https://doi.org/10.1039/d2bm00525e

[183] Yan, Fang., Yanyan, Zheng., Chongyi, Chi., Sai, Jiang., Wanbang, Qin., Yicheng, Zhang., Haiqing, Liu., Qinhui, Chen. (2024). (1) PAA-PU Janus Hydrogels Stabilized by JANUS Particles and its Interfacial Performance During Hemostatic Processing.. Advanced Healthcare Materials. https://doi.org/10.1002/adhm.202303802

[184] Wanxin, Guo., Bin, Zhao., Muhammad, Shafiq., Xiao, Yu., Yihong, Shen., Jie, Cui., Yujie, Chen., Pengfei, Cai., Zhengchao, Yuan., Mohamed, H., El-Newehy., Hany, El-Hamshary., Yosry, Morsi., Binbin, Sun., Jian-feng, Pan., Xiumei, Mo. (2023). (2) On the development of modular polyurethane-based bioelastomers for rapid hemostasis and wound healing. Regenerative Biomaterials. https://doi.org/10.1093/rb/rbad019

[185] Yang, Bo., Wang, Yiliang., He, Jianxiong. (2018). (4) Polyurethane expansive film material for hemostasis and preparation method thereof.

[186] Aaron, C, Wilson., Shih-Feng, Chou., Roberto, Lozano., Jonathan, Y., Chen., Pierre, F., Neuenschwander. (2019). (6) Thermal and Physico-Mechanical Characterizations of Thromboresistant Polyurethane Films. Bioengineering. https://doi.org/10.3390/BIOENGINEERING6030069

[187] Xiaotian, Ge., Li, Zhang., Xianshun, Wei., Xi, Long., Yingchao, Han. (2024). (1) Plasma Surface Treatment and Application of Polyvinyl Alcohol/Polylactic Acid Electrospun Fibrous Hemostatic Membrane. Polymers. https://doi.org/10.3390/polym16121635

[188] Erfan, Dorkhani., Ali, Faryabi., Yasmin, Noorafkan., Asieh, Heirani., Behnam, Behboudi., Mohammad, Sadegh, Fazeli., Alireza, Kazemeini., Mohammad, Reza, Keramati., Amir, Keshvari., Seyed, Mohsen, Ahmadi, Tafti. (2023). (3) Biomedical properties and hemostatic efficacy of polyvinyl alcohol (PVA) based hydrogel in experimental rat liver injury model.. Journal of Applied Biomaterials & Biomechanics. https://doi.org/10.1177/22808000231198803

[189] Mohammad, Ashfaq., Tepparit, Wongpakham., Neetu, Talreja., Divya, Chauhan., Theerawat, Tharasanit., Werayut, Srituravanich. (2022). (2) Synthesis of polymeric composite

grafted with mineral particles/graphene oxide-based biomaterial: A promising robust hemostatic bandage. Materials today communications. https://doi.org/10.1016/j.mtcomm.2022.104786

[190] Tomoko, Ito., Shingo, Yamaguchi., Daisuke, Soga., Keisuke, Ueda., Takayuki, Yoshimoto., Toshiyuki, Koyama. (2022). (4) Water-Absorbing Bioadhesive Poly(Acrylic Acid)/Polyvinylpyrrolidone Complex Sponge for Hemostatic Agents. Bioengineering. https://doi.org/10.3390/bioengineering9120755

[191] Li, Q., YangMasood, U., Zhang, Z., Feng, G., Yang, H., & Song, Y. (2023). The efficacy and safety of thrombin-based flowable hemostatic agents in spine surgery: a protocol for systematic review and meta-analysis.

[192] Desiree, D., Rosselli., Benjamin, M., Brainard., Chad, W., Schmiedt. (2015). (2) Efficacy of a topical bovine-derived thrombin solution as a hemostatic agent in a rodent model of hepatic injury.. Canadian Journal of Veterinary Research-revue Canadienne De Recherche Veterinaire,

[193] Christine, M., Cheng., Carla, Meyer-Massetti., Steven, R., Kayser. (2009). (4) A review of three stand-alone topical thrombins for surgical hemostasis. Clinical Therapeutics. https://doi.org/10.1016/J.CLINTHERA.2009.01.005

[194] Yuanxing, Zhou., Xiaochi, Ma., Zhonghai, Li., Bo, Wang. (2021). (1) Efficacy, safety, and physicochemical properties of a flowable hemostatic agent made from absorbable gelatin sponge via vacuum pressure steam sterilization.. Journal of Biomaterials Applications. https://doi.org/10.1177/0885328220950899

[195] Roberto, Gazzeri., Marcelo, Galarza., Marika, Morabito., Alex, Alfieri. (2018). (5) Clinical Use and Hemostatic Application of Gelatin. doi: 10.1007/978-981-10-6083-0_3

[196] Roland, Z, White., Roland, Z, White., Lachlan, Kerr., Lachlan, Kerr., Tyler, J, White., Matthew, J, Sampson. (2021). (4) Review of topical gelatin-based haemostatic agents; an insidious culprit of intraoperative anaphylaxis?. Anz Journal of Surgery. https://doi.org/10.1111/ANS.16716

[197] Andrey, Karpenko., A., V., Cheban., A, A, Rabtsun., German, Y, Sokurenko., K.A., Andreychuk., Igor, I, Prostov., Natalia, Germanovna, Sapronova., N.D., Mzhavanadze., A, A, Kamaev., I, A, Suchkov. (2023). 1. Fibrin Sealant TISSEEL Lyo as a haemostatic agent in vascular surgery: Results of randomized, controlled, patient-blinded, multicentre clinical study in the Russian population.. Science Progress. https://doi.org/10.1177/00368504231182834

[198] W., Danker., Ashley, Deanglis., Nicole, Ferko., David, Garcia., A., Hogan. (2021). 2. Comparison of fibrin sealants in peripheral vascular surgery: A systematic review and network meta-analysis. Annals of medicine and surgery. https://doi.org/10.1016/J.AMSU.2020.12.003

[199] Ehsan, Natour., Michael, Suedkamp., Otto, Dapunt. (2012). (3) Assessment of the effect on blood loss and transfusion requirements when adding a polyethylene glycol sealant to the anastomotic closure of aortic procedures: a case–control analysis of 102 patients undergoing Bentall procedures. Journal of Cardiothoracic Surgery. https://doi.org/10.1186/1749-8090-7-105

Materials Research Foundations 172 (2025) 413-460
https://doi.org/10.21741/9781644903353-18

[200] William, M., Stone., David, L., Cull., Samuel, R., Money. (2012). (4) A randomized prospective multicenter trial of a novel vascular sealant. Annals of Vascular Surgery. https://doi.org/10.1016/J.AVSG.2012.02.013

[201] Quentin, Pellenc., Joseph, Touma., Raphaël, Coscas., Grégoire, Edorh., Maria, J., N., Pereira., Jeffrey, M., Karp., Yves, Castier., Pascal, Desgranges., Jean-Marc, Alsac., Jean-Marc, Alsac., Jean-Marc, Alsac. (2019). (1) Preclinical and clinical evaluation of a novel synthetic bioresorbable, on-demand, light-activated sealant in vascular reconstruction.. Journal of Cardiovascular Surgery. https://doi.org/10.23736/S0021-9509.19.10783-5

[202] Tetsushi, Taguchi., Mizuta, Ryo., Temmei, Ito., Keiko, Yoshizawa., Mikio, Kajiyama. (2016). (2) Robust Sealing of Blood Vessels with Cholesteryl Group-Modified, Alaska Pollock-Derived Gelatin-Based Biodegradable Sealant Under Wet Conditions.. Journal of Biomedical Nanotechnology. https://doi.org/10.1166/JBN.2016.2210

[203] Alan, B., Lumsden., Eugene, R., Heyman. (2006). (3) Prospective randomized study evaluating an absorbable cyanoacrylate for use in vascular reconstructions.. Journal of Vascular Surgery. https://doi.org/10.1016/J.JVS.2006.06.039

[204] Jan, Brunkwall., Ruemenapf, G., Florek, Hj., Werner, Lang., Schmitz-Rixen, T. (2007). (2) A single arm, prospective study of an absorbable cyanoacrylate surgical sealant for use in vascular reconstructions as an adjunct to conventional techniques to achieve haemostasis.. Journal of Cardiovascular Surgery,

[205] Bastiaan, P., Vierhout., Alewijn, Ott., Michel, M.P.J., Reijnen., Jacques, Oskam., Jan, J.A.M., van, den, Dungen., Clark, J., Zeebregts. (2014). (1) Cyanoacrylate Skin Microsealant for Preventing Surgical Site Infection after Vascular Surgery: A Discontinued Randomized Clinical Trial. Surgical Infections. https://doi.org/10.1089/SUR.2013.191

[206] Hamit, Serdar, Başbuğ. (2016). (4) Ultrasound-guided perforator vein sealing with cyanoacrylate glue. Turkish Journal of Thoracic and Cardiovascular Surgery. https://doi.org/10.5606/TGKDC.DERGISI.2016.12883

[207] Junichiro, Sageshima., Junichiro, Sageshima., Gaetano, Ciancio., Gaetano, Ciancio., K., Uchida., K., Uchida., Antonio, Romano., Antonio, Romano., Z., Acun., Z., Acun., L., Chen., L., Chen., George, W., Burke., George, W., Burke. (2011). (5) Absorbable Cyanoacrylate Surgical Sealant in Kidney Transplantation. doi: 10.1016/J.TRANSPROCEED.2011.03.094

[208] Regina, Moura., Francisco, Humberto, de, Abreu, Maffei., L., Mattar., Viciany, Erique, Fabris., P., R., Cury., S., Lastoria., E., A., Gregorio. (2009). (1) Glutaraldehyde-treated homologous vein graft as a vein substitute: experimental study in rabbits.. International Angiology. https://doi.org/10.4172/2157-7552-S1.002

[209] Herbert, Dardik. (1986). (5) Lower Extremity Revascularization with the Glutaraldehyde Stabilized Human Umbilical Cord Vein Graft. ASTM special technical publications. https://doi.org/10.1520/STP33281S

[210] Herbert, Dardik., Ibrahim, M., Ibrahim., Barry, Sussman., Mamoon, Jarrah., Irving, Dardik. (1979). (3) Glutaraldehyde-stabilized umbilical vein prosthesis for revascularization of the legs. Three year results by life table analysis.. American Journal of Surgery. https://doi.org/10.1016/0002-9610(79)90376-3

[211] Herbert, Dardik., Ibrahim, M., Ibrahim., Mamoon, Jarrah., Barry, C., Sussman., Irving, Dardik. (2005). (4) Three-year experience with glutaraldehyde-stabilized umbilical vein for limb salvage.. British Journal of Surgery. https://doi.org/10.1002/BJS.1800670402

[212] Zeng, Q., Zhou, J., Ji, Y., & Wang, H. (2024). A semiparametric Gaussian mixture model for chest CT-based 3D blood vessel reconstruction. Biostatistics, kxae013.

[213] Song, Xiaokun., Zhou, Long., Xie, Weiguo., Wang, Lei., Xu, Ning. (2018). (3) Blood vessel image reconstruction method and reconstruction apparatus.

[214] Yih, Yang, Chen., Benjamin, R., Kingston., Warren, C., W., Chan. (2020). (4) Transcribing In Vivo Blood Vessel Networks into In Vitro Perfusable Microfluidic Devices. Advanced materials and technologies. https://doi.org/10.1002/ADMT.202000103

[215] Salazar, H. F., O'Connell, G. M., Corpuz, G. S., Salingaros, S., Chopoorrian, A. H., Carrion, K., ... & Spector, J. A. (2023). 146. A novel 3D printed sacrificial material for the development of perfused intricate blood vessel structures. Plastic and Reconstructive Surgery–Global Open, 11(5S), 91-92.

[216] Nazanin, Amiryaghoubi., Marziyeh, Fathi., Jaleh, Barar., Hossein, Omidian., Yadollah, Omidi. (2023). (1) Hybrid polymer-grafted graphene scaffolds for microvascular tissue engineering and regeneration. European Polymer Journal. https://doi.org/10.1016/j.eurpolymj.2023.112095

[217] Rumiana, Tzoneva., Claudia, Weckwerth., Barbara, Seifert., Marc, Behl., Matthias, Heuchel., Iana, Tsoneva., Andreas, Lendlein. (2011). (3) In vitro evaluation of elastic multiblock co-polymers as a scaffold material for reconstruction of blood vessels.. Journal of Biomaterials Science-polymer Edition. https://doi.org/10.1163/092050610X537147

[218] Halima, Kerdjoudj., Halima, Kerdjoudj., Fouzia, Boulmedais., Nicolas, Berthelemy., H., Mjahed., H., Louis., Pierre, Schaaf., Jean-Claude, Voegel., Jean-Claude, Voegel., Patrick, Menu. (2011). (5) Cellularized alginate sheets for blood vessel reconstruction. Soft Matter. https://doi.org/10.1039/C0SM00998A

[219] Alida, Abruzzo., Calogero, Fiorica., Vincenzo, Davide, Palumbo., Roberta, Altomare., Giuseppe, Damiano., Maria, Concetta, Gioviale., Giovanni, Tomasello., Mariano, Licciardi., Fabio, Salvatore, Palumbo., Gaetano, Giammona., Attilio, Ignazio, Lo, Monte. (2014). (4) Using Polymeric Scaffolds for Vascular Tissue Engineering. International Journal of Polymer Science. https://doi.org/10.1155/2014/689390

[220] Mina, Shahriari-Khalaji., Muhammad, Shafiq., Haitao, Cui., Ran, Cao., Meifang, Zhu. (2023). (1) Advancements in the fabrication technologies and biomaterials for small diameter vascular grafts: A fine-tuning of physicochemical and biological properties. Applied Materials Today. https://doi.org/10.1016/j.apmt.2023.101778

[221] Panitporn, Laowpanitchakorn., Marie, Piantino., Kentaro, Uchida., Misa, Katsuyama., Michiya, Matsusaki. (2024). (4) Biofabrication of engineered blood vessels for biomedical applications. Science and Technology of Advanced Materials. https://doi.org/10.1080/14686996.2024.2330339

[222] Georgia, Papavasiliou., Ming-Huei, Cheng., Eric, M., Brey. (2010). (2) Strategies for Vascularization of Polymer Scaffolds. Journal of Investigative Medicine. https://doi.org/10.2310/JIM.0B013E3181F18E38

[223] Alida, Abruzzo., Calogero, Fiorica., Vincenzo, Davide, Palumbo., Roberta, Altomare., Giuseppe, Damiano., Maria, Concetta, Gioviale., Giovanni, Tomasello., Mariano, Licciardi., Fabio, Salvatore, Palumbo., Gaetano, Giammona., Attilio, Ignazio, Lo, Monte. (2014). (5) Using Polymeric Scaffolds for Vascular Tissue Engineering. International Journal of Polymer Science. https://doi.org/10.1155/2014/689390

[224] Elena, López-Ruiz., Seshasailam, Venkateswaran., Macarena, Perán., Macarena, Perán., Gema, Jiménez., Salvatore, Pernagallo., Juan, J., Díaz-Mochón., Olga, Tura-Ceide., Francisco, Arrebola., Juan, Melchor., J., I., Soto., Guillermo, Rus., Pedro, J., Real., María, Diaz-Ricart., Antonio, Conde-González., Mark, Bradley., Juan, A., Marchal. (2017). (1) Poly(ethylmethacrylate-co-diethylaminoethyl acrylate) coating improves endothelial re-population, bio-mechanical and anti-thrombogenic properties of decellularized carotid arteries for blood vessel replacement. Scientific Reports. https://doi.org/10.1038/S41598-017-00294-6

[225] E., A., Velikanova., Vera, G., Matveeva., E., O., Krivkina., V., V., Sevostianova., M., Yu., Khanova., T., V., Glushkova., Yu., A., Kudryavtseva., Larisa, V., Antonova. (2019). (3) Development of a polymeric scaffold for vascular tissue engineering. doi: 10.1063/1.5132248

[226] Soodabeh, Gorgani., Anousheh, Zargar, Kharazi., Shaghayegh, Haghjooy, Javanmard., Mohammad, Rafiinia. (2020). (4) Improvement of Endothelial Cell Performance in an Optimized Electrospun Pre-polyglycerol Sebacate-Poly Lactic Acid Scaffold for Reconstruction of Intima in Coronary Arteries. Journal of Polymers and The Environment. https://doi.org/10.1007/S10924-020-01749-0

About the Editor

Dr. Inamuddin is working as an Assistant Professor at the Department of Applied Chemistry, Aligarh Muslim University, Aligarh, India. He obtained a Master of Science degree in Organic Chemistry from Chaudhary Charan Singh (CCS) University, Meerut, India, in 2002. He received his Master of Philosophy and Doctor of Philosophy degrees in Applied Chemistry from Aligarh Muslim University (AMU), India, in 2004 and 2007, respectively. He has extensive research experience in multidisciplinary fields of Analytical Chemistry, Materials Chemistry, and Electrochemistry and, more specifically, Renewable Energy and Environment. He has worked on different research projects as a project fellow and senior research fellow funded by the University Grants Commission (UGC), Government of India, and the Council of Scientific and Industrial Research (CSIR), Government of India. He has received the Fast Track Young Scientist Award from the Department of Science and Technology, India, to work in the area of bending actuators and artificial muscles. He has also received the Sir Syed Young Researcher of the Year Award 2020 from Aligarh Muslim University. He has completed four major research projects sanctioned by the University Grant Commission, Department of Science and Technology, Council of Scientific and Industrial Research, and Council of Science and Technology, India. He has published 216 research articles in international journals of repute and nineteen book chapters in knowledge-based book editions published by renowned international publishers. He has published 200 edited books with Springer (U.K.), Elsevier, Nova Science Publishers, Inc. (U.S.A.), CRC Press Taylor & Francis Asia Pacific, Trans Tech Publications Ltd. (Switzerland), IntechOpen Limited (U.K.), Wiley-Scrivener, (U.S.A.) and Materials Research Forum LLC (U.S.A). He is a member of various journals' editorial boards. He has served as Associate Editor for journals (Environmental Chemistry Letter, Applied Water Science and Euro-Mediterranean Journal for Environmental Integration, Springer-Nature), Frontiers Section Editor (Current Analytical Chemistry, Bentham Science Publishers), Editorial Board Member (Scientific Reports-Nature) and Review Editor (Frontiers in Chemistry, Frontiers, U.K.) He has also guest-edited various special thematic issues for the journals of Elsevier, Bentham Science Publishers, and John Wiley & Sons, Inc. He has attended as well as chaired sessions at various international and national conferences. He has worked as a Postdoctoral Fellow, leading a research team at the Creative Research Initiative Center for Bio-Artificial Muscle, Hanyang University, South Korea, in the field of renewable energy, especially biofuel cells. He has also worked as a Postdoctoral Fellow at the Center of Research Excellence in Renewable Energy, King Fahd University of Petroleum and Minerals, Saudi Arabia, in the field of polymer electrolyte membrane fuel cells and computational fluid dynamics of polymer electrolyte membrane fuel cells. He is a life member of the Journal of the Indian Chemical Society. His research interest includes ion exchange materials, a sensor for heavy metal ions, biofuel cells, supercapacitors and bending actuators.

Dr. Arwa Alrooqi is an Assistant Professor in the Department of Chemistry, Faculty of Science, at Al Baha University, Saudi Arabia. She has published several research articles in internationally recognized journals. Dr. Alrooqi is actively engaged in multidisciplinary research, focusing on the synthesis and engineering of organic semiconductors and nanomaterials, their characterization, and their application in smart windows.

Dr. Amal Alshehri is an Assistant Professor of Physical Chemistry in the Department of Chemistry, Faculty of Science, at Al Baha University, Saudi Arabia. She earned her Ph.D. with a focus on nanomaterials, catalysis, and adsorption. Her research expertise encompasses green synthesis approaches for nanomaterials, water treatment, heterogeneous catalysis, photocatalysis, and the selective oxidation of cyclohexane. Dr. Alshehri has published several research articles in internationally recognized journals, contributing to fundamental advancements in the fields of nanomaterials and catalysis.

Dr. Hind Alluqmani is an Assistant Professor of Analytical Chemistry in the Department of Chemistry, Faculty of Science, at Al Baha University, Saudi Arabia. She earned her Ph.D. with a focus on the "Analytical Study of Physical Behavior in Clay Hydrogel Media." Her research expertise includes surface analyses, such as scanning electron microscopy (SEM) and associated techniques, including surface area (BET) measurements, energy-dispersive X-ray spectroscopy (EDX), and powder X-ray diffraction (PXRD). Additionally, she specializes in thermal analysis, including thermogravimetric analysis (TGA) and differential scanning calorimetry (DSC), as well as rheological analysis.

Dr. Mohammad A. Jafar Mazumder has been serving as a Professor of Chemistry at King Fahd University of Petroleum & Minerals (KFUPM), Saudi Arabia. He has extensive experience in designing, synthesizing, and characterizing various organic compounds, ionic and thermo-responsive polymers for corrosion, water treatment, and biomedical applications. Dr. Jafar Mazumder obtained his B.Sc (Hons.), M.Sc (Chemistry) from Aligarh Muslim University, India, MS (Chemistry) from KFUPM, Saudi Arabia, and Ph.D. in Chemistry (2009) from McMaster University, Canada.

In more than 20 years of academic research, Dr. Jafar Mazumder has had the opportunity to work with several international collaborative research groups and has exposed himself to a broad range of research areas. Dr. Jafar Mazumder secured 8 US patents, published more than 85 articles in peer-reviewed journals, 37 conference proceedings, 9 book chapters, and co-edited 4 books with Springers and Trans Tech publications. He is awarded as a Fellow of the Royal Society of Chemistry and Chartered Chemist, Association of Chemical Profession of Ontario, Canada. Besides, Dr. Jafar Mazumder is a member of the American Chemical Society (ACS), Canadian Society for Chemistry (CSC), Canadian Biomaterial Society (CBS), and a life member of the Bangladesh Chemical Society (BCS). In his academic career, he was awarded numerous national and international scholarships and awards including the prestigious Indian Council for Cultural Relations (ICCR) Scholarship from Govt. of India for undergraduate studies in India, Aligarh Muslim University undergraduate & graduate Gold medal, and certificate of excellence from the Ministry of Human Resource Development, Govt. of India, and MITACS postdoctoral fellowship (Canada) for pursuing postdoctoral research in Chemical and Biomedical Engineering.

Currently, Dr. Jafar Mazumder is actively involved in several ongoing university (KFUPM), government (KACST, NSTIP), and client (Saudi Aramco) funded projects in the capacity of principal and co-investigators. His current research interest includes the design, synthesis, and characterization of various modified monomers and polymers for potential use in the inhibition of mild steel corrosion in oil and gas industries and the preparation of multilayered polyelectrolyte coated membranes for the removal of heavy metals and organic contaminants from aqueous water samples. The long-term scientific goal of Dr. Jafar Mazumder is not merely to make science fun and entertaining for people. It is to engage them with a multidisciplinary scientific mission at a deeper level to create a space through which they can interact with scientific ideas, develop connections between science, engineering, and biology, and thoughts of their own to contribute to society. He feels this goal and engaging personality make him a pleasant person to work with and help inspire his co-workers in any professional setting.

www.ingramcontent.com/pod-product-compliance
Lightning Source LLC
LaVergne TN
LVHW052216200926
848481LV00001B/126

* 9 7 8 1 6 4 4 9 0 3 3 4 6 *